INTRODUCTION TO BUSINESS AND ECONOMIC STATISTICS

FIFTH EDITION

JOHN R. STOCKTON

Professor Emeritus of Business Statistics
Former Director, Bureau of Business Research
The University of Texas at Austin

CHARLES T. CLARK

Associate Professor of Business Statistics
The University of Texas at Austin

M24

Published by

SOUTH-WESTERN PUBLISHING CO.

CINCINNATI WEST CHICAGO, ILL. DALLAS PELHAM MANOR, N.Y.
PALO ALTO, CALIF. BRIGHTON, ENGLAND

ISBN: 0-538-13240-X

Library of Congress Catalog Card Number: 74-24472

1234567Ki1098765

Printed in the United States of America

Preface

The major objective of a textbook in elementary business statistics should be to inform students how people in business use statistics in making decisions and controlling the operations of a business. Information is needed on the sources of statistical data useful to the businessperson and on the methods of analysis that will make these data of maximum usefulness. This text attempts to describe in detail the important basic principles used by the business statistician and to supplement this description with numerous examples of how the methods have been used.

It has been kept in mind throughout the discussion that many students majoring in business administration are not highly trained mathematicians. Whenever possible a simple explanation of methods and theory has been used in preference to mathematical terminology. The sophisticated methods of mathematics may be preferred by those students with an adequate background, but a fairly large percentage of business students have gone no farther than an introductory course in calculus. Some have not even been introduced to calculus.

The coverage of the methods of analysis of data, published sources of data, and the collection of data are more complete than in the majority of statistics texts. This has been done in the belief that without an understanding of these basic principles the business student will not be equipped to make effective use of these data. Special emphasis is given to the analysis of time series and index numbers. The present turmoil in the world economy has increased to a considerable degree the interest of businesspersons and government officials in statistics of this nature. Unless he has a thorough understanding of the methods of measuring and forecasting changes in the various segments of the economy, the student of business has not received a well-rounded education.

It has been assumed that not all students taking a course in business statistics are familiar with probability. For the benefit of students who have not studied probability, a simple explanation of the most important theorems is given as a basis for the study of sampling and statistical inference. For students who are adequately prepared in probability, this chapter can serve as a brief review.

The major probability distributions are covered in expectation that the majority of business students will have had only a limited introduction to them.

Chapters 5 and 6 will serve as the theoretical basis for understanding the remaining chapters in Part 2, which are devoted to an introduction to the collection of sample data and to some of the most generally used methods of statistical inference. It is impossible to do more than introduce the student to statistical inference in a first course. The methods described were selected on the basis of their frequency of use and the fact that they are typical of the type of analysis that can be carried out in this manner. This subject matter is taught much more satisfactorily in an advanced course for students who have a considerable interest in statistical methods and who are attracted to the field of research.

Chapter 7 has been retained without much change to give the basic theory for estimates from random samples and the computation of confidence intervals. This chapter does not go into the widely used methods of stratified and cluster sampling since it is believed that this subject matter is more properly covered in an advanced text.

A new feature of this edition is the increased emphasis on nonparametric methods. These methods of testing for statistical significance are becoming increasingly important in certain types of business research and a few of the most useful methods have been described in Chapter 9 to supplement the coverage of the more traditional parametric methods in Chapter 8.

One of the most widely used applications of statistical inference is in the use of sampling to control the quality of manufactured product. Some of the techniques of statistical quality control are described as an introduction to a subject that has become a highly specialized use of the methods of statistical inference in making industrial decisions.

Regression and correlation analysis is given fairly complete coverage in the belief that even the most elementary course in statistics should explain the theory and methods of this very useful tool with its many applications to business problems.

Various types of study material are given at the end of each chapter to aid the student in applying the principles discussed in the text. This material has been grouped into two categories: study questions and problems. The study questions require the student to give some thought to the meaning of the various topics discussed in the chapter. The problems are to be used in developing skill in performing the calculations needed for the various methods of analysis. Various aids to calculation are provided in the tables in the appendixes.

We are indebted to the Literary Executor of the late Sir Ronald A. Fisher, F.R.S., Cambridge; to Dr. Frank Yates, F.R.S., Rothamsted; and to Messrs. Oliver & Boyd, Ltd., Edinburgh, for permission to reprint in Appendixes K and L the tables from their book *Statistical Tables for Biological, Agricultural, and Medical Research.* We are also indebted to Professor Stephen P. Shao, Old Dominion University, and the publisher for permission to reprint in Appendix E the table from his book *Mathematics for Management and Finance;* and to Professor Charles T. Clark, The University of Texas at Austin, Professor

Lawrence L. Schkade, The University of Texas at Arlington, and the publisher for permission to reprint in Appendixes F and J the tables from their book *Statistical Analysis for Administrative Decisions.*

J.R.S.
C.T.C.

Contents

INTRODUCTION

Part 1 DATA ANALYSIS

Part 2 STATISTICAL INFERENCE

Part 3 ANALYSIS OF RELATIONSHIP

Part 4 ANALYSIS OF BUSINESS CHANGE

APPENDIXES

1 Statistics in Business Decision Making

Statistical techniques are used today in almost every area of human enterprise. Whether the concern is with predicting the weather, fighting disease with new drugs, preventing crime, evaluating a new food product, or forecasting population growth, there is an underlying core of uncertainty which must be dealt with in an objective and scientific manner to obtain the best results. The use of statistics provides some capability for dealing with this uncertainty.

The application of statistical techniques to business decisions are many and result from the fact that virtually all important business decisions are made under conditions of uncertainty. Business has become increasingly complex, and the responsibilities of the business executive have become correspondingly greater. Often the business executive must choose one definite course of action from all those open to him, although the consequences of each course of action cannot be fully known at the time the decision must be made.

To make such difficult decisions in the face of uncertainty demands two things. First, there must be an elaborate information system to gather and to supply all the facts needed by the executive. Second, there must be available a tested, and often sophisticated, set of tools for evaluating those facts. The nature and application of this set of tools is the subject of this book.

When most business units were small, the facts needed by the manager were not only fewer in number but were also more easily obtained. Since the market was close at hand and the customers were, for the most part, personal friends of the owner of the business, there was little need for an elaborate analysis of the market. The owner of the business could find out what his customers thought about his product and service simply by listening to their comments or asking them questions. However, in today's large business enterprises, goods are manufactured long before they are offered for sale. Producers and consumers are strangers to each other, and manufacturers must try to anticipate not only what consumers will buy but also the quantity they will buy. In order to get the information needed for intelligent planning of the output of a factory, management must rely on a systematic method of securing information from consumers.

The personnel problems of the small business may be similar in nature to those of the large concern, but in a small enterprise the owner or manager is close to his employees and knows a great deal about them. The information that must now be collected in elaborate personnel records was at one time a matter of personal knowledge of the manager of the business. In the same way the owner of a small factory could keep informed on the progress of production in an informal manner without the elaborate system of records and reports that is necessary in a large organization.

The increasing reliance on information collected and summarized in reports has increased the time required in getting the information to the businessmen who are to use it. Not only has the preparation time for completing reports increased the cost of the information, but the collection time has reduced its value in making decisions and controlling the business. The pressure for timely information has revolutionized the techniques and the machinery for collecting and summarizing business data.

NATURE OF STATISTICAL DATA

Statistical data are facts expressed in quantitative form. They are the building blocks of statistical analysis. As business has grown more complex and the demand for information needed to make decisions has increased, a change has come about in the nature of information required. Greater emphasis is being placed on information expressed precisely as quantities rather than in more general qualitative terms.

In many cases the characteristics of the item for which information is desired can be expressed in measurable or countable quantities such as dollar sales, tons of coal, or bushels of wheat. Statistical data of this quantitative type can be analyzed in many ways to make the facts more meaningful and more useful to the business executive.

Some kinds of facts are not directly quantifiable; that is, they cannot be measured on any kind of a scale. This type of statistical data is produced by grouping individual observations into qualitative categories or classes. Good examples are hospital patients classified according to their ailments or machined parts classified as either acceptable or not acceptable. The kinds of statistical analysis possible with qualitative data are not the same as with quantitative data, but useful information can still be obtained.

Confusion frequently arises over the use of the word *statistics* because it has three meanings: (1) it is used to refer to statistical data; (2) it is used to refer to a body of methods and techniques for dealing with statistical data; and (3) the term "statistic," used in the singular, refers to a measure derived from a sample of data. The third meaning of the term will be discussed in Chapter 7, which deals with statistical sampling.

Anything that exhibits differences of magnitude or number is called a *variable* and serves as the raw material for statistical analysis. A company's retirement and pension program probably requires the age of each employee on the payroll. The value of this variable is obtained by measuring the age of each employee. The sales department may be concerned with the cost of processing small orders and may ask for information about the size of the orders received in a selected period of time. This characteristic may be measured by computing the amount of each order in either physical quantities or in total cost of the goods bought. Either of these units is a value of a variable.

It is also possible to derive the values of a variable by computing the ratio between two amounts such as the number of deaths per thousand population. This is a ratio between the number of deaths in a given period and the thousands of persons in the population. Derived variables may be in many forms. A few examples are per capita income for each state, average yield of cotton per acre for various cotton-growing regions of the south, and fatalities on airlines per million passenger-miles flown.

ANALYSIS OF STATISTICAL DATA

After a mass of measurements has been made to secure the statistical data related to a given problem, the data must be processed in order to be in the most useful form for management. Generally the individual using the data does not want all the details that have been collected; he usually wants summaries and as much analysis of the figures as possible. The methods used to measure the characteristics of the data, to put the individual measurements into some kind of summary form, and to analyze the data to bring out the greatest amount of information are called *statistical methods*.

The theory underlying the methods of analysis used on statistical data is referred to as *statistical theory*. Any discussion of the methods of analyzing data will also include a discussion of the logic underlying the methods, but the rigorous development of this logic is mathematical in nature. Courses entitled mathematical statistics stress the mathematical theory underlying the methods of analysis.

The technical statistical methods used to make the needed measurements of the characteristics that are important to the problem and to bring them together in a summary form may be called *descriptive statistics*. This term is appropriate because the purpose of these methods is to make effective use of statistical data in communicating with the executive making a decision. The better the data describe the situation facing the executive the more useful they are. Methods that have been developed to collect, summarize, and in other ways manipulate the various types of data needed to describe situations are discussed in detail in Part 1.

It is not always necessary to compile all the data concerning a particular problem in order to secure valuable information for decision making. Under certain conditions the analysis of a sample taken from a large group will give the information needed at a small fraction of the cost required to compile complete data. The methods used in selecting the sample and in drawing inferences from sample data are powerful tools of research and are used with great effectiveness in making business decisions. Reaching conclusions from a sample is known as *statistical induction* and will be discussed in Part 2.

Parts 1 and 2 are concerned with methods of analyzing data involving only one variable at a time. That is to say, an observation consists of only one measurement or count on the variable under study.

Part 3 of this text discusses problems dealing with the relationship between two or more variables considered simultaneously. An observation consists of a measurement on several variables at one time. When it can be shown that variations in one or more independent variables are related to the variations in another variable called the dependent variable, this fact may be useful in predicting the action of the dependent variable based on prior knowledge of the independent variables to which it is related.

For example, if it is known that gasoline consumption in any state is related to the number of cars registered and the number of miles of highway in that state, this fact can be used to estimate gasoline consumption based on a knowledge of the other two factors long before the gasoline is purchased for use by the consumer. How this estimate is made and how accurate it may be is a matter of statistical analysis.

Part 4 deals with the analysis of business change. Because much of the data with which the businessman works are observations measured at various points in time, a special kind of statistical analysis is required to understand them.

For example, a personnel director keeps records of the number of employees on the payroll each month. The statistical data with which he deals are part of a time series, and the observations are not independent of each other. While the numbers may change from month to month, each month is related to what was observed the previous month. By careful statistical analysis it may be possible to measure the trend or average growth in the work force, the seasonal pattern if one exists, and the effect of swings in the business cycle, and perhaps forecast future changes in the size of the working group.

MODERN DECISION MAKING

Developments in the use of statistical data in business decision making have created some new and powerful tools that can be used by the businessman. The pressure of more complex business problems and the scale of operation have stimulated research in methods of analyzing data to meet these growing problems. Many fields of knowledge have contributed to the improvement in business decision making, but the advances in science on many fronts during the twentieth century have had an unusually significant influence on the thinking about the decision-making process in business.

At first glance it might appear that the business executive faces problems that bear no resemblance to those with which the scientist deals. However, the use of objective, unbiased methods that attempt to make use of all the facts in a given situation may justify the executive in believing that he is using the methods of science in making his decisions. The use of more objective methods and more accurate information may be expected to improve the quality of business decisions.

World War II gave tremendous impetus to efforts to improve decision-making techniques, and this trend has continued at an accelerated pace since the end of the war. The increasing use of statistical methods of analysis has provided a larger volume of precise information to use in the analysis of business problems, and in recent years a great deal of experimentation has been carried on in the hope of finding better methods of analysis in making decisions. During World War II the methods used by scientists were applied with outstanding success to certain military problems, such as determining the best program for strategic bombing, for mining enemy waters, and for searching out enemy ships. Business organizations now apply these same principles to such problems as determining the best policy to follow with respect to product mix, inventory fluctuations, and shipments of goods to various markets from different warehouses. The names commonly applied to this scientific or mathematical approach to the solution of business problems are *management science* or *operations research.*

Operations Research

Operations research is the application of scientific methods of analysis to executive-type problems involving the operations of man-machine systems in industrial organizations in order to provide those in control of the operations with optimum solutions to problems. Because business is so complex, operations research has usually been able to make effective use of a team whose members have been drawn from a number of special fields, including mathematics, statistics, physics, economics, and engineering. The methods used are those employed by scientists, and in general the approach is quantitative. The methods involve the prediction of various courses of action and thus provide a basis that management can use in choosing between alternative courses of action. The complex problems for which solutions are sought involve many important aspects of a firm's operations.

One of the most effective techniques used in operations research is *linear programming,* a body of techniques for solving problems dealing with many variables in which an objective that can be expressed quantitatively is attained, subject to certain restraining conditions. This mathematical method has succeeded in giving the best solution to problems that previously had no solution except through the judgment of the executive. It is reasonable to conclude that progress will continue to be made in the solution of more complex problems by techniques of this type, although this does not mean that it will be possible to optimize all decisions in the near future. Fairly simple methods of analysis of statistical data in making decisions will continue to be important, and as the use of quantitative methods by business executives becomes more widespread, better decisions will be made.

The Role of the Computer

The computer is involved in analyses such as operations research, for without the computer most of the calculations needed could not be made.

During the last two decades the businessman has found the computer to be a valuable aid to decision making, has revamped his organization and his paperwork routines to adapt to the computer, and is now digesting the gains and looking ahead to the changes a new generation of computers will bring. The advanced systems currently being developed are likely to have an impact on management planning and control equal to the revolution of paperwork processing by early computers.

The computer has been readily and enthusiastically accepted to perform mathematical computations with great speed and accuracy. Some business problems which are routinely solved via computer in just a few seconds would require years of manual computation time. A few of the jobs performed by computers are controlling inventories, preparing payrolls, handling checkouts in a supermarket, making airplane reservations, and handling credit card billing. In recent years decreasing costs and the use of remote terminals have enabled even small businesses to take advantage of the savings of automated data processing that were once available only to large firms.

OUTLINE OF STEPS IN STATISTICAL METHODS

The methods of analysis of quantitative data may be broken into the following six steps:

1. Definition of the problem
2. Assembly of information
3. Collection of original data
4. Classification
5. Presentation
6. Analysis.

A brief description of these steps is given in the following paragraphs.

Definition of the Problem

Before starting to assemble the information needed to solve a problem, the problem should be clearly defined so that quantitative methods can be used whenever possible in reaching a solution. If a problem is defined using quantitative analysis, there is a strong possibility that a better decision can be reached than would otherwise have been possible. Use of newer techniques, such as operations research, to make the decision requires a precise statement of the problem in quantitative terms.

Assembly of Available Information

The first step to take after defining the problem is to make a search for all readily available information that relates to the problem. Although sometimes the facts will be predominantly quantitative, on other occasions the most important information may be in other than numerical form. In a large percentage of problems, however, the statistical information will be an important part of the total information assembled.

In assembling the statistical data on a problem, one should first determine what information has already been collected. This information, if it deals with internal operations, may be in the records of the business. Unfortunately, it is not uncommon for a businessman to make decisions without using information that could be readily obtained from the records of the business. One reason for this is that records may be kept in one part of the business for a specific purpose without other executives knowing that such records exist. Even if the information is not available in a suitable form it often can be easily assembled into the appropriate form for making the decision.

Data on conditions outside the business may already have been collected and published. It is a fundamental principle that the businessman seeking information on phenomena outside the organization should rely as far as possible on data already available for use rather than make the collection himself. There are many fact-gathering agencies, both public and private, which publish information of value to executives. These agencies include federal and state governments, trade associations, banks, newspapers, business periodicals, universities, and private business concerns. Business executives who fail to make use of available information are neglecting an important aid to better decision making.

In using published data, two types of sources, primary and secondary, should be recognized. A *primary source* is the agency that makes the compilation and first publishes it; a *secondary source* is any republication by another agency. For example, the Bureau of the Census collects and publishes the census of population; but many other agencies, both governmental and private, republish certain material in the original volumes. The census publications are primary sources and all subsequent publications are secondary sources.

It is generally preferable to use the primary source, since it usually contains a detailed description of the exact meaning and limitations of the information. For example, a recent Census of Manufactures presents 30 pages of detailed explanation of the data. This includes a detailed statement of the establishments covered by the census and the type of establishments excluded, the definition of the industries used in the classification, the degree of comparability of this census with other enumerations, and many other facts that are important in the use of the data. Secondary sources almost invariably give less explanation about the meaning of the statistics, and frequently present no explanation other than the captions, stubs, and footnotes in the tables.

The advantage of using secondary sources lies in their convenience. Many of them, such as the *Statistical Abstract of the United States,* collect data from a large number of primary sources. Accordingly, the user finds many of the facts he wants in one volume instead of having to consult a number of primary sources. In searching for information on an unfamiliar subject, it is good practice to look in a comprehensive secondary source such as the *Statistical Abstract.* The data wanted may not be there, but the secondary source may suggest primary sources that do contain the facts. The bibliography published in the *Statistical Abstract* is a valuable guide to data readily available to the businessman.

Collection of Original Data

A careful statement of the problem will provide the basis for deciding what information will be needed in its solution. After assembling the internal data from business records and published information, the need for additional information must be determined. If the needed information is important enough to the decision to justify the cost of collecting it, an original investigation is the next logical step. In such an investigation data are collected firsthand, as opposed to utilizing data already collected and tabulated. Typically, some kind of survey is used to secure the data.

It is unlikely that an executive concerned with the data used in a problem will make the survey himself, but he should know a great deal about the problems of collecting data, the different methods used, and the advantages and disadvantages of each method. It will be his decision as to whether the investigation should be made or whether the problem should be settled on the basis of the facts already available.

To collect original data from a large group, it is possible to draw reasonably accurate conclusions from a study of a small portion of the group called a *sample*. A sample is selected from the total of the individual units of the group, which is called the *universe* or the *population*. The sample is a small proportion of these individual units. There will always be some difference between the estimate from a sample and the value that would be obtained by enumerating all the units in a group, but this difference can be controlled and reduced to as small an amount as desired. (The control of sample variance is discussed in Chapter 7.)

The ability to make estimates from a sample representing only a small percentage of the universe is a powerful device for providing useful information on business problems. If it were necessary to rely solely on complete enumerations of data, much of the information that is now available for making business decisions could not be afforded. For example, consider the cost of surveying soft drink consumption in this country. It would be wasteful to spend the money necessary to make a complete enumeration when an approximation from a sample will adequately serve the purpose for which the information is to be used.

The actual procedures used in collecting data are essentially the same whether all the items are to be included or only a sample. Making estimates from samples and computing measures of the reliability of these estimates represent an important part of statistical methods. A discussion of the methods used is presented in Chapter 7.

Classification

The process of grouping a large number of individual facts or observations on the basis of similarity among the items is called *classification*. The items put into one group have certain characteristics in common that differentiate them from the items put in other groups. For example, if sales are classified on the basis of salesmen, all the invoices in one group would be alike in that they were all sales of one man. If the invoices are classified on the basis of territories, the characteristic common to all members of a class would be the geographic location of the dealers buying the goods. The executive of a business or a government agency will normally make use of statistical data in classified form.

A mass of unclassified information cannot be used effectively until it has been organized. The following example emphasizes the importance of classification of data and illustrates that a great deal of classification is taken for granted. Assume that a concern manufacturing cosmetics and selling to wholesalers and retailers made 11,765 individual sales during a certain month. If the sales manager had in front of him the 11,765 sales invoices, he would have all the facts available about the sales of the company for that month, but in this form the data would be practically worthless to the sales manager as far as giving him significant information about the business for that month. On the other hand, properly classified and organized, this same mass of information would tell the sales manager a great deal he would want to know. The following list suggests some of the classifications of the data the sales manager would ordinarily want to have:

1. Amount of sales by sales territories
2. Amount of sales by salesmen
3. Amount of sales by price groups
4. Comparison of all of these with other periods of time.

Most sales managers would want more classifications than are given here if they could be obtained without too much work and expense. However, the sales manager probably would not want to see the individual invoices unless there was something unusual about certain ones. The significant facts about a mass of detailed information are seen more easily in a classification of these facts than in the original data.

Because classification is the first step in analyzing a mass of statistical data, the manner in which the classification is made largely determines what further analysis is possible. There are several recognized basic types of statistical classification, and the methods of analyzing them represent a major part of statistical method. These various classification methods are discussed in Chapter 2.

Presentation

Tabular form, an orderly arrangement of data in columns and rows, is the method most commonly used for presenting statistical data. This form of arrangement reduces a large mass of individual items to a relatively limited number of rows and columns, thus compressing it into a manageable form. If the number of classes is large, the tabular form is the only practicable method of presentation. If the number of classes is very small, the data may be presented in *paragraph form*.

Anyone who deals with statistical data must be able to organize and present data in a compact, logical table. The construction of ordinary tables is not a highly technical problem, although it does require some experience and a thorough understanding of the data and what the table should show. Every businessman should be familiar enough with tabular forms to be able to set up ordinary tables and particularly to be able to read and easily understand any correctly designed table. The construction of very complex analytical tables is properly the function of the statistician, but most tables used by business executives are not this type. Table construction is discussed in Chapter 2.

Analysis

The last step in the use of statistical methods is the analysis of statistical data. Since tabulation of data usually does not yield all the information that it is possible to extract from a given collection of data, the data must be manipulated into its most useful form. It should be emphasized that the collection of data is closely related to the analysis of the data, since the information collected frequently is determined by the analysis intended. Because of this, there is no definite line separating the collection and the analysis of statistical data. In a sense collection and tabulation of data become a part of the analysis.

Most of the chapters in this text are devoted to methods of analyzing statistical data after they have been collected and tabulated. Since this is only an introduction, the methods most generally used will be discussed.

Methods of analysis differ greatly in various situations. A growing proportion of routine decisions are being made automatically according to a criterion set up in advance. A well-known example of this kind of decision making is the automatic reorder system employed by many firms to monitor the inventory levels of thousands of items and to issue a purchase order automatically when the stock of any item drops below a predetermined level.

A disadvantage of such an automatic decision system is that up to now it has been possible to apply it only to relatively simple business problems. A large percentage of the problems on which the business executive makes decisions are too complex to be incorporated into such simple decision models. Although new techniques for problem solving are being developed constantly, the judgment of an experienced executive, supported by as much objective information as can be accumulated, must still be relied upon for the majority of decisions.

STATISTICAL PITFALLS

The utilization of facts expressed precisely as measurable quantities can bring about great improvements in solving problems that beset management, but there are pitfalls to avoid when using the techniques of statistics. Many of the methods of analysis are so complex that, unless the user is aware of all the complexities, he can easily misinterpret the data's meaning. There is a growing danger that the user of the results of sophisticated analyses may be deluded by the sophistication and place too much confidence in the solutions suggested. Because of the abstract nature of much of the methodology of statistics, it is possible to make serious errors, particularly in the logic underlying the selection of the technique. These errors may never become apparent to the people using the results. When a simple set of facts is observed by the person making a decision, he has a certain amount of protection against drawing incorrect conclusions. The more elaborate the analysis of the data on which he must base his conclusions, the less opportunity he has to know the reliability of the information he is using.

A great many examples of the incorrect use of statistical data have been published,[1] but understanding what is wrong with the analysis usually requires some knowledge of the correct method of making the analysis. Throughout this text the various errors that commonly arise in the use of different methods of analysis will be pointed out. The following paragraphs present a few simple but all too common examples of fallacious reasoning as a general warning of the pitfalls that are present when using statistical data. Statistical pitfalls include mistakes in recording data and making calculations, since both produce incorrect information. However, the subject is limited here to improper analysis that is not immediately or easily recognizable but leads to faulty conclusions.

Bias

It is almost impossible to be completely objective or to have no preconceived ideas on a subject, and there are many ways in which such a state of mind can influence the results of any collection or analysis of data. In its simplest form, *bias,* as it is used in this discussion, means that one gives more weight to the facts that support his opinion than to conflicting data. The failure of the election polls to forecast the election of former President Truman in 1948 probably received more publicity than any other statistical error in history. Although the reasons for the error were complex and technical, to a certain degree the overwhelming belief on the part of most analysts that President Truman would not be elected made them fail to recognize the significance of the information they were analyzing. After the election it was not difficult to see that the polls had given a strong warning that the election would be close and that any forecast would be risky.

An extreme case of bias would be a situation in which a course of action is chosen and then statistical analyses are used to find reasons for doing what has already been decided. Some business statisticians have been likened to the boy in school who looked up the answers to his problems in the back of the book, and who now as a businessman works to find reasons to support the decisions made by his boss. The cynical pronouncement that "figures don't

[1] For an entertaining treatment of this subject, see Darrell Huff, *How to Lie with Statistics* (New York: W. W. Norton and Company, Inc., 1954).

lie but liars figure" reflects the same belief that decisions are based on preconceived ideas and that statistics are used only as a pretense that the decision had a logical basis. When statistical data or the results of statistical analyses are used in selling, there is always the suspicion that the facts have been manipulated to tell the most favorable story.

Noncomparable Data

Comparisons are an important part of statistical analysis, but it is extremely important that they be made only between comparable sets of data. Comparing the cost of living at the present time with the cost of living 50 years ago raises questions of comparability, since many of the items in the present-day budget either did not exist or were relatively unimportant 50 years ago. Comparison from one period to another of the number of deaths attributed to a particular disease may show an increasing rate in the more recent year due to more accurate reporting of the causes of death; thus, the figures for the two periods are not comparable.

One of the most important problems of comparing changes in economic activity over a period of time is determining whether the statistical data mean the same thing in each period. In the Census of Population prior to 1950, students and military personnel were counted in the population of their hometowns; but in 1950, students and military personnel were counted in the population of the city in which they were attending school or in which they were stationed. This change in classification caused a considerable degree of noncomparability in cities with either large military establishments or educational institutions with a large number of students who were not residents of the community. Totally wrong conclusions might be drawn from the comparison of population in 1940 and 1950. It is imperative that data always be comparable or the conclusions drawn may be highly inaccurate.

Uncritical Projection of Trends

An uncritical projection of past trends into the future is a pitfall that has done much to discredit the use of statistical analysis. It is true that the only information on which to base a forecast is the record of the past, but many decisions are made by accepting a naive projection of a past trend without determining whether it may reasonably be expected to continue in the same manner in the future. For example, the population of a city may be forecast by projecting the average rate of growth for the past ten years without making any study of the factors that might have caused that growth. A careful analysis might indicate that the factors involved in the past growth were unusual, and changes already taking place might foretell an entirely different rate of growth in the future. An appreciable amount of space is devoted in future chapters to the techniques used in making forecasts. An important part of any such discussion is convincing warning of the dangers of using these techniques and the precautions that should be taken.

Improper Assumptions Regarding Causation

It is much easier to measure the relationship between two events than it is to explain the cause-and-effect relationships that exist. Even if accurate statistical data are compiled to describe what happened, it is not a simple matter to determine why it happened. It is a fallacy to conclude that because two events occur together, one was the cause of the other. Often it is possible to make a case for either factor being the cause, and it may be equally likely that both events were the result of some third factor.

One of the simplest examples of this type of fallacious reasoning would be to measure two factors that are both related to growing population and then designate one factor as the cause of the change in the other. An obviously incorrect analysis would be to conclude that because church attendance in a growing city had increased by about the same percentage as the number of arrests, that either the increase in church attendance was the cause of the increase in arrests or the increase in arrests was the cause of the increase in church attendance. There probably would be little argument with the statement that both arrests and church attendance increased as the population increased, and that neither was the cause of the increase in the other. However, all such relationships are not so simple. A study showed that areas of a city with the highest degree of crowding (largest ratio of persons to square feet of floor space in their homes) also showed the highest incidence of tuberculosis. One might conclude that overcrowding was a cause of tuberculosis, although a more careful analysis might suggest that the economic factors that resulted in overcrowding were also responsible for the high frequency of tuberculosis.

The most direct method of determining the existence or direction of a causal relationship is through experimentation. Simply stated, the variable hypothesized to be the "cause" is deliberately manipulated and the variable thought to be the "effect" is observed. If the hypothesized cause and effect relationship exists, the effect variable takes on a new value. Studies which are not experimental, and most surveys are not, must be carefully evaluated, for a causal influence may be asserted or implied with no basis for doing so.

Comparison with an Abnormal Base

In making comparisons of one period with another, strange distortions can result if the base of the comparison is abnormal, particularly if it is abnormally small. The most satisfactory base for comparison of business and economic data over a period of time is one that is approximately average or normal. Completely erroneous conclusions may be drawn if the base used is abnormal in any respect.

If a salesman reported that farm income in a portion of his territory was 400% higher this year than last, it might be interpreted to mean that business had been unusually good. However, if income last year was very low because of drought, the comparison would not be as favorable as the statement seemed to indicate. An extreme case of an abnormal base is a very small number, such as one or two, used as the base for a percentage. Very large percentage increases will occur in such cases, but these increases generally will not be as significant as they appear.

Improper Sampling

The volume of data available in most studies is so immense that a great deal of emphasis is placed on the study of samples to draw conclusions regarding the larger group from which the sample was drawn. Under the right circumstances sampling is a powerful tool of the statistician. If the sample is properly selected, it will have essentially the same characteristics as the group from which it is taken; but if the sample is not selected properly, the results may be meaningless. Since it is not easy to select a sample correctly, there is great danger that a sample study will not give accurate information. Because the methods to be used are rather technical, the discussion of sampling will be deferred to Chapter 7.

Uncritical Use of the Computer

The computer has become so useful in the field of statistics that there is some tendency to forget its limitations. Like any other powerful device, it should be used with care. The computer will do only what it is instructed to do, and the validity of the results it generates depends on the quality of the information fed into it and the skill with which it is instructed to analyze the data. If faulty methods of analysis are employed, the results may not improve decisions no matter how rapidly the computations are made. If inaccurate data are used in making analyses, the decisions based on the complex analysis of these data may be worse than if only simple methods of analysis were used. The statistician using the computer must use the same care in selecting data and deciding on the methods of analysis to be used as if he were making the calculations on a desk calculator.

The preceding discussion has been general, since specific examples of fallacious reasoning with statistical data are usually subtle and require a fairly complete discussion of the techniques of analysis involved. Further examples of the misuses of statistical methods will be given more detailed treatment in the discussions of the different methods of analysis.

STUDY QUESTIONS

1-1. From your reading and experience list ten areas in which statistics are useful in making business decisions.

1-2. Differentiate the various meanings of the word "statistics."

1-3. Why are statistical data more important to the modern business executive than they were 15 or 20 years ago?

1-4. Describe the differences in the amount and kind of information needed to run the following types of grocery outlets:

(a) A small, one-owner grocery store in a rural community

(b) A supermarket food chain operating 250 stores in 11 states.

1–5. Assume you have compiled a table from the *Statistical Abstract.* When you present the table in a report, the argument is advanced that the data have been taken from a secondary source and are not as accurate as if you had secured them from a primary source. Justify the use of the secondary source, explaining the advantages of using it instead of the primary source.

1–6. How do the problems of data collection by a trade association differ from the problems encountered by a governmental agency collecting industry data?

1–7. What is a variable and what part does it play in statistical analysis? Give two concrete examples to demonstrate what you mean.

1–8. In compiling data on the production of automobile tires, the unit most likely to be used is the number of tires. However, technological progress in the manufacture of tires has improved so much in the past 30 years that a tire may now be expected to be usable for several times the number of miles that it could be used in the early days of the industry. This seems to mean that the unit (one tire) has not remained the same over the years. In other words, the production of 1,000 tires in 1973 has a different significance than the production of 1,000 tires in 1932, in terms of the amount of usage in the tires.

Does the fact that the unit of measurement has changed its significance over the years make it improper to show data on the number of tires produced? Can you suggest any way to take into account the change in the unit of measurement?

1–9. Assume that you are making a survey to determine the number of families living in a given area. Write a definition of "family" for the persons collecting the information.

1–10. What are the advantages of using sample data instead of a complete enumeration? What disadvantages, if any, would this procedure have?

1–11. What advantages do you consider might result from the team approach to problem solving as used in operations research?

1–12. Why is a precise definition of a problem important in making business decisions?

1–13. It is generally a mistake to fail to make use of internal business data or to fail to discover published external data when these two types of information will be of value to an executive in making a decision. Which mistake do you consider is the more likely to be made?

1–14. The 1969 Census of Agriculture was conducted primarily by mail. The mailings were made just before January 1, 1970, and the information collected was intended to cover the full year of 1969. The data-collection phase lasted through September, 1970.

Prior censuses were taken by enumerators; the field work for 1964 was largely completed in November and December of the census year. The Census of Agriculture gave the total of farms in Texas as 205,110 in 1964 and as 213,550 in 1969.

Comment on whether or not you think these two differences (use of mail and timing) affect the comparability of data between the two censuses.

1–15. At a congressional hearing, the comptroller of a large corporation stated that his company wanted the government statistical agencies to reduce the amount of data they were collecting from business firms. He stated that it would not only save the government money but it would also reduce the cost of filling out the reports, which had become a large expense for his company. A few days later a sales representative from the same company appeared at the hearing and protested any cuts being made in the amount of information collected by the government statistical agencies.

Comment on this. What criteria can an appropriations committee use in deciding how much information should be collected by a government agency?

1–16. Do you think it is a proper function of government to collect statistical information on the economy of the country and furnish it free or at a price that merely covers the cost of printing and distributing the publications? If you think it is a proper function of government, how would you decide what information should be collected? It is obvious that all the external data businessmen want cannot be furnished.

1–17. Reports submitted to the Bureau of the Census by business organizations are confidential, and the law prohibits the use of the reports in any way that will reveal information about an individual business concern. This sometimes results in information not being published for certain small areas, since the number of business concerns is so small that such publication would reveal information about an individual concern. Not publishing some information leaves gaps in the data on certain areas, and considerable objection to the law on nondisclosure of census data has occasionally been raised. Do you think the law is wise or do you think it should be changed to permit the publication of data even though they reveal information about individual concerns?

PROBLEMS

1–1. The doctors belonging to a county medical association are often in disagreement with insurance company representatives over the customary and usual fees charged in that community for various kinds of medical services. If you were asked by the organization to set up a statistical office to gather data to use in arbitrating these disagreements, how would you carry out such as assignment?

1–2. Suppose you are working for a wholesale grocery business and the question arises as to whether it would be profitable to close out a large number of very small accounts. One group argues that the cost of keeping the records, delivering the small orders, and collecting the accounts absorb all the profit on the sales. Another group counterargues that the small customers may become large customers, and when they do their business will be profitable. Neither group has any information on which to base its arguments.

You are instructed to assemble enough facts to settle the argument. Make a list of the items of information you think will be needed to make the best decision. In this problem it is not necessary to develop methods of securing the information. Whether the information can be obtained at a reasonable cost can be decided later.

1–3. A business office is considering air conditioning. One of the advocates of installing air conditioning insists that the increased efficiency of the office staff would pay for the cost. The annual cost of air conditioning has been compiled, but measuring the increased efficiency of the office employees raises many questions.

Outline a method for measuring the increase in efficiency that could be expected as a result of air conditioning the office. The measurement should make possible the computation of the total monetary savings of the air conditioning installation.

1–4. A research organization publishes a monthly magazine in which business and economic conditions in a certain section of the United States are analyzed. For years the magazine has presented the data in tables and repeated it in discussions, in the belief that some people would read the text instead of the tables.

Recently a proposal was made to eliminate all repetition of the data in the tables and confine the text to analysis of business conditions. The argument supporting this change was that anyone interested in the statistical data could read the tables, and repeating the information in the text was unnecessary. State and support your recommendations for the policy you feel should be used to present the statistical data.

1–5. A local businessman, who lives in a large timber production area, has asked his banker for a loan to finance the construction of a plant which would produce charcoal for barbecue pits and other forms of outdoor cooking. Assemble all the information you can find on the production and sale of this kind of charcoal, and outline a checklist to follow so no information is forgotten that might be valuable to the banker's decision concerning the loan. It is not necessary to compile the data, just describe the sources.

2 Statistical Data

Many statistical data needed by businessmen have already been collected and are available in published sources. When data are published, the information is presented in tables and frequently is shown in charts and graphs. If the data needed are not available, it is necessary to collect, summarize, and present the information needed in properly designed tables and charts. This chapter presents a brief discussion of the procedures used in performing these operations.

COLLECTING STATISTICAL DATA

It is convenient to distinguish according to their source the kinds of statistical data used by the businessman. The chief distinction is between data that originate inside a particular business organization and data that originate outside it. Information that relates to the operations of an individual business is classified as *internal data* for that business. The source of the data is a record or a report originating inside the business. Information that relates to some activity outside the individual business is classified as *external data*. This classification is significant only with respect to an individual concern, since internal data for one concern are external data for another. For example, the sales of Sears, Roebuck & Company are internal data for that company but external data for any other organization.

Internal data normally must be collected by the firm, since no one else has access to them. External data, however, can be collected by an outside agency. A substantial amount of external data is collected and published by nonprofit agencies, such as governmental departments, trade associations, and universities, and by private business concerns that specialize in collecting and selling information of value to businessmen. Banks, newspapers, business periodicals, and public utilities also collect and publish business data as a service to their customers. The business executive should take full advantage of all sources of external data in assembling the information needed in making a decision, since such information can usually be acquired at much less cost than if the business tried to collect it. However, if the information cannot be obtained from a published source, there may be no recourse but to make a survey to collect the data needed or to employ a statistical organization to make the collection. Normally data acquired in this manner is much more expensive than published data, and such a method is usually considered a last resort.

Sample Surveys Versus Complete Enumerations

Generally whenever the information to be collected exists in great volume, there is an incentive to substitute a sample survey for a complete enumeration of all the information. For example, the information furnished by the inspection department of a manufacturing concern may be so voluminous that it is unwise, as a practical matter, to inspect every component and piece of material used in the factory. There is also the situation in inspection where the test is destructive, such as an abrasion test on rubber soles and heels, which means that a sample *must* be used in determining the characteristics of the material being examined. It becomes important under these circumstances to be able to collect a sample of the information and to use this limited amount of information as a substitute for complete data.

Because of its great savings, sampling has become one of the distinctive characteristics of statistical methods. The problems of designing a sample of units that is representative of the entire group of items and the related problem of determining the degree of accuracy with which the sample measures the characteristics of the larger group are discussed in Part 2. Since the problems of collecting sample data are not significantly different from those of collecting complete data, the discussion in this chapter will not distinguish between them. The problems associated with the use of sample data to estimate values of the whole group from which the sample was taken are also covered in Part 2.

Problems of Collecting Data

Since internal data originate inside the business, collecting the desired information is fairly easy, even though it is frequently scattered among several departments of the business. The main consideration usually relates to the efficient preparation and maintenance of records in order to facilitate speedy low-cost access to information. When collecting external data, however, the statistician must secure information from sources outside the organization. Seldom are these sources especially willing to cooperate by responding with the needed information.

The problems of securing external data that are free of error are particularly critical, since the information is not under the control of the business concern collecting the data. As a result, the possibility of error is generally greater for external data than for internal data.

Since external data deal with a great volume of information that is scattered geographically, wide use has been made of sampling procedures to collect external data. Samples have also been used whenever the internal data are extremely numerous, as in the inspection of raw materials or manufactured products.

Methods of Collecting Statistical Data

In many respects the processes of collecting and tabulating internal statistical data are so interrelated that they are essentially one process. Internal data are usually recorded currently, and the data are organized and made ready for use in the process of keeping records. External data, however, must be secured either by making observations or inquiries. In either case, the records which are gathered become the original documents from which the statistician prepares his tabulations. In collecting external data, therefore, the processes of collection and tabulation are more distinctly separated than in internal data.

When the observation method is used for data collection, the observer must have some systematic method of recording his observations. When the inquiry method is used, the questions are listed on a form, called a *questionnaire* or *schedule,* which provides space for recording the answers.

Examples of data which can be collected via observation methods are: (1) traffic studies, which provide useful information for planning streets and highways; (2) store location studies which measure the potential "walk-in" business by counting the number of people walking past a particular location; (3) fashion analyses, which are made by fashion writers who observe the types of clothing worn in order to secure information on style trends; (4) size of the audience viewing a given television program, measured by a recording device such as the Audimeter which when attached to individual receivers indicates when the set is turned on and to which station it is tuned; and (5) information about the brands of food products used, which is obtained by securing permission to take an inventory of people's pantry shelves.

The amount and type of data that the business statistician can obtain by observation are limited, since so many of the facts needed can be obtained only by questioning someone who has the desired information. When it is possible to secure the information by observation, this method is usually used in preference to inquiry. For example, the Audimeter will generally give more accurate information on the televiewing habits of a family than could be obtained by asking them what programs they watch.

Because of the limited application of the observation method, inquiry is the most widely used method of collecting external data. The standard questioning methods are mail, telephone, and personal interview. The advantages of the different methods vary with circumstances. The best method to use is the one that will secure the information needed with the greatest degree of accuracy and at the lowest cost, with cost being determined in terms of dollars and time.

Using the mail is often the least expensive method, but it is notoriously weak in that the return rate is usually very low. Personal interviews are much more successful in securing respondent cooperation, but they are considerably more expensive than the mail. The telephone method is less expensive than personal interviews, but many kinds of information cannot be obtained accurately by telephone. In some situations the person to be questioned cannot be reached by telephone.

Form in Which Facts Exist

The statistician collecting data from individuals or from business concerns should distinguish between facts that can be supplied from the respondent's records and facts, opinions, or estimates that are not recorded anywhere. The records may be public records or private records of individuals or business concerns. Records of individual businesses will not often be made available to other concerns. On the other hand, a great deal of information, such as data on sales, production, employment, and profits, may be collected from individual concerns by public agencies or trade associations. When such information is given out, the facts are confidential and will be released only as part of a total to prevent linking the information to individual businesses.

Most information collected by business concerns is from individuals and not records. Information given by consumers on commodities used and reasons for their use, reading and televiewing habits, and many other types of data collected for use by marketing executives represent facts supplied from memory and existing in no formal record.

Errors in Surveys

The accuracy of the final data depends to a large degree on the precision of the records from which the data have been collected. A simple example is the age of the respondent. Almost everyone's date of birth is recorded with an official agency, and they may have a certified copy of this record. It might be specified in a questionnaire that the information on age be secured from official records or copies of these records, since this would probably be the most accurate information that could be obtained on age; but in many situations it cannot be obtained. When applying for a passport, a citizen must furnish a copy of his birth certificate, but this would not be practicable in taking a census. As an alternative, the interviewer asks age at last birthday, age at nearest birthday, or date of birth. Sometimes the respondent is asked both age and date of birth to serve as a check. A substantial amount of evidence indicates that individuals frequently do not report their ages accurately, either accidentally or deliberately. However, inquiry is often the only way the information can be secured.

Even though recorded information is preferable in terms of accuracy to unrecorded data, the information desired is not always recorded, and unrecorded data must be used regularly. The accuracy of the findings based on unrecorded data is quite dependent upon the skill with which the questions are asked. This is particularly true when asking for opinions and motives. The wording of questions in marketing research and public-opinion research can have a sizable effect on the responses. The proper wording of questions capable of producing unbiased responses is a job for a skilled psychologist.

TABULATING STATISTICAL DATA

Tabulation is the arrangement of individual items into summary or condensed form and is the first step in the analysis of statistical data. Tabulating statistical data consists essentially of grouping similar items into classes and summarizing each group, usually by counting the number of items or computing totals for each class. The various methods and mechanical devices employed comprise the methods of statistical tabulation.

Because of the large volume of repetitive clerical work involved in tabulating large external surveys and masses of internal data, there are strong incentives to improve these operations. Many mechanical methods have been developed and the speed of processing data has been increased tremendously. For practical purposes, the tabulation of data within any large business or as part of any large statistical survey uses some form of automated data processing.

Assume that a classification is made of sales by departments by months. At the end of each month the sales manager wants the dollar value of sales for each department. The *columnar record* is designed to record each sale and to divide the total amount of each sale among the departments. Each sales invoice is analyzed, and the amount of each departmental sale is entered in the proper column. After the entry is made, the amounts are added and the total checked against the invoice total. After the last invoice for the month is entered, the columns are totaled, giving a classification of sales for the month by department.

When a classification is made by counting the number of items in each group, it is usually done with a *tally sheet,* shown on page 23, instead of a columnar record. A tally mark is recorded on the tally sheet instead of writing an amount as in the columnar work sheet. The classes are set up on the tally sheet by allowing a line rather than a column for each class.

Table 2–1 gives the strength of 125 lots of cotton yarn, represented by the pounds per square inch required to break a skein of yarn. The tally sheet in Figure 2–1 classifies these measurements.

STATISTICAL TABLES

It is difficult for the average person to comprehend a mass of numerical facts unless they are organized in a meaningful way. A *statistical table* is an organized and logical presentation of quantitative data in vertical columns and horizontal rows. A complete table includes titles, headings, and explanatory notes, all of which clarify the full meaning of the data presented. Table 2–2 is a typical statistical table.

TABLE 2-1

STRENGTH OF 125 LOTS OF COTTON YARN

(Pounds per Square inch Required to Break a Skein of 22s)

Lot No.	Pounds	Lot No.	Pounds	Lot No.	Pounds	Lot No.	Pounds	Lot No.	Pounds
1	98	26	112	51	93	76	108	101	86
2	97	27	111	52	80	77	117	102	111
3	99	28	90	53	86	78	93	103	124
4	93	29	89	54	88	79	94	104	92
5	95	30	88	55	80	80	97	105	102
6	93	31	90	56	83	81	88	106	91
7	94	32	85	57	86	82	100	107	95
8	95	33	91	58	88	83	87	108	104
9	98	34	84	59	90	84	95	109	89
10	102	35	88	60	88	85	93	110	127
11	102	36	111	61	90	86	90	111	83
12	108	37	109	62	87	87	80	112	98
13	87	38	84	63	85	88	86	113	92
14	95	39	85	64	84	89	76	114	99
15	107	40	84	65	89	90	80	115	113
16	80	41	110	66	91	91	70	116	93
17	82	42	107	67	86	92	75	117	87
18	79	43	105	68	87	93	85	118	80
19	85	44	103	69	95	94	77	119	83
20	81	45	102	70	98	95	103	120	78
21	86	46	100	71	91	96	106	121	116
22	82	47	100	72	87	97	96	122	122
23	86	48	101	73	89	98	93	123	79
24	83	49	99	74	109	99	106	124	100
25	88	50	102	75	119	100	107	125	95

Source: Cotton Economic Research, The University of Texas at Austin.

TABLE 2-2

SALES AND INCOME OF UNITED STATES STEEL CORPORATION, 1967 TO 1972

(Amounts in Millions of Dollars)

Year	Sales of Products and Services	Income			
		Amount	Percentage of Sales	Dividends Declared	Reinvestment in Business
1967	4,067.2	172.5	4.2	129.9	42.6
1968	4,609.2	253.7	5.5	129.9	123.8
1969	4,825.1	217.2	4.5	129.8	87.4
1970	4,883.2	147.5	3.0	130.0	17.5
1971	4,963.2	154.5	3.1	97.5	57.0
1972	5,428.9	157.0	2.9	86.7	70.3

Source: 1972 Annual Report of United States Steel Corporation.

Strength of 125 Lots of Cotton Yarn
(Pounds per Square Inch Required
to Break a Skein of 22s)

Pounds Required	Tally	No. of Lots
70-74	/	1
75-79	卌 /	6
80-84	卌 卌 卌 //	17
85-89	卌 卌 卌 卌 卌 ////	29
90-94	卌 卌 卌 卌	20
95-99	卌 卌 卌 //	17
100-104	卌 卌 ///	13
105-109	卌 卌	10
110-114	卌 /	6
115-119	///	3
120-124	//	2
125-129	/	1
Total		125

FIGURE 2-1

TALLY SHEET

Table Construction

The basic problem of constructing a statistical table is to formulate the data classification scheme. Once this scheme and its classes have been defined, the task of separating the items into their proper classes is usually a routine operation. In classifications based on differences of kind, the classes are set up on the basis of qualitative differences. An example of this type of classification is separating total sales into sales made to wholesalers, to retailers, and directly to users. When the classes are defined by differences in degree of a given characteristic, the table is based on a quantitative classification instead of qualitative differences of kind. Examples are classifying survey respondents by age or classifying companies by number of employees. If the classes show the geographic location of the items being measured or counted, the result is a geographic classification; and if the data are separated into groups on the basis of time intervals, the table is a time series.

Frequency Distributions

Special problems occur when data to be entered represent the number of items in each class. This type of classification, called a *frequency distribution,* may cause problems to arise because the definition of the classes cannot make use of generally recognized or existing groups. In such cases the determination of class boundaries is always somewhat arbitrary. The following principles should be considered when setting up the classes for a frequency distribution.

Number of Classes. *The number of classes should not be so large as to destroy the advantages of summarization; yet condensation must not be carried too far.* The classes must provide enough groups to show the chief characteristics of the data. At the same time, the classes must not be so numerous that it is difficult to comprehend the distribution as a whole.

The correct number of classes depends somewhat on the number of figures to be classified. A small number of items justifies a small number of classes. If 100 items are classified into 25 classes, there are likely to be groups with few or no frequencies. On the other hand, 25 classes might not be too many for several thousand items. A rule often cited is that an interval that casts the variable into 15 to 25 classes is most satisfactory.

Size of Classes. *Whenever possible, all classes should be the same size,* because this simplifies any subsequent analyses. However, when the items being classified contain a few extremely large or extremely small items, it is usually impossible to set up equal class intervals. Table 2–3 is an example of a classification of this kind. The size of assets ranges from under $1 million to over $300 million. If the data were divided in 12 equal classes, the class intervals might be: under $25 million, $25 million and under $50 million, $50 million and under $75 million, and so on. With such a distribution 5,308 or 95.7% of the associations would fall in the first six classes, and only 236 or 4.3% would fall into the last six classes. The first class alone would contain almost 70% of the associations, leaving the remaining 30% to be distributed among the other 11 classes. Much of the information wanted about the distribution would be lost in the first class. Since a more detailed breakdown of this group is needed, class intervals below $25 million should be much smaller than those for that part of the distribution above $25 million. The principle that class intervals should be equal does not mean that they should be equal if the characteristics of the data can be shown better by using classes of unequal size.

When one or both extremes of a distribution are a considerable distance from the center, it may be necessary to use an *open-end class,* as in Table 2–3. Using this type of class eliminates the need for a large number of classes that may have few or no frequencies. The disadvantage is that there is no way of telling how large the items in the open-end class actually are unless a special note is made of the facts. Therefore, when making a tabulation it helps a great deal to give the total sum of the items going into an open-end class in addition to their number. This simple procedure removes the most striking disadvantage of the open-end class, but many classifications are published without this information, as is Table 2–3.

TABLE 2-3

DISTRIBUTION OF SAVINGS ASSOCIATIONS BY ASSET SIZE
DECEMBER 31, 1971

Asset Size (Millions of Dollars)	Number of Associations	Percentage of Total
Under $1	706	12.7
$ 1 and under $ 5	860	15.5
$ 5 and under $ 10	823	14.8
$ 10 and under $ 25	1,465	26.5
$ 25 and under $ 50	817	14.7
$ 50 and under $100	462	8.3
$100 and under $150	175	3.2
$150 and under $200	64	1.2
$200 and under $300	78	1.4
$300 and over	94	1.7
Total	5,544	100.0

Source: Federal Home Loan Bank Board and United States Savings and Loan League.

Continuous and Discontinuous Variables. Quantities that can take on any value within a given range are called *continuous variables.* All measurements are examples of continuous variables, although they are always rounded to a certain number of places, depending on the degree of accuracy desired in a particular situation. Quantities that occur only at certain values are called *discontinuous,* or *discrete, variables.* These variables are associated with counting. Examples of discontinuous variables are size of families, number of rooms in a house, or number of employees. Since continuous data are rounded, they are often expressed in the same manner as discontinuous data and may easily be mistaken for a truly discontinuous distribution. For this reason the two types should be carefully distinguished.

For discontinuous data, the process of setting the class limits is simple once the width of the class is decided. If each value falls in a different class, the value itself designates the class boundaries. An example of this situation is the classification of families according to the number in the family. When several values of the variable are included in one class, the class limits are designated by stating the largest and the smallest units allowed in the class. For example, the duration of overnight automobile trips might be classified by the number of nights spent away from home. The first class includes those trips involving one night; the second class includes those trips involving two nights away from home. The third class includes those trips requiring at least three nights or as many as five. The first three classes are stated as:

1 night
2 nights
3–5 nights.

All other classes are stated according to the same principle, except that the last class is open-end, such as 21 or more nights.

Real Class Limits. When the variable is continuous, the designation of the class limits cannot be accomplished so simply. The data presented in Table 2-1 on page 22 are shown as a frequency distribution. In Table 2-4, the class intervals are expressed in pounds. The individual amounts given in Table 2-1 are expressed in even pounds since the individual observations were rounded at the time the tests were made. In setting up this tabulation, the determination of class limits was not a problem since they could also be rounded to even pounds. A value of 74 was put in the first class and the next largest observation, which is 75, belonged in the next class. Since the data were rounded to the nearest pound, no value between 74 and 75 was recorded.

TABLE 2-4

STRENGTH OF 125 LOTS OF COTTON YARN

(Pounds per Square Inch Required to Break a Skein of 22s)

Pounds per Square Inch	Number of Lots
70– 74	1
75– 79	6
80– 84	17
85– 89	29
90– 94	20
95– 99	17
100–104	13
105–109	10
110–114	6
115–119	3
120–124	2
125–129	1
Total	125

Source: Table 2-1.

The real upper limit of the first class is not 74 pounds, since pressures as great as 74.5 pounds were rounded to 74. In other words, 74.5 pounds is the real upper limit of the class since 74.5 is the largest value of the variable that can be assigned to this class. The real upper limit of each class in Table 2-4 is .5 greater than the stated limit, since pressures as much as .5 pounds larger than the stated limit are included in each class. By the same reasoning each of the real lower limits is .5 pounds less than the stated limits.

The real class limits for the classes in Table 2-4 are:

69.5–74.5

74.5–79.5

79.5–84.5, and so on.

It is immediately apparent that the class intervals overlap; for example, 74.5 pounds may be assigned to either the first class or the second class. If the rule is to round to the nearest pound, there is no logical method of rounding a value that comes halfway between them except to round half of them to the larger value and half to the smaller.

The most commonly used method of rounding is to carry fractions of more than one half up and those under one half down, although in some cases the fraction is simply dropped regardless of whether it is larger or smaller than one half. This is called *truncating* a number. For example, a common method of asking for age is obtaining the age at the last birthday, which is in effect dropping the fraction of a year since the last birthday. While this method of rounding is less accurate than reporting age at the nearest birthday, it seems to be more easily understood.

When age at the last birthday is given, determination of the real limits of the classes must take the method of rounding the data into account. The class intervals might be stated as:

21–24
25–29
30–34, and so on.

If ages were reported at the last birthday, the real lower limit would be 21 because this is the smallest age that could be entered in this class. The largest value that may be included in the class is not 24, but any value between 24 and 25. Any value less than 25, no matter how small an amount it is less than 25, is recorded as 24. The real limits of the three classes shown above may be stated as follows:

21 and under 25
25 and under 30
30 and under 35, and so on.

It is probably clearer in this case to state that the limits are 21 and under 25 rather than 21 to 24.

CHARTS

Any statistical classification can be expressed in charts as well as in tables. However, most graphs or charts are used for reasons other than the mere presentation of data. A graph is inferior to a table as a method of presenting the data, since one can get only approximate quantities from the graph; however, a graph has the capacity to emphasize certain facts or relationships existing in the data. Tables, especially if there are numerous entries, can make these relationships more difficult to recognize. The graph or chart is not a substitute for a table; it is a different form of data analysis.

Bar Charts

A basic form of graphic presentation is the *bar chart,* which can be employed when the classification is one of kind, degree, geographic, or temporal. The basic principle is to set up a scale and draw bars of the correct length, according to the scale, to represent the different classes.

Figures 2–2 and 2–3 illustrate the following important points to consider in the construction of bar charts:

1. *Title.* The title should tell exactly what information the chart contains.
2. *Scale.* A scale must be provided to allow the reader to interpret the significance of the length of the bars. Guidelines should be drawn across the chart at major points of the scale to aid in estimating the lengths of the bars.
3. *Class designations.* The classes represented by the bars should be clearly indicated at the left of the bars for horizontal bar charts, and at the bottom of vertical bar charts.
4. *Source.* The source of the chart, if previously published, or the source of the data from which the chart was constructed, should be indicated below the chart.

Figure 2–2 shows the 1972 Annual Report of International Harvester Company, which gives the sales of the company classified into major product groups. The length of each bar represents the amount of sales in the designated group.

Geographical distributions may be shown on a bar chart the same manner as qualitative distributions, except the designations for the bars are geographical units instead of differences of kind. Bar charts may also be used to present

Classification of Sales by Major Product Group, 1972

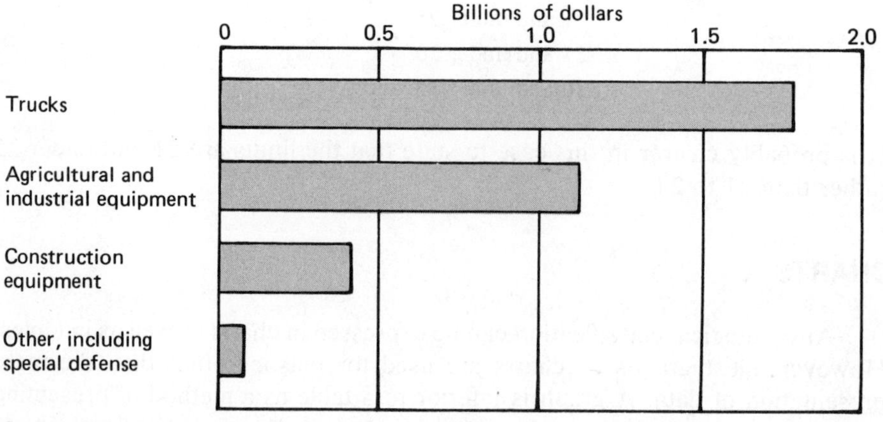

Source: Adapted from 1972 Annual Report of International Harvester Company.

FIGURE 2–2

INTERNATIONAL HARVESTER COMPANY

time series, but it is customary to use vertical bars rather than horizontal bars. The time period represented by each bar is shown at the bottom of the bar, but otherwise the charts follow the same principles as the other bar charts described on page 28.

Figure 2–3 shows the frequency distribution of the data in Table 2–4 (page 26). Charts of frequency distributions always use vertical bars, which are usually identified by a scale that labels the class boundaries. Since a bar is drawn to represent each class, this scale identifies the bars. In this type of chart, both the vertical and the horizontal scales should be carefully labeled.

Table 2–4 is an example of continuous series of data and for this reason there is no space between the bars in Figure 2–3. A bar chart of continuous data is called a *histogram.*

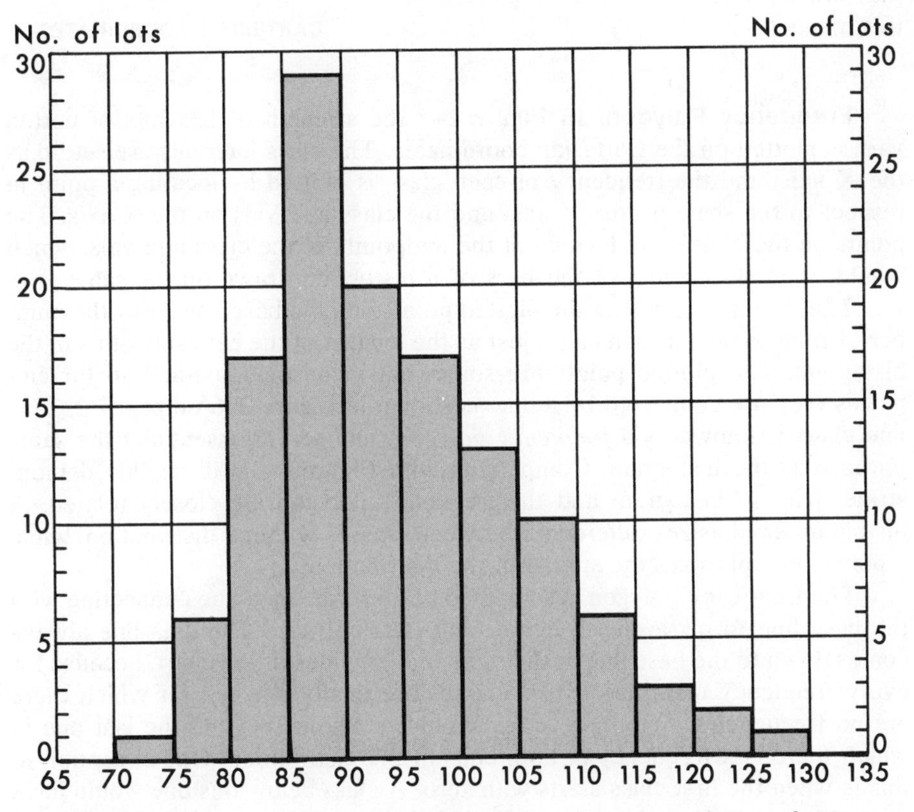

Source: Table 2-4.

FIGURE 2–3

STRENGTH OF 125 LOTS OF COTTON YARN

Line Charts

Most statistical classifications can be shown graphically by a bar chart. Some may also be shown on a line chart. In constructing a line chart, points are located on a grid in relation to two scales on the X and Y axes that intersect at right angles. These scales are known as *Cartesian coordinates;* the points plotted on them are connected by straight lines, as shown in Figure 2–4. Since the two scales are quantitative, it follows that only classifications that are in serial order can be plotted in this manner.

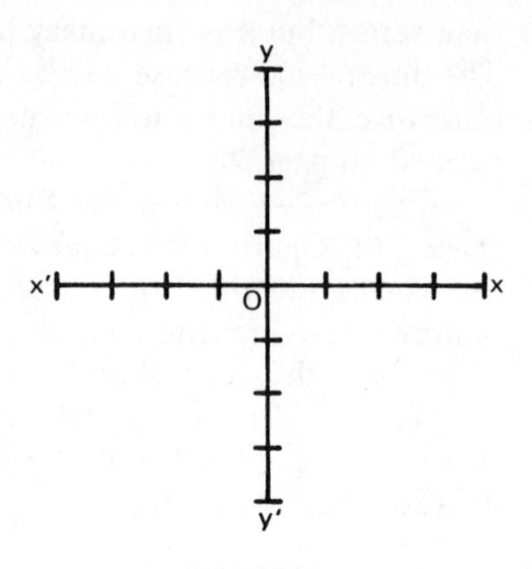

FIGURE 2–4
CARTESIAN COORDINATES

Frequency Polygon. In Figure 2–5 the strength of 125 lots of cotton yarn is plotted on the Cartesian coordinates. The class intervals are set up as the X axis, and the frequency of each class is plotted by locating a point in respect to the scale on the Y axis and the class intervals on the X axis. The points on the X axis are located at the midpoints of the class intervals, which would be at the middle of the bars of a histogram drawn on the same grid.

The distances between the plotted points and the base line show the number of frequencies in each class, just as the lengths of the bars show this in the histogram. The plotted points are somewhat inconspicuous, and so for emphasis they are connected by a line as shown in Figure 2–6 on page 32. This line chart is known as a *frequency polygon,* and serves essentially the same purpose as the histogram. Comparison with Figure 2–3 will quickly demonstrate that the histogram and the frequency polygon are closely related. A histogram emphasizes differences between classes within a distribution, while a polygon emphasizes the spread of the distribution.

The frequency polygon is so named because the data line connecting with the base line forms a closed figure; that is, a polygon. The data line always comes down to the base line at the right and left sides of the chart, because for every frequency distribution two classes are finally reached for which there are no frequencies. One more class should be shown beyond the last one in which there is a frequency at both ends of the distribution. One exception to this is when the first class starts with zero. A class below this one would have minus values, and the interpretation of this would be meaningless in most cases. In this case, bring the plotted line down to the base line at zero.

Instead of plotting the *number* of frequencies, it is possible to construct a frequency polygon with the frequencies expressed as *percentages of the total number of items.* This is treated as any other frequency distribution in drawing a frequency polygon, except that the Y axis is labeled "percentage of" instead of "number of." Two or more frequency distributions containing a

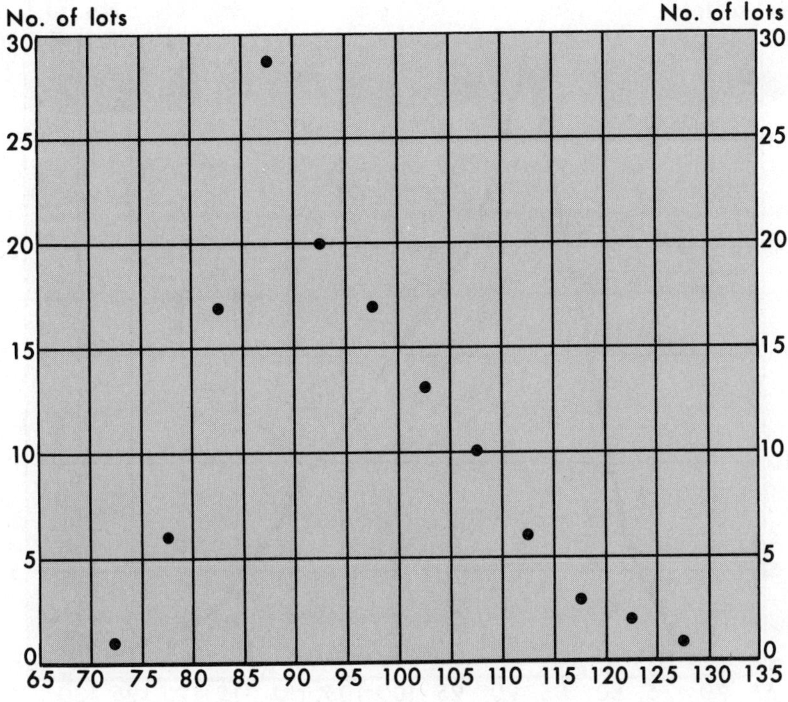

Source: Table 2-4.

FIGURE 2–5

STRENGTH OF 125 LOTS OF COTTON YARN

different number of items can be plotted on the same chart, after expressing the frequencies for each distribution as percentages of the total number in the distribution. Figure 2–7 on page 33 compares the average monthly salaries of file clerks and typists in the state of Tennessee as of June, 1971. Each plotted point represents the percentage of employees receiving a salary which falls within a particular salary class.

Time Series. In plotting time series as line charts, the time classification is always laid out on the horizontal axis, with the data scale on the vertical axis. The time scale on the horizontal axis is a continuous scale of the period covered by the data. The vertical rulings always indicate specific dates, but the number of rulings varies considerably for different charts.

Several different methods of designating the time scale are shown in Figure 2–8 on page 34. In general the scale designation is written in the middle of the respective time period, as demonstrated in Example A. The middle of each space is July 1 of the year indicated. In Example B the solid vertical line above the year is July 1 of that year. Thus, January 1, 1970, is between the first and

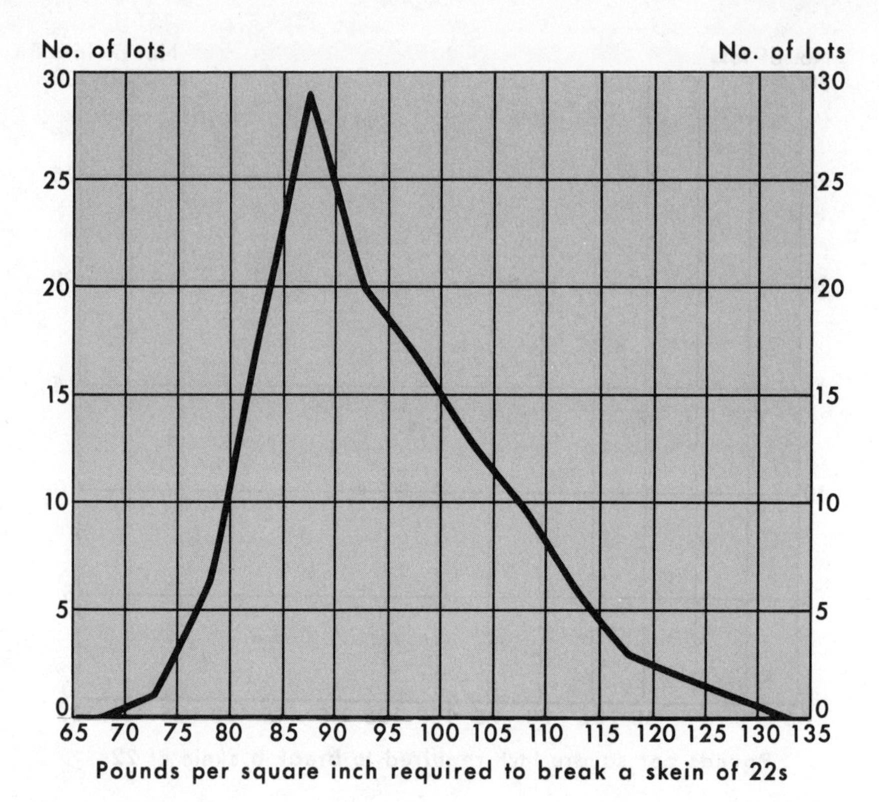

Source: Table 2-4.

FIGURE 2–6
STRENGTH OF 125 LOTS OF COTTON YARN

the second solid vertical lines, and January 1, 1971, is between the second and third lines. The space between the first and second dotted lines represents the whole year 1970.

In Example C every fifth year is represented by a guideline, and the intervening years are shown with short lines. In this example the guidelines and the short lines represent the middle of the years.

Example D shows another time scale with a guideline every five years, and short lines to mark the intervening years. The scale designations are written in the middle of the spaces as in Example A, in this case labeling every fifth space.

Example E is designed for monthly data and parallels Example A for annual data in that the scale designations are written directly below the spaces rather than below the vertical lines. Some graph paper that is preprinted to handle monthly data is laid out as in Example F. This design may cause confusion since the vertical lines represent the middle, rather than the end, of a time interval; in those cases a year ends between and not on a vertical rule.

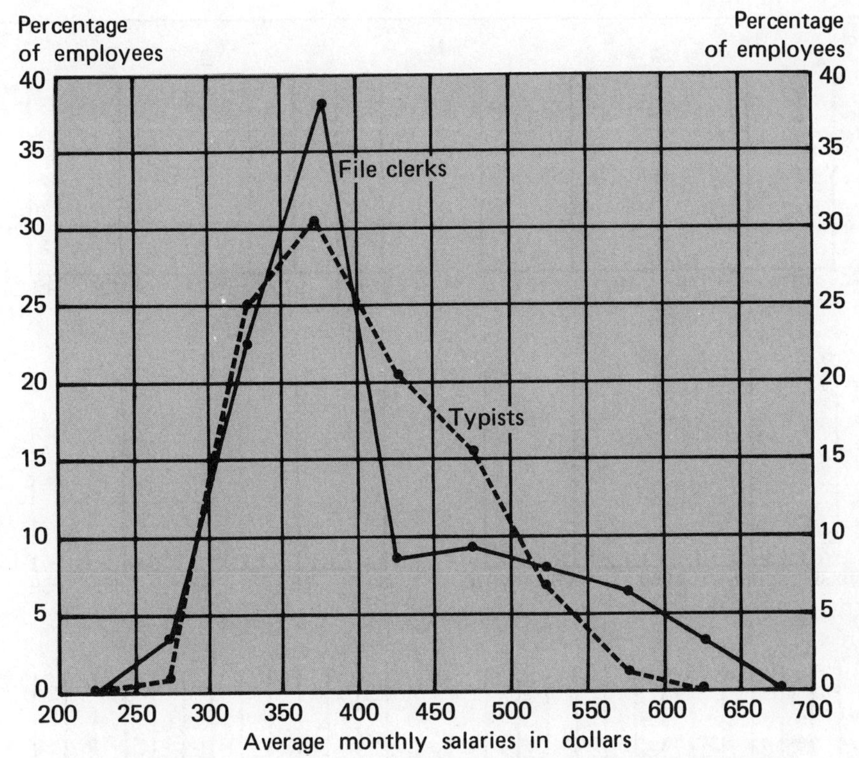

Source: Adapted from U.S. Department of Labor, *Regional Report 17* (May, 1972).

FIGURE 2–7

AVERAGE MONTHLY SALARIES OF FILE CLERKS AND TYPISTS IN THE STATE OF TENNESSEE, JUNE, 1971

Locating Point and Period Data on the Time Scale (X Axis). Data classified on the basis of time periods may be either point or period data. *Point data* are measurements of some characteristic *at a given point of time,* such as prices, inventories, or number of employees. *Period data* are accumulations of amounts *over a period of time,* such as the number of cars sold, tons of coal mined, or amount paid out in wages during a week, month, or year. Figure 2–9 (page 35) is an example of point data, and Figure 2–10 (page 36) is an example of period data. This distinction is also found in accounting, where point data is balance sheet information, representing the status of the business at a given date. Period data is the information from the income statement, which covers operations over a period of time.

When point data are taken at frequent intervals, it may be desirable to combine the numerous individual values. For example, the price of wheat in Chicago for the month of January might be the price at a given date in the month or the average of the prices of wheat taken at different dates during the month. If the price of wheat at the close of the day were compiled for each day

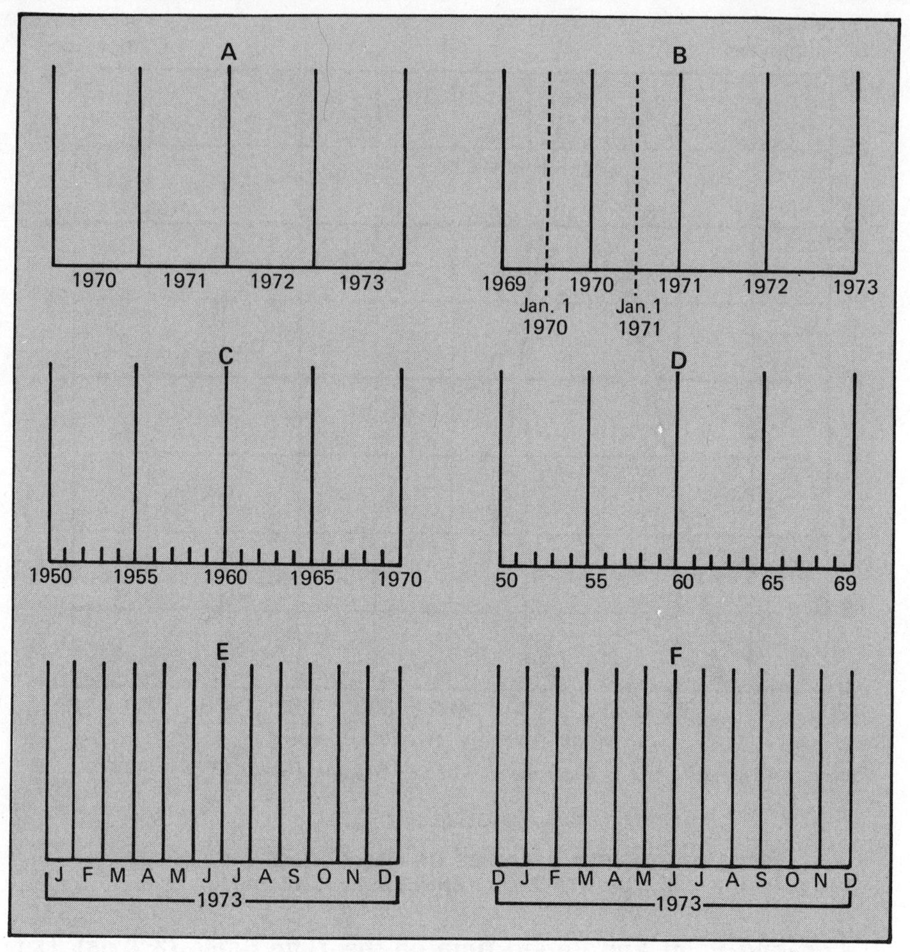

FIGURE 2–8

EXAMPLES OF TIME SCALES

of January, an average of these prices could be used as the January price of wheat. The individual prices would be point data, and the average of the daily prices would be period data.

Locating point data on the time scale is simple and logical. If inventory data as of January 1 of each year are plotted, a point is located at the beginning of the space designated to represent each year and the points are connected by straight lines. If data represent the number of employees on the payroll on the fifteenth of each month, the points are plotted in the middle of the space representing each month. In every case, point data are located according to this rule.

Plotting period data for a line chart is slightly different, since the line chart must represent the data for a period of time by the location of one point. It is standard practice to plot the point for a period in the *middle of the space representing that period.* If the vertical guidelines are erected at the middle of the

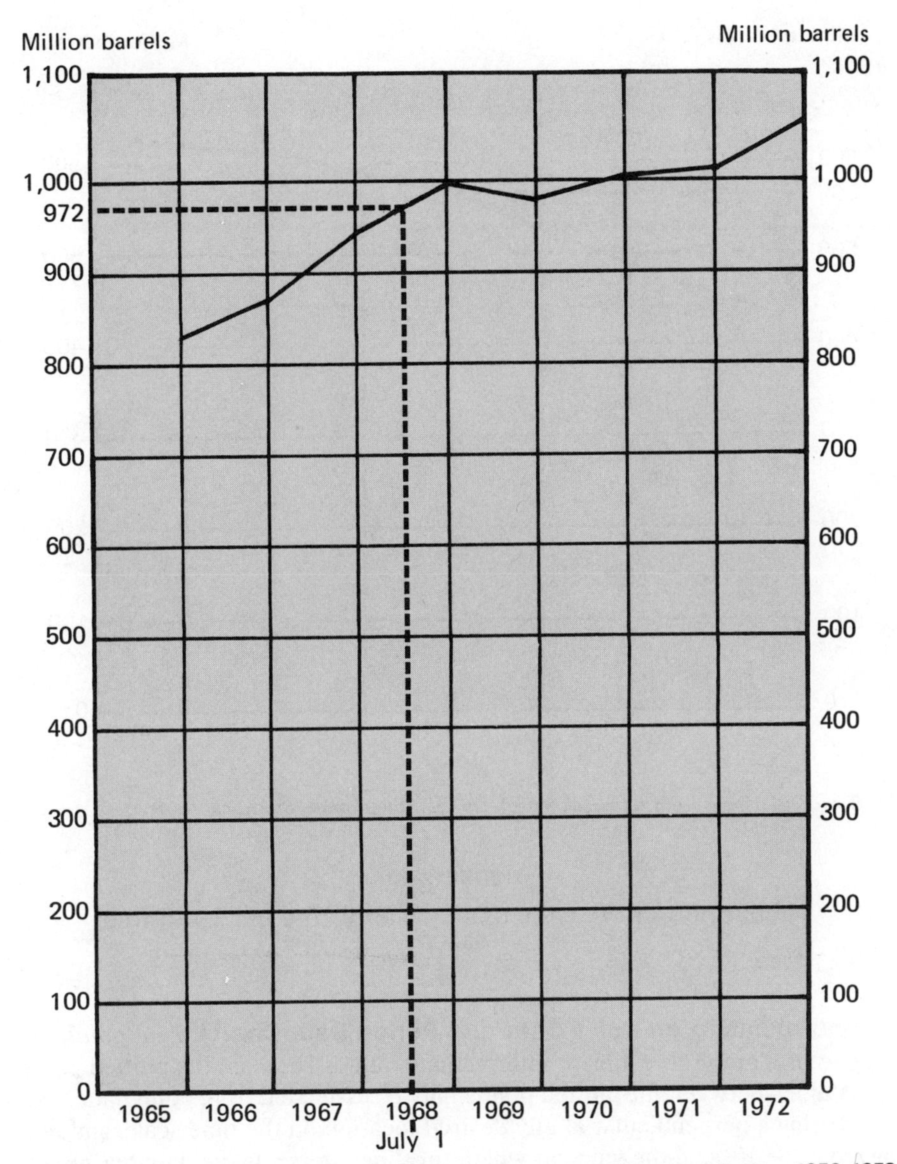

Source: Adapted from American Petroleum Institute, *Annual Statistical Review, 1956–1972.*

FIGURE 2–9

STOCKS OF CRUDE OIL IN THE UNITED STATES AT END OF YEAR, 1965–1972

time period, as in Example B on page 34, the points are plotted on the vertical guidelines. If the guidelines are at the beginning and end of the period, as in Example A, the data are plotted between the lines. If care is taken to designate the beginning and end of the time periods, no trouble is experienced with this type of chart.

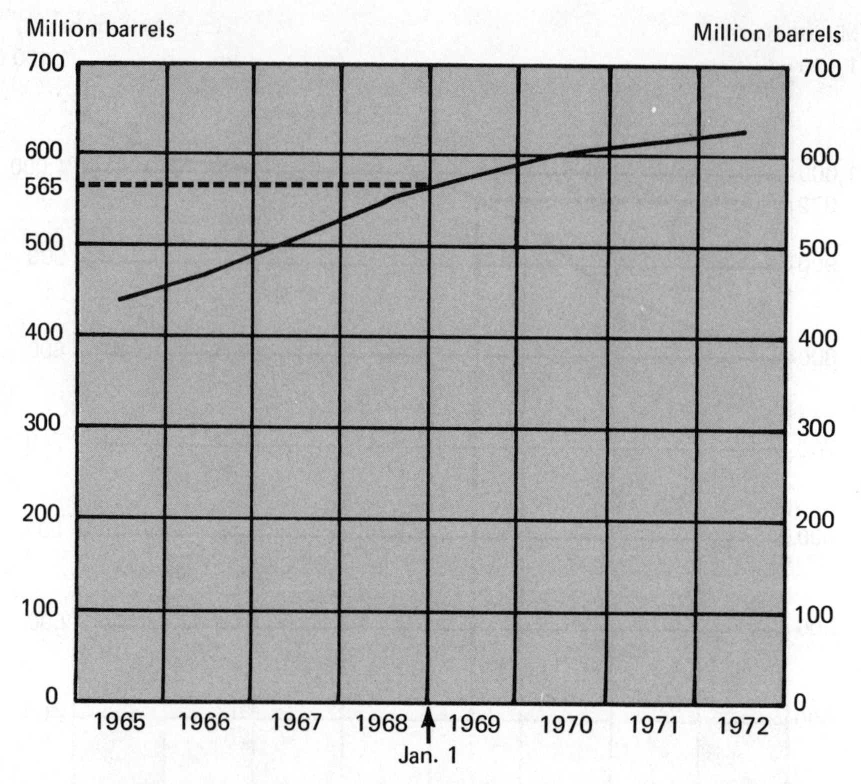

Source: Adapted from American Petroleum Institute, *Annual Statistical Review, 1956–1972.*

FIGURE 2–10

PRODUCTION OF NATURAL GAS LIQUIDS IN THE UNITED STATES 1965–1972

Interpolating on Point Data and Period Data Charts. *Interpolation* is the use of a graph to estimate intermediate values between the plotted points.

Values between the plotted points may be estimated from point data charts by erecting a perpendicular at any desired location on the time scale, and reading from the *Y* scale the value at which this line intersects the data line. Figure 2–9 shows stocks of crude oil in the United States at the end of each year, with the point plotted at the end of each space representing a year. The lines connecting the plotted points may be taken as estimates of the changes in stocks of crude oil. For example, in Figure 2–9 the stocks of crude oil on July 1, 1968, is estimated to be about 972 million barrels. This amount is read from the Y scale as the distance of the plotted line above the base line at July 1 on the time scale. This method of estimating amounts between two points is known as *straight line interpolation.* Such an estimate assumes that the number of producing wells changed at a constant amount between the two dates. If this is not true, the estimate will not be correct; but this method is commonly used in the absence of any other information. If information is available on how the change actually occurred, these facts should be used instead of the straight line interpolation.

Interpolating on a graph of period data is done in the same manner, although at first it may appear to be different. In Figure 2–10 each plotted point represents the number of barrels of natural gas liquids produced in the designated year. The data are plotted in the midpoint of the year and indicate the production for a one-year period beginning six months before the plotted point and ending six months after it. By interpolation it is determined that January 1, 1969, has a value of 565 million barrels, which is an estimate of the natural gas liquids produced in a one-year period beginning six months before and ending six months after January 1, 1969. This would be the 12-month period from July 1, 1968, to June 30, 1969. The production for any 12-month period can be estimated this way. Any data estimated from this chart will represent an annual figure with the point located in the middle.

STUDY QUESTIONS

2–1. The statement has been made that in making a decision it is extremely wasteful to overlook internal data. Is it any more wasteful to overlook internal data than to fail to use external data related to the solution of the problem? Give reasons for your answer.

2–2. What are the advantages of collecting external data by observation rather than by inquiry? What are the advantages of inquiry over observation? Which of the two methods is more widely used in collecting external data for business? Why?

2–3. What are the factors that determine whether personal interviews, mail, or the telephone should be used in a survey?

2–4. In studies of consumer buying habits and related subjects, it is often helpful to seek the advice of a psychologist in designing a questionnaire that will be used to collect data. On the other hand, it is much less likely that a psychologist would be employed to design a schedule that would be used to collect data from a business concern. What characteristics of the two situations mentioned would influence the usefulness of a psychologist in the design of a questionnaire?

2–5. Comment on the statement that the last half of the twentieth century will be known as the "age of the computer." Do you agree or disagree with it?

2–6. A business starts a campaign to reduce the number of reports prepared and circulated, since there is some evidence that many of the reports no longer serve any useful purpose. In the course of the discussion there is some argument about the difference between reports and records. Discuss this difference and point out how you distinguish between reports and records.

2–7. Why is it sometimes not convenient to have all the classes of a frequency distribution the same size? What precautions must be taken in using the data in a frequency distribution with unequal class intervals?

2–8. What characteristics of a distribution make it necessary to use open-end classes?

2–9. Explain the difference between a continuous variable and a discrete variable. Give two examples of both types of variables that might be used by a business statistician.

2–10. A group of students is instructed on a questionnaire to report their age at their last birthday.

(a) What are the real limits of the following classes?

21 to 24 years
25 to 28 years

(b) What are the midvalues of these two classes?

2–11. Employees are asked on a questionnaire to report their age at their nearest birthday.

(a) What are the real limits of the following classes?

21 to 24 years
25 to 28 years

(b) What are the midvalues of these two classes?

2–12. What are the advantages of presenting statistical data graphically? What are the advantages of the tabular form of presentation?

2–13. For a given frequency distribution, how would you decide whether to draw a bar chart (histogram) or a line chart (frequency polygon)?

2–14. When point data for a number of dates are averaged, is the average point data or period data? (An example of this situation is an average of the number of employees on the payroll of a business on the fifteenth of each month for a particular year.)

PROBLEMS

2–1. Assume that you are employed in a large city by a dairy that has been manufacturing *Mellorine*—a frozen dessert made of vegetable fat instead of butterfat. *Mellorine* has been sold, at a substantial reduction in price, in competition with your brand of ice cream. You have analyzed sales figures for the past three years and find that the percentage of *Mellorine* sales has increased steadily. You have tabulated the sales data by areas of the city and find that the increase in the percentage of *Mellorine* sales has been greatest in the areas with the lowest income. It is felt that the popularity of *Mellorine* in low-income areas can easily be attributed to the fact that the price of *Mellorine* is about 40% below the price of the best grade of ice cream that you market.

Questions have regularly arisen regarding the consumer acceptance of *Mellorine*—such questions as whether the increase in sales has been due entirely to price, and whether consumers can tell any difference between *Mellorine* and other brands of ice cream even if they know what the difference is. The label explains what *Mellorine* is made of, but there is some question as to whether consumers read the labels. Finally, a decision is made to collect information about the consumer acceptance of *Mellorine*, in the hope that questions such as those given above can be answered.

You are instructed to design a questionnaire that will supply management with the information needed about the sale of *Mellorine* and ice cream. It is considered important that each product be fitted to its proper market, but management does not know whether it possesses all the facts. The purpose of your survey is to supply all the facts that can be collected from consumers about *Mellorine* and its acceptance.

2–2. The table at the top of page 39 gives the ages of 210 employees of a governmental agency employing a large number of field workers.

Empl. No.	Age (Yrs.)	Empl. No.	Age (Yrs.)	Empl. No.	Age (Yrs.)	Empl. No.	Age (Yrs.)	Empl. No.	Age (Yrs.)	Empl. No.	Age (Yrs.)
1	67	73	35	133	59	206	30	290	29	380	41
3	34	74	38	135	53	207	44	293	28	382	30
6	36	75	36	140	36	211	40	294	41	383	39
7	48	76	46	144	61	212	32	299	58	384	26
12	49	79	43	147	46	213	51	301	38	386	26
13	31	80	34	148	41	215	35	303	29	387	33
14	61	81	62	150	29	220	34	305	34	389	28
15	34	83	69	151	27	221	33	308	42	390	30
16	43	84	50	152	58	222	39	313	35	392	31
21	45	85	28	154	41	223	37	317	42	393	43
22	38	86	44	157	50	224	29	322	42	395	43
23	32	87	43	158	55	225	37	324	41	396	26
24	27	90	60	159	60	230	68	327	64	397	28
28	61	91	39	160	34	232	26	340	26	401	44
34	29	92	47	161	30	235	51	342	31	402	26
35	47	95	42	162	42	236	45	343	50	403	28
38	36	99	49	163	51	237	37	344	29	405	35
39	50	101	44	165	43	239	31	345	56	406	30
40	46	102	37	167	43	241	42	350	25	407	40
41	30	103	30	168	27	243	39	351	41	408	49
42	46	107	42	169	33	250	44	354	31	411	38
44	32	108	42	170	56	251	37	355	27	412	25
45	30	109	31	172	43	254	55	356	29	414	34
47	33	111	34	173	43	257	44	358	30	415	26
48	45	113	24	174	37	260	47	359	37	416	33
50	49	114	38	175	42	261	31	362	30	417	37
52	48	116	59	176	33	264	32	363	53	418	33
53	41	119	41	181	50	265	37	365	33	419	37
56	53	121	43	183	44	266	30	366	31	420	26
57	36	122	50	184	29	267	32	367	40	421	39
61	37	123	40	189	39	268	30	368	47	422	39
62	47	124	41	191	34	272	42	369	35	423	24
63	47	125	46	193	39	273	56	370	42	424	32
64	30	126	43	195	33	275	32	372	37	426	24
66	46	127	48	196	35	278	56	373	35	432	34

Source: Records of the agency.

(a) Set up a tally sheet and classify the employees according to age.
(b) From the tally sheet prepare a table in proper form.
(c) Summarize briefly what this table shows about the distribution of ages of the employees in the agency.
(d) Construct a histogram of the distribution.
(e) Construct a frequency polygon of the distribution.

2–3. The following table shows the weight in ounces of 84 cans of tomatoes that were filled by an automatic filling machine. All of the cans vary slightly when weighed, although the company intended to have cans of equal weight.

(a) Set up a tally sheet and classify the cans of tomatoes according to weight.
(b) From the tally sheet prepare a table in proper form.
(c) Summarize briefly what this table shows about the distribution of the weights of the cans of tomatoes.
(d) Construct a histogram of the distribution.
(e) Construct a frequency polygon of the distribution.

14.96	14.78	14.77	14.71	14.83	14.74	14.85
14.81	14.84	14.88	14.89	14.79	14.76	14.89
14.93	14.88	14.72	14.78	14.92	14.97	14.67
14.84	14.68	14.70	14.83	14.91	14.80	14.79
14.70	14.75	14.80	14.98	14.72	14.85	14.78
14.88	14.73	14.76	14.84	14.82	14.99	14.73
14.82	14.67	14.75	14.86	14.89	14.79	14.81
14.78	14.86	14.81	14.96	14.76	14.79	14.93
14.89	14.92	14.82	14.84	14.72	14.85	14.73
14.67	14.83	14.78	14.70	14.73	14.61	14.94
14.79	14.97	14.96	14.63	14.95	14.83	14.80
14.60	14.63	14.77	14.65	14.70	14.74	14.82

Source: Company records.

2–4. The following table shows the number of member associations of the Federal Home Loan Bank System from 1963 through 1971. Show these data as a bar chart.

Year	Number
1963	4,960
1964	4,985
1965	5,006
1966	4,982
1967	4,919
1968	4,849
1969	4,747
1970	4,601
1971	4,471

Source: Federal Home Loan Bank Board, *Savings and Home Financing: 1972 Source Book* (July, 1973).

2–5. The following table gives net assets of mutual funds in the United States from 1961 through 1971. From these data construct a line chart showing the total net assets at the end of each calendar year.

(a) Is the line chart an example of point data or period data?
(b) Explain the significance of the values represented by the portion of the line between the plotted points.
(c) Estimate the mutual fund assets on July 1, 1970.

Year	Millions of Dollars
1961	22,789
1962	21,271
1963	25,214
1964	29,116
1965	35,220
1966	34,829
1967	44,701
1968	52,677
1969	48,291
1970	47,618
1971	55,045

Source: Investment Company Institute, *1972 Mutual Fund Fact Book* (October, 1973).

2-6. The following table shows private housing starts in the United States, 1961–1971. Construct a line chart showing this data.

(a) Is the line chart an example of point data or period data?

(b) What would be the meaning of the point on the line halfway between the 1,507,700 starts in 1968 and the 1,466,800 starts in 1969?

Year	Number of Units
1961	1,313,000
1962	1,462,700
1963	1,610,300
1964	1,528,800
1965	1,472,900
1966	1,165,000
1967	1,291,600
1968	1,507,700
1969	1,466,800
1970	1,433,600
1971	2,048,200

Source: United States Savings and Loan League, *1972 Savings and Loan Fact Book.*

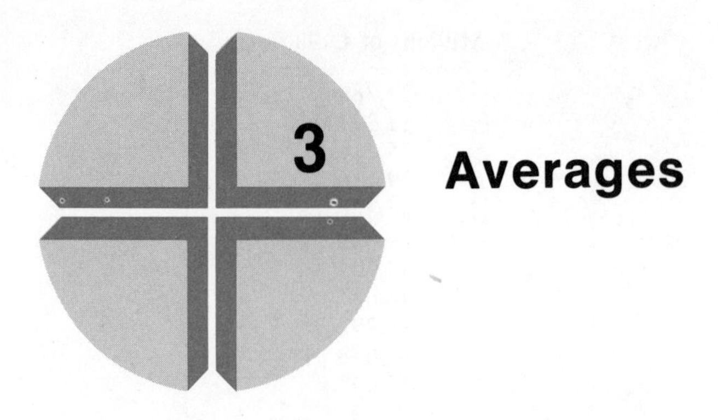

Averages

Chapter 2 shows how information provided by a large number of observations can be condensed into a relatively small number of classes. This chapter discusses an even more concise value, the average.

USES OF AVERAGES

An *average* summarizes a mass of individual observations with only one value. This high degree of summarization results in some information loss, but it has the advantage of providing a very concise summary of the information. Usually more than one summary figure is provided to compensate for the loss. The summary measures used in addition to averages are described in Chapter 4.

A Common Denominator

One important use of averages is the reduction of two or more aggregates to a common denominator in order to make direct comparisons between them. Comparing the totals for time periods of varying lengths, for example, the production of crude oil during different months, demonstrates this use. Total production in the United States in January, 1973, was 284.6 million barrels, and in February, 1973, 262.5 million barrels. Since January has 31 days and February, 28, the two aggregates are not strictly comparable. By computing the average production per day for each month, the monthly production is reduced to the common denominator of days. The resulting daily production rates of 9.18 million barrels per day in January and 9.38 million barrels per day in February more accurately indicate the level of activity in the industry during the two months. The total oil production in February was 7.8% lower than in January, but the production per day in February was 2.2% greater than in January.

Many similar uses of the average to provide a common denominator occur in analyzing data over periods of time when the time intervals are of unequal length. For example, the factory sales of automobiles for the first six months of 1973 totaled 7,001,900 cars, as compared with total factory sales during 1972 of 11,270,700. Expressing each of these aggregates as the number of automobiles per month makes them more nearly comparable. Thus, production in 1972 was at a level of 939,225 per month, and for the first six months of 1973, at a level of 1,166,983 per month.

Many ratios used for making comparisons between aggregates fall into the category of reducing data to a common denominator. The well-known batting average is actually the ratio between the number of hits and the number of times a player was at bat. Simply to compare the number of hits of different players during the season would be rather meaningless, since much of the variation among players is related to the number of times each is at bat.

As another example, a common denominator is needed to express the increasing number of deaths in traffic accidents, which rose from 34,763 in 1950 to 54,700 in 1971. This represents an increase of 57% in 21 years. However, the population grew considerably during the period. For a significant comparison the data should be reduced to a common base by expressing the ratio of deaths in traffic accidents to population. Traffic deaths were 23.0 per 100,000 population in 1950 and 26.7 per 100,000 population in 1971. In other words, the rate of traffic fatalities, taking into account the increase in population, increased at a much lower rate between 1950 and 1972 than the number of accidents. For some purposes such a comparison is more meaningful than the change in the total number of traffic deaths; however, comparisons made for other reasons, such as planning emergency equipment needs, would be most meaningful if the total number were used.

When making comparisons between different time periods and geographic areas, the data are frequently reduced to a comparable basis with respect to population. For example, it may be meaningful to compare total income received by individuals in the various states, but for other uses it is more significant to reduce the income in the different states to a common denominator. Expressing the ratio of personal income in a state to the population of that state allows for the variations in population in the various states. For example, in 1972 the total personal income in New York was $97.69 billion, and in Nevada, $2.75 billion. Personal income per person in New York, however, was $5,319 compared with $5,215 per person in Nevada. The ratios of total income to population in the two states represent significant information that is quite different from the data represented by total income.

Ratios such as those in the preceding paragraph involve: (a) the amount of the numerator and the unit in which it is expressed; and (b) the amount of the denominator and the unit in which it is expressed. Any unit used for comparison must be stated in the ratio. Whenever the denominator represents the number of persons, the phrase "per capita," which means "per person," is used. The term per capita can be used only when the common denominator is the number of persons. If some other unit, such as the number of families, is used as the common denominator, the ratio is computed by dividing the numerator by the number of families in each state. The ratio is then stated as a given amount of income per family.

Birth rates and death rates are computed by finding the ratio between the number of births or deaths and the population. The population per square mile is a ratio which serves as a measure of the density of population in different geographic divisions of the country. These examples use ratios having like denominators to make data, which otherwise would be noncomparable, suitable for direct comparison.

A Measure of Typical Size

A second important use of averages is to provide one summary figure that is typical of all the items in a distribution of individual items that are essentially different. The greater the diversity or variability among the items, the less satisfactory such a measure will be; but many situations need this type of summary statement. An average is used when the items are, in general, different. When they are all the same size, an average is not needed to summarize them.

In popular usage the word average is frequently considered to be a synonym for typical, as when one says that a certain employee's ability is "about average." If a sales manager were asked the average age of his salesmen, he might think of the most common age, demonstrating the concept of the average as the value most typical of the whole group.

The significance of an average used as a measure of typical size should be distinguished from an average used as a common denominator. Per capita personal income in Nevada was computed to be $5,215 in 1972. It is possible to compare the level of income in Nevada with that in New York by making allowance for the difference in population of the two states. It does not necessarily follow that such a ratio may also be used as a measure of typical size of individual incomes in either state. In using averages as typical of a whole group of individual items, there are very definite precautions that must be taken to insure that the average is actually typical of the individual items in the group it is to represent.

A popular magazine appealing primarily to homeowners published a table giving the cost of operating water heaters of various types. The table stated that 4,000 cubic feet of gas were used monthly by the average family with an automatic-type heater. The cost of operating this heater for a month was $3.78 when natural gas was used. Reader reaction to the table was mixed; some wrote that the amount of gas consumed was too much, and others wrote that the amount was not enough. There was too much diversity among the individual figures averaged to find any one value that was typical of the whole group; that is, typical in the sense that a substantial proportion of the group was at or near the average amount. This resulted basically from the lack of homogeneity of the data—too many different kinds of families and too many different gas rates were represented. The situation could have been improved if the data had been broken up into groups and averages had been computed for the different groups. Each average would have been more typical of its group than the one average was of all the families included.

When using an average as typical of all the values of a variable, the data should be homogeneous for the purpose at hand. The use of homogeneous data usually results in a small enough dispersion of the individual items so that an average that is really typical can be computed. The dispersion of the data, which is an important characteristic of the data themselves, is discussed in Chapter 4.

THE PRINCIPAL AVERAGES

The three most common averages are the arithmetic mean, the median, and the mode. The arithmetic mean is the best known and most widely used in statistical work. The geometric mean and the harmonic mean, though not as well known, have important specialized uses.

Arithmetic Mean

The *arithmetic mean* of a series of values of a variable $X_1, X_2, X_3, \ldots, X_N$ is the sum of the values divided by the number of values. Expressed as an equation, this operation may be written

$$\mu = \frac{\Sigma X}{N} \tag{3-1}$$

where

μ (mu) = the arithmetic mean[1]
Σ (sigma) = the sum of or the summation of
X = the value of an individual variable
N = the numbers of values of the variable X.

The formula is read: The arithmetic mean is equal to the sum of the values divided by the number of values.

The arithmetic mean is usually referred to as the *average*. It is also called the *common average* and the *mean*. Any of these terms used without qualification ordinarily refers to the arithmetic mean.

The mean strength of the 125 lots of cotton yarn in Table 2–1, page 22, is computed by summing the 125 values and dividing the sum by 125.

$$\mu = \frac{11{,}715}{125} = 93.72 \text{ pounds per square inch.}$$

[1] The terminology used in this chapter will apply when the data represent a complete enumeration. If the arithmetic mean is computed from a sample, a different symbol will be used to represent the mean. The subject of samples is introduced in Chapter 7.

Since the arithmetic mean is based on the sum of the values of the variable, it can be computed even if the individual values are not available. For example, the Institute of Life Insurance reports that the total amount of individual ordinary life insurance policies sold in 1971 was $132,676,000,000, consisting of 11,372,000 policies. The average size policy is computed

$$\mu = \frac{132{,}676{,}000{,}000}{11{,}372{,}000} = \$11{,}667.$$

Computing the Mean from a Frequency Distribution—Long Method. In situations where the individual values of a variable are not given, it is important to have a method of finding the mean. Before computers were available the method of making an approximation of the mean from a frequency distribution was used even when all the variables were known. Since this method is reasonably accurate and is a considerable time-saver when making the computations, it is often used even when all the values of the variable are known.

Computing the mean from a frequency distribution assumes that the average value of all the items (f) falling in a class is the midpoint of that class. For example, in Table 3–1 it is assumed that the one lot of yarn in the first class has a strength of 72 pounds per square inch, the midpoint of the class. The six lots of yarn in the second class are assumed to have an average strength of

TABLE 3–1

COMPUTATION OF THE MEAN FROM A FREQUENCY DISTRIBUTION (LONG METHOD)

Strength of 125 Lots of Cotton Yarn
(Pounds per Square Inch Required to Break a Skein of 22s)

Pounds per Square Inch (Class Interval)	Number of Lots f	Pounds per Square Inch (Midpoints) m	Total Pounds of Yarn Strength fm
70– 74	1	72	72
75– 79	6	77	462
80– 84	17	82	1,394
85– 89	29	87	2,523
90– 94	20	92	1,840
95– 99	17	97	1,649
100–104	13	102	1,326
105–109	10	107	1,070
100–114	6	112	672
115–119	3	117	351
120–124	2	122	244
125–129	1	127	127
Total	125		11,730

Source: Table 2–1.

$$\mu = \frac{\Sigma fm}{N} = \frac{11{,}730}{125} = 93.84 \text{ pounds per square inch.}$$

77 pounds. The product of 77 times 6 (fm) gives the total number of pounds of yarn strength represented by the six items. Each of the values in the fm column, computed in the same manner, represents an estimate of the total pounds of yarn strength of all the lots in the class. The sum of the amounts in the fm column is an estimate of the total of 11,715 pounds obtained by adding all the amounts in Table 2–1.

In the following formula the estimated total Σfm, which is 11,730 pounds per square inch, is divided by 125 giving a mean of 93.84 pounds per square inch.

$$\mu = \frac{\Sigma fm}{N}. \tag{3-2}$$

The 93.84 compares favorably to the 93.72 found when Formula 3–1 was used to compute the mean on page 45. In this case the approximation of ΣX by Σfm is very good and supports the general rule that Σfm is a good estimate of ΣX. When there is a large number of frequencies with a class interval that is relatively small, the mean computed from the frequency distribution can be accepted as a good estimate of the actual mean of the individual observations.

Computing the Mean from a Frequency Distribution—Short Method. When the class interval is uniform throughout the classification, the procedure represented by Formula 3–2 is ordinarily abbreviated in practice by computing the deviations of the midpoints from an arbitrarily chosen value of m, designated as A.

If $\quad d = m - A,$

then multiplying both sides by Σf yields $\quad \Sigma fd = \Sigma fm - \Sigma fA.$

Or, since A is a constant, $\quad \Sigma fd = \Sigma fm - NA.$ (Note that $\Sigma f = N$.)

and $\quad \dfrac{\Sigma fd}{N} = \dfrac{\Sigma fm}{N} - A.$

Or $\quad A + \dfrac{\Sigma fd}{N} = \dfrac{\Sigma fm}{N}$

but $\quad \mu = \dfrac{\Sigma fm}{N}.$

Therefore, $\quad \mu = A + \dfrac{\Sigma fd}{N}.$

Using the class interval (i) as the unit, $\quad d' = \dfrac{m - A}{i}$

since $\quad d = m - A$

$$d = id'.$$

Substituting id' for d gives $\mu = A + \frac{\Sigma fid'}{N}$.

Since i is a constant $\mu = A + \frac{\Sigma fd'}{N} i,$ **(3–3)**

letting $c = \frac{\Sigma fd'}{N} i.$ **(3–4)**

Then $\mu = A + c$ **(3–5)**

where

A = the arbitrary origin (midpoint of an arbitrarily chosen class interval)
m = the midpoint of any class
i = the number of units in the class interval.

The computation of the mean by the use of this formula is illustrated in Table 3–2.

Formula 3–3 is usually used in computing the arithmetic mean from a frequency distribution with a uniform class interval, but Formula 3–5 is

TABLE 3–2
COMPUTATION OF THE MEAN FROM A FREQUENCY DISTRIBUTION (SHORT METHOD)

Strength of 125 Lots of Cotton Yarn
(Pounds per Square Inch Required to Break a Skein of 22s)

Pounds per Square Inch (Class Interval)	Number of Lots f	Deviation of Midpoint from Arbitrary Digit d'	fd'
70– 74	1	−4	− 4
75– 79	6	−3	−18
80– 84	17	−2	−34
85– 89	29	−1	−29
90– 94	20	0	0
95– 99	17	1	17
100–104	13	2	26
105–109	10	3	30
110–114	6	4	24
115–119	3	5	15
120–124	2	6	12
125–129	1	7	7
Total	125		46

Source: Table 2–1.

$$A = 92 \qquad \mu = A + \frac{\Sigma fd'}{N} i = 92 + \frac{46}{125} 5$$

$$i = 5 \qquad \mu = 92 + 1.84 = 93.84 \text{ pounds per square inch.}$$

identical except that it emphasizes the two steps in the operation. The value of A is a first approximation of the mean. Computing c gives a measure of the amount this approximation differs from the mean. This difference may be positive or negative, but when it is added to the value of A it *corrects for the difference between the arbitrary origin and the value of the mean.* The method works whether the arbitrary origin is close to the mean or far from it. The only reason for attempting to locate A fairly close to the value of the mean is that the closer it is to the mean the smaller the numbers that have to be used in the calculations. No matter what value of A is chosen, the value of c corrects exactly for the deviation and gives the same answer as given by Formula 3–2, which is an approximation of the true value of the mean given by Formula 3–1. By taking a different arbitrary origin and performing the computations a second time, the arithmetic can be checked.

The short method of computing the mean is feasible only when the class intervals are all the same size. With unequal class intervals Formula 3–2, which is probably the simplest method, may always be used. An open-end class renders a close approximation of the mean impossible, or at least uncertain, by either the long or the short method, unless the sum of the items in the open-end class is known.

Weighted Arithmetic Mean. In the computation of the mean on page 48, each item was included in the total only once. A method of increasing the influence of a particular item on the average is to include this item more than once in the total. Each item can be given a different influence, or weight, by including each one a different number of times when finding the sum of the items. The result is called a *weighted arithmetic mean.*

The calculation of a weighted arithmetic mean is performed by multiplying each item to be averaged by the weight assigned it, totaling the products, and dividing the total by the sum of all the weights used. These operations may be summarized by

$$\mu = \frac{\Sigma wX}{\Sigma w} \qquad \textbf{(3–6)}$$

where

w = the weight assigned to each item, or X value.

Students are familiar with the use of weighted averages to combine several grades that are not equally important. For example, assume that the grades representing a semester of work consist of 1 final examination, 4 one-hour examinations, and 10 ten-minute tests—a total of 15 grades. It is logical that these grades should be given different weights in computing the final average. Possibly a ten-minute test might be given a weight of 2; an hour examination, a weight of 10; and the final examination, a weight of 40. The procedure for compiling the average is to multiply each grade by its weight, add these products, and then divide the total of the products by the sum of the weights. This calculation is illustrated in Table 3–3.

TABLE 3–3

COMPUTATION OF THE WEIGHTED ARITHMETIC MEAN

Semester Grades of One Student

Tests and Examinations	Grade X	Weight w	wX
Ten-minute tests			
1	81	2	162
2	75	2	150
3	68	2	136
4	90	2	180
5	40	2	80
6	71	2	142
7	60	2	120
8	58	2	116
9	90	2	180
10	73	2	146
Hour examinations			
1	75	10	750
2	84	10	840
3	79	10	790
4	87	10	870
Final examination	91	40	3,640
Total		100	8,302

Source: Hypothetical data.

$$\mu = \frac{\Sigma wX}{\Sigma w} = \frac{8{,}302}{100} = 83.$$

The unweighted average, in reality, is the weighted average obtained by giving the same weight to each item. If each item were given a weight of 1, the sum of the weights would simply be the number of items, or N. The formula for the unweighted average becomes the formula already given for the arithmetic mean

$$\mu = \frac{\Sigma X}{N}.$$

The calculation of the arithmetic mean from a frequency distribution, illustrated in Table 3–1, is a weighted arithmetic mean of the midpoints of the classes. The midpoint of a class is weighted by the number of items in the class. The total of the products divided by the sum of the frequencies gives a weighted average of the midpoints, with the frequencies being the weights.

Averaging Ratios. In computing an average of ratios, each ratio must be given its proper weight. It is not uncommon to compute ratios for a number of different classes, and then compute a ratio for the total of all classes. Unless the classes are of equal importance, however, an unweighted average of the individual ratios will not give the correct result.

Table 3–4 shows the unemployment rate for June, 1973, in each of the five Pacific states. Each percentage is the ratio of the number unemployed to the total number in the labor force multiplied by 100 to make it a percentage. To compute the average percentage of unemployment in the Pacific states, the percentages for the individual states must be averaged carefully. The unweighted arithmetic mean of the percentages for the five states is 4.22%. If there were the same number in the labor force in each state, this would be the correct mean. But there are 1,160,200 in the labor force in California, and 112,200 in the labor force in Alaska. Thus, it is not logical to give both states the same weight in computing the mean.

TABLE 3–4

PERCENTAGE OF UNEMPLOYMENT IN NONAGRICULTURAL INDUSTRIES, JUNE, 1973

Pacific States

State	Percentage Unemployed
Washington	5.2
Oregon	3.1
California	3.5
Alaska	5.9
Hawaii	3.4

Source: U.S. Department of Labor, *Employment and Earnings* (August, 1973).

The weighted arithmetic mean of these percentages is computed in Table 3–5, in which the percentage of the labor force unemployed in each state is weighted by the size of the work force in each state. The weighted mean of 3.7% is the correct value because the weights applied to the individual percentage reflect the importance of each state.

In this case another method of achieving the same result is to compute the relationship between the total number in the labor force and the total number of unemployed in the five states. The total number in the labor force in the five states is 9,968,100 and the actual number of unemployed is 367,664 or 3.7% of the labor force. This is the percentage found by taking a weighted arithmetic mean of the five individual percentages.

Generally an average of ratios can best be computed by combining the data from which the ratios were computed, and computing a ratio from the total data. In the example above, the data from which the ratios were derived were used in weighting the ratios, and the two computations were identical. Since the computations from the total data are easier to make and to understand than those in Table 3–5, the individual ratios should be weighted only when the original total data are not available.

TABLE 3–5

COMPUTATION OF THE WEIGHTED ARITHMETIC MEAN

Proportion of the Labor Force Unemployed, June, 1973
Pacific States

State	Proportion Unemployed X	Number of Employees w	wX
Washington	.052	1,160,200	60,330
Oregon	.031	828,000	25,668
California	.035	7,544,000	264,040
Alaska	.059	112,200	6,620
Hawaii	.034	323,700	11,006
Total		9,968,100	367,664

Source: U.S. Department of Labor, *Employment and Earnings* (August, 1973).

$$\mu = \frac{\Sigma wX}{\Sigma w} = \frac{367{,}664}{9{,}968{,}100} = .037 \text{ or } 3.7\%.$$

As another example of computing an average of ratios, a brick company has five trucks hauling shale from a pit two miles from the brickyard. The trucks are loaded by a power shovel, driven to the brickyard to dump their load, and returned to the pit to be loaded again. The computation in Table 3–6 shows the time required per load for each truck on April 12 and the average of these figures.

TABLE 3–6

COMPUTATION OF ARITHMETIC MEAN

Time Required per Load for Five Trucks

Truck	Minutes per Load
1	48.0
2	40.0
3	53.3
4	30.0
5	26.7
Total	198.0

Source: Hypothetical data.

$$\mu = \frac{198.0}{5} = 39.6 \text{ minutes.}$$

The average can be computed another way. Each of the five trucks operated eight hours on April 12. According to the computation in Table 3–7, the five trucks hauled an average of 13 loads on that eight-hour day. Since there are 480 minutes in an eight-hour day, the average length of time per load was 36.9 minutes (480 divided by 13). However, the computation in Table 3–6 showed that the average length of time required to haul one load was 39.6 minutes. The difference between the two averages is due to the difference in weighting. The number of minutes per load in Table 3–6 are ratios, and it is incorrect to give each ratio the same weight as in the first computation. Each of the trucks worked the same length of time, but *each one hauled a different number of loads*. Truck 3 hauled 9 loads in an eight-hour day, while Truck 5 hauled 18 loads. The time per load for each truck should be weighted by the number of loads hauled, as shown in Table 3–8. This computation gives the average length of time required to haul one load as 36.9 minutes, which agrees with the average time computed by dividing 480 minutes by an average of 13 loads per day.

TABLE 3–7

COMPUTATION OF ARITHMETIC MEAN
Number of Loads Hauled in Eight-Hour Day

Truck	Number of Loads Hauled
1	10
2	12
3	9
4	16
5	18
Total	65

Source: Hypothetical data.

$$\mu = \frac{65}{5} = 13 \text{ loads.}$$

TABLE 3–8

COMPUTATION OF WEIGHTED ARITHMETIC MEAN
Minutes per Load Weighted by Number of Loads per Eight-Hour Day

Truck	Minutes per Load X	Number of Loads in Eight Hours w	wX
1	48.0	10	480
2	40.0	12	480
3	53.3	9	480
4	30.0	16	480
5	26.7	18	480
Total		65	2,400

Source: Tables 3–6 and 3–7.

$$\mu = \frac{\Sigma wX}{\Sigma w} = \frac{2{,}400}{65} = 36.9 \text{ minutes.}$$

If each truck had hauled one load and we wanted to know the average time required, it would be correct to average the number of minutes, giving each load equal weight, for we would not be averaging ratios. However, since each truck worked the same amount of time, the minutes per load must be weighted by the number of loads hauled in a day.

Harmonic Mean

The *harmonic mean* is computed by averaging the reciprocals of the values of the variable and finding the reciprocal of this average. Under certain circumstances this calculation automatically gives the correct weight to the values of the variable and can be used instead of weighting the arithmetic mean as was done in Table 3–8. The reciprocal of an individual item is written $\frac{1}{X}$ and the mean of these reciprocals is $\frac{\Sigma\frac{1}{X}}{N}$. This computation gives the reciprocal of the harmonic mean *(H)*, which may be written

$$\frac{1}{H} = \frac{\Sigma\frac{1}{X}}{N}$$

or

$$H = \frac{N}{\Sigma\frac{1}{X}}. \tag{3-7}$$

The computations represented by Formula 3–7 are shown in Table 3–9, where the harmonic mean is found to be the same as the weighted mean of the ratios computed in Table 3–8. In Table 3–9 the harmonic mean is used in averaging ratios when all the trucks operated the same length of time (8 hours or 480 minutes). If, however, the problem is to find the average length of time per load when all trucks hauled the same number of loads, the unweighted arithmetic mean of the minutes per load is the appropriate average to use.

Since a ratio is computed from two numbers, one of two situations is encountered when averaging ratios: (1) All the ratios may have the same value in the denominator, which in the preceding example means that all trucks hauled the same number of loads with different elapsed times. Each truck would have a ratio consisting of the number of minutes of time elapsed for all loads divided by the number of loads. When all ratios have the same value in the denominator, the arithmetic mean is the appropriate mean for averaging the ratios. (2) All the ratios may have the same value in the numerator. In the preceding example all the trucks worked the same number of minutes during the day, but the number of loads hauled varied. When all the ratios have the same value in the numerator, the harmonic mean should be used to compute the mean of the ratios.

TABLE 3–9

COMPUTATION OF THE HARMONIC MEAN

Minutes Required per Load for Five Trucks

Truck	Minutes per Load X	$\frac{1}{X}$
1	48.0	.0208333
2	40.0	.0250000
3	53.3	.0187617
4	30.0	.0333333
5	26.7	.0374532
Total	198.0	.1353815

Source: Table 3–6.

$$H = \frac{5}{.1353815} = 36.9 \text{ minutes.}$$

A more general case is encountered when the number of loads hauled and the amount of time worked are different for each truck. In this situation neither the unweighted arithmetic mean nor the unweighted harmonic mean is the appropriate average to use. The ratios can be averaged by computing the total number of minutes worked by all the trucks and the total number of loads hauled by all the trucks. The average length of time will be the ratio between the total number of minutes worked and the total number of loads hauled, as computed in Table 3–8. This is the general method of averaging ratios that can be used in all cases *if the original values from which the ratios were computed are available.* Since this information is not always available, it is often necessary to average the ratios; in such cases it must be determined whether the harmonic mean or the arithmetic mean is the proper average to use.

Geometric Mean

The *geometric mean (G)* of a series of values such as $X_1, X_2, X_3, \ldots, X_N$ is the Nth root of the product of the values. It may be expressed as

$$G = (X_1 \cdot X_2 \cdot X_3 \cdot \ldots \cdot X_N)^{\frac{1}{N}}. \tag{3–8}$$

The simplest method of computing the geometric mean, at least when N is more than two, is by the use of logarithms.[2] In terms of logarithms, Formula 3–8 becomes

$$\log G = \frac{\Sigma \log X}{N}. \tag{3–9}$$

[2] The use of logarithms is discussed in Appendix D.

The geometric mean is necessarily zero if any value of X is zero, and may become imaginary if negative values occur. Except for these cases, the geometric mean is always determinate and is rigidly defined. The geometric mean is always less than the arithmetic mean of the same variables, unless all the values are the same.

Table 3–10 illustrates the computation of the geometric mean of the percentages of increase in the population of Los Angeles over six decades. The population in 1920 was 180.7% of the 1910 population; the 1930 population was 214.7% of the 1920 population; and so on. The problem is to determine the average rate of increase over the five decades. The procedure is to find the logarithms of the five percentages and then to find the arithmetic mean of the logarithms, which is the logarithm of the geometric mean. The average percentage of population of one census date to the preceding census date is 143.75. The average percentage increase, therefore, is 43.75.

TABLE 3–10

COMPUTATION OF THE GEOMETRIC MEAN

Percentage Change from Previous Census in Population of Los Angeles, California, 1920 to 1970

Census Date	Population	Percentage of Population on Previous Census Date X	Log X
1910	319,198	. . .	. . .
1920	576,673	180.7	2.256958
1930	1,238,048	214.7	2.331832
1940	1,504,277	121.5	2.084576
1950	1,970,358	131.0	2.117271
1960	2,479,015	125.8	2.099681
1970	2,815,998	113.6	2.055378
Total			12.945696

Source: U.S. Department of Commerce, Bureau of the Census, *Census of Population (1960)*, and *idem., Statistical Abstract of the United States (1972)*.

$$\log G = \frac{\Sigma \log X}{N} = \frac{12.945696}{6} = 2.157616$$

$$\text{antilog } 2.157615 = 143.7$$

$$G = 143.7$$

$$\text{Average rate of increase} = 143.7 - 100.00\%$$

$$= 43.7.$$

By adding the percentages in the third column of Table 3–10 and dividing by 6, the arithmetic mean of the percentages equals 147.9%. If this average rate of increase per decade is applied to the 1910 population to compute an estimate of the 1920 population, then applied to the 1920 population to compute the estimated 1930 population, and so on, it should end up with the correct 1970 population, even though the individual decade estimates would not agree with the actual population. If the average rate of increase per decade is applied each 10 years and the 1970 population does not come out at the number that is known to be correct, something is wrong with the average.

When the arithmetic mean percentage computed for the 60-year period is applied as described above, the population for 1970 is estimated to be 3,340,947. The calculations are

1920: 147.9% of 319,198 = 472,094
1930: 147.9% of 472,094 = 698,227
1940: 147.9% of 698,227 = 1,032,678
1950: 147.9% of 1,032,678 = 1,527,331
1960: 147.9% of 1,527,331 = 2,258,923
1970: 147.9% of 2,258,923 = 3,340,947.

The computed population is considerably higher than it should be, due to the fact that the geometric mean should have been used instead of the arithmetic mean. If the average *amount* of change per decade is wanted, the arithmetic mean is the proper measure to use, but whenever the average *percentage* change is desired, the geometric mean is the correct average.

The geometric mean of percentages of increase can be computed without knowing the percentage increase in the individual decades, just as the arithmetic mean can be computed without knowing the individual items being averaged. The computations in Table 3–10 may be abbreviated by letting P_0, P_1, P_2, P_3, P_4, P_5, and P_6 be the population in 1910, 1920, 1930, 1940, 1950, 1960, and 1970, respectively. The percentages in Table 3–10 may then be represented

$$1920: \frac{P_1}{P_0} \times 100 \qquad 1940: \frac{P_3}{P_2} \times 100 \qquad 1960: \frac{P_5}{P_4} \times 100$$

$$1930: \frac{P_2}{P_1} \times 100 \qquad 1950: \frac{P_4}{P_3} \times 100 \qquad 1970: \frac{P_6}{P_5} \times 100$$

$$G = \sqrt[6]{\frac{P_1}{P_0} \times \frac{P_2}{P_1} \times \frac{P_3}{P_2} \times \frac{P_4}{P_3} \times \frac{P_5}{P_4} \times \frac{P_6}{P_5} \times (100)^6} = \sqrt[6]{\frac{P_6}{P_0}} \times 100.$$

The population was 319,198 in 1910 and 2,815,998 in 1970. In other words, the ratio of the 1970 population to the 1910 population is 8.822104. Since six decades are covered, the average increase per decade is found by taking the sixth root of 8.822104 as shown in the following computations. This is the geometric mean of the different rates of increase over the period.

$$G = \sqrt[6]{\frac{2,815,998}{319,198}}$$

$$G = \sqrt[6]{8.822104}$$

$$\log G = \frac{\log 8.822104}{6} = \frac{.945572}{6} = .157595$$

$$\text{antilog } .157595 = 1.437$$

$$G = 1.437 \times 100 = 143.7$$

$$\text{average rate of increase} = 143.7\% - 100.0\% = 43.7\%.$$

The geometric mean may be used to find the average monthly rate of increase between two census dates, such as 1960 and 1970. The population was 2,479,015 on April 1, 1960, and 2,815,998 on April 1, 1970. The 1970 population was 1.136 times the 1960 population. Since 120 months elapsed between the two census dates, the average monthly increase may be found by extracting the 120th root of 1.136, which is the geometric mean of the monthly rates of increase.

$$\log G = \frac{\log 1.136}{120} = \frac{.055378}{120} = .0004615$$

$$\text{antilog } .0004615 = 1.00106$$

$$G = 1.00106$$

$$\text{average monthly rate of increase} = 1.00106 - 1.0000 = .00106 \text{ or } .106\%.$$

The population increased .106% each month between the 1960 and 1970 censuses. If we wish to estimate the population at any date between these two census dates, it can be done by using the average monthly percentage of increase. For example, July 1, 1965, fell 63 months after the 1960 census date. Since the population increased at the average of .106% for the 120 months, it is logical to assume that this average rate of increase prevailed for the 63 months ending July 1, 1965. Each month was 1.00106 times the preceding month, and in 63 months the population would have increased at this rate 63 times. Therefore, the July 1, 1965, population was 1.00106^{63} times the population on April 1, 1960. The computation of 1.00106^{63} is

$$\log 1.00106^{63} = (\log 1.00106)(63) = (.000460)(63) = .028980$$
$$\text{antilog } .028980 = 1.069.$$

The population on July 1, 1965, may be estimated to be 1.069 times 2,479,015, which is the population on April 1, 1960. This multiplication gives an estimated population of 2,650,000 on July 1, 1965.

Median

The *median* is the value of the middle item in an array. An *array* is an arrangement of values in the order of their size from the smallest to the largest, or from the largest to the smallest. The location of the median of the grades of a student on five one-hour examinations is illustrated in the following array:

93%
89%
76% ← The middle grade is 76%. This is the median.
74%
50%

If there are an even number of items, there is no middle value; any amount between the two middle items of the array might be considered the median, since there are an equal number of items on each side. Although the median is to a certain extent indeterminate in this situation, it is generally defined as the arithmetic mean of the two middle items. This is a practical solution to a situation in which the definition does not always give a unique value. Another case in which the median is indeterminate is illustrated by the following values:

93%
91%
89% } This is the median.
89% }
80%

If 89% is selected as the median, there is one grade that is smaller and two that are larger. It is impossible to find a value of the variable that has the same number of items larger as smaller; so, strictly speaking, the median is indeterminate. In such cases, especially where there are a large number of items, the value nearest the middle value would be considered to be the median. In this case, it would be 89%.

In comparing the arithmetic mean with the median, the value of the arithmetic mean is always strictly determinate as long as all the values of the variable are known. This grows out of the fact that the arithmetic mean is a computed average, while the median is a position average. In other words, the median is the position on the scale of the variable that divides the array into two equal parts. The position averages have certain characteristics in common that differentiate them in certain respects from the computed averages, as we shall see later in this chapter.

Approximating the Median from a Frequency Distribution. If the individual values of the variable are recorded in punched cards or on magnetic tape, it is not difficult to arrange them in an array. However, unless the information is in machine readable form, the median is commonly estimated from a frequency distribution, except when the number of values of the variable is very small. Since there is no other important use for the array, the sorting is normally not considered worth the time required, especially if it must be done by hand. Furthermore, it may be that only the frequency distribution is available, in which case the median must be approximated from the frequency distribution.

The location of the median is facilitated by the use of a cumulative frequency distribution, used in connection with the simple frequency distribution, as shown in Table 3–11. Column 3 of Table 3–11 shows that there are 53 lots out of 125 lots of yarns that require less than 90 pounds per square inch to break a skein of yarn. Column 2 shows that 20 lots require between 90 and 94 pounds. The median pounds per square inch must lie between 90 and 94 pounds, since fewer than half of the lots are less than 90 but more than half are less than 94.

TABLE 3–11

LOCATION OF THE MEDIAN

Strength of 125 Lots of Cotton Yarn

Pounds per Square Inch (1)	Number of Lots (2)	Cumulative Number of Lots (3)
70– 74	1	1
75– 79	6	7
80– 84	17	24
85– 89	29	53
90– 94	20	73
95– 99	17	90
100–104	13	103
105–109	10	113
110–114	6	119
115–119	3	122
120–124	2	124
125–129	1	125
Total	125	

Source: Table 2–1.

$$Md = L_{Md} + \frac{\frac{N}{2} - F_{L_{Md}}}{f_{Md}} i_{Md}$$

$$Md = 89.5 + \frac{\frac{125}{2} - 53}{20} 5$$

$$Md = 89.5 + \frac{9.5}{20} 5 = 89.5 + 2.4 = 91.9 \text{ pounds.}$$

At this point it becomes necessary to distinguish carefully between the real limits of the class and the stated limits, 90 to 94. In making computations it is more precise to use the real lower limit, 89.5, even though the class is generally referred to by the stated limits.

The formula for locating the median within the class bounded by the limits 90–94 is written

$$Md = L_{Md} + \frac{\frac{N}{2} - F_{L_{Md}}}{f_{Md}} i_{Md} \tag{3-10}$$

where

Md = median
L_{Md} = real lower limit of the class in which the median falls
FL_{Md} = cumulative frequencies less than the lower limit of the class in which the median falls
f_{Md} = frequency of the class in which the median falls
N = number of frequencies in the distribution
i_{Md} = class interval of the class in which the median falls.

Using Formula 3–10, the median is located as shown in Table 3–11.

According to the definition on page 59, the median in Table 3–11 is the 63d item. If the original data are not available, it would be impossible to know the exact value of this item, but its value can be estimated fairly accurately. There are 20 items between 89.5 and 94.5, and it is assumed that they are distributed uniformly between these two class limits. This assumption is not entirely valid, but it gives results that are reasonably accurate. The 20 items distributed uniformly within the class are shown graphically in Figure 3–1.

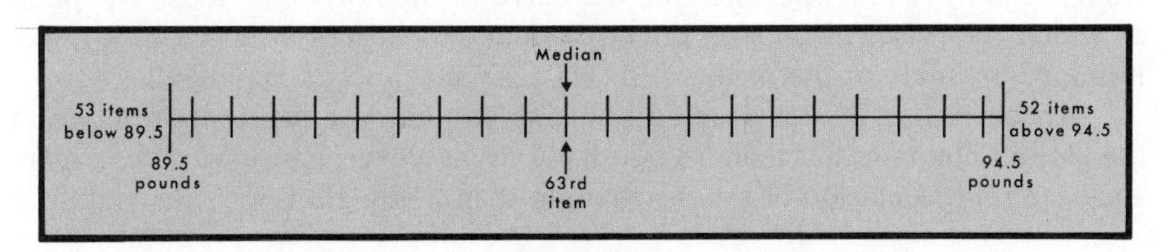

FIGURE 3–1

ASSUMED DISTRIBUTION OF 20 ITEMS WITHIN CLASS INTERVAL

Not only are there 20 items between 89.5 pounds and 94.5 pounds, but there are also 20 subspaces between the items. Actually there are 19 whole spaces and one half-space at each end. Since the items are all the same distance apart, by dividing the number of spaces, 20, into the class interval, 5 pounds, the size of each space is found to be .25 pounds.

As shown in Figure 3–1, the median is located 9.5 spaces from the lower boundary of the class, and 10.5 spaces from the upper boundary. This means that the median is 89.5 pounds + (9.5) (.25), or 91.9 pounds. The steps above may be summarized

$$Md = 89.5 + \frac{9.5}{20}5 = 89.5 + 2.4 = 91.9 \text{ pounds.}$$

Since the preceding method of calculation is presented as an approximation, it is interesting to compare the value with the correct value of the median derived from an array of the data. If the data on strength of the 125 lots in Table 2–1 on page 22 are put into an array, the 63d item is 92 pounds. Thus, the value of the median interpolated from the frequency distribution gives the same value when rounded to the nearest pound. Since the original data in Table 2–1 are rounded to the nearest pound, the median is significant only to the same number of places.

In computing the median, it is not necessary to construct a diagram such as that shown in Figure 3–1 in order to find the values for use in the computations or to compute the number of the item in the array that is the median. It is necessary to find the various values specified in Formula 3–10 and compute accordingly. Table 3–11 is an example of this procedure.

Graphic Interpolation of the Median. Graphic interpolation may be substituted successfully for the approximating method described in the previous section. Figure 3–2 is an *ogive,* which is a graphical curve of the cumulative frequencies. The curve was drawn from the cumulative frequencies data in Table 3–11. Once the ogive is plotted, a point representing $\frac{N}{2}$ is located on the vertical axis. In this case $\frac{N}{2} = \frac{125}{2} = 62.5$. A horizontal line is extended from 62.5, located on the vertical axis, to the ogive. A vertical line is drawn from the point where this line cuts the ogive to the horizontal axis. The point at which this vertical crosses the horizontal axis serves as a graphically determined estimate of the median. In this case, it is approximately 92.

An ogive can be constructed by using straight line segments to connect the plotted points rather than a smooth curve; however, the curve gives a more accurate representation of the distribution than a straight line.

Mode

The value of the variable that occurs most frequently is called the *mode.* It is the position of greatest density, the predominant or most common value—the value that is the fashion *(la mode).*

Although the concept of the mode is simple, it is not always easy to locate. All methods of locating the mode of continuous data give only approximations. When data are discrete, such as the size of family, the mode is the value of the variable that occurs most frequently.

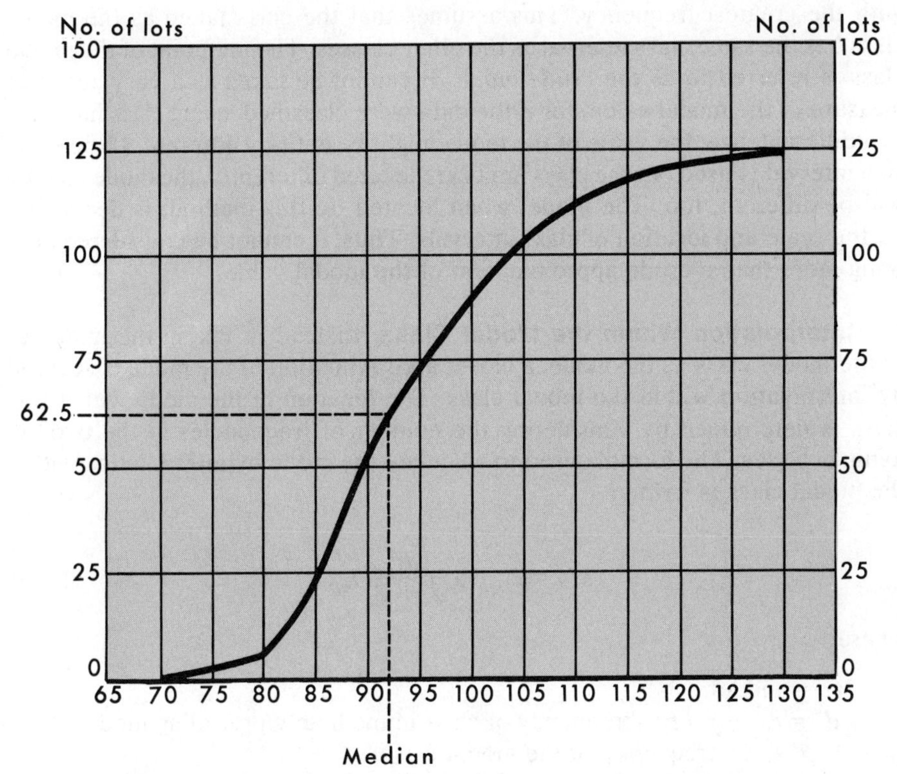

Source: Table 3-11.

FIGURE 3-2

LOCATING THE MEDIAN GRAPHICALLY

Bimodal Distribution. Some frequency distributions, called *bimodal,* have two classes that are considerably larger than the adjoining classes, thus giving the chart of the distribution two peaks. Such a distribution with two modes is usually the result of a distribution that is not homogeneous. For example, if wage rates in an industry show considerable difference between wages paid to skilled and unskilled workers, the distribution of wages paid to all wage earners might show two modes. One would be the point of concentration of wage rates of skilled workers and the other would be the wage rates of unskilled workers. When a bimodal distribution is encountered, one normally attempts to classify the data into two homogeneous distributions, each with its own mode.

Crude Mode. The simplest value for the mode is the midpoint of the class with the greatest frequency. This assumes that the class taken as the modal class has the same class interval as the other classes. The midpoint of the modal class is referred to as the *crude mode*. It cannot be taken as a very accurate measure of the modal value, for if the data were classified, using class intervals of a different size, the value of the mode might be entirely different. If the same size interval is used but the class limits are located differently, the mode usually will be different, too. The mode, when located by this method, is dependent on the scale and location of class intervals. Thus, it cannot be considered anything more than a crude approximation of the modal value.

Interpolation Within the Modal Class. Instead of taking the midpoint of the modal class as the mode, a closer approximation of the mode is secured by interpolation within the modal class. The location of the mode within this class is determined by considering the number of frequencies in the two adjoining classes. The formula used to calculate the mode by interpolating within the modal class is written

$$Mo = L_{Mo} + \frac{d_1}{d_1 + d_2} i_{Mo} \qquad \textbf{(3–11)}$$

where

Mo = mode
$d_1 = f_{Mo} - f_1$ (f_1 = frequency of class immediately preceding modal class, f_{Mo} = frequency of the modal class)
$d_2 = f_{Mo} - f_2$ (f_2 = frequency of class immediately following modal class)
L_{Mo} = real lower limit of the modal class
i_{Mo} = class interval of the class in which the mode falls.

The modal class in Table 3–11 is 85 to 89 pounds, with 29 frequencies. To interpolate within the modal class, the difference between the frequency of the modal class and the frequency of the two adjoining classes is computed

$$d_1 = 29 - 17 = 12$$
$$d_2 = 29 - 20 = 9.$$

Using Formula 3–11, the mode is then calculated

$$Mo = 84.5 + \frac{12}{12 + 9} 5 = 84.5 + 2.9 = 87.4.$$

When the frequencies on each side of the modal class are the same, the mode is the midpoint of the modal class. When the frequency is the same for two classes that both have the greatest frequency, the mode falls on the class limit that divides them.

Graphic Interpolation of the Mode. The mode may also be computed graphically on a histogram, as illustrated in Figure 3–3. Using the data in Table 3–11, lines are drawn diagonally from the upper corners of the rectangle representing the modal class to the upper corners of the adjacent rectangles. The diagonal lines intersect at the modal ordinate; that is, a line through this point and perpendicular to the X axis intersects the X axis at the modal value. If the chart is drawn accurately, the modal value will agree with the value computed using Formula 3–11.

Empirical Mode (Mo_E). In a perfectly symmetrical distribution the mean, the median, and the mode are the same. As the distribution departs from perfect symmetry, these three values separate and maintain a rather definite relationship to each other. However, as the asymmetry becomes more pronounced, this relationship becomes increasingly distant.

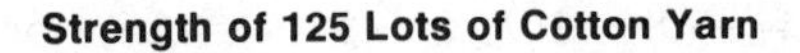

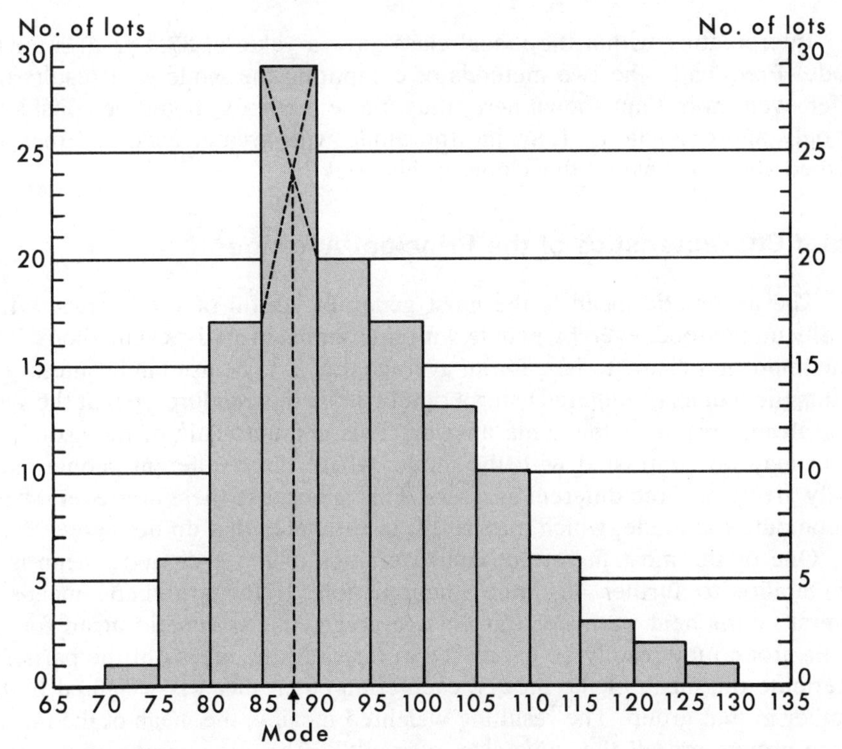

Source: Table 3-11.

FIGURE 3–3

LOCATING THE MODE GRAPHICALLY

When dealing with distributions that are nearly symmetrical, it is possible to find an approximation of the mode based upon the relationship between the values of the mean, the median, and the mode, since the mean and the median may usually be determined. The relationship between the three averages is expressed

$$Mo_E = \mu - 3(\mu - Md) \qquad \textbf{(3–12)}$$

The *empirical mode* may be defined, therefore, as the value that lies a distance from the mean equal to three times the distance of the median from the mean, in the same direction as the median deviates from the mean.

Applying Formula 3–12 to the data in Table 3–11 for the strength of cotton yarn, the value of the mode is computed

$$\mu = 93.8$$
$$Md = 91.9$$
$$Mo_E = 93.8 - 3(93.8 - 91.9) = 88.1 \text{ pounds.}$$

Interpolating within the modal class gave a value of 87.4 pounds for the mode. Frequently the two methods of computing the mode give results that differ even more than shown here; thus, these methods should be considered as only approximations. Locating the mode with greater accuracy than described above is beyond the scope of this book.

Chief Characteristics of the Principal Averages

The arithmetic mean is the most generally useful of the averages. It is readily understood, even by people with no training in statistical methods. This is an important characteristic for an average that is to be used in business. The arithmetic mean is computed using a rigidly defined procedure, so that the same data always result in the same answer. This characteristic of the arithmetic mean may be contrasted with the mode, where three different people might easily arrive at three different answers. This is because there are several ways to compute the mode, which may result in estimates that do not agree.

One of the most important characteristics of the arithmetic mean is its adaptability to further algebraic manipulation. If the arithmetic means of several component parts of a group are given, the arithmetic mean for the whole group may readily be expressed in terms of the means of the parts. To determine the mean of the means, each component mean is weighted by the number in that group. The resulting weighted mean is the mean of the means. If the groups are all the same size, a simple arithmetic average of the component means is all that is required to find the mean of the means.

Neither the median nor the mode is easily adaptable to further algebraic manipulation. Table 3–12 gives the median 1973 income of households classified by age of the head of the household. It shows the computation of the median income of all households, weighting the median for each group by the

TABLE 3–12

COMPUTATION OF WEIGHTED MEDIAN

Median Income of Households in 1973 Classified by Age of Head of Households

Age of Head of Household	Median Family Income (Dollars) X	Number of Households w	wX
14–24	7,061	5,476	38,666,036
25–34	10,877	13,562	147,513,874
35–44	12,688	11,721	148,716,048
45–54	13,125	12,805	168,065,625
55–64	10,034	11,212	112,501,208
65 and over	4,169	13,473	56,168,937
Total		68,249	671,631,728

Source: U.S. Department of Commerce, Bureau of the Census, *Current Population Reports,* Series P-60, No. 87 (June, 1963).

$$\text{Weighted median} = \frac{\Sigma wX}{\Sigma w} = \frac{671{,}631{,}728}{68{,}249} = \$9{,}841.$$

number of households in the group. The median income computed by weighting the group medians is $9,841. The correct median income disregarding the age of the head of the household is $9,698. The median computed by weighting the median income of each group is not correct because the median is a positional measure and must be determined by forming a combined distribution and locating the midpoint. In this case the weighted median is not far wrong, but there is no way of computing it with complete accuracy for the whole group from the medians of the subgroups.

Another consideration is how easily an average may be calculated. When the original data are available, there is no significant difference in the ease with which the mean and median may be calculated. But when the total of the values and the number are known, and the individual values are not known, the mean can be computed but the median cannot. In an open-end class, the mean cannot be computed accurately unless the total or average value of the items in the open-end class is known. The open-end class does not interfere, however, with the location of the median. The location of the mode does not depend on the values of the variable at the extremes of the series, so open-end classes make no difference in its location.

The calculation of the geometric mean requires more work than either the arithmetic mean or the median; but if the conditions warrant the use of the geometric mean, it should be used. The harmonic mean requires about the same amount of computation as the geometric mean.

The arithmetic mean is based on all the values of the variable. A few extremely large or extremely small values of the variable have an undue influence on the mean. This tendency of the mean is demonstrated in the distribution of five grades of a student:

```
     89%
     83%
     81%  ← The median is the middle item of the array, 81%.
     77%
     20%
  5)350

     70%  ← The mean is the total, 350, divided by 5, or 70%.
```

The mean is 11 percentage points less than the median. This is because the median is influenced only by the *position* of the values being averaged, while the mean is affected by the *size* of the values, since the sum of the values is used in its computation. If instead of being 20% the lowest grade were 75%, the median would still be 81% and the mean would also be 81%.

```
     89%
     83%
     81%  ← The median is the middle item of the array, 81%.
     77%
     75%
  5)405

     81%  ← The mean is the total, 405, divided by 5, or 81%.
```

The important problem at this point is determining which average is more typical of the five grades ranging from 89% to 20%. The mean is 70% and the median is 81%. They cannot both be considered equally representative of the student's performance. When performance has been fairly uniform throughout a course, it is not as important which average is used to summarize the work, since each average will give approximately the same results. But when an extremely low grade is made in one or two tests, it makes a considerable difference in the final grade.

If 70% is taken as the typical grade, the nearest individual grade is 77%; and there are three higher grades. If the median, 81%, is taken as the typical grade, the nearest individual grade is 83%, and the next removed is four less or 77%. There is one higher than 83%, and then the lowest grade, 20%. In other words, the median grade comes much nearer the individual items, and, therefore, better reflects central tendency. The 20% grade is given less influence, since it differs so greatly from the other four grades. The mean attempts to be representative of all the grades, with the result that it is pulled away from the four grades between 77% and 89% toward the 20% grade. However, it is representative of neither the lowest grade, nor the four higher grades. The median is more representative of the four grades between 77% and 89%, since it is not influenced by the one extremely small amount.

Many distributions of business data, particularly dollar figures, have extremely large amounts that have an undue influence on the mean. In these cases the median is the better average to use in summarizing the data. Table

3–13 classifies according to size of deposits the insured banks in the United States that required disbursements to protect depositors between 1934 and 1966. The deposits of the 466 banks totaled $804,300,000 and the mean deposit was $1,725,966. The median of the data, computed by the methods illustrated on pages 59 to 62, is $319,672.

Since the deposit size of many banks concentrated below $250,000, the median deposit size of $319,672 is more representative of the banks than the mean amount of $1,725,966. The large deposits of a few banks exert an undue influence on the mean. Five banks had deposits between $25 million and $50 million and one bank over $50 million. The interpretation of the median is that there were the same number of banks with deposits of less than $319,672 as there were banks with more than $319,672.

TABLE 3–13

INSURED BANKS REQUIRING DISBURSEMENTS TO PROTECT DEPOSITORS BETWEEN 1934 AND 1966

Classified by Size of Deposits

Size of Deposits	Number of Banks
$100,000 or less *	107
$100,000 and under $250,000	109
$250,000 and under $500,000	61
$500,000 and under $1,000,000	71
$1,000,000 and under $2,000,000	52
$2,000,000 and under $5,000,000	38
$5,000,000 and under $10,000,000	16
$10,000,000 and under $25,000,000	6
$25,000,000 and under $50,000,000	5
Over $50,000,000	1
Total	466

* The first class limit, $100,000, overlaps the lower limit of the second class. Although this classification violates the principles given in Chapter 2, the table is presented as it appears in the source.

Source: U.S. Department of Commerce, Bureau of the Census, *Statistical Abstract of the United States (1967).*

The following summary briefly outlines the chief characteristics of the arithmetic mean, harmonic mean, geometric mean, median, and mode.

Arithmetic Mean

1. The value of the arithmetic mean is based on all the observations, and thus is affected by all the values of the variable. This may result at times in giving certain extreme values too much influence.

2. The arithmetic mean is rigidly defined and is always determinate if the individual values of the variable are available.

3. The arithmetic mean may be calculated if the individual items are known, if the total value and the number of items are known, or if a frequency distribution is available from which it is possible to estimate closely the average value of the items falling in each class.
4. The arithmetic mean lends itself to further algebraic manipulation.

Harmonic Mean

1. The harmonic mean is based on all the observations, and thus is affected by all the values of the variable. It gives even less weight to the extremely large values than does the geometric mean, and gives considerably more weight to the small values than either the arithmetic mean or the geometric mean.
2. The harmonic mean is indeterminate if any value of the variable is zero. Finding the reciprocal of a zero value of the variable would require division by zero, which is not valid. Except when a value of the variable is zero, the harmonic mean is always determinate and is rigidly defined.
3. The harmonic mean is the average to use when ratios are being averaged and the numerators of the fractions from which the ratios were computed are the same for all ratios.
4. The harmonic mean lends itself to further algebraic manipulation.

Geometric Mean

1. The geometric mean is based on all the observations, and thus is affected by all the values of the variable. However, it gives less weight to extremely large values than does the arithmetic mean.
2. The geometric mean is zero if any value of the variable is zero, and it may become imaginary if negative values occur. Except for these cases, the value is always determinate and is rigidly defined.
3. The geometric mean is the average to use when rates of change or ratios are being averaged, and *it is intended to give equal weight to equal rates of change.*
4. The geometric mean lends itself to further algebraic manipulation.

Median

1. The median is affected by the position of each item in the series but not by the value of the items. This means that extreme deviations from the central part of the distribution affect the median less than the arithmetic mean.
2. The median may be located when the data are incomplete, e.g., if the actual values of the variable at the extremes are unknown, but their general location and frequency are known.
3. Strictly speaking, the median is indeterminate for an even number of cases, although by general agreement it is the mean of the two central values of the

variable. When several items in the center of the distribution are the same size, the median may be indeterminate to a slight degree.

4. The median does not lend itself to algebraic treatment in as satisfactory a manner as the arithmetic mean, the harmonic mean, or the geometric mean.

Mode

1. The value of the mode is determined by the items at the point of greatest concentration, and is not affected by the remaining values of the variable.
2. Since the mode is the point of greatest concentration, it is typical of the distribution. When a distribution is bimodal, it should be broken into more than one distribution to secure greater homogeneity.
3. The true mode is difficult to compute, although an approximate value is easily found.
4. The mode does not lend itself to algebraic treatment.

STUDY QUESTIONS

3–1. The average-size farm in the United States in 1971 was 389 acres. Explain how this figure might be considered a ratio.

3–2. What danger is there in using an arithmetic mean without knowing anything about the size of the individual items included in the total?

3–3. If a salesman says that he must call on three prospects "on the average" to make one sale, which average do you think he means? Explain.

3–4. Why is the arithmetic mean the most generally used of the averages? Do you think it should be the most widely used average?

3–5. The fact that the arithmetic mean can be used logically in further calculations is given as an advantage over the use of the median and the mode. Why is this factor considered to be an important advantage?

3–6. Do you consider the fact that the geometric mean cannot be used if one of the items is zero to be an important disadvantage of the geometric mean? Give a reason for your answer.

3–7. Explain why the averaging of percentages, ratios, and rates must be done with unusual care.

3–8. Why is it impossible to compute the harmonic mean of a distribution if one of the values of the variable is zero?

PROBLEMS

3–1. Assume that you ride a commuter train to work and that you live ten miles from the railway station. You allow 20 minutes to drive to the station and average 30 miles per hour. One morning, because of street repairs, your speed is reduced to 15 miles per hour for half of the distance. How fast will you need to drive for the last half of the distance in order to catch the train? You had allowed yourself the usual 20 minutes, thinking that you could drive at 30 miles per hour all the way, so you must average 30 miles per hour.

3–2. The average hourly earnings of 8,669,000 production employees in durable-goods manufacturing concerns in the United States in June, 1973, was $4.30. The average hourly earnings of 6,083,000 production employees in nondurable-goods manufacturing concerns was $3.65. Compute the average hourly earnings of all production workers in manufacturing in June, 1973.

3–3. The following table gives the distribution of hourly wage rates in the four departments of a manufacturing plant. Compute the arithmetic mean of the hourly rates for each of the four departments. *Retain the solutions to Problems 3–3 through 3–7 for use in later chapters.*

Hourly Wage Rates (Dollars)	Number of Employees				
	Dept. A	Dept. B	Dept. C	Dept. D	Total
4.00 and under 4.20	8	4	5	4	21
4.20 and under 4.40	15	10	12	10	47
4.40 and under 4.60	40	15	14	17	86
4.60 and under 4.80	30	7	10	20	67
4.80 and under 5.00	4	4	6	10	24
5.00 and under 5.20	3	—	3	9	15
Total	100	40	50	70	260

Source: Hypothetical data.

3–4. Compute the median of the hourly rates for each of the four departments in Problem 3–3.

3–5. Compute the mode by interpolating within the modal class for each of the four departments in Problem 3–3.

3–6. Compute the empirical mode of the hourly rates for each of the four departments in Problem 3–3.

3–7. Compute the following measures for total production workers in Problem 3–3:

(a) Arithmetic mean
(b) Median
(c) Mode by interpolating within the modal class
(d) Empirical mode.

3–8. Demonstrate that the arithmetic mean for total hourly wage rates for production workers given in Problem 3–3 can be computed accurately from the measures for the individual departments.

3–9. Demonstrate that the median for total hourly wage rates for production workers given in Problem 3–3 cannot be computed accurately from the measures for the individual departments.

3–10. The following three cars, driven 500 miles on a test track, obtained the following gasoline mileages. What was the average number of miles per gallon for the three cars?

Car	Miles per Gallon
A	12.5
B	15.6
C	19.4

3–11. Assume that each of the cars in Problem 3–10 had ten gallons of gasoline in the tank. The cars were driven until all the gasoline had been used, and the miles per gallon were the same as in Problem 3–10. What was the average number of miles per gallon for the three cars? Compare this answer with the one obtained in Problem 3–10 and explain.

3–12. Each of five typewriting students copied ten pages of printed matter at the following speeds. Each page contained 300 words.

Words per Minute
43
64
75
52
31

What was the average speed of the five typists (in words per minute)?

3–13. Assume that each of the five students in Problem 3–12 typed for ten minutes at the speeds given. What would be the average speed (in words per minute)? How does this average compare with the one computed in Exercise 3–12? Explain.

3–14. A fund of $30,000 was invested in 1963, and for ten years all dividends and interest payments received were reinvested. At the end of the ten years, the total of the fund was $49,783.64. What was the average rate of return, compounded annually, on the original investment?

3–15. The following tabulation gives the retail prices in dollars of four commodities in Dallas, Texas, September, 1971, and January, 1972.

Commodity	September 1971	January 1972
Oranges, dozen	1.333	1.173
Sugar, five pounds	.331	.664
Tomatoes, pound	.409	.548
Pork chops, pound	1.112	1.032

Source: U.S. Department of Labor, Bureau of Labor Statistics, Region 6, *Consumer Prices* (March, 1973).

(a) Compute the ratio of January, 1972, prices to September, 1971, prices for each commodity.

(b) Compute the geometric mean and the arithmetic mean of these ratios.

(c) Explain the significance of each of these averages.

3–16. Compute the harmonic mean of the ratios of Problem 15.

3–17. The number of people living in families in the United States was 187,137,000 in March, 1971. There were 51,948,000 families, which gives an average family size of 3.60 people (arithmetic mean). How do you interpret such a number? In other words, how could a family have .60 of a person in it?

3–18. The table at the top of page 74 gives the distribution of families by size in the United States on March 1, 1971.

(a) What difficulty would be encountered in trying to compute the arithmetic mean for this distribution?

(b) Compute the median for the distribution. If the median does not come out a whole number, how do you interpret it? By definition, the median is the value of the variable that divides the distribution into two equal parts. Is it possible to find any value of the variable that divides the frequencies into two equal parts?

Number of People	Number of Families
2	18,282
3	10,724
4	9,899
5	6,528
6	3,381
7 or more	3,133
Total	51,947

Source: U.S. Department of Commerce, Bureau of the Census, *Current Population Reports,* Series P-20, No. 233 (February, 1972).

3–19. A supervisor in the circulation department of a daily newspaper was interested in the ages of the boys who delivered papers in his district, and compiled the following information, based on the ages of the boys at their last birthday.

Age (Years)	Number of Carriers
14	7
15	15
16	23
17	27
18	20
19	8
Total	100

What is the arithmetic mean age? The median age?

3–20. Find the arithmetic mean, median, and mode of the following set of ungrouped data.

72	107	523
892	703	14
707	29	408
93	85	211
680	90	85

3–21. Given the following average income and number of workers for each union local, compute the mean income of all workers in the union.

Local Union	Number of Members	Average Hourly Earnings
A	27	$3.29
B	108	2.98
C	54	3.04
D	82	3.19
E	220	3.42

3–22. Refer to the table in Problem 2–3, page 40, and compute the arithmetic mean weight of the 84 cans of tomatoes, using Formula 3–1.

3–23. If Problem 2–3 has not been solved, step (a) must be completed before working this problem and the three following problems.

(a) Set up a tally sheet and classify the 84 cans of tomatoes in Problem 2–3 according to weight.

(b) Compute the arithmetic mean weight of the 84 cans of tomatoes, using Formula 3–2. How does this compare with that secured in Problem 2–3?

3–24. Compute the arithmetic mean weight of the frequency distribution of the 84 cans of tomatoes in Problem 3–23, using deviations from an arbitrary origin. (If uniform class intervals were used, deviations in class interval units may be used.) *Retain the solutions to Problems 3–24, 3–25, and 3–26 for use in later chapters.*

3–25. Compute the median weight of the 84 cans of tomatoes in Problem 2–2. Is the median or the arithmetic mean the better average to use for this distribution? Why?

3–26. Compute the mode for the weight of the 84 cans of tomatoes in Problem 2–2 by two methods. Which method do you consider to be better? Does the mode have any distinct advantage over the mean and the median for this distribution?

3–27. Refer to the table in Problem 2–2, page 39, and compute the arithmetic mean age of the 210 employees, using Formula 3–1.

3–28. The following table shows the number of automobiles in the United States in 1971 by their estimated age.

(a) What figure divides the distribution so that half the cars are newer and the other half are older? What is this measure called?

(b) What percentage of the cars are between three and six years old?

(c) Seventy-five percent of the cars are newer than what age?

(d) What two ages include the middle 50% of the cars in terms of age?

Age (Years)	Number (Thousands)
Under 1	5,927
1 and under 2	8,888
2 and under 3	9,280
3 and under 4	8,802
4 and under 5	7,772
5 and under 6	8,313
6 and under 7	8,171
7 and under 8	6,651
8 and under 9	5,624
9 and under 10	4,274
10 and under 11	2,525
11 and under 11	2,035
12 and under 13	1,183
13 and under 14	563
14 and older	3,114
Total	83,122

Source: Automobile Manufacturers Association, *Automobile Facts and Figures* (1972), p. 31.

3–29. If Problem 2–2 has not been solved, step (a) must be completed before working this problem and the following three problems.

(a) Set up a tally sheet and classify the employees in Problem 2–2 according to age.
(b) Compute the arithmetic mean age of the 210 employees, using Formula 3–2. How does this mean compare with that secured in Problem 3–27?

3–30. Compute the arithmetic mean age of the frequency distribution of the 210 employees in Problem 3–29, using deviations from an arbitrary origin. (If uniform class intervals were used, deviations in class interval units may be used.) *Retain the solutions to Problems 3–30, 3–31, and 3–32 for use in later chapters.*

3–31. Compute the median age of the frequency distribution of the 210 employees in Problem 3–29. Is the median or the arithmetic mean the better average to use for this distribution? Why?

3–32. Compute the mode for the ages of the frequency distribution of the 210 employees in Problem 3–29 by two methods. Which method do you consider to be better? Does the mode have any distinct advantage over the mean and the median for this distribution?

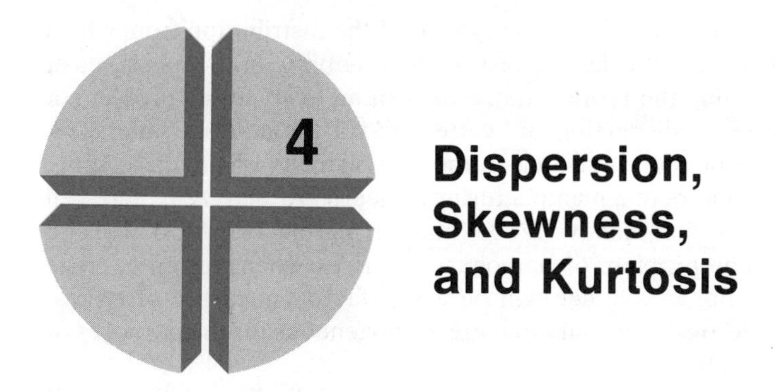

4 Dispersion, Skewness, and Kurtosis

The principal methods of summarizing individual observations by computing measures of typical size are described in Chapter 3. This chapter discusses three other important characteristics of frequency distributions that should be studied in order to give a complete description of the distribution.

IMPORTANCE OF DISPERSION

The tendency of individual values of a variable to scatter about the average is known as *dispersion*. Dispersion is an important characteristic that should be measured for the information it gives about the data in a frequency distribution. Since dispersion may not be the same in both directions, it is important to measure the *skewness,* or lack of symmetry, of the distribution. The *kurtosis,* or degree of peakedness of the frequency distribution, has specific applications.

The decision to use an average as a value typical of a distribution should be made carefully. Before a mean may be considered typical, the dispersion of the items around it should be examined. The highest degree of concentration would be to have all the individual items the same size. The scatter in such a case would be zero, and the mean would be exactly the same as the individual values of the variable. The more the individual amounts differ from each other, the less typical of the whole distribution an average will be.

Assume that a class of 20 college students is made up of 18 who come from the town where the college is located, while the other 2 are foreign students from France and Japan. One might ask, "What is the average distance the 20 students traveled to attend the college?" The arithmetic mean is 174 miles. In this case the mean is much larger than the distances traveled by the 18 hometown students and much smaller than the distances traveled by the other 2. Thus, it is difficult to think of the mean as representative of any one of the 20 students.

Since an average may not always be typical of the distribution from which it is computed, a method is needed for testing the reliability of the measures of typical size. Determining the representativeness of an average is a problem of measuring the degree of dispersion or scatter among the amounts averaged.

The degree of dispersion is related to the homogeneity of the data. If the wages of unskilled laborers in a manufacturing concern are analyzed, there will ordinarily be less dispersion than would be found in a group of both skilled and unskilled laborers. The problem of homogeneity arises whenever an average is computed. As a rule, an average will be a satisfactory measure of typical size only when it is derived from data that are homogeneous for the purposes of the particular investigation.

Dispersion is also important when the scatter in a distribution may itself be significant. In evaluating the performance of a student, it might be significant to measure the *consistency* of his work. If the grades made on different tests show a wide dispersion, it means that his work was sometimes good and sometimes poor. Two students might have the same median grade on their work for the year; yet all of one student's grades might be close to the median, while the grades of the second might vary from 100% to 20%. A manufacturer interested in controlling the quality of his product tries to prevent variations between individual units. A manufacturer of light bulbs tries to produce bulbs that will burn a long time, but it is also desirable that there be as little variation as possible in the length of life between individual bulbs. Uniformly high quality in a product is better than a high average in quality with wide variations between units.

MEASURES OF ABSOLUTE DISPERSION

If measures of dispersion are to be used in further calculations, they need to be precise and logical mathematically; but if an approximation is not needed for further calculations, a simple method of measuring dispersion may be used. Each of the following methods is appropriate for use under certain circumstances.

Range

The simplest measure of dispersion is the *range* of the data, that is, the distance between the smallest and the largest amounts. Frequently the range is expressed by giving the smallest and the largest amounts. For example, a production manager might say that the average daily wage in a certain department is $23.98 and that the individual daily wages range from $14 to $41. This would give a rough measure of scatter that could be compared with other departments. A second department might have a mean daily wage of $24.61, with daily wages ranging between $19 and $30. The average of the second department is probably more representative of the wage distribution than the average of the first, since there is less scatter in the wages received by employees in the second department. The high and the low prices for which securities or commodities sell on an exchange in a given day are reported regularly in financial periodicals. These quotations serve as a reasonably accurate measure of the dispersion in prices for each day.

A serious weakness of such a measure of dispersion is that it is based on only two items and tells nothing about the scatter of the other items. The range is a simple and easily understood measure of scatter, but it is less informative than other measures.

Since the range can be computed with very little work, it is used extensively to measure dispersion in the construction of control charts for variables used in statistical quality control. Maintaining control charts requires the measurement of the dispersion in many distributions, and the saving of time resulting from the use of the range becomes an important factor. The types of distributions used in this work do not distort the range as much as some distributions, so it serves as a reasonably accurate measure of dispersion. The use of the range in connection with statistical quality control is described in detail in Chapter 10.

Quartile Deviation (Q)

The median is defined as the point that divides the distribution into two equal parts, half the items falling on one side and half on the other. By exactly the same process, it is possible to divide an array or a frequency distribution into four equal parts. Three points must be located: (1) the *first quartile* (Q_1) is the point that has one fourth of the frequencies smaller and three fourths larger; (2) the *second quartile* is the median; (3) the *third quartile* (Q_3) has one fourth of the frequencies larger and three fourths smaller.

Half the frequencies fall between the first and third quartiles, since one fourth fall between the first and second quartiles, and one fourth between the second and third. The distance between the first and the third quartiles is the *interquartile range*; the smaller this distance, the greater the degree of concentration of the middle half of the distribution. A measure of dispersion based on the interquartile range does not give any influence to the items above and below the middle half of the distribution, but it is much superior to the range, which is based on only the largest and the smallest amounts of the array.

The measure of dispersion based on the interquartile range is called the *quartile deviation* or *semi-interquartile range* and the formula for its computation is

$$Q = \frac{Q_3 - Q_1}{2}. \tag{4-1}$$

The formulas used to interpolate for the quartiles after locating the class in which each quartile falls are

$$Q_1 = L_{Q_1} + \frac{\frac{N}{4} - F_{L_{Q_1}}}{f_{Q_1}} i_{Q_1} \tag{4-2}$$

$$Q_3 = L_{Q_3} + \frac{\frac{3N}{4} - F_{L_{Q_3}}}{f_{Q_3}} i_{Q_3} \tag{4-3}$$

where

L_{Q_1} = real lower limit of the class in which the first quartile falls

L_{Q_3} = real lower limit of the class in which the third quartile falls

FL_{Q_1} = cumulative frequencies less than the lower limit of the class in which the first quartile falls

FL_{Q_3} = cumulative frequencies less than the lower limit of the class in which the third quartile falls

f_{Q_1} = frequency of the class in which the first quartile falls

f_{Q_3} = frequency of the class in which the third quartile falls

N = number of frequencies in the distribution

i_{Q_1} = class interval of the class in which the first quartile falls

i_{Q_3} = class interval of the class in which the third quartile falls.

The computation of the quartiles is basically similar to the computation of the median. For the data in Table 3–11, page 60, $\frac{N}{4} = \frac{125}{4} = 31.25$. There are 24 frequencies less than 84.5 pounds (24 lots had a strength that required less than 84.5 pounds per square inch to break a skein of yarn), and 53 lots required less than 89.5 pounds. Therefore, Q_1 falls between 84.5 and 89.5 pounds. Substituting the data from Table 3–11, the first quartile is computed

$$Q_1 = 84.5 + \frac{\frac{125}{4} - 24}{29} 5 = 84.5 + \frac{31.25 - 24}{29} 5$$

$$Q_1 = 84.5 + \frac{(7.25)(5)}{29} = 84.5 + 1.2 = 85.7.$$

In the same table $\frac{3N}{4} = 93.75$. Cumulating the frequencies, 90 are found to be less than 99.5 pounds and 103 are less than 104.5 pounds. Therefore, Q_3 falls between 99.5 pounds and 104.5 pounds. Substituting the data from Table 3–11, the third quartile is computed

$$Q_3 = 99.5 + \frac{\frac{(3)(125)}{4} - 90}{13} 5 = 99.5 + \frac{93.75 - 90}{13} 5$$

$$Q_3 = 99.5 + \frac{(3.75)(5)}{13} = 99.5 + 1.4 = 100.9.$$

Since the first quartile is 85.7 pounds and the third quartile is 100.9 pounds, the interquartile range is 15.2 pounds (the distance between the two quartiles). Half the interquartile range (the semi-interquartile range) is the quartile deviation, which is computed

$$Q = \frac{100.9 - 85.7}{2} = \frac{15.2}{2} = 7.6.$$

Other measures of a similar nature may be constructed by finding the points that divide the distribution into tenths *(deciles)* or hundredths *(percentiles).*

Average Deviation *(AD)*

The measures of dispersion described in the previous section are simple to compute and understand, but they ignore some of the information furnished by the distribution of the individual items. A better measure of dispersion is based on an average of the deviations of the individual items from a central value of the distribution. The *average deviation* is the arithmetic mean of the deviations from the mean or the median, in which all the deviations are treated as positive regardless of sign.

Computing the Average Deviation from Ungrouped Data. The computation of the average deviation from ungrouped data is illustrated by the semimonthly expenditures for food given in Table 4–1. The third column in Table 4–1 shows the deviation of each of the individual items from the median $(X - Md)$. The median is \$111.50, the mean of the two middle items in an array of the ten values. Since the average deviation is found by computing the mean of the deviations without regard to sign, the signs of the $X - Md$ values are not entered in Table 4–1. The computation of the average deviation is

$$AD = \frac{\Sigma|X - Md|}{N}. \qquad \textbf{(4–4)}$$

The parallel lines $|\quad|$ around the term $X - Md$ indicate that the absolute sum, rather than the algebraic sum, is taken. Dividing this absolute sum by the number of deviations gives the average deviation.

The average deviation summarizes in one figure the whole group of deviations and is a measure of the typical amount of dispersion among the values.

TABLE 4–1

COMPUTATION OF THE AVERAGE DEVIATION FROM UNGROUPED DATA

Semimonthly Expenditure for Food by Ten Families

Family	Semimonthly Expenditures (Dollars)	$X-Md$ (Disregarding Signs)
1	124.00	12.50
2	120.00	8.50
3	133.00	21.50
4	97.00	14.50
5	112.00	0.50
6	110.00	1.50
7	117.00	5.50
8	99.00	12.50
9	103.00	8.50
10	111.00	0.50
Total		86.00

Source: Hypothetical data.

Median = $111.50

$$AD = \frac{\Sigma|X - Md|}{N}$$

$$AD = \frac{86.00}{10} = \$8.60.$$

Since interest lies in the magnitude of the deviations rather than in their direction, the simple expedient of ignoring the signs gives a measure of the average size of the deviations. If the algebraic sum of the deviations were computed, the sum would be zero when the deviations are taken from the arithmetic mean. This would, of course, give an average deviation of zero. When deviations are taken from the median, their sum will not be zero (unless the median and the mean have the same value); but the total will always be quite small since half of the deviations from the median will be positive and half will be negative.

Computing the Average Deviation from a Frequency Distribution. It is common to have data in the form of a frequency distribution rather than in the form of individual amounts. Table 4–2 shows the computation of the average deviation of items grouped into a frequency distribution.

The first step in computing the average deviation is to determine the deviation of the midpoint of each class from the median. It is assumed that this shows the amount of deviation for *each value in the class*. The total deviation of all the items in a class is found by multiplying the deviation by the frequency of the class. The sum of these products, disregarding the signs, is the sum of the deviations from the median. Dividing this sum by the number of frequencies in the distribution gives the average deviation. The computation of the average deviation from a frequency distribution is

$$AD = \frac{\Sigma f|m - Md|}{N}. \tag{4-5}$$

TABLE 4-2

COMPUTATION OF THE AVERAGE DEVIATION FROM A FREQUENCY DISTRIBUTION

Strength of 125 Lots of Cotton Yarn

Pounds per Square Inch (Class Interval)	Number of Lots f	Pounds per Square Inch (Midpoints) m	Deviation of Midpoints from Median (91.9) $\|m - Md\|$	Frequency Times Deviation $f\|m - Md\|$
70– 74	1	72	19.9	19.9
75– 79	6	77	14.9	89.4
80– 84	17	82	9.9	168.3
85– 89	29	87	4.9	142.1
90– 94	20	92	.1	2.0
95– 99	17	97	5.1	86.7
100–104	13	102	10.1	131.3
105–109	10	107	15.1	151.0
110–114	6	112	20.1	120.6
115–119	3	117	25.1	75.3
120–124	2	122	30.1	60.2
125–129	1	127	35.1	35.1
Total	125			1,081.9

Source: Table 2–1.

$$AD = \frac{\Sigma f|m - Md|}{N}$$

$$AD = \frac{1{,}081.9}{125} = 8.66.$$

Standard Deviation and Variance

The most widely used measure of dispersion is the *standard deviation,* which resembles the average deviation in that it is based on the deviations of all the values of the variable from a measure of typical size. The standard deviation differs from the average deviation in the method of averaging the deviations. The standard deviation is always computed from the mean and is a minimum when measured from this value. Although the average deviation may be computed from either the mean or the median, the fact that it is a minimum when measured from the median has resulted in a tendency to use it as a measure of dispersion when the median is used as a measure of typical size.

In computing the standard deviation, the deviations from the mean are first squared. Next, the squared deviations are averaged by dividing their total by the number of deviations. The average of the squared deviations is the *variance,* which is the square of the standard deviation. For some purposes the variance is more useful than the standard deviation, but both of the measures have many important uses in statistical analysis. They are both computed measures and may be used in further calculations, which makes them particularly valuable when a measure of dispersion is needed in a formula.

In computing the average deviation, the problem of getting rid of the minus signs in averaging the deviations is handled by ignoring the signs and dealing only with the absolute size of the deviations. The standard deviation overcomes this problem by squaring the deviations and making them all positive. This is a logical mathematical step, and as a result, the variance and the standard deviation may be used in further calculations.

The standard deviation will be represented by the small Greek letter sigma (σ), and the variance, by σ^2. In this chapter the discussion is concerned only with measures computed from complete enumerations. Samples will be discussed in Chapter 7, where a different symbol for the standard deviation of a sample will be introduced.

Computing the Standard Deviation from Ungrouped Data. The computations of the variance and the standard deviation are

$$\sigma^2 = \frac{\Sigma(X-\mu)^2}{N} \tag{4-6}$$

$$\sigma = \sqrt{\frac{\Sigma(X-\mu)^2}{N}} \tag{4-7}$$

where

$X - \mu =$ the deviations from the mean.

The computation of the variance and the standard deviation is illustrated in Table 4–3. The mean is \$112.60 and the deviation of each observation from the mean is written in the third column. The squared deviations are entered in the fourth column, the sum of which is 1,150.40. Therefore, the value of the variance is 115.04 and the standard deviation is \$10.73.

Formula 4–6 is based on the definition of the variance, but it is not generally used for calculation purposes except when a computer is available. If a desk calculator is used, it is possible to modify this equation to a form that requires fewer computations. The numerator of Formula 4–6, the sum of the squared deviations from the arithmetic mean, is an important concept in statistical analysis which will be used in many situations. An easy method of calculation will be a valuable tool for analysis. Several computational formulas for the variance and the standard deviation follow.

TABLE 4–3

COMPUTATION OF THE STANDARD DEVIATION AND THE VARIANCE FROM UNGROUPED DATA

Semimonthly Expenditure for Food by Ten Families

Family	Semimonthly Expenditures (Dollars) X	$X - \mu$	$(X - \mu)^2$
1	124.00	11.40	129.96
2	120.00	7.40	54.76
3	133.00	20.40	416.16
4	97.00	−15.60	243.36
5	112.00	−0.60	0.36
6	110.00	−2.60	6.76
7	117.00	4.40	19.36
8	99.00	−13.60	184.96
9	103.00	−9.60	92.16
10	111.00	−1.60	2.56
Total	1,126.00	0	1,150.40

Source: Table 4–1.

$$\mu = \frac{1{,}126.00}{10} = \$112.60$$

$$\sigma^2 = \frac{\Sigma(X - \mu)^2}{N} \qquad \sigma = \sqrt{\frac{\Sigma(X - \mu)^2}{N}}$$

$$\sigma^2 = \frac{1{,}150.40}{10} = 115.04 \qquad \sigma = \$10.73.$$

Squaring the term $\Sigma(X - \mu)$ gives

$$\Sigma(X - \mu)^2 = \Sigma(X^2 - 2X\mu + \mu^2).$$

Applying the summation sign to each term gives

$$\Sigma(X - \mu)^2 = \Sigma X^2 - \Sigma 2X\mu + \Sigma\mu^2.$$

Interchanging the constant and the summation sign in the middle term gives

$$\Sigma(X - \mu)^2 = \Sigma X^2 - 2\Sigma X\mu + \Sigma\mu^2.$$

The summation sign associated with μ^2 can be changed to an N, since there are as many μ^2's being summed as there are items in the distribution. This gives

$$\Sigma(X - \mu)^2 = \Sigma X^2 - 2\Sigma X\mu + N\mu^2.$$

Substitution of the definitional formula for μ gives

$$\Sigma(X-\mu)^2 = \Sigma X^2 - 2\Sigma X\left(\frac{\Sigma X}{N}\right) + N\left(\frac{\Sigma X}{N}\right)^2.$$

Collecting terms gives

$$\Sigma(X-\mu)^2 = \Sigma X^2 - \frac{(\Sigma X)^2}{N}. \tag{4-8}$$

This formula will be used in various places, but the use here is computing the variance by dividing by N to give

$$\sigma^2 = \frac{\Sigma(X-\mu)^2}{N} = \frac{\Sigma X^2 - \frac{(\Sigma X)^2}{N}}{N}. \tag{4-9}$$

The calculations may be simplified by multiplying Formula 4–9 by $\frac{N}{N}$, which reduces to

$$\sigma^2 = \frac{N\Sigma X^2 - (\Sigma X)^2}{N^2}. \tag{4-10}$$

The standard deviation can be found from Formula 4–10 simply by taking the square root of the variance, which gives

$$\sigma = \sqrt{\frac{N\Sigma X^2 - (\Sigma X)^2}{N^2}}. \tag{4-11}$$

The square root of the numerator and the denominator may be taken separately to give

$$\sigma = \frac{\sqrt{N\Sigma X^2 - (\Sigma X)^2}}{N}. \tag{4-12}$$

The computation of the variance and the standard deviation by Formulas 4–10 and 4–12 is shown in Table 4–4. The values computed by this method are identical to those computed on page 85.

Computing the Standard Deviation from a Frequency Distribution. When data are available only in the form of a frequency distribution, it is necessary to have some method of computing the standard deviation from the distribution instead of from the individual items. In fact, the computation from a frequency distribution is so much shorter when the number of items is large that if a computer is not available, time can be saved by making a frequency distribution to use in computing the mean and the standard deviation.

TABLE 4–4

COMPUTATION OF THE STANDARD DEVIATION AND THE VARIANCE FROM UNGROUPED DATA

Semimonthly Expenditure for Food by Ten Families

Family	Semimonthly Expenditures (Dollars) X	X^2
1	124	15,376
2	120	14,400
3	133	17,689
4	97	9,409
5	112	12,544
6	110	12,100
7	117	13,689
8	99	9,801
9	103	10,609
10	111	12,321
Total	1,126	127,938

Source: Table 4–1.

$$\sigma^2 = \frac{N\Sigma X^2 - (\Sigma X)^2}{N^2} = \frac{(10)(127{,}938) - (1{,}126)^2}{10^2}$$

$$\sigma^2 = \frac{1{,}279{,}380 - 1{,}267{,}876}{100} = \frac{11{,}504}{100}$$

$$\sigma^2 = 115.04$$

$$\sigma = \$10.73$$

or by Formula 4–12, $\sigma = \frac{\sqrt{11{,}504}}{10} = \frac{107.3}{10} = \$10.73.$

The basic principle employed in this method is the same as that used to approximate the mean and the average deviation from a frequency distribution. The midpoint of each class is assumed to be the value of all the items in the class, and the standard deviation is based on the deviation of these midpoints from the mean. This assumption is not entirely correct, but since the grouping error is usually relatively small, the approximation is widely used without attempting to compute the value more accurately.

The variance may be defined as

$$\sigma^2 = \frac{\Sigma f(m - \mu)^2}{N} \quad \textbf{(4–13)}$$

where

m = the midpoints of the classes of a frequency distribution.

Assuming that the deviations of all the values in a class are the same, the squared deviation of each class must be multiplied by the frequency of the class to get the total squared deviations. The variance is then computed by finding the average of the squared deviations from the mean, and the standard deviation is the square root of the variance.

The calculations are similar to those made in Table 4–2 for the average deviation except that the deviations of the midpoints are computed from the arithmetic mean instead of from the median, and the deviation for each class is squared before being multiplied by the frequency of the class. To compute the mean squared deviation, which is the variance, $\Sigma f(m-\mu)^2$ is divided by N.

When the class interval is uniform throughout the frequency distribution, the calculations for the variance can be shortened by using the same device described in computing the arithmetic mean in Table 3–2. The midpoint of one of the classes is designated as A and deviations of the midpoints from this arbitrary origin are computed in class interval units. These computations are made in Table 4–5, which merely extends Table 3–2 by the addition of one column, $f(d')^2$. Each of the values in the column headed fd' is multiplied by the value on the same line in column d' to give the value of $f(d')^2$. The sum of $f(d')^2$ is the sum of the squared deviations of all the 125 lots from the arbitrarily chosen value A, which equals 92, the midpoint of the class 90–94. The quantity $\Sigma f(d')^2$ when divided by N is analogous to the variance, except it is in class interval units and is measured from A instead of from μ. It may be written

$$\sigma_a^2 = \frac{\Sigma f(d')^2}{N}.$$

Since the deviations used in computing σ_a^2 were from the arbitrary origin A, instead of from the arithmetic mean, the value of σ_a^2 is always larger than σ^2, except when the arithmetic mean equals A. This is true because the sum of the squared deviations from the arithmetic mean is less than the sum of the squared deviations from any other value. Furthermore, it can be demonstrated that the difference between σ_a^2 and σ^2 is equal to the square of c, given by Formula 3–5. Therefore,

$$\sigma^2 = \sigma_a^2 - c^2$$

and

$$\sigma = \sqrt{\sigma_a^2 - c^2}.$$

The values of σ_a^2 computed above in class-interval units, and c^2, also expressed in class-interval units (Formula 3–4, page 48), are substituted to give

$$\sigma = i\sqrt{\frac{\Sigma f(d')^2}{N} - \left(\frac{\Sigma fd'}{N}\right)^2}. \qquad \textbf{(4–14)}$$

The standard deviation is converted to original units by multiplying by i.

The computation of the standard deviation from a frequency distribution is shown in Table 4–5, which gives the values for Formula 4–14.

TABLE 4–5

COMPUTATION OF THE STANDARD DEVIATION FROM A FREQUENCY DISTRIBUTION

Strength of 125 Lots of Cotton Yarn

(Pounds per Square Inch Required to Break a Skein of 22s)

Pounds per Square Inch (Class Interval)	Number of Lots f	d'	fd'	$f(d')^2$
70– 74	1	−4	−4	16
75– 79	6	−3	−18	54
80– 84	17	−2	−34	68
85– 89	29	−1	−29	29
90– 94	20	0	0	0
95– 99	17	1	17	17
100–104	13	2	26	52
105–109	10	3	30	90
110–114	6	4	24	96
115–119	3	5	15	75
120–124	2	6	12	72
125–129	1	7	7	49
Total	125		46	618

Source: Table 2–1.

$$\sigma = i\sqrt{\frac{\Sigma f(d')^2}{N} - \left(\frac{\Sigma fd'}{N}\right)^2} = 5\sqrt{\frac{618}{125} - \left(\frac{46}{125}\right)^2}$$

$$\sigma = 5\sqrt{4.9440 - .1354} = 5\sqrt{4.8086} = (5)(2.1928)$$

$$\sigma = 10.96.$$

Since one of the chief reasons for computing a measure of dispersion is to use it in further computations, the standard deviation holds an important place in the discussion of dispersion. The quartile deviation and the average deviation are good measures of dispersion when they are not to be used in further calculations.

Two of the most important uses of a measure of dispersion in further calculations occur in the measurement of the reliability of computations from samples and in the computation of measures of relationship. Measures of the precision of estimates made from samples are discussed in Part 2, and the measures of relationship are discussed in Part 3. Such computations make use of a measure of dispersion, and in every case the standard deviation is used in preference to any other.

CHIEF CHARACTERISTICS OF MEASURES OF DISPERSION

The following summary briefly outlines the chief characteristics of the measures of absolute dispersion described in this chapter—the range, the quartile deviation, the average deviation, and the standard deviation.

Range

1. The range is the simplest measure of dispersion with respect to its calculation and its significance, since it is merely the distance between the largest and the smallest items in a distribution.
2. Since the range is based on the two extreme values, it tends to be erratic. A few very large or very small values are not unusual in distributions of business data. When these items occur, the range measures only their dispersion and ignores the remaining values of the variable.
3. The fact that the range is not influenced by any of the values of the variable except the two extremes is its chief weakness. There is always a danger that it will give an inaccurate description of the dispersion in the distribution.
4. Extensive use of the range is found in statistical quality control when the distributions used do not distort the range, and the saving in computation time is an important factor.

Quartile Deviation

1. The quartile deviation is fundamentally the same type of measure as the range, since it is based on the range over which the middle half of the distribution scatters. Like the range, it is not affected by the dispersion of all the individual values of the variable.
2. Basing the quartile deviation on the range of the middle half of the distribution is an improvement over the range of the entire distribution because extreme deviations in the largest and the smallest values cannot distort the quartile deviation to the extent that they distort the range.
3. The quartile deviation requires more work to compute than the range but generally less work than the average deviation or the standard deviation.

Average Deviation

1. The average deviation is the average of the absolute size of the deviations of the individual items from their average value (either arithmetic mean or median).
2. By reflecting the dispersion of every item in the distribution, the average deviation is superior to the range and the quartile deviation as a measure of dispersion.
3. Since the average deviation is the average variation of the individual items from their average value, it has a precise meaning. Its significance is logical and is not difficult to understand.

4. The calculation of the average deviation from individual values of a variable or from a frequency distribution is not difficult.
5. The major flaw of the average deviation arises from the fact that it averages the absolute deviations; that is, it ignores the signs of the deviations. This fact makes its use in further calculations less satisfactory than the standard deviation.

Standard Deviation

1. The standard deviation is the square root of the arithmetic mean of the squared deviations of the individual items from the arithmetic mean. Squaring the deviations makes all of them positive so there is no need to ignore the signs as in the computation of the average deviation.
2. The standard deviation is affected by the deviation of every item.
3. The need to square each deviation increases the amount of work required in comparison with the calculation of the average deviation, but the standard deviation can be computed from the individual values of the variable or from a frequency distribution.
4. The standard deviation resembles the average deviation in that it has a precise meaning, but it has a further advantage since there is no mathematical flaw in its calculation.
5. The fact that the standard deviation is logical mathematically means that it can be used satisfactorily in further calculations. This is its outstanding superiority over the other measures of dispersion.

RELATIVE DISPERSION

Measures of absolute dispersion furnish methods for expressing definitely the amount of scatter in a distribution. At times, however, it is necessary to compare the scatter in one distribution with that of another; and if the items in one distribution are decidedly different in size from the other, it is difficult to compare the degree of scatter in the two. For example, assume that an executive wants to compare the variation in the salaries of clerical workers in the organization with the salaries paid the sales force. The clerks are paid weekly and the salesmen monthly. The mean clerical salary is $123.60 per week, and the mean salary of the salesmen is $1,094.66 per month. The standard deviation of the clerical salaries is $24.10, and of the salesmen's salaries, $367.81.

Because of the differences in the original data, it is impossible to decide whether $367.81 is a more important amount of scatter for the salaries paid to the salesmen than $24.10 is for the clerical salaries. Since the salaries of salesmen are larger, partly because they are expressed on a monthly rather than a weekly basis, they will show more dispersion than the smaller weekly wages. Some method is needed to express the amount of dispersion in the two series on a *relative* basis.

The most common method of comparing amounts of different sizes is to reduce them to a comparable percentage basis. In this case each standard deviation is expressed as a percentage of the mean of the data from which it was computed. The percentage computed in this way is called the *coefficient of variation* and is represented by V. The formula for computing V is

$$V = \frac{\sigma}{\mu} \cdot 100. \tag{4-15}$$

Dividing the standard deviation by the mean and multiplying by 100 expresses the standard deviation as a percentage of the mean.

For salesmen's salaries the coefficient of variation is computed

$$V = \frac{\$367.81}{\$1{,}094.66} \cdot 100 = 33.6\%.$$

For clerical salaries the coefficient of variation is computed

$$V = \frac{\$24.10}{\$123.60} \cdot 100 = 19.5\%.$$

The coefficient of variation for salesmen's salaries may now be compared logically with the coefficient of variation for clerical salaries. These two measures show that there was somewhat more scatter among the salaries of salesmen than among the clerical salaries. The two percentages can be compared, whereas the comparison of the two standard deviations means nothing, since the individual items in one distribution differ so much from the items in the other.

Other measures of relative dispersion might be based on the average deviation and the quartile deviation, but the coefficient of variation as described here is the only one generally used. For measuring the dispersion of statistical data used in business, the average deviation and the quartile deviation are highly satisfactory and more easily understood than the standard deviation. If they become more popular as measures of dispersion, it would be advisable to compute measures of relative dispersion based on them.

SKEWNESS

Sometimes it is desirable to compute a measure that shows the direction of the dispersion about the center of the distribution. Measures of dispersion indicate only the amount of the scatter, and give no information about the direction in which the scatter occurs. Measures of *skewness* indicate the lack of symmetry in a distribution and show the direction of the skewness.

Computing Skewness

The distribution of the strength of 125 lots of cotton yarn in Table 4–5 shows a moderate amount of skewness to the larger values, which is reflected in the difference between the arithmetic mean, median, and mode. Figure 4–1 shows the location of the three averages, with the mean the largest, the mode the smallest, and the median falling between them.

A measure of skewness can be based on the difference between the mean and the mode. The greater the amount of skewness, the more the mean and the mode differ because of the influence of extreme items. Dividing this difference by the standard deviation gives a coefficient of skewness. The formula is

$$\text{Skewness} = \frac{\mu - Mo}{\sigma}. \qquad \textbf{(4–16)}$$

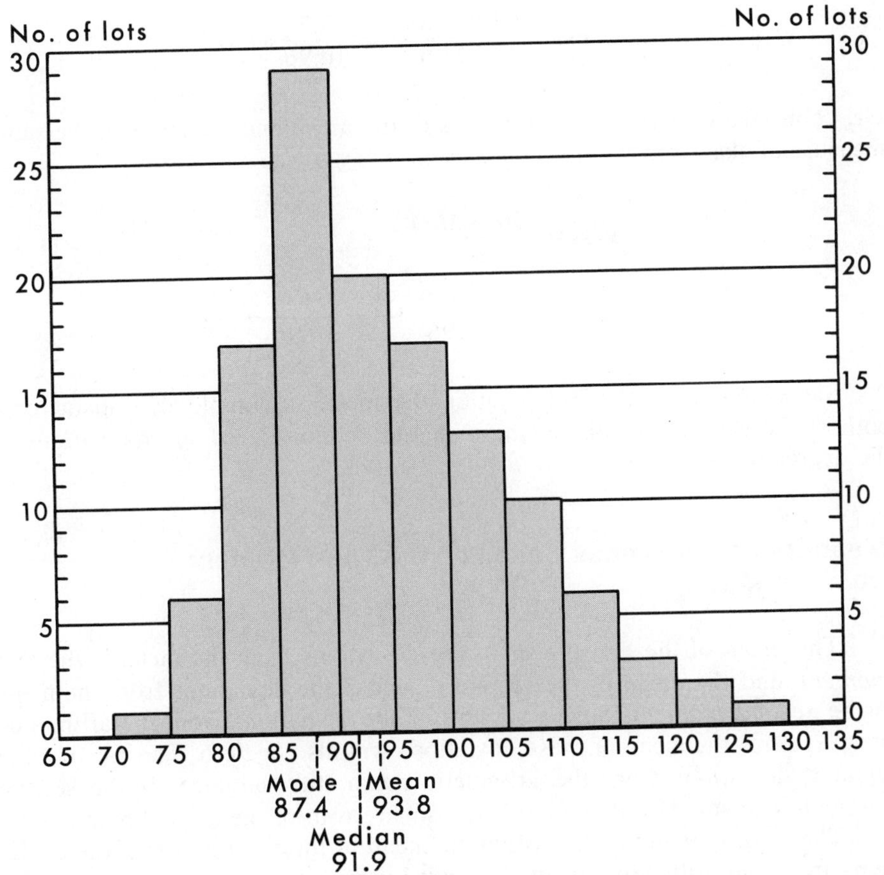

Source: Table 4-5.

FIGURE 4–1

HISTOGRAM SHOWING LOCATION OF MEAN, MEDIAN, AND MODE

When the distribution is skewed to the larger values, the mean is larger than the mode; therefore, the measure of skewness will be positive. When the skewness is to the smaller values, the mean will be smaller than the mode, and the measure of skewness will be negative.

Using Formula 4–16, the calculation of the measure of skewness for the distribution of strength of cotton yarn is

$$\text{Skewness} = \frac{93.8 - 87.4}{10.96} = \frac{6.4}{10.96} = .58.$$

If the empirical mode computed by Formula 3–12 on page 66 is used instead of the mode located within the modal class, the value of skewness is

$$\text{Skewness} = \frac{93.8 - 88.1}{10.96} = \frac{5.7}{10.96} = .52.$$

Or, computed directly from the values of the mean and the median, the same measure of skewness is

$$\text{Skewness} = \frac{3(\mu - Md)}{\sigma} \tag{4–17}$$

$$\text{Skewness} = \frac{3(93.8 - 91.9)}{10.96} = \frac{3(1.9)}{10.96} = .52.$$

Since neither method of locating the mode is completely satisfactory, both of these measures of skewness should be considered approximations of the degree of skewness.

Measures of Skewness Based on the Third Moment from the Mean

The mean of the first power of the deviations from the mean is the *first moment,* and the mean of the second power of the deviations from the mean is the *second moment.* Since the sum of the deviations from the arithmetic mean is zero, the mean of these deviations will also be zero. The mean of the squared deviations from the arithmetic mean, the variance, is the second moment. The standard deviation is the square root of the second moment.

The *third moment* is the arithmetic mean of the third power of the deviations from the arithmetic mean. The third moment serves as a more precise measure of skewness than those described previously. The third power of a deviation will retain the original sign of the deviation; thus, if the minus deviations to the third power are greater in total than the plus deviations, it indicates skewness toward the smaller values. The sign of the sum of the deviations cubed always indicates the direction of the skewness.

For Ungrouped Data. Using the data from Table 4–6, the computation of the third moment for the semimonthly expenditures of ten families is

$$M_3 = \frac{\Sigma(X-\mu)^3}{N} \qquad \textbf{(4–18)}$$

$$M_3 = \frac{3{,}243.12}{10} = 324.312.$$

TABLE 4–6 *

COMPUTATION OF THE FIRST FOUR MOMENTS FROM THE ARITHMETIC MEAN (UNGROUPED DATA)

Semimonthly Expenditure for Food by Ten Families

Family	Semimonthly Expenditures (Dollars) X	$X-\mu$	$(X-\mu)^2$	$(X-\mu)^3$	$(X-\mu)^4$
1	124.00	11.40	129.96	1,481.544	16,889.6016
2	120.00	7.40	54.76	405.224	2,998.6576
3	133.00	20.40	416.16	8,489.664	173,189.1456
4	97.00	−15.60	243.36	−3,796.416	59,224.0896
5	112.00	− 0.60	0.36	−0.216	0.1296
6	110.00	− 2.60	6.76	−17.576	45.6976
7	117.00	4.40	19.36	85.184	374.8096
8	99.00	−13.60	184.96	−2,515.456	34,210.2016
9	103.00	− 9.60	92.16	−884.736	8,493.4656
10	111.00	− 1.60	2.56	4.096	6.5536
Total	1,126.00	0	1,150.40	3,243.120	295,432.3520

* The last column shows the computation of the fourth power of the deviations, which is used in computing the fourth moment discussed on page 98.

Source: Table 4–1.

The value of the third moment is a measure of the absolute amount of skewness and β_1 is a measure of relative skewness, based on the formula

$$\beta_1 = \frac{M_3^2}{M_2^3}. \qquad \textbf{(4–19)}$$

The second moment, which is the variance, is needed in computing β_1. Using Formula 4–6, the second moment is computed

$$M_2 = \frac{\Sigma(X-\mu)^2}{N}$$

$$M_2 = \frac{1{,}150.40}{10} = 115.04$$

$$\beta_1 = \frac{324.312^2}{115.04^3} = .069.$$

The positive sign of the value of the third moment means that the distribution is skewed to the larger values. The larger the value of β_1, the greater the skewness, either to the right or to the left. For a symmetrical distribution, the value of β_1 is zero.

From a Frequency Distribution. Since the third and the fourth moments require more calculations than the mean or the standard deviation, a computer is even more necessary in finding these measures. However, if a computer is not available, the moments can be computed from a frequency distribution by an extension of the device used in Table 4–5, where the first two moments were computed. The first five columns in Table 4–7 are computed in the same manner as the five columns in Table 4–5, but Column 6 in Table 4–7 provides the data needed for the computation of the third moment. Column 7 will be used to compute the fourth moment in the following section.

TABLE 4–7

COMPUTATION OF THE MOMENTS FROM THE MEAN FOR A FREQUENCY DISTRIBUTION

Strength of 125 Lots of Cotton Yarn (Pounds per Square Inch to Break a Skein of 22s)

Pounds per Square Inch (Class Interval) (1)	Number of Lots f (2)	d' (3)	fd' (4)	$f(d')^2$ (5)	$f(d')^3$ (6)	$f(d')^4$ (7)
70– 74	1	−4	− 4	16	− 64	256
75– 79	6	−3	−18	54	−162	486
80– 84	17	−2	−34	68	−136	272
85– 89	29	−1	−29	29	− 29	29
90– 94	20	0	0	0	0	0
95– 99	17	1	17	17	17	17
100–104	13	2	26	52	104	208
105–109	10	3	30	90	270	810
110–114	6	4	24	96	384	1,536
115–119	3	5	15	75	375	1,875
120–124	2	6	12	72	432	2,592
125–129	1	7	7	49	343	2,401
Total	125		46	618	1,534	10,482

Source: Table 2–1.

The computation of the third moment from a frequency distribution uses the deviations from an arbitrary origin in class interval units, as was done in Table 4–5. The values of $f(d')^3$ in Column 6 are computed by multiplying the values of $f(d')^2$ in Column 5 by the values of d'. The algebraic sum of these values is used to compute the value of the third moment in class-interval units in the formula

$$M_3 = \frac{\Sigma f(d')^3}{N} - 3\,\frac{\Sigma fd'}{N}\,\frac{\Sigma f(d')^2}{N} + 2\left(\frac{\Sigma fd'}{N}\right)^3. \tag{4-20}$$

Substituting the values from Table 4–7 in Formula 4–20, the value of the third moment is computed

$$M_3 = \frac{1{,}534}{125} - 3\,\frac{46}{125}\,\frac{618}{125} + 2\left(\frac{46}{125}\right)^3$$

$$M_3 = 12.2720 - 5.4582 + .0997$$

$$M_3 = 6.9135 \text{ (in class-interval units).}$$

Formula 4–14, for the standard deviation, may be modified to give M_2, which is the variance in class-interval units, and is written

$$M_2 = \frac{\Sigma f(d')^2}{N} - \left(\frac{\Sigma fd'}{N}\right)^2. \tag{4-21}$$

Using the values in Table 4–7,

$$M_2 = \frac{618}{125} - \left(\frac{46}{125}\right)^2 = 4.9440 - .1354 = 4.8086 \text{ (in class-interval units).}$$

From the values of M_2 and M_3, the value of β_1 from Formula 4–19 is computed

$$\beta_1 = \frac{M_3^2}{M_2^3} = \frac{6.9135^2}{4.8086^3} = \frac{47.7965}{111.1875} = .430.$$

Since the sign of the third moment is positive, the skewness is to the larger values. The small value of β_1 shows that the skewness is not pronounced.

It should be noted that the value of β_1 may be computed with the values of M_2 and M_3 expressed in class-interval units or in original units. This is possible because the value of i in both the numerator and the denominator is raised to the same power and therefore cancels out. In practice, the values of β_1 and β_2 are normally computed from the moments expressed in class-interval units.

KURTOSIS

Another important characteristic of a frequency distribution is *kurtosis,* the degree of peakedness of a given distribution. The fourth moment of a frequency distribution is a measure of the amount of kurtosis in a frequency distribution, and a relative measure β_2 is based on the fourth moment. The smaller the value of β_2, the flatter the distribution; the larger the value of β_2, the more peaked the distribution.

The fourth moment is the arithmetic mean of the fourth power of the deviations from the arithmetic mean. Using the data in Table 4–6, the fourth moment for the semimonthly expenditures for food by ten families is computed

$$M_4 = \frac{\Sigma(X-\mu)^4}{N} \quad \textbf{(4–22)}$$

$$M_4 = \frac{295{,}432.352}{10} = 29{,}543.2352.$$

The value of β_2, a relative measure of kurtosis, is computed

$$\beta_2 = \frac{M_4}{M_2^2} \quad \textbf{(4–23)}$$

$$\beta_2 = \frac{29{,}543.2352}{115.04^2} = \frac{29{,}543.2352}{13{,}234.2016} = 2.23.$$

The sign of β_2 is always plus, since the powers of deviations involved in both the numerator and the denominator are even. The fourth moment (the numerator) is based on the fourth power of the deviations and the second moment squared is the denominator.

In Table 4–7 the deviations from the arbitrary origin are raised to the fourth power to find the value $\Sigma f(d')^4$. This sum is used with other values calculated in Table 4–7 to compute the fourth moment from the mean by the formula

$$M_4 = \frac{\Sigma f(d')^4}{N} - 4\frac{\Sigma fd'}{N}\frac{\Sigma f(d')^3}{N} + 6\left(\frac{\Sigma fd'}{N}\right)^2\frac{\Sigma f(d')^2}{N} - 3\left(\frac{\Sigma fd'}{N}\right)^4 \quad \textbf{(4–24)}$$

$$M_4 = \frac{10{,}482}{125} - 4\frac{46}{125}\frac{1{,}534}{125} + 6\left(\frac{46}{125}\right)^2\frac{618}{125} - 3\left(\frac{46}{125}\right)^4$$

$$M_4 = 83.8560 - 18.0644 + 4.0172 - .0550$$

$$M_4 = 69.7538 \text{ (in class-interval units).}$$

The value of β_2 is computed below by Formula 4–23 from the values of M_4 and M_2 in class-interval units (calculated on the preceding pages).

$$\beta_2 = \frac{69.7538}{4.8086^2} = \frac{69.7538}{23.1226} = 3.02.$$

STUDY QUESTIONS

4–1. Why is it important that something be known about the shape of the distribution of the values of a variable if one is to use an average as a measure of typical size?

4–2. Why should some knowledge of the dispersion be available when using an average as typical of the values of the variable?

4–3. Why is it important to know something about the skewness of a distribution?

4–4. Although the range is considered the simplest measure of dispersion, it is not a very reliable measure. What is the chief weakness of the range?

4–5. Why are the quartiles and the quartile deviation generally better measures of dispersion than the range?

4–6. Explain why deciles and percentiles are called by these names.

4–7. The average deviation and the standard deviation are the two measures of dispersion that are affected by all the values of the variable. Are the average deviation and the standard deviation thus better measures of dispersion than other measures? Why?

4–8. In computing the average deviation and the standard deviation from a frequency distribution, it is necessary to assume that the midpoint of a class is the average value of the items in the class. Does this assumption introduce any error into the computation of these measures of dispersion? If it does introduce error, does it detract significantly from the usefulness of these measures? Give reasons for your answer.

4–9. Why is the standard deviation considered a better measure of dispersion than the average deviation?

4–10. Explain the advantages of using a measure of relative dispersion.

4–11. Why does the third moment serve as a measure of skewness?

4–12. What advantages does β_1 have as a measure of skewness in comparison with the third moment?

PROBLEMS

4–1. The following table gives the union basic hourly wage rates for selected building trades in April, 1973. Determine which building trade has the greatest variation in basic rates and which building trade has the least variation. These comparisons should be made using the coefficient of variation.

Occupation	Albuquerque, New Mexico	Dallas, Texas	El Paso, Texas	Houston, Texas
Bricklayers	$6.290	$7.375	$5.750	$7.500
Carpenters	5.910	7.020	5.370	6.660
Electricians	7.500	7.600	6.700	7.335
Painters	5.170	6.735	4.750	6.110
Plasterers	5.920	7.045	5.940	6.825
Plumbers	8.000	7.450	6.250	7.080
Building Laborers	4.020	4.780	3.180	4.700

Source: Southwest Regional Office, Bureau of Labor Statistics, Dallas, Texas (April, 1973).

4–2. Two students made the same average (arithmetic mean) on the three one-hour examinations given in a course, but the individual grades on the examinations differed as shown in the table at the top of page 100.

(a) Summarize briefly the difference in the performance of the two students as seen from simply looking at the grades.

(b) Compute the average deviation from the mean for the two sets of grades. What information do the two average deviations give with regard to the performance of the two students?

	Student A	Student B
First examination	84	58
Second examination	84	98
Third examination	85	93
Arithmetic mean	83	83

4–3. For the following observations compute the variance, standard deviation, average deviation about the median, and the coefficient of variation.

63 45 39 55 69 21 50 25 33 25

4–4. The following frequency distribution shows the time required to wrap 127 mail-order packages. Compute the variance, standard deviation, and coefficient of variation.

Time (Minutes)	Number of Packages
0.5 and under 1.0	6
1.0 and under 1.5	12
1.5 and under 2.0	30
2.0 and under 2.5	42
2.5 and under 3.0	28
3.0 and under 3.5	12
Total	130

4–5. Measure the skewness (β_1) and kurtosis (β_2) of the distribution of values in Problem 4–3 using moments about the arithmetic mean.

4–6. Assume that 20 clerks are paid the following monthly salaries:

600	490	425	627	675	450	475	590	520	480
500	480	580	511	490	330	338	410	459	605

(a) Compute the range.

(b) Compute Q_1, Q_3, and determine the interquartile range.

(c) Compute the quartile deviation.

(d) Compute a measure of skewness using the mean, median, and standard deviation.

4–7. The following frequency distribution shows the number of borrowers at a farm-community bank by age of the farm borrower, rounded to the nearest birthday.

Age	Number of Borrowers
18–25	3
26–35	8
36–45	14
46–55	25
56–75	37
Total	87

(a) Compute the variance of this distribution.

(b) Compute the quartile deviation and compare it with the standard deviation.

(c) Compute a measure of skewness using the mean, mode, and standard deviation.

(d) Compute the average deviation about the median.

4–8. The following table gives the distribution of hourly wage rates of production workers in the four departments of a manufacturing plant.

Hourly Wage Rates (Dollars per Hour)	Number of Employees				
	Dept. A	Dept. B	Dept. C	Dept. D	Total
4.00 and under 4.20	8	4	5	4	21
4.20 and under 4.40	15	10	12	10	47
4.40 and under 4.60	40	15	14	17	86
4.60 and under 4.80	30	7	10	20	67
4.80 and under 5.00	4	4	6	10	24
5.00 and under 5.20	3		3	9	15
Total	100	40	50	70	260

Source: Hypothetical data.

Compute the standard deviation and the coefficient of variation for hourly wage rates in each of the four departments.

4–9. Compute the measure of skewness by Formula 4–17 for the hourly wage rates of the four departments given in Problem 4–8.

4–10. Compute the measure of skewness by Formula 4–17 for the hourly wage rates of all production workers given in Problem 4–8.

4–11. Compute β_1 for the hourly wage rates of the four departments given in Problem 4–8.

4–12. Compute the values of Q_1, Q_3, and Q for the distribution of times shown in Problem 4–4.

4–13. Compute β_1 and β_2 for the distribution of wrapping times in Problem 4–4.

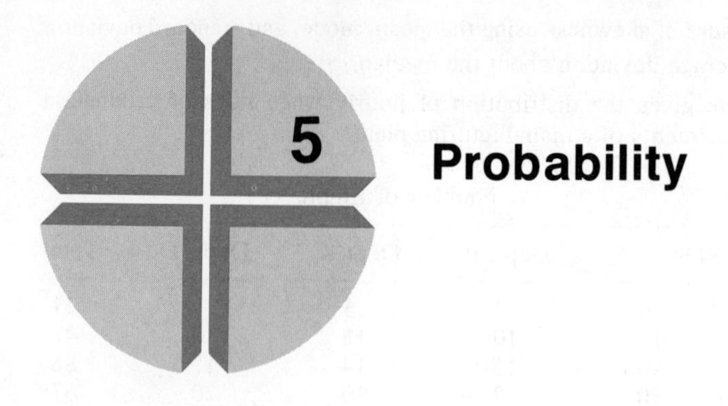

5 Probability

The discussions in Part 1 dealt with statistical description. In most cases the observations available were assumed to be complete sets of data from populations so that averages, dispersion, skewness, and other descriptive measures were parameters. When descriptive statistical measures are compiled from populations, they can be used directly in making business decisions. However, when only part of a population is observed, when descriptive measures are sample statistics and business decisions must be made on the basis of partial information, the kind of analysis employed is known as *statistical induction.* Part 2 of this text deals with the fundamental areas of statistical induction. This chapter introduces the concept of probability and Chapter 6 discusses some of the most common probability distributions. Probability and probability distributions are applied in Chapter 7 to basic sampling theory and in Chapters 8 and 9 to the testing of hypotheses about universe parameters using sample data. In Chapter 10 probability and sampling theory are applied to industrial sampling problems.

SOME PRELIMINARY IDEAS

A few preliminary ideas must be discussed before it can be demonstrated how probabilities are derived and used and before precise definitions of the concepts of probability can be made.

Events

An *event* is the outcome of some experiment. There are several kinds of events; a few examples are discussed in the following paragraphs.

Suppose the experiment is the tossing of a fair, six-sided die. There are six possible outcomes, which are labeled x_1, x_2, x_3, x_4, x_5, and x_6. Each of these outcomes is called a *simple event* because it cannot be further subdivided into more than one event. If the simple events are designated E_i, then E_1 represents the event of rolling the die so that the 1 appears on top, E_2 represents the outcome of the 2 on top, and so forth.

Since it is not possible for both the 1 and the 2 to appear on top in a single roll of the die, the events E_1 and E_2 are said to be *mutually exclusive events.* In mutually exclusive events the occurrence of one event precludes the occurrence of any other. In the single flip of a coin, the events heads and tails are mutually exclusive events.

Now consider an experiment in which two dice are thrown simultaneously. The events E_i might be defined as the sum of the values of the faces which land on top for the two dice. The values of i are 2, 3, . . . , 12. The event E_7 is not a simple event as it is composed of two simple events which may be the event of obtaining a 6 and the event of obtaining a 1. The compound event E_7 could also occur as a result of obtaining a 1 and 6; 5 and 2; 2 and 5; 4 and 3; or 3 and 4. Any event that can be decomposed into two or more simple events is called a *compound event.*

Another type of event can be illustrated using the experiment of tossing two fair dice. The likelihood that any one number will appear on top in any one throw is the same for all numbers. If the toss of the first die produces a 4, this does not in any way affect the likelihood that the toss of the second die will or will not produce a 4. When the occurrence or nonoccurrence of one event does not in any way affect the likelihood of the occurrence or nonoccurrence of another event, the two events are said to be *independent events.*

Sample Space and Sample Points

Events are also called *sample points.* The set or collection of all sample points, each corresponding to only one possible event, is called the *sample space.*

Take again the experiment of tossing a single die. The sample space, designated as S, can be defined as the set of the simple events $E_1, \ldots, E_6$.

$$S = \{E_1, E_2, E_3, E_4, E_5, E_6\}.$$

The sample space is represented graphically as the one-dimensional space shown in Part A of Figure 5–1.

In Figure 5–1, Part B represents the sample space for one toss of two dice, shown as a two-dimensional space. The sample space for three tosses (or one toss of three dice) might be represented as a three-dimensional space, and so on.

The sample points in B are all possible compound events, and the compound events which represent a sum of 7 are shown within the dashed boundary. The sample points which represent a sum of 4 are shown within the dotted boundary. There are fewer sample points for event 4 than there are for event 7. As will be shown later, this means that the probability of getting a 4 is less than the probability of getting a 7. The concept of sample space is related to the concept of probability.

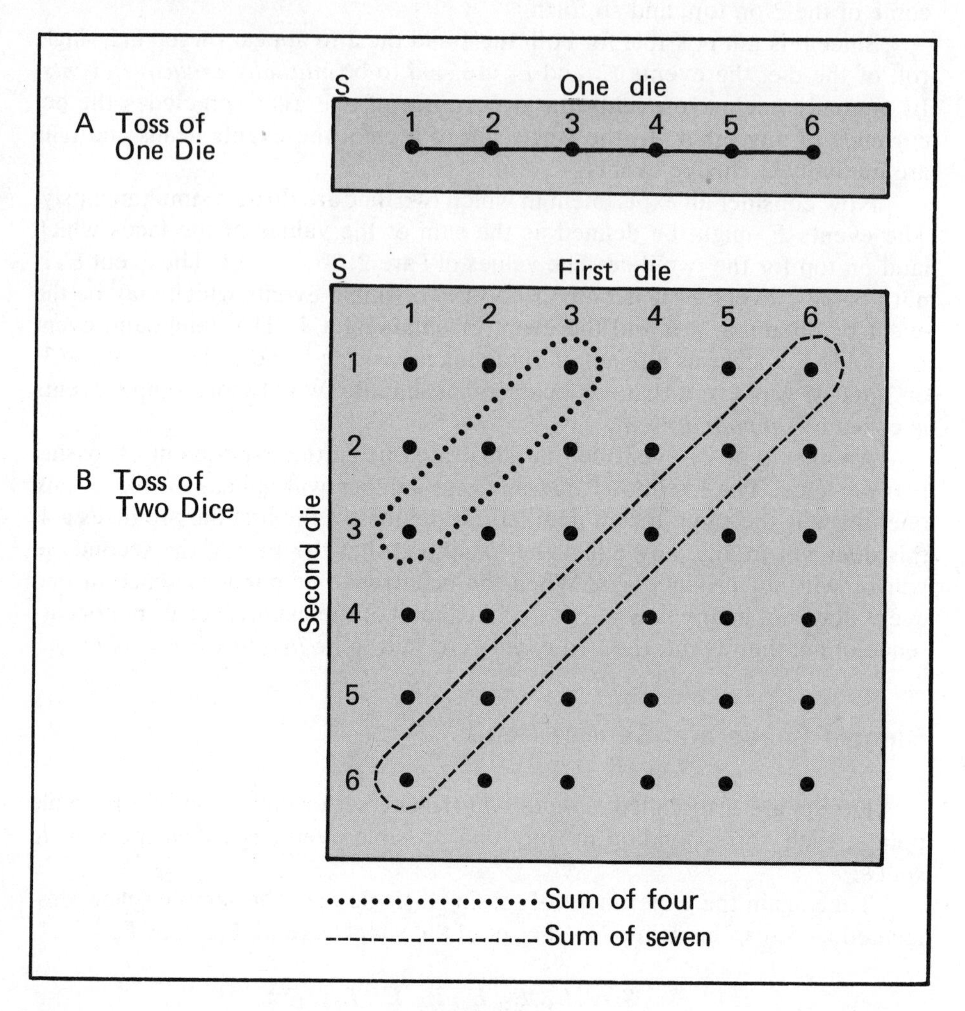

FIGURE 5–1

SAMPLE SPACE IN TOSSING ONE OR TWO SIX-SIDED DICE

Venn Diagrams

The concepts of events, sample points, and sample spaces are often represented graphically by a specialized type of chart known as a *Venn diagram.* A Venn diagram is constructed by drawing an enclosed shape which depicts the sample space, *S,* representing all possible events. The individual events need not be shown but are assumed to be distributed throughout the sample space. In Figure 5–2 the rectangle marked *S* represents the sample space. The circle within the rectangle marked *A* represents all the events that have the

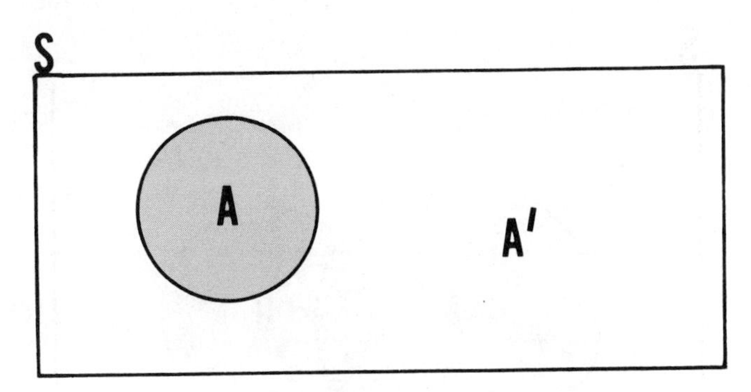

FIGURE 5-2

VENN DIAGRAM SHOWING $A + A' = S$

particular characteristic A. For example, S might represent the set of all possible sums of the top faces of two randomly thrown dice, and A might represent the subset of all sums of 7. The area designated A' would represent the remainder of the sample space or the subset of all values not 7.

Figure 5-3 shows two mutually exclusive events. These might be the sum of 7 (represented by A) and the sum of 4 (represented by B) in an experiment consisting of a toss of two dice. Mutually exclusive events have no common sample point. Therefore, in Figure 5-3 a sample point cannot be in A and in B; that is, a toss of the dice cannot come up both a 7 and a 4.

Using the same experiment of rolling two dice, assume that A represents all events having an even number sum, and B represents all of the events that have a sum greater than 8. The sample points represented in A and B are no longer mutually exclusive; they have some points in common. For example, the sum 10 is both even and greater than 8.

The crosshatched area in Figure 5-4 represents the events which have both characteristics A and B. This area is called the *intersection of A and B*, usually denoted $A \cap B$ and read "A and B." The entire shaded area in the

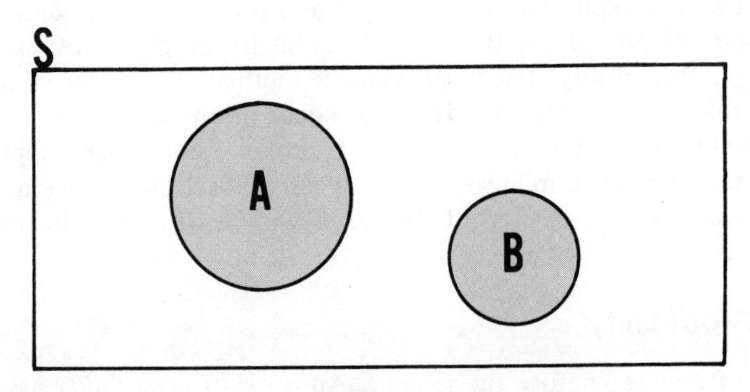

FIGURE 5-3

VENN DIAGRAM OF TWO MUTUALLY EXCLUSIVE EVENTS

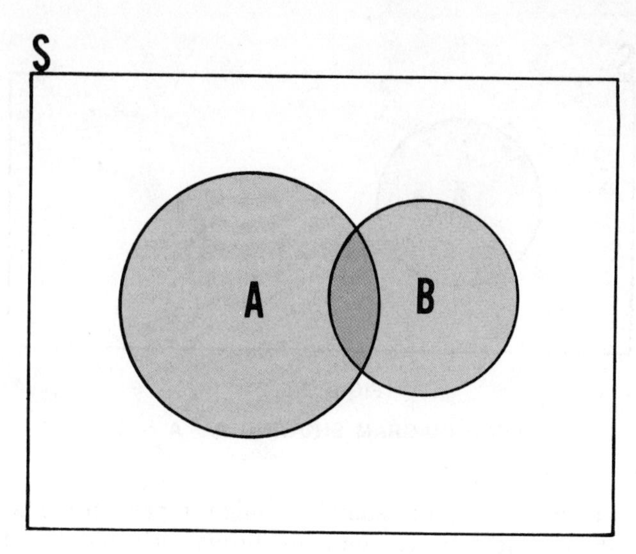

FIGURE 5-4

VENN DIAGRAM SHOWING TWO EVENTS NOT MUTUALLY EXCLUSIVE

figure formed by the two sample spaces represents events having either characteristic A, characteristic B, or both. This is the *union of A and B,* which is denoted $A \cup B$ and read "A union B" or just "A or B."

The concepts of sample space, intersections, and unions, as well as their graphic counterparts, will be instrumental later in this chapter in developing probability concepts.

KINDS OF PROBABILITY

The concept of probability has many applications in everyday life, and there are many practical applications in business decision making when the outcome of a particular event is uncertain. A businessman may weigh the probability that his business will succeed at a particular location; a promoter may reflect on the likelihood that a new stage show will be a hit; or a contractor may estimate the probability that he will be able to complete a construction job in 100 days. Probability theory provides a mathematical way of stating the likelihood that some particular event will occur in the future.

Probability is part of everyone's vocabulary, but most people are not precise as to what the term "probability" means. There are three kinds of probability relevant to this discussion and each will be discussed in the sections which follow.

A Priori Probability

A priori means "before the event." An *a priori probability* assumes that all possible events, E_i, are known and that they all have equal likelihood of occurrence.

If an event E can happen in A different ways and cannot happen in A' different ways, the probability that event E will occur is

$$P(E) = \frac{A}{A + A'},$$

and the probability that event E will not occur is

$$P(E') = \frac{A'}{A + A'}.$$

Since the sum of the ways something can happen plus the sum of the ways something cannot happen equals the total events,

$$A + A' = n.$$

Thus,

$$P(E) = \frac{A}{n}$$

and

$$P(E') = \frac{A'}{n}.$$

It is also true that

$$P(E) + P(E') = 1.$$

The probability of drawing a certain card, such as the king of clubs, from a deck of 52 playing cards is 1/52, since there are 52 cards but only one king of clubs. The probability of drawing a heart is 13/52 as there are 13 hearts in the deck. If a box of poker chips contains 100 white chips, 150 blue chips, and 50 red chips, the probability of drawing a white chip is $\frac{100}{100 + 150 + 50}$ or 1/3.

In the situations above, it is assumed that the player knows the contents of the deck of cards, that it is a fair deck and the cards are drawn by chance. In the drawing of the chips the number of chips of each color must be known, and they must all have the same chance of selection. Unfortunately such a precise knowledge of the universe is seldom part of a real business situation.

The preceding discussion illustrates the *axioms of probability*, which may be stated as follows:

1. $0 \leq P(E) \leq 1$. No probability can be greater than 1, and no probability can be negative. Something which is impossible is given a probability zero, and something that is certain to occur is given a probability of 1.

2. $P(E_1) + P(E_2) + \ldots + P(E_k) = 1$. The sum of the probabilities of all k possible mutually exclusive events is equal to 1. In other words, each time an experiment is conducted, one of the several possible outcomes must occur.

3. $P(E') = 1 - P(E)$. The probability that event E will not occur is 1 minus the probability that it will occur. The occurrence and the nonoccurrence of an event forms a complete and mutually exclusive sample space. Hence, the probability occurrence and the probability of nonoccurrence must sum to 1. If the probability of occurrence is $P(E)$, then the probability of nonoccurrence, $P(E')$, must be $1 - P(E)$.

Experimental Probability

If an experiment is performed n times under the same conditions, and if there are A outcomes of an event E, the estimate of the probability of E as n approaches infinity is

$$P(E) = \lim_{n \to \infty} \frac{A}{n}.$$

The estimate of a probability obtained in this manner is an *experimental probability*. As a practical matter n does not have to be infinitely large. Very good estimates of probabilities may often be secured with reasonably small samples, as will be shown in Chapter 7.

An example of experimental probability being used to estimate the probability of an event might be determining the probability of picking a white chip from a box of chips. If one does not know the number of chips of each color, then there is no way to compute the a priori probability. One could, however, make repeated trials of an experiment by randomly selecting a single chip from the box, noting its color, and replacing it. By noting the frequency of occurrence of white chips one could estimate the probability of drawing a white.

After 5 trials the results may show 2 white and three nonwhite. Based upon 5 trials the probability of drawing a white would be estimated to be $2/5 = .40$. After 10 trials the results may show 3 white and 7 nonwhite, or $3/10 = .30$ probability of white. After 100 trials the results would likely be very near the a priori probability, which is 33 white and 67 nonwhite. As the number of trials increases, the estimate of the probability of an event based upon the relative frequency becomes very near the a priori value. This can be verified by using the relative frequency method to estimate the probability of obtaining a head on a coin toss. When probability of a head is estimated after 2, 4, 8, 16, and 32 tosses, the estimates move nearer to the a priori value of .5 as the number of trials increase.

Subjective Probability

On some occasions, it is impossible to compute an experimental probability through a series of trials or by drawing a sample of observations from some universe. If one wished to express the probability that science can produce a cheap substitute for gasoline by 1980, one could not compute the a priori probability of this, nor could the experiment be completed a certain number of times to arrive at a relative frequency. The probability of that event could only be expressed as one's personal confidence that it would occur.

Since individuals may differ in their degrees of confidence in the outcome of some future event even when offered the same evidence, their opinions, expressed as probabilities, will differ. Statements of opinion regarding the likelihood that an event will occur when expressed as probabilities are called *subjective probabilities*.

While the use of a subjective probability is always open to challenge and criticism, the judgment of an experienced businessman or a qualified expert as to what he expects to happen provides useful information which is often treated as if it were a probability in the classical sense. For instance, a manager may estimate the probability of a competitor's price reduction to be .10, or may estimate the probability of a raw material price increase to be .90. Having made these estimates the manager will be able to incorporate them into a formalized statistical decision process to be introduced later. In general, though, subjective probabilities are criticized as being guesses; the manager's opinion is already present in a nonquantitative form and in the absence of other information he may act upon this opinion. Expressing this opinion quantitatively gives no more credibility to the opinion; it merely enables the opinion to be included into a more formal quantitative analysis of the situation.

MATHEMATICS OF PROBABILITY

The last part of this chapter uses the known or estimated probabilities introduced in the previous section to secure other probabilities. These calculations involve the mathematics or calculus of probability and are based on four rules—two addition rules and two multiplication rules.

Addition of Probabilities

Assume that one wishes to know the probability that a card drawn from an ordinary deck will be a king or a black card (spade or club). The probability is the sum of the probability of a king, plus the probability of a black card, minus the probability that the card is both black and a king. The probability of the two black kings must be subtracted to keep from counting them twice.

The *general rule of addition of probabilities* is written

$$P(A \cup B) = P(A) + P(B) - P(A \cap B). \tag{5-1}$$

In the card example $P(\text{king} \cup \text{black}) = \frac{4}{52} + \frac{26}{52} - \frac{2}{52} = \frac{28}{52}$ or .5385.

If the two events are mutually exclusive, Formula 5–1 is simplified by the elimination of the last term, giving the *special rule of addition,* which is written

$$P(A \cup B) = P(A) + P(B). \tag{5-2}$$

In other words, the probability of the occurrence of at least one of two mutually exclusive events is the sum of the probabilities that the individual events will occur. It is called the special rule because it can be used only in the case of mutually exclusive events.

For example, the probability of drawing the ace of spades from a deck of 52 cards is $\frac{1}{52}$. The probability of drawing the ace of hearts is also $\frac{1}{52}$. If the probability of drawing the ace of spades is designated $P(A)$ and the probability of drawing the ace of hearts is $P(B)$, the probability of at least one of these events occurring may be written

$$P(A \cup B) = P(A) + P(B) = \frac{1}{52} + \frac{1}{52} = \frac{2}{52}.$$

The events are mutually exclusive since both of them cannot occur. If the ace of hearts is drawn, the ace of spades cannot be drawn. This rule can be extended to more than two events and the probability of drawing any one of the four aces is

$$P(A \cup B \cup C \cup D) = P(A) + P(B) + P(C) + P(D)$$

$$P(A \cup B \cup C \cup D) = \frac{1}{52} + \frac{1}{52} + \frac{1}{52} + \frac{1}{52} = \frac{4}{52}.$$

Assume that a box contains 100 white balls, 150 black balls, and 50 red balls. If the probability of drawing a white ball is written $P(A)$, the probability of *not* drawing a white ball is $P(A')$, and $P(A') = 1 - P(A)$. Since $P(A) = \frac{100}{300}$, $P(A') = 1 - \frac{100}{300} = \frac{200}{300} = \frac{2}{3}$. The probability of not drawing a white ball is the same as the probability of drawing a red ball or a black ball, which may be designated $P(B)$ and $P(C)$ respectively.

$$P(B \cup C) = P(B) + P(C) = \frac{150}{300} + \frac{50}{300} = \frac{200}{300} = \frac{2}{3}.$$

Multiplication of Probabilities

The probability that two independent events will both occur is the product of the probabilities of the separate events. The probability of throwing one die and securing a 5, $P(A)$, and then throwing again and securing a 4, $P(B)$, is

$$P(A \cap B) = P(A)P(B) = \frac{1}{6} \cdot \frac{1}{6} = \frac{1}{36}$$

where

$A = 5$ on the first throw
$B = 4$ on the second throw.

Likewise, the probability of tossing a coin and getting a head three times in succession is

$$P(A \cap B \cap C) = P(A)P(B)P(C) = \frac{1}{2} \cdot \frac{1}{2} \cdot \frac{1}{2} = \frac{1}{8}$$

where

$A =$ the first head
$B =$ the second head
$C =$ the third head.

In the preceding examples the outcomes of the successive experiments are in no way related to the outcomes of the earlier events. It is frequently the case, however, that two events are not independent. Assume that a manufacturer buys 80% of a given article used in his plant from Company K, which has been able to supply the product with an average of 4% defective. Since Company K cannot supply all that is needed, 20% of the items used are purchased from Company L, which supplies the product, with an average of 6% defective. The probability that a given item selected at random from inventory will be defective is .04 if it is known for certain that the item came from Company K. The probability that an item is supplied by Company K and is also defective is the product of the probability of being supplied by that company, written $P(A)$, and the probability of its being defective, which is the probability of that company's product being defective. This probability is written $P(B|A)$, read "the probability of B given A." In this example it reads "the probability of a defective product given its production by Company K." The formula for the *general rule of multiplication* is

$$P(A \cap B) = P(A)P(B|A). \tag{5-3}$$

This can also be written

$$P(A \cap B) = P(B)P(A|B).$$

In this case the probability of being supplied by Company K is .80, so $P(A) =$ .80. The probability of a defect, $P(B|A)$, is .04, since this is the relative frequency with which a defective item appears in the product supplied by the company. Thus, the probability that the item was supplied by Company K and is defective is

$$P(A \cap B) = (.80)(.04) = .032.$$

Since Company L supplies 20% of the product purchased, the probability that the item was supplied by Company L is 20%, so in this case $P(A') = .20$. It is known from past experience that this company's product is 6% defective, so $P(B|A') = .06$. The probability that an item was supplied by Company L and is defective is

$$P(A' \cap B) = (.20)(.06) = .012.$$

The probabilities just computed may be entered in Table 5–1, which is a cross classification of the product based on the supplying company and on whether or not the product was defective. In the column headed "Defective" the probability .032 is entered on the line with the stub "Company K," and the probability .012 is entered in the second line, which represents Company L. In the total column the probabilities .800 and .200 are entered for the two companies. These reflect the probabilities that an item has been produced by Companies L and K respectively. The probability that a nondefective item is supplied by the company is found by subtracting the probability of a defective item from the probability in the total column. This is .768 for Company K and .188 for Company L. The probability of a defective item regardless of manufacturer is the total of the probabilities of defective items for the two companies, which is found using the addition theorem given previously. In the same way, the probability that an item will not be defective is the total of the probabilities of nondefective items. Both the total column and the total row of the table add to 1.000.

TABLE 5–1

FRACTION DEFECTIVE AND FRACTION NOT DEFECTIVE

Classified by Supplying Company

Supplier	Defective (B)	Not Defective (B′)	Total
Company K (A)	.032	.768	.800
Company L (A′)	.012	.188	.200
Total	.044	.956	1.000

The totals of the columns and of the rows are usually referred to as *marginal probabilities*. If an item is selected at random without knowing which company supplied it, the probability of a defective item would be .044. The probability of a nondefective item would be 1.000 − .044 = .956. The probability that an item selected at random was supplied by a particular company is given in the right-hand column of Table 5–1.

Since Formula 5–3 states that $P(A \cap B) = P(A)P(B|A)$, it is equally true that $P(A \cap B) = P(B)P(A|B)$. The conditional probabilities are written

$$P(A|B) = \frac{P(A \cap B)}{P(B)}$$

and

$$P(B|A) = \frac{P(A \cap B)}{P(A)}.$$

When $P(A) = P(A|B)$, A and B are independent events. For example, event A is defined as drawing a king from a deck of playing cards. Event B is drawing a black card from the same deck. The probability of drawing a king, $P(A)$, is $4/52 = 1/13$; and the probability of drawing a king given that a black card has been drawn, $P(A|B)$, is $2/26 = 1/13$. Thus, $P(A) = P(A|B)$, which indicates that events A and B are independent. The formula for the *special rule of multiplication* is

$$P(A \cap B) = P(A)P(B) \qquad \textbf{(5–4)}$$

given the special requirement that A and B are independent events.

In the preceding example the probability of drawing a card which is both black and a king is $\left(\frac{26}{52}\right)\left(\frac{4}{52}\right) = \left(\frac{1}{2}\right)\left(\frac{1}{13}\right) = \frac{2}{52}$, which is the proportion of black kings in the deck.

Another Example of the Multiplication of Probabilities

Assume that a lot being tested contains 50 items, 5 defective and 45 effective. If 3 of these 50 items are selected at random and tested, what is the probability of getting zero defectives, that is, 3 effective items?

The probability of getting an effective item on the first draw would be $\frac{45}{50}$, since all items have the same chance of being selected and 45 out of the 50 are effective. The probability of getting an effective on the second draw, provided an effective had been drawn on the first draw, would be $\frac{44}{49}$, since there would be only 44 effectives in the reduced lot of 49 items. The 5 defective items would still be in the lot since none was drawn in the first two draws. Since the probability of getting an effective item on the second draw would be $\frac{44}{49}$, the probability of getting two effective items in succession would be $\frac{45}{50} \times \frac{44}{49} = .80816$.

If two effective items were drawn, there would be 43 effectives in the remaining 48, so the probability of getting an effective would be $\frac{43}{48}$. The probability of getting two effectives in succession was computed earlier as $.80816 \times \frac{43}{48} = .72398$. It may, therefore, be written that $P(0) = \frac{45}{50}\frac{44}{49}\frac{43}{48} = \frac{85{,}140}{117{,}600} = .72398$.

BAYES' THEOREM

Bayes' theorem is a formula for determining conditional probabilities, developed by Thomas Bayes, an eighteenth century British clergyman. To illustrate, suppose that in a particular statistics course given by a large university, 60% of the students have good grades and 40% have poor grades. It is known that, on the average, the good students are late to class only 5% of the time while poor students are late about 15% of the time. If a student selected at random from the course comes in late on a particular day, what is the probability that he is a good student?

To solve the problem, let E represent the event that a student is late. Let H_1 represent a good student, and let H_2 represent a poor student. The answer requires the computation of $P(H_1|E)$, which is the probability that the student is a good student, given that he has come to class late. This can be done using Bayes' theorem, which is written

$$P(H_i|E) = \frac{P(H_i) \cdot P(E|H_i)}{\Sigma[P(H_i) \cdot P(E|H_i)]} \qquad \textbf{(5–5)}$$

where

H_i = the possible outcomes
$P(H_i)$ = the *prior probability,* the probability of each possible outcome prior to consideration of any other information
$P(E|H_i)$ = the *likelihood,* the conditional probability that the event E will happen under each possible outcome, H_i
$P(H_i) \cdot P(E|H_i)$ = the *joint probability,* the probability of $(E \cap H_i)$ determined by the general rule of multiplication
$P(H_i|E)$ = the *posterior probability,* which combines the information given in the prior distribution with that provided by the likelihoods to give the final conditional probability.

The probability that the late student is one with good grades is

$$P(H_1|E) = \frac{P(H_1) \cdot P(E|H_1)}{P(H_1) \cdot P(E|H_1) + P(H_2) \cdot P(E|H_2)} = \frac{(.6)\,(.05)}{(.6)\,(.05) + (.4)\,(.15)}$$

$$= \frac{.03}{.09} = .33.$$

The probability that the late student has poor grades is

$$P(H_2|E) = \frac{P(H_2) \cdot P(E|H_2)}{P(H_1) \cdot P(E|H_1) + P(H_2) \cdot P(E|H_2)} = \frac{(.4)\,(.15)}{(.6)\,(.05) + (.4)\,(.15)}$$

$$= \frac{.06}{.09} = .67.$$

Note that $P(H_1|E) + P(H_2|E) = .33 + .67 = 1.0$.

In applying Bayes' theorem it is sometimes easier to understand if presented in tabular form. For example, a credit manager who deals with three types of credit risks: prompt pay, slow pay, and no pay. The proportions of people falling into each group are determined from past records to be .75, .20, and .05 respectively. The credit manager has also learned from experience that 90% of the prompt-pay group are homeowners, while 50% of the slow pay, and 20% of the no-pay groups are homeowners. Bayes' theorem would allow the credit manager to determine the probability that a new applicant for credit would fall into each of the pay categories if he first knew whether the applicant was a homeowner. The computations are shown in Table 5–2.

In Column 1 the three payment outcomes are H_1, H_2, and H_3. Column 2 indicates the prior probabilities of each outcome. Column 3 indicates the likelihood of homeowning given each possible outcome. Column 4 contains the joint probabilities formed by multiplying respective elements of Columns 2 and 3. The sum of Column 4 is the value of the denominator of Bayes' theorem. The posterior probabilities in Column 5 are found by dividing the appropriate joint probability in Column 4 by the sum of Column 4.

TABLE 5–2

APPLICATION OF BAYES' THEOREM TO CREDIT RISK AS RELATED TO HOME OWNERSHIP

Pay Group	Outcomes H_i (1)	Prior $P(H_i)$ (2)	Likelihood $P(E\|H_i)$ (3)	Joint $P(H_i) \cdot P(E\|H_i)$ (4)	Posterior $P(H_i\|E)$ (5)
Prompt	H_1	.75	.90	.675	.860
Slow	H_2	.20	.50	.100	.127
No pay	H_3	.05	.20	.010	.013
Total		1.00		.785	1.000

Note: *E* is the event of being a homeowner.

$$P(H_1|E) = \frac{P(H_1) \cdot P(E|H_1)}{P(H_1) \cdot P(E|H_1) + P(H_2) \cdot P(E|H_2) + P(H_3) \cdot P(E|H_3)} = \frac{.675}{.785} = .860$$

$$P(H_2|E) = \frac{P(H_2) \cdot P(E|H_2)}{P(H_1) \cdot P(E|H_1) + P(H_2) \cdot P(E|H_2) + P(H_3) \cdot P(E|H_3)} = \frac{.100}{.785} = .127$$

$$P(H_3|E) = \frac{P(H_3) \cdot P(E|H_3)}{P(H_1) \cdot P(E|H_1) + P(H_2) \cdot P(E|H_2) + P(H_3) \cdot P(E|H_3)} = \frac{.010}{.785} = .013.$$

Thus, the probability that a homeowner will be a prompt pay is .860, that he will be a slow pay is .127, and a no pay is .013. Such information is quite helpful in deciding whether to approve an application.

USING PROBABILITY IN DECISION MAKING

A graphic presentation of conditional probabilities, called a *tree diagram,* is often used to show the probability of various events occurring. An example of the use of a tree diagram is making the decision to accept or reject a shipment of product received by a manufacturer from a supplier. The inspection plan used to make a decision may be based on a sample rather than inspecting all the items in a lot. This is commonly done to save the expense of a complete screening of a lot, although if the test of the product destroys it, a sample is the only method of determining whether or not a lot meets specifications.

Assume that the buyer decides to inspect a random sample of 25 items from a large lot. He classifies each item inspected as acceptable or not acceptable. If none of the items tested is defective, the lot is accepted. If 4 or more of the items are defective, the lot is rejected. If 1, 2, or 3 items are defective, an additional sample of 50 items is selected and tested, and the total number of defective items in the two samples is determined. If 4 or more defectives are found in the sample of 75, the lot is rejected. If the total number of defectives in the 2 samples is 3 or fewer, the lot is accepted.

The information needed to use this sampling plan is the probability of a defective lot being accepted and the probability of its being rejected. For example, if the lot has 4% defective items, what is the probability that a sample of 25 will contain 0 defectives? This will be the probability of accepting the lot on the first sample. Likewise, the probability of 4 or more defective items is the probability of rejecting the lot on the first sample. If 1, 2, or 3 defectives are found, the second sample must be taken, and the probability of getting 4 or more in the 2 samples will be the probability of rejecting the lot on the second sample. Assume that there were 2 defectives in the first sample and a second sample was taken. This is an example of the use of conditional probabilities; the probability of rejecting the lot on the second sample depends upon the probability of getting 2 defectives in the first sample times the probability of getting 2 or more defectives in the second sample. The probability of accepting the lot on the second sample depends on the probability of getting 2 in the first sample times the probability of fewer than 2 defectives in the second sample.

Administering this sampling plan requires that one know the probability of getting each number of defective items in a sample of a given size from a lot with a known fraction defective. The determination of these probabilities is discussed in Chapter 6, but without explaining how they were derived, the various probabilities of getting r defectives in a sample of 25 from a lot that is 4% defective are

$$\begin{aligned} P(r = 0) &= .368 \\ P(r = 1) &= .368 \\ P(r = 2) &= .184 \\ P(r = 3) &= .061 \\ P(r \geq 4) &= \underline{.019} \\ \text{Total} \quad & \;\; 1.000 \end{aligned}$$

The probability that one of these events occurs is 1.000. The probability of accepting the lot on the first sample is .368, and the probability of rejecting it on the first sample is .019. The probability of 1, 2, or 3 defectives is also shown.

The tree diagram shown in Figure 5–5 is constructed by drawing 5 branches of the tree, each branch representing 1 of the events that can occur with the first sample. The probability of each event occurring is shown on the appropriate branch. The branch representing 4 or more defectives is marked "reject"; the branch representing 0 defectives is marked "accept." No decision is made for the other number of defectives since a second sample must be taken.

The second sample consists of 50 items, so the probability of getting different numbers of defectives is not the same as for the sample of 25. The probability of each number of defective items occurring in the second sample is given below, again deferring the explanation of how they were determined for Chapter 6.

$$\begin{aligned} P(r' = 0) &= .135 \\ P(r' = 1) &= .271 \\ P(r' = 2) &= .271 \\ P(r' = 3) &= .180 \\ P(r' \geq 4) &= \underline{.143} \\ \text{Total} \quad & \;\; 1.000 \end{aligned}$$

The proper probabilities are entered on the second set of branches according to the following reasoning of the conditional probabilities. If the first sample contained 3 defectives, the lot will be rejected on the second sample if 1 or more defectives are found. If 1 defective is found, the cumulative number of defectives would be 4 and a rejection would follow. If more than 1 defective is found in the second sample, the lot would, of course, be rejected. The probability of a rejection on the second sample is the probability of getting 3 defectives on the first sample times the probability of 1 or more defectives on the second sample. This product is $.061 \times .865 = .052765$. The decision is to reject.

If the second sample has no defective items, the lot will be accepted. The probability of acceptance will be the probability of 3 defectives on the first sample times the probability of zero defectives on the second sample, which is $.061 \times .135 = .008235$.

Each branch of the tree diagram is extended in the same manner, and the probability of acceptance or rejection is computed as the product of the

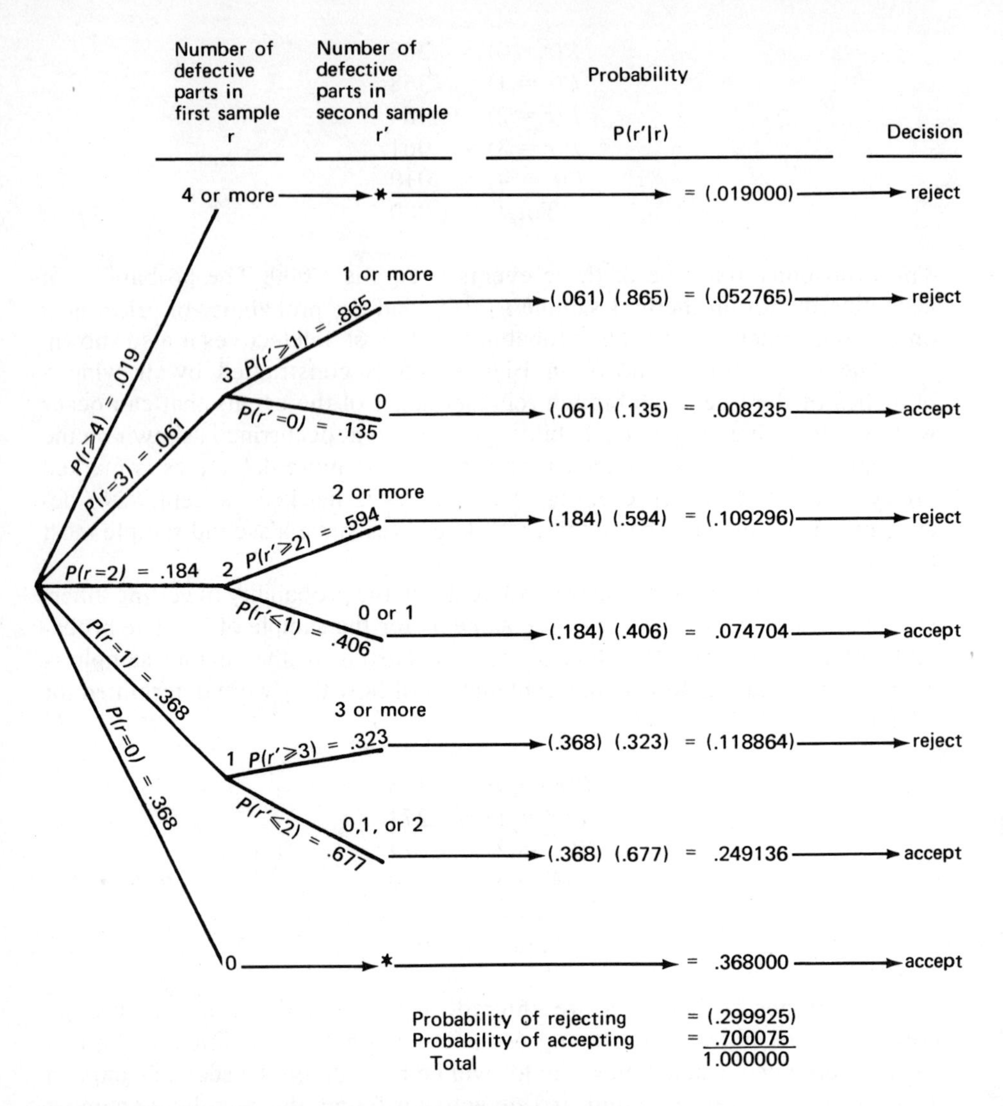

* Indicates that it was not necessary to draw the second sample.
Note: Rejection probabilities are shown in parentheses.

FIGURE 5–5

USE OF A TREE DIAGRAM IN DECISION MAKING

probability of a specific number of defectives in the second sample, given the probability of the number of defectives in the first sample. There are four ways in which an acceptance can be secured and four ways in which a rejection can occur. The total probability of an acceptance is .700075, and the total probability of a rejection is .299925. This knowledge of the probability of accepting a lot that has 4% defective items can be used to decide whether or not to use this sampling plan. If the buyer wants a smaller probability of accepting a lot with 4% defective items, it will be necessary to take larger samples.

STUDY QUESTIONS

5-1. Distinguish between a compound event and a simple event.

5-2. Give two examples of mutually exclusive events.

5-3. What are independent events?

5-4. If two events are mutually exclusive, are they also independent? Explain.

5-5. Distinguish between a sample point and sample space.

5-6. What is a Venn diagram and how can it be used to explain probability concepts?

5-7. Distinguish between experimental probabilities and a priori probabilities.

5-8. Are the majority of probabilities known in advance, or is it necessary to estimate them from an experiment?

5-9. What is a subjective probability? Give an example.

5-10. Define the following terms that are used in connection with Bayes' theorem:

(a) Posterior probabilities.

(b) Likelihood.

(c) Prior probabilities.

(d) Joint probabilities.

5-11. What is a tree diagram and what part does it play in determining probabilities and in probability decision theory?

5-12. A card is drawn at random from a deck of playing cards. What rule of probability would you use to secure the probabilities of each of the following?

(a) King or a jack.

(b) Four or a spade.

(c) Five and a heart.

(d) Queen and a black card.

5-13. A penny is tossed in such a manner that pure chance determines whether heads or tails will come up. If 5 successive tosses have been heads, what is the probability that the sixth toss will be heads?

5-14. If 2 fair dice are thrown simultaneously, compute the probability of:

(a) A total of 3 on the faces which are up.

(b) A total of 5 on the faces which are up.

(c) A total greater than 6.

(d) A total of 7 or 11.

(e) A total of 8 or less.

5-15. What is the probability of securing a 2, 3, or 4 from a throw of 1 die?

5-16. A new keypunch operator is hired in the tabulating department of a business. How can you determine the probability that a card punched by this operator would have an error on it?

5–17. If items are drawn from a finite universe but each item drawn is replaced before the next one is drawn, why is the effect the same as if the items had been drawn from an infinite universe?

PROBLEMS

5–1. Compute the probabilities listed in Study Question 5–12.

5–2. Draw Venn diagrams to illustrate the probabilities listed in Study Question 5–12.

5–3. What are the respective probabilities of getting 0, 1, 2, 3, 4, 5, 6, 7, and 8 heads in a toss of 8 coins?

5–4. A product is assembled from Components *A* and *B*. The probability of *A* being defective is .03, and the probability of *B* being defective is .02. What is the probability that the assembled product will not be defective?

5–5. A salesman has observed that if he makes a sale to Customer A, the probability is .50 that he will also make a sale to Customer B. If he does not make a sale to A, the probability that he will make a sale to B is .10. If he decides on making a call to A and he has a probability of .32 of making a sale to A, what is the probability that he will make a sale to B that day? (Assume that his opinion of the probabilities is correct.)

5–6. Mr. Jones can come home from the office in one of three ways. He drives (1) through town 30% of the time, (2) around town on a bypass 25% of the time, and (3) by way of a shortcut through the industrial district 45% of the time. When he takes the first route, he is late to supper 25% of the time. When he takes the second and third routes, he is late 20% and 30% of the time, respectively. If on a particular night Mr. Jones is late to supper, what is the probability he took the shortcut?

5–7. The output of a given process is classified as follows:

Major defect only	1,627
Minor defect only	5,860
Both major and minor defects	185
No defects	23,603
Total output	31,275

(a) What is the probability that an item chosen at random from this output will have a major defect? A minor defect? No defects?

(b) What is the probability that an item chosen at random will have a defect, either minor or major?

(c) In the next 1,000 items produced, how many would you expect to be without defects?

(d) If 2 items are chosen at random, what is the probability that both will have a major defect? That both will have a defect, either major or minor? That both will have no defects?

5–8. The otherwise identical urns have the following number of red and green balls:

	Urn 1	Urn 2	Urn 3
Red Balls	4	3	8
Green Balls	6	7	2

(a) If 1 ball is drawn at random from each urn, what is the probability that all 3 balls will be red?

(b) If 2 balls are drawn at random (without replacement) from an unidentified urn, and both are green, what is the probability that the balls came from Urn 1?

(c) If in (b) the 2 balls are both red, what is the probability that they were drawn from Urn 3?

5-9. A ball is drawn from a box containing 4 black, 5 red, and 11 white balls. The balls are mixed thoroughly and 1 is drawn in such a manner that each ball has the same chance of being selected. What is the probability that the ball will be either black or red?

5-10. A large shipment of parts contains 5% defective parts. A random sample of 30 parts is selected and inspected. If there are no defective parts in the sample, the shipment is accepted. If there are 4 or more defective parts, the shipment is rejected. If there are 1, 2, or 3 defective parts, a second sample of 60 parts is drawn and inspected. If there are 4 or more defective parts in the total of 90 in the sample, the shipment is rejected. Otherwise, the shipment is accepted. The probabilities of defective parts in the 2 samples of 30 and 60 items are shown in the following table:

Number of Defective Parts in the Sample	Probability	
	Sample of 30	Sample of 60
0	.2231	.0498
1	.3347	.1494
2	.2510	.2240
3	.1255	.2240
4	.0471	.1680
5 or more	.0186	.1848
Total	1.0000	1.0000

(a) Draw a tree diagram to show all possible sample combinations that could lead to the acceptance or rejection of the shipment.

(b) What is the probability that the shipment will be rejected?

(c) What is the probability that the shipment will be accepted?

(d) If the probability of accepting the shipment with 5% defective is too great, how can this probability be reduced?

5-11. Given that events A and B are independent and $P(A) = .30$ and $P(B) = .60$, evaluate the following probabilities:

(a) $P(A \cap B)$ (c) $P(A')$ (e) $P(A|B)$

(b) $P(A \cup B)$ (d) $P(A' \cap B')$

5-12. A box contains an unknown number of red and blue balls. If 2 balls are drawn at random (without replacement), the probability that both are red is .5. What is the fewest number of balls that could be in the box?

5-13. If A, B, and C are independent events, evaluate the following:

(a) $P(A)$ if $P(A \cap B) = 1/2$, and $P(B) = 3/4$.

(b) $P(B)$ if $P(A \cap B \cap C) = 1/8$, and $P(A \cap C) = 1/2$.

(c) $P(B)$ if $P(A|B) = .4$, and $P(A \cap B) = .2$.

5–14. The following table shows numbers of men and women students in a sample of 100 who own cars on a college campus:

	Car	No Car
Men	40	20
Women	30	10

Let C denote car and M denote man. Evaluate:

(a) $P(C)$
(b) $P(M)$
(c) $P(C \cap M)$
(d) $P(C \cup M)$
(e) $P(C'|M)$
(f) $P(M|C')$
(g) $P(M' \cap C')$

6 Probability Distributions

Chapters 3 and 4 contained several examples in which the observations of some variable had been grouped into a frequency distribution. The data in these distributions were actual observations of the variables under study, and the distributions could be called *observed distributions* or *empirical distributions*. Chapter 5 discussed some of the fundamentals of probability. In this chapter these two concepts are merged in a discussion of probability distributions.

A *probability distribution* can be defined as a theoretical distribution of all the possible values of some variable and the probabilities associated with the occurrence of each value. It is a systematic arrangement of the probabilities associated with mutually exclusive and exhaustive simple events of some experiment. When the probabilities associated with each value of the distribution are multiplied by the total number of observations in the experiment, the result is an *expected distribution.*

A very simple example designed to illustrate the differences in the three types of distributions is shown in Table 6–1. Suppose that a fair coin is tossed 20 times. The probability of heads is .5 and the probability of tails is .5. These two probabilities constitute a probability distribution. Before the coins are

TABLE 6–1

DIFFERENCE BETWEEN A PROBABILITY DISTRIBUTION, AN EXPECTED FREQUENCY DISTRIBUTION, AND AN OBSERVED FREQUENCY DISTRIBUTION

20 Tosses of a Fair Coin

Event	Probability Distribution	Expected Distribution	Observed Distribution
Heads	.5	10	12
Tails	.5	10	8
Total	1.0	20	20

tossed, the anticipated number of heads and tails is each 10. This is the expected distribution. If the 20 tosses of the coin result in 12 heads and 8 tails, these results constitute the observed distribution. The fact that the tosses are random explains the fact that the expected and the observed distributions are not exactly the same. In Chapter 9 a chi-square test will be used to determine whether the differences between expected and observed distributions are too large to be attributed to chance or whether they are so great as to represent a significant difference.

THE HYPERGEOMETRIC DISTRIBUTION

The discussion on page 113 was concerned with a lot of 50 manufactured items, 5 of them defective and 45 good. The general rule of multiplication was used to determine the probability of getting zero defectives in a random sample of 3 of the items drawn from the 50.

Another method of finding the probability of drawing zero defectives would be to use Formula 6–1 to compute the number of combinations, C, of n items taken r at a time. There are 19,600 different combinations of 3 that can be made from 50 different items. For 50 items taken 3 at a time

$$_{n}C_{r} = \frac{n!}{r!(n-r)!} \qquad \textbf{(6–1)}[1]$$

$$_{50}C_{3} = \frac{50!}{3!47!} = 19{,}600.$$

Since 47! is the product of all the integers from 47 to 1, and 50! is the product of all the integers from 50 to 1, 47! cancels into 50! leaving only $50 \cdot 49 \cdot 48$. Calculation of $_{50}C_{3}$ is therefore

$$_{50}C_{3} = \frac{50 \cdot 49 \cdot 48}{3 \cdot 2 \cdot 1} = \frac{117{,}600}{6} = 19{,}600.$$

Since 45 of the items in the population are not defective, the number of different samples of 3 that can be drawn from these nondefective items is computed

$$_{45}C_{3} = \frac{45!}{3!42!} = 14{,}190.$$

Out of the total of 19,600 samples that may be drawn, 14,190 will have no defective items, so the probability of such a sample being drawn is computed

[1] In Formula 6–1 the symbol ! represents the factorial of a number. The factorial of a number is the product of the integer multiplied by all the lower integers; the factorial of 4 (written 4!) is $4 \times 3 \times 2 \times 1 = 24$. It is important to remember that $0! = 1$.

$$P(0) = \frac{14,190}{19,600} = .72398.$$

This calculation gives the same value as the calculation on page 114.

The probability of drawing a sample with 1 defective is computed by determining the total number of samples of 3 that will contain only 1 defective item and 2 that are not defective. Since there are 45 good items in the lot and each sample will contain 2 of these items, the first step is to compute the number of different samples of 2 that can be made from the 45 good items. This number is computed by finding the number of combinations of 45 taken 2 at a time, or ${}_{45}C_2$.

$${}_{45}C_2 = \frac{45!}{2!43!} = 990.$$

Each of these 2 nondefective items must now be combined with a defective item to make a sample of 3 with 1 defective. There are 5 defective items in the lot, and the number of combinations of these 5 items taken 1 at a time is

$${}_5C_1 = \frac{5!}{1!4!} = 5.$$

Each of these 5 defective items can be included in each of the 990 samples containing 2 nondefective items, giving a total of 5×990 or 4,950 different combinations. The probability of drawing a sample at random with 1 defective is, therefore, computed

$$P(1) = \frac{{}_{45}C_2\,{}_5C_1}{{}_{50}C_3} = \frac{(990)(5)}{19,600} = \frac{4,950}{19,600} = .25255.$$

The probability that a sample will contain 2 defective items and 1 nondefective is computed

$$P(2) = \frac{{}_{45}C_1\,{}_5C_2}{{}_{50}C_3} = \frac{(45)(10)}{19,600} = \frac{450}{19,600} = .02296.$$

The probability that all three will be defective is

$$P(3) = \frac{{}_5C_3}{{}_{50}C_3} = \frac{10}{19,600} = .00051.$$

Since these 4 situations cover all possible combinations of defective and nondefective items in a sample of 3, the sum of these probabilities must be 1.00000, as shown in Table 6–2. The distribution shown in this table is known as the *hypergeometric* and is the correct distribution to use when the probabilities of an event change as successive events occur.

TABLE 6–2

PROBABILITY OF 0, 1, 2, AND 3 DEFECTIVE ITEMS IN A SAMPLE OF 3 FROM A LOT OF 50 (BASED ON THE HYPERGEOMETRIC)

Fraction Defective = .10
Computation of Mean and Standard Deviation

Number of Defective Items r	Probability of Each Number Defective f	fr	$(r-\mu)$	$(r-\mu)^2$	$f(r-\mu)^2$
0	.72398	0	−.3	.09	.0651582
1	.25255	.25255	.7	.49	.1237495
2	.02296	.04592	1.7	2.89	.0663544
3	.00051	.00153	2.7	7.29	.0037179
Total	1.00000	.30000			.2589800

(Formula 3–2) $$\mu = \frac{\Sigma fm}{N}$$

Substituting r for m, and Σf for N $$\mu = \frac{\Sigma fr}{\Sigma f} = \frac{.30000}{1.00000} = .30$$

(Formula 4–13) $$\sigma^2 = \frac{\Sigma f(m-\mu)^2}{N}$$

Substituting r for m, and Σf for N $$\sigma = \sqrt{\frac{\Sigma f(r-\mu)^2}{\Sigma f}} = \sqrt{\frac{.258980}{1.00000}} = \sqrt{.258980} = .509$$

The mean and the standard deviation of the distribution are computed in Table 6–2 using Formulas 3–2 and 4–13. It is not necessary, however, to make the computations from these formulas since both measures can be computed from the values of N, n, and π.

$$\mu = n\pi \tag{6–2}$$

$$\mu = (3)(.10) = .30$$

$$\sigma = \sqrt{n\pi(1-\pi)\frac{N-n}{N-1}} \tag{6–3}$$

$$\sigma = \sqrt{(3)(.1)(.9)\frac{50-3}{50-1}} = \sqrt{.258980} = .509$$

where

$N = 50$, the total number of items
$n = 3$, the number of items in the sample
$\pi = .10$, the fraction defective $\frac{5}{50}$.

THE BINOMIAL DISTRIBUTION

The *binomial theorem* is a theoretical discrete frequency distribution which has many practical applications in business statistics. It is also known as the *Bernoulli distribution,* after James Bernoulli, a Swiss mathematician who died in 1705.

The hypergeometric is the correct distribution to use when the probability of an event is not constant, but this situation does not always exist. For example, after an item has been selected and tested it might be replaced and be subject to selection again. If this were done, it would mean that the probability of drawing a defective item would remain the same for each article selected for testing. It would be rather unusual to replace an article after it has been selected in a sample, but such a practice is called *sampling with replacement* and has the effect of the universe being infinite, since it makes possible drawing an unlimited number of samples of a given size.

The above discussion suggests that if the universe is very large, it might be practicable to assume that the probability of drawing a defective article changes so slightly when one unit is selected that it might be considered to remain unchanged. This is commonly done when the sample selected is small in relation to the size of the universe.

Problems in which the probability of occurrence of an event remains constant or can be assumed to remain approximately constant may be solved by using the binomial distribution, also called the Bernoulli distribution, which is based on the expansion of the binomial. The general term of the binomial expansion applying to all terms is

$$P(r|n,\pi) = {}_nC_r\pi^r(1-\pi)^{n-r} = \frac{n!}{r!(n-r)!}\pi^r(1-\pi)^{n-r} \qquad \textbf{(6–4)}$$

where

$P(r|n,\pi)$ = the probability of r out of n events occurring, given π.

Using the problem of the manufactured product that on the average is 10% defective, Formula 6–4 may be used to compute the probability of each of the number of defective articles that might occur in a sample of 3. It was stated above that $\pi = .10$ and $n = 3$. The probability of 0 defectives can be found by setting $r = 0$.[2]

When $r = 0$

$$P(0|3,.10) = {}_3C_0\pi^0(1-\pi)^{3-0}$$

$$= \frac{3!}{0!3!}(.10)^0(.90)^{3-0}$$

$$= \frac{3 \cdot 2 \cdot 1}{3 \cdot 2 \cdot 1}(.10)^0(.90)^3$$

[2] $(b)^0$ is 1 for every nonzero b.

$$= (1)(1)(.729)$$

$$= .729.$$

When $r = 1$

$$P(1|3,.10) = {}_3C_1(.10)^1(.90)^{3-1}$$

$$= \frac{3!}{1!2!}(.10)^1(.90)^{3-1}$$

$$= \frac{3 \cdot 2 \cdot 1}{1 \cdot 2 \cdot 1}(.10)^1(.90)^2$$

$$= (3)(.10)(.81)$$

$$= .243.$$

When $r = 2$

$$P(2|3,.10) = {}_3C_2(.10)^2(.90)^{3-2}$$

$$= \frac{3!}{2!1!}(.10)^2(.90)^{3-2}$$

$$= \frac{3 \cdot 2 \cdot 1}{2 \cdot 1 \cdot 1}(.10)^2(.90)^1$$

$$= (3)(.01)(.90)$$

$$= .027.$$

When $r = 3$

$$P(3|3,.10) = {}_3C_3(.10)^3(.90)^{3-3}$$

$$= \frac{3!}{3!0!}(.10)^3(.90)^{3-3}$$

$$= \frac{3 \cdot 2 \cdot 1}{3 \cdot 2 \cdot 1}(.10)^3(.90)^0$$

$$= (1)(.001)(1)$$

$$= .001.$$

When the probabilities computed from the binomial are compared with those computed from the hypergeometric, it is found that they are somewhat different, and the smaller the universe the more important it is to recognize that the probabilities change after each item is selected.

Table 6–3 gives the values of the binomial distribution, which is the probability of 0, 1, 2, and 3 defectives in a sample of 3 from a very large lot that is 10% defective. (Theoretically the size of the lot should be infinite or sampling should be with replacement.) The total of the probabilities is 1.000, since the probability is 0 to 3 defectives equals the sum of the individual probabilities. Also shown is the computation of the arithmetic mean and the standard deviation of the distribution, substituting r for m, and Σf for N.

TABLE 6-3

PROBABILITY OF 0, 1, 2, AND 3 DEFECTIVE ITEMS IN A SAMPLE OF THREE FROM A LARGE LOT (BASED ON THE BINOMIAL)

Fraction Defective = .10
Computation of Mean and Standard Deviation

Number of Defective Items in Sample r	Fraction of Defective Items in Sample r/n	Probability of Occurrence f	fr	$r - \mu$	$(r - \mu)^2$	$f(r - \mu)^2$
0	0/3	.729	0	−.3	.09	.06561
1	1/3	.243	.243	.7	.49	.11907
2	2/3	.027	.054	1.7	2.89	.07803
3	3/3	.001	.003	2.7	7.29	.00729
Total		1.000	.300			.27000

$$\mu = \frac{\Sigma fr}{\Sigma f} = \frac{.300}{1.000} = .30$$

$$\sigma = \sqrt{\frac{\Sigma f(r - \mu)^2}{\Sigma f}} = \sqrt{\frac{.27000}{1.000}} = \sqrt{.27} = .52$$

The arithmetic mean and standard deviation of the binomial may be computed from the values of n and π as follows without using the formulas from Chapters 3 and 4. The formulas for the arithmetic mean of the hypergeometric (Formula 6–2) and the arithmetic mean of the binomial (Formula 6–5) are the same.

$$\mu = n\pi \tag{6-5}$$

$$\mu = (3)(.10) = .30$$

$$\sigma = \sqrt{n\pi(1 - \pi)} \tag{6-6}$$

$$\sigma = \sqrt{(3)(.10)(1 - \pi)} = \sqrt{.27} = .52.$$

If the distribution is stated in terms of the fraction defective instead of the number of items defective, the number of defectives in each sample is divided by n so the mean of the fraction defective will be

$$\mu = \frac{n\pi}{n} = \pi. \tag{6-7}$$

The standard deviation of the distribution is

$$\sigma = \sqrt{\frac{\pi(1 - \pi)}{n}}. \tag{6-8}$$

When $\pi = .50$ the binomial is symmetrical. Figure 6–1 shows graphically the binomial distribution for $n = 20$ and $\pi = .50$. The mean and the standard deviation of this distribution are computed

$$\mu = n\pi = (20)(.50) = 10$$

$$\sigma = \sqrt{n\pi(1-\pi)} = \sqrt{(20)(.50)(.50)} = \sqrt{5} = 2.236.$$

Since the distribution is symmetrical, the arithmetic mean, the median, and the mode are the same value.

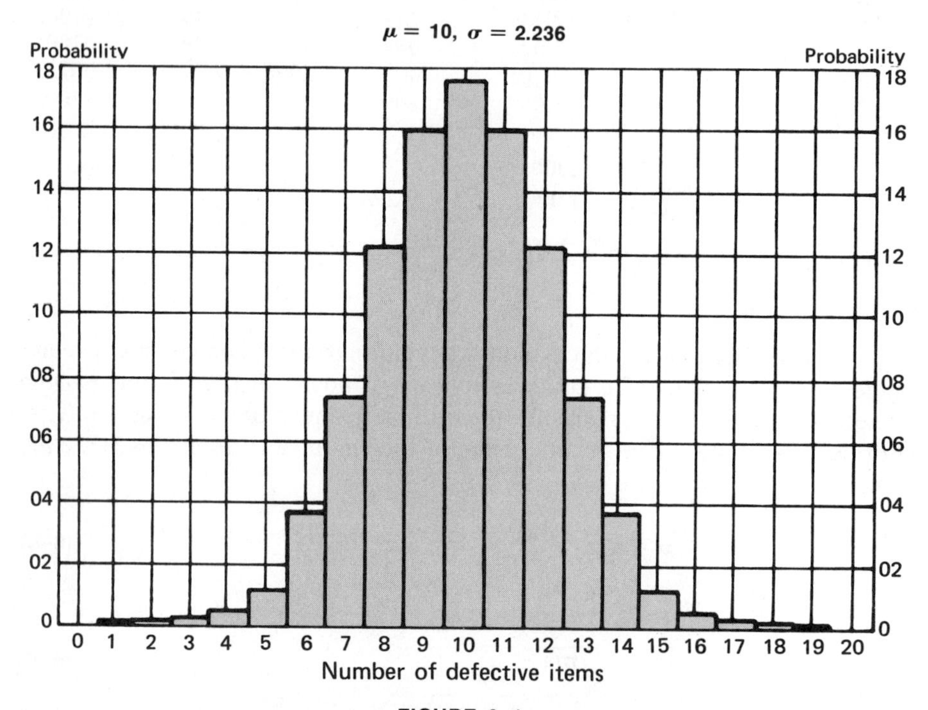

FIGURE 6–1

BINOMIAL DISTRIBUTION WITH $n = 20$ AND FRACTION DEFECTIVE = .50

In solving any problem which involves a binomial distribution of probabilities, considerable time and effort can be saved by using a table of the binomial probability distribution. For example, suppose that an automobile rental agency has available both sedans and station wagons. Records show that 3 persons in 4 prefer sedans. The agency manager would like to know the probability that of the next 10 calls, exactly 4 will request a station wagon. Using Formula 6–4, this probability is computed

$$P(r=4|n=10,\pi=.25)=\frac{10!}{4!6!}(.25)^4(.75)^6=(210)(.0039062)(.1779785)$$
$$=.1460.$$

This same value can be secured from the table in Appendix F by locating the table value associated with $n=10$, $r=4$, and $\pi=.25$.

If, however, the agency manager would like to know the probability that the next 10 customers will request exactly 4 sedans, the problem is a bit more complicated. Since the probability that a single customer will want a sedan is $\pi=.75$, this value exceeds those listed in the table. This difficulty can be circumvented by determining from the table the probability that 6 out of 10 will want station wagons when the probability that anyone will want a station wagon is .25. Note that the probability of the 4 requests for sedans by the next 10 people is exactly the same as 6 requests for station wagons by the next 10 people. In fact, requesting 6 station wagons and requesting 4 sedans are actually the same event. From the table it is determined that

$$P(r=6|n=10,\pi=.25)=.0162.$$

THE NORMAL DISTRIBUTION

The *normal distribution* is a theoretical continuous distribution discovered over 200 years ago. It was then considered to be the "law" to which distributions of observations of natural phenomena were believed to conform. This belief has been modified as other distributions were discovered, but the normal distribution is still one of the most important in statistics.

The binomial distribution is symmetrical when $\pi = 1/2$. But when π is greater or less than one half, the distribution is skewed. However, as the size of n increases, the distribution becomes more nearly symmetrical. Regardless of the value of π, as n increases indefinitely, the binomial distribution approaches the symmetrical curve known as the normal distribution. As n increases, the number of terms in the binomial increases; and the larger the number of terms, the smoother the graph of the distribution. If in the preceding example of the binomial n had been equal to 1,000 rather than 10, the probability of drawing zero defective items would be very small, as would be the probability of drawing 1,000 defectives. The frequency polygon of a binomial distribution with a large value of n resembles closely the chart of the normal distribution, called the *normal curve.* For many purposes, the normal distribution can be substituted for the binomial distribution without any significant loss of accuracy.

In plotting the normal curve, the origin of the X scale is located in the middle of the distribution. Since the distribution is symmetrical, this midpoint is the mean of the distribution. All the X values are measured as deviations from the mean in standard deviation units. The values on the X scale in terms of z units are represented by $\frac{X-\mu}{\sigma}$. Since the mean is at the middle of the distribution, the values of $\frac{X-\mu}{\sigma}$ will be either plus or minus values. A normal distribution with a mean of $400 is shown in Figure 6–2; the values on the X scale are shown both in the original X units and in $\frac{X-\mu}{\sigma}$ units.

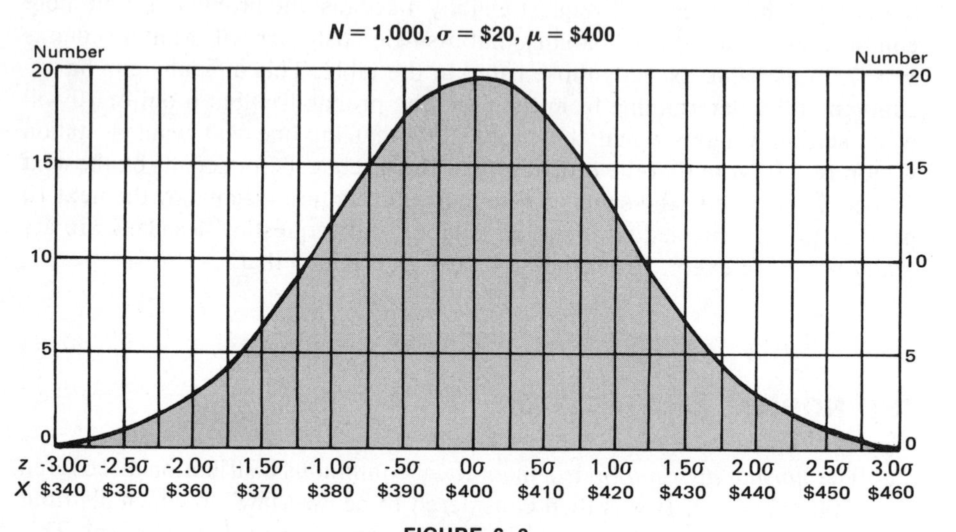

FIGURE 6–2

NORMAL DISTRIBUTION

Ordinates of the Normal Distribution

The height of the *maximum ordinate* of any given normal distribution depends on the total number on the distribution, N, and the value of the standard deviation, σ. The maximum ordinate is represented by Y_o, since it is the value of Y when $X-\mu=0$. The value of the maximum ordinate of the normal distribution is

$$Y_o = \frac{N}{\sigma\sqrt{2\pi}} \quad \text{or} \quad Y_o = \frac{N}{\sigma}.39894. \tag{6-9}$$

For a normal distribution with total frequencies of 1,000 ($N = 1{,}000$), a standard deviation of $20 ($\sigma = \$20$), and an ordinate computed for each integral value of X, the value of the maximum ordinate Y_o is computed

$$Y_o = \frac{1{,}000}{20}.39894 = (50)(.39894)$$

$$Y_o = 19.947.$$

From the values of $X - \mu$, other ordinates of the normal distribution may be computed[3]

$$Y = Y_o e^{-\frac{1}{2}\left(\frac{X-\mu}{\sigma}\right)^2}. \tag{6-10}$$

Since $Y_o = \frac{N}{\sigma}.39894$, the formula of the ordinates may be written

$$Y = \frac{N}{\sigma}.39894e^{-\frac{1}{2}\left(\frac{X-\mu}{\sigma}\right)^2}$$

When $X - \mu = \$40$, Y is computed

$$Y = \frac{1{,}000}{20}(.39894)(2.718282)^{-\frac{1}{2}\left(\frac{40}{20}\right)^2} = \frac{1{,}000}{20}(.39894)(2.718282)^{-2}$$

$$= \frac{1{,}000}{20}(.39894)\frac{1}{(2.718282)^2} = \frac{1{,}000}{20}(.39894)(.13535) = 2.6995.$$

The value .05399, which is the product of .39894 times .13535, can be secured from the table of Ordinates of the Normal Distribution (Appendix G). For the ordinate computed above, $X - \mu = \$40$; and since $\sigma = \$20$, $\frac{X-\mu}{\sigma} = \frac{40}{20} = 2$. It is customary to let $\frac{X-\mu}{\sigma} = z$, and these two terms will be used interchangeably. The value from Appendix G is .05399 when $z = 2.0$. This value is multiplied by the value of $\frac{N}{\sigma}$ to secure the ordinate of the normal distribution when $N = 1{,}000$, $\sigma = \$20$, and $z = 2.0$

$$Y = \frac{N}{\sigma}(.05399) = \frac{1{,}000}{20}(.05399) = 2.6995.$$

Note that this is the same value computed by Formula 6–10 above.

The table of Ordinates of the Normal Distribution can be used to compute the maximum ordinate (Y_o) in the same manner as any other ordinate. When $z = 0$, the value in Appendix G is .39894; thus,

$$Y_o = \frac{1{,}000}{20}(.39894) = 19.947.$$

By computing a number of ordinates of the normal distribution when $N = 1{,}000$ and $\sigma = \$20$, the distribution may be plotted to give the curve in Figure 6–2. The ordinates used to plot this curve were computed by obtaining

[3] In Formula 6–10 the symbol e represents the base of the Naperian or natural logarithms. The value of e to six decimal places is 2.718282.

the values in Appendix G; these values, which give the same amounts as would be computed by the formula, are generally used to save time. In Figure 6–2 the ordinate was computed for each integral value of X from \$340 to \$460. If a frequency distribution is to be plotted at the midpoint of each class, the frequency of each class would be i times the ordinate at the midpoint of the class (letting i represent the width of the class interval). This can be accomplished by expressing the standard deviation in class-interval units $\left(\frac{\sigma}{i}\right)$ in all calculations.

Areas Under the Normal Curve

It is frequently important to know the proportion of the total frequencies of the normal distribution that falls between two points on the X axis, instead of knowing the ordinate at these points. Figure 6–3 shows three shaded segments of the total area. The area in the center of the curve is the portion that lies between the ordinate with a value of $+1\sigma$ and the ordinate with a value of -1σ on the X axis. The shaded area at the right of the curve is the portion that lies above $+1.96\sigma$, while the shaded area at the left is the portion that lies below -1.96σ. Each tail of the curve is asymptotic to the base line; that is, the tail approaches the base line but never touches it.

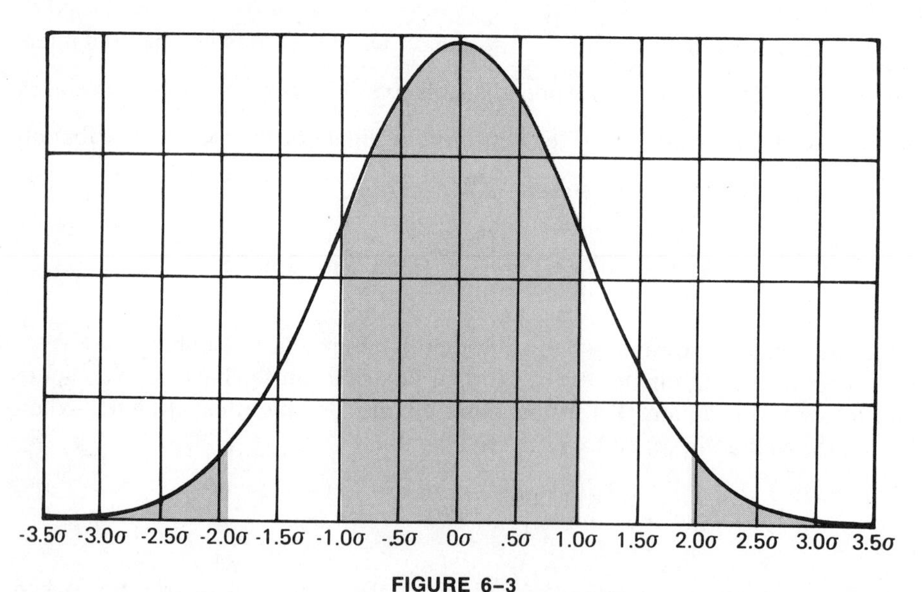

FIGURE 6–3

NORMAL DISTRIBUTION

It is possible to compute the area between any two ordinates by using integral calculus and the equation of the normal distribution, but it is not necessary to make these computations. Because the value of these areas is used so frequently in calculations, a table has been prepared to show the areas under the normal curve between different ordinates expressed in units of z. The proportion of the total area lying between any two ordinates may be determined from this table, which is given in Appendix H. The values in this table represent the proportion of the curve between the maximum ordinate (the mean) and the ordinates at given values on the X axis expressed as deviations from the maximum ordinate in units of the standard deviation. The values on the X axis are the same as those used in the table of ordinates. For example, 34.13% of the area under the normal curve falls between the maximum ordinate ($X - \mu = 0$) and $+ 1\sigma$. Since the curve is symmetrical, the same proportions apply to the plus and the minus values of z. Therefore 68.26% of the area under the curve falls between $+ 1\sigma$ and $- 1\sigma$.

From Appendix H the proportion of the curve that falls above $+ 1.96\sigma$ is found to be .02500, by the following calculation:

> The proportion of the normal curve between the maximum ordinate and the 1.96σ point on the X axis is shown in the table to be .47500. Since the normal curve is symmetrical, .50000 of the area lies above the maximum ordinate. Since .47500 of the curve is below 1.96σ, subtraction gives .02500 of the curve above 1.96σ. The same percentage of the curve falls below $- 1.96\sigma$. Thus, .05000 of the curve falls beyond the points $- 1.96\sigma$ and $+ 1.96\sigma$.

Similar calculations can be made to find the proportion of the curve falling between any two points on the X axis. It is first necessary to express the value of the X axis as a deviation from the mean in standard deviation units, that is, as a value of z. Appendix H gives the proportion of the normal curve between the maximum ordinate and selected values of z. If it is desired to compute the proportion of the curve between two X values, neither of which is the maximum ordinate, it is necessary first to compute the proportion of the curve between each value of X and the maximum ordinate. The proportion between these two points can then be found as the difference between the two proportions, if both points are on the same side of the mean. The proportion of the curve between $+ 1.0\sigma$ and $.5\sigma$ is found as follows:

> The proportion of the curve between the maximum ordinate and $z = +1.0$ is .34134, and the proportion between the maximum ordinate and $z = +.5$ is .19146. Therefore, the proportion of the curve between $z = +1.0$ and $z = +.5$ is the *difference* between these two proportions, or $.34134 - .19146 = .14988$.

If the two points fall on different sides of the mean, the proportion of the curve on each side of the mean is computed as was just shown. The proportion of the curve between the maximum ordinate and $z = -1.0$ is .34134. The proportion between the maximum ordinate and $z = +.5$ is .19146. The proportion of the curve between $z = 1.0$ and $z = +.5$ is the sum of the two proportions, or $.34134 + .19146 = .53280$.

The Normal Distribution as an Approximation of the Binomial

It was stated on page 131 that as the size of n increases the binomial approaches the normal distribution. Since this is true, in many situations the normal distribution may be used as an approximation of the binomial. When the value of π in the binomial is near .50, the distribution is approximately symmetrical, but regardless of the value of π, the distribution approaches the normal distribution as n increases without limit.

Figure 6–4 shows graphically the histogram of the binomial distribution with $\pi = .25$ and $n = 100$ given in Table 6–4. Added to the histogram of the binomial distribution is a smooth curve showing the normal distribution that has the same arithmetic mean and standard deviation drawn through the points plotted in the middle of the bars. The binomial distribution follows closely the shape of the normal distribution. The fact that the binomial is a discrete distribution instead of being continuous accounts for most of the difference between the two distributions. Discrete distributions can take on only certain values and have no meaning for other values. Continuous distributions, on the other hand, can take on an infinite number of values between any two given values.

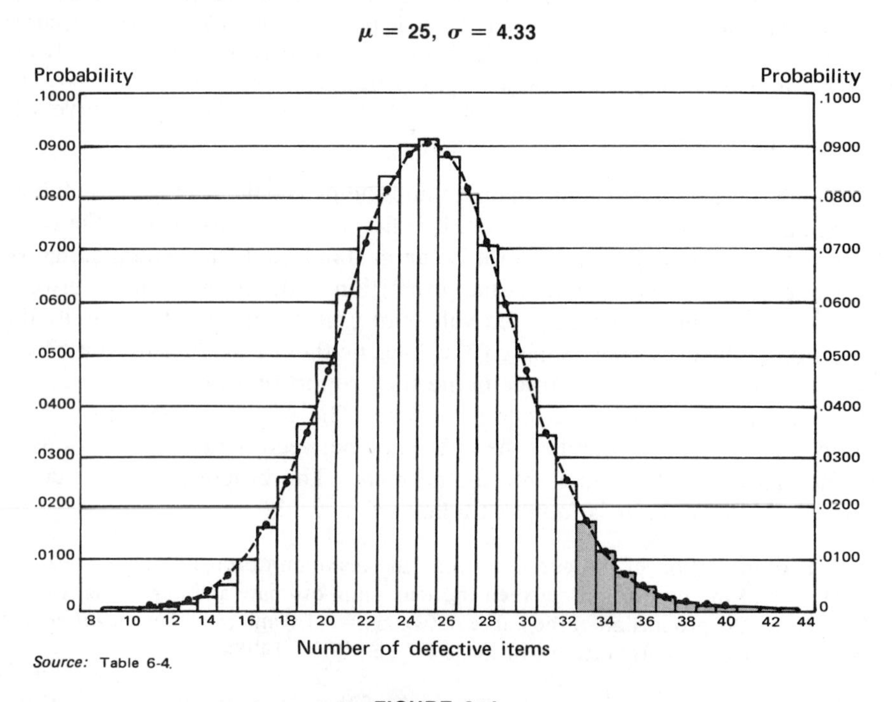

Source: Table 6-4.

FIGURE 6–4

BINOMIAL DISTRIBUTION AND NORMAL DISTRIBUTION WITH SAME μ AND σ

Since the areas under various portions of the normal curve have been computed, it is possible to determine the fraction of the distribution above or below any given value much easier than can be done for the binomial. Reference to the equation of the binomial will show the amount of work needed to compute the probability of a sample of 100 containing 33 or more defective

TABLE 6–4

BINOMIAL DISTRIBUTION WITH n = 100 AND FRACTION DEFECTIVE = .25, AND NORMAL DISTRIBUTION WITH SAME μ AND σ

$\mu = 25 \qquad \sigma = 4.33$

Number of Defectives	Probability of Occurrence	
	Binomial	Normal
9	.0000	.0001
10	.0001	.0002
11	.0003	.0005
12	.0006	.0008
13	.0014	.0019
14	.0030	.0037
15	.0057	.0064
16	.0100	.0106
17	.0165	.0167
18	.0254	.0248
19	.0365	.0351
20	.0493	.0471
21	.0626	.0597
22	.0749	.0727
23	.0847	.0830
24	.0906	.0898
25	.0918	.0923
26	.0883	.0898
27	.0806	.0830
28	.0701	.0727
29	.0580	.0597
30	.0458	.0471
31	.0344	.0351
32	.0248	.0248
33	.0170	.0167
34	.0112	.0106
35	.0070	.0064
36	.0042	.0037
37	.0024	.0019
38	.0013	.0008
39	.0007	.0005
40	.0004	.0002
41	.0002	.0001
42	.0001	.0001
43	.0000	.0000

items. Table 6–4 gives the probability of each number of defective items occurring in a sample of 100, and the probability of 33 or more defectives can be found by adding the probability of 33 defectives and of each number above 33. For the binomial distribution the sum of these probabilities is .0445. It will be noted that they are computed to four decimal places, which means that there is a slight probability of more than 42 defective items, but rounded to four decimal places it is .0000.

The fraction of the normal distribution that is 33 and above can be derived from the table of areas under the normal curve by computing the value of z. It has been shown that

$$z = \frac{X - \mu}{\sigma}.$$

The value of μ for the normal distribution is $n\pi = (100)(.25) = 25$. It is necessary to compute the area of the distribution to the right of the point halfway between 32 and 33, since the value 33 is located in the middle of the bar representing 33 defectives. The standard deviation is equal to 4.33, so

$$z = \frac{X - \mu}{\sigma} = \frac{32.5 - 25}{4.33} = \frac{7.5}{4.33} = 1.73.$$

The value of 1.73 for z is found in Appendix H to include .45818 of the area in the upper half of the distribution. Therefore, .50000 − .45818 = .04182, which represents the area of the curve that is to the right of the point 32.5 on the X axis. This is a good approximation of the probability of 33 and more defects occurring in a sample of 100, which according to the binomial is .0445.

Care must be taken not to use the normal distribution as an approximation of the binomial when the value of π is very small, unless n is very large. For $n = 100$ and $\pi = .25$, Figure 6–4 demonstrates that the binomial is nearly symmetrical and resembles the normal distribution very closely. Since the areas of the normal distribution are readily available in tables, the computations are much simpler than for the binomial.

The Normal Distribution as an Approximation of the Hypergeometric

It is possible to use the normal distribution to estimate the values of the hypergeometric, and this use is important because the computation of the values of the hypergeometric is even more burdensome than for the binomial. When the mean and standard deviation of a hypergeometric distribution have been computed, using Formulas 6–2 and 6–3, it is possible to use them exactly as the mean and standard deviation of the binomial distribution were used to find the values of the approximate normal distribution. Except when the value of π or n is very small, the normal distribution serves as a good approximation of the hypergeometric, just as it does for the binomial. The only difference between using the normal distribution as an approximation of the binomial and using it as an approximation of the hypergeometric is that in the latter case the term $\sqrt{\frac{N-n}{N-1}}$ is used in computing the standard deviation. The hypergeometric must be used when the probabilities change substantially as items are drawn from a relatively small universe, and in such a situation the term $\sqrt{\frac{N-n}{N-1}}$ will be substantially less than 1. When the universe is large in relation to the sample being used, this term approaches 1, and since as it approaches 1 it has less and less influence on the result, it may be ignored.

THE POISSON DISTRIBUTION

The *Poisson distribution* is another theoretical discrete probability distribution with many business applications. It is named after a French mathematician, Siméon Poisson, who derived the distribution in 1837 as a special (limiting) case of the binomial distribution when the number of trials is very large and the probability of success on any one trial is very small. The Poisson was later derived as a distribution in its own right. It serves as a model for situations where one is concerned with the number of occurrences per unit of observation, such as the number of trucks arriving at a loading dock per hour, flaws per 100 feet of cable, or the number of imperfections per square yard of cloth.

For example, assume that a manufacturing process has been turning out woolen cloth with an average (arithmetic mean) of 3.6 defects per 10 yards of cloth (one yard wide). If the distribution of the defects is random, the fractions of random samples of 10 yards having no defects, one defect, two defects, etc., will be a Poisson distribution. The relative frequency shows the probability of a lot having zero defects, one defect, two defects, etc.

In a situation in which the fraction of the items that will be defective is known, as in the example on page 27 where the fraction defective is 10%, the number of defective items will be distributed as the binomial. In the case of woolen cloth, however, it is impossible to determine the total number of defects possible in 10 yards of cloth, so the *percentage* of total defects present in a given lot of cloth cannot be computed. Perhaps the number of possible defects per 10 square yards is infinite, or at least very large. In this case only the *number* of defects per 10 square yards can be determined, rather than the percentage of items that are defective.

The formula for the Poisson distribution is

$$P(r|\lambda) = \frac{\lambda^r}{r!} e^{-\lambda}. \qquad \textbf{(6–11)}$$

where

e = the base of the natural logarithms, having a value of 2.71828+
λ = the mean number of occurrences in some continuum
r = a variable that takes on the values 0, 1, 2, . . . , and represents the number of occurrences that may take place.

The relative frequency for each value of r can be computed from Formula 6–11 and the value of $e^{-\lambda}$ can be computed by logarithms.

The probability of zero defects in one lot of cloth is computed

$$P(0|3.6) = \frac{3.6^0}{0!}(2.71828)^{-3.6} = \frac{(1)(.02732)}{1} = .02732.$$

The value of $2.71828^{-3.6}$ equals .02732 (see footnote 4). The probability of one defect in a lot is

$$P(1|3.6) = \frac{3.6^1}{1!}(2.71828)^{-3.6} = \frac{(3.6)(.02732)}{1} = .09835.$$

The probability of two defects is

$$P(2|3.6) = \frac{3.6^2}{2!}(2.71828)^{-3.6} = \frac{(3.6)(3.6)(.02732)}{(2)(1)} = .17703.$$

The computations for the Poisson are much simpler than for the binomial. Note that the value $2.71828^{-3.6}$ in the numerator is the same in each calculation. The term is computed only once, using logarithms, or it can be read from Appendix H where the value of $e^{-\lambda}$ is given for selected values of λ. After the probability for $r = 0$ has been computed, the value for $r = 1$ can be found by simply multiplying the probability for $r = 0$ by 3.6 and dividing by the next r value, that is, 1. Each successive value can be computed in the same way.

Table 6–5 gives the probability of each number of defects from 0 to 10 and also "11 and over." The total of these probabilities is 1.0000, so the value of "11 and over" can be found by subtracting the total of the probabilities from 0 to 10 from 1.0000. The probability of 11 or more defects is so small that there is no need to go further than the value of $r = 11$ or more.

The Poisson can be used to describe occurrences of many kinds. Sampling for number of defects has been given as an example, and this application is widely used in sampling inspection. This use has become so common that extensive tables of the values of the Poisson for many values of λ have been computed and can be used to solve problems with a minimum of computation. A table giving certain values of the Poisson is shown in Appendix J. The values in this table may be used to check the values of the Poisson computed in Table 6–5. It is possible that a discrepancy may occur in the last decimal place due to a different number of decimals being carried.

When the possible number of occurrences of an event is very large but the percentage of these potential occurrences that actually occur is so small that the average occurrence is very small, the Poisson distribution gives an excellent description of the data.

---

[4] $e^{-3.6} = \frac{1}{e^{3.6}}$ (The minus sign in the exponent indicates the reciprocal.)

$\log e^{3.6} = (3.6)(\log 2.71828) = (3.6)(.434294) = 1.563458$

$e^{3.6} = \text{antilog } 1.563458 = 36.60$

$e^{-3.6} = \frac{1}{36.60} = .02732$

TABLE 6–5

PROBABILITY OF OCCURRENCE OF VARIOUS NUMBERS OF DEFECTS PER 10 SQUARE YARDS OF WOOLEN CLOTH

Number of Defects	Probability of Occurrence
0	.0273
1	.0984
2	.1770
3	.2125
4	.1912
5	.1377
6	.0826
7	.0425
8	.0191
9	.0076
10	.0028
11 and over	.0013
Total	1.0000

An Example Application of the Poisson

A study published in 1936 on the number of vacancies filled on the United States Supreme Court by years over the 96-year period of 1837–1932 illustrates the use of the Poisson.[5] In these 96 years there were 48 vacancies filled, an average of .5 per year. It can be assumed that these vacancies would be distributed throughout the years in the form of the Poisson, since the total number of vacancies conceivably could be very large, but the actual number of vacancies occurring in any year must have been small, since the average was only .5 per year. The probabilities of a year having zero vacancies, 1 vacancy, 2 vacancies, etc., are computed

$$P(r|\lambda) = \frac{\lambda^r}{r!}\, e^{-\lambda}$$

$$P(0|.5) = \frac{.5^0}{0!}\,(2.71828)^{-.5} = \frac{(1)(.60653)}{1} = .60653$$

$$P(1|.5) = \frac{.5^1}{1!}\,(2.71828)^{-.5} = \frac{(.5)(.60653)}{1} = .30327$$

$$P(2|.5) = \frac{.5^2}{2!}\,(2.71828)^{-.5} = \frac{(.5)(.5)(.60653)}{(2)(1)} = .07582$$

$$P(3|.5) = \frac{.5^3}{3!}\,(2.71828)^{-.5} = \frac{(.5)(.5)(.5)(.60653)}{(3)(2)(1)} = .01264$$

[5] W. Allen Wallis, "The Poisson Distribution and the Supreme Court," *Journal of the American Statistical Association,* Vol. 31, No. 194 (1936), pp. 376–380.

The probability of 3 or fewer vacancies in a year is the sum of the probabilities at the bottom of page 141, or .99826. The probability of over 3 vacancies is 1 − .99826 = .00174, since the sum of all probabilities must equal 1.0000. The probabilities computed above can be applied to the 96 years covered by the study to show the expected number of years with zero vacancies, 1 vacancy, 2 vacancies, etc. The relative frequencies multiplied by 96 will give the expected number of years for which there were zero vacancies, 1 vacancy, 2 vacancies, etc. These expected frequencies are shown in Table 6–6, where they are carried out to one decimal place. In the adjoining column is given the actual frequency of occurrence of each number of vacancies as reported in the original study. It appears that the theoretical Poisson distributed describes fairly accurately what happened.

TABLE 6–6

NUMBER OF YEARS IN WHICH SPECIFIED NUMBERS OF VACANCIES ON THE SUPREME COURT WERE FILLED, 1837–1932

Number of Vacancies	Number of Years	
	Expected Frequency	Actual Frequency
0	58.2	59
1	29.1	27
2	7.3	9
3	1.2	1
Over 3	.2	0
Total	96.0	96

Since 48 vacancies were filled over the 96-year period, the average (arithmetic mean) number of vacancies filled per year was computed as $\frac{48}{96} = .50$. The mean can also be computed by Formula 3–2, and the standard deviation can be computed by Formula 4–13. These calculations are shown in Table 6–7.

The variance of the Poisson distribution is equal to the mean of the distribution, and the standard deviation is equal to $\sqrt{\mu}$. Since the theoretical frequencies computed for the Poisson were computed for a distribution with a mean equal to .5, the variance of the theoretical distribution is also equal to .5. It can be demonstrated by computing the variance for the theoretical frequencies given in Table 6–7 that the mean and the variance of this Poisson distribution are equal.

Use of a Poisson Distribution in a Statistical Decision Problem

In Chapter 5 a tree diagram was used to illustrate the use of conditional probabilities in statistical decision making. The problem was to test an industrial sampling plan to determine the probability that it would reject a shipment of items that was 4% defective. The probability distribution which was

TABLE 6–7

NUMBER OF YEARS IN WHICH SPECIFIED NUMBERS OF VACANCIES ON THE SUPREME COURT WERE FILLED, 1837–1932

Theoretical Frequencies

Number of Vacancies m	Number of Years f	fm	$m-\mu$	$(m-\mu)^2$	$f(m-\mu)^2$
0	58.2	0	−.50	.25	14.550
1	29.1	29.1	.50	.25	7.275
2	7.3	14.6	1.50	2.25	16.425
3	1.2	3.6	2.50	6.25	7.500
4	.2	.8	3.50	12.25	2.450
Total	96.0	48.1			48.200

$$\mu = \frac{48.1}{96} = .50 \qquad \sigma^2 = \frac{48.2}{96} = .50 \qquad \sigma = .71$$

used, but not explained, in that example was the Poisson distribution. Since the probability that any one item was defective was 4%, and the size of the first sample was 25 parts, the average number of defective items per sample was

$$\mu = np = 25(.04) = 1 = \lambda.$$

The probability of the number of defective items in a sample of 25 was then computed

$$P(r|\lambda = 1),$$

and for $r = 2$ defective items this would be

$$P(r = 2|\lambda = 1)\frac{1^2}{2!}(2.71828)^{-1}$$

$$= \frac{1}{2}(.367879)$$

$$= .1839 \text{ or } .184.$$

From Appendix I, $e^{-1} = .367879$; this probability can also be read from Appendix J.

The second sample used in the illustration on page 117 was 50 items. Therefore, the average number of defective items per sample was

$$50(.04) = 2 = \lambda$$

and

$$P(r' = 2|\lambda = 2) = .2707 \text{ or } .271.$$

The Poisson as an Approximation of the Binomial

Not only does the Poisson describe the distribution of many types of statistical data, but it is also useful in approximating the values of the binomial under certain conditions. When the mean of the binomial is small, the Poisson gives a fairly close estimate since the binomial approaches the Poisson distribution as a limit as the value of π decreases and n increases in size. Because the binomial behaves in this manner, it is logical to set the value of λ in the equation of the Poisson equal to $n\pi$ and use the Poisson as an approximation of the binomial. Since the calculations required for the Poisson are much less burdensome than for the binomial, this approximation is widely used.

The closeness with which the Poisson follows the binomial can be illustrated by comparing the two distributions with the same mean of .5 defective items. Assume that a sample of 25 is selected from a very large lot that has a fraction defective .02. The probability of getting various numbers of defective items in a sample of 25 would be the binomial $(.02 + .98)^{25}$. Using Formula 6–4 the probability of 0 defectives ($r=0$) is computed

$$P(r|n, \pi) = {}_nC_r\pi^r(1-\pi)^{n-r}$$

$$P(0|25, .02) = {}_{25}C_0(.02)^0(1-.02)^{25-0}$$

$$= \frac{25!}{0!25!}(1)(.98)^{25}$$

$$= (1)(1)(.6035) = .6035.$$

The probabilities of the successive values of r are computed by substituting the value of r ($r=1, r=2$, etc.) in the equation. When the probabilities are carried to four decimal places, the probability when $r=6$ is zero, although if carried out enough places there would be a small value. This means that the last probability in a table of this binomial should be labeled "5 and over."

The probability of zero defectives can be estimated by the Poisson when the average number of defectives is small. In this case the average number of defectives is .5 since the fraction of items in a lot is .02 and the number in each sample is 25. ($\lambda = n\pi = (.02)(25) = .5$). This value of λ is substituted in the equation for the Poisson with $r=0$.

$$P(r|\lambda) = \frac{\lambda^r}{r!}e^{-\lambda}$$

$$P(0|.5) = \frac{.5^0}{0!}(2.72828)^{-5}$$

$$= \frac{(1)(.60653)}{1} = .6065.$$

The probabilities for the remaining values of r are computed and entered in Table 6–8. The approximation for all values appears to be close. The cumulative probabilities are given in the last two columns of Table 6–8 to show that the agreement of the cumulatives is even closer than for the individual number of defectives. For example, the probability of securing 1 or fewer defectives in a sample of 25 from a universe with 2% defective is .91 when computed by the use of the binomial, and also .91 when estimated by the Poisson.

TABLE 6–8

PROBABILITY DISTRIBUTION OF NUMBER DEFECTIVE (BINOMIAL AND POISSON)

$\pi = .02 \quad n = 25 \quad \mu = .5$

Number of Defective Items	Binomial	Poisson	Cumulative Binomial	Cumulative Poisson
0	.6035	.6065	.6035	.6065
1	.3079	.3033	.9114	.9098
2	.0754	.0758	.9868	.9856
3	.0118	.0126	.9986	.9982
4	.0013	.0016	.9999	.9998
5 and over	.0001	.0002	1.0000	1.0000
Total	1.0000	1.0000		

The Poisson is widely used in sample inspection as an approximation of the binomial whenever the number of defective items ($n\pi$) is small. Any value of $n\pi$ under 5 will usually give a very close approximation of the binomial and the Poisson is frequently used with larger values than 5.

Whenever the Poisson is a reasonably close approximation of the binomial, the saving in computation is a good reason for using it. The development of more extensive tables of the binomial has reduced somewhat the dependence on the Poisson as an approximation. However, tables of the Poisson are also readily available and are much briefer than the tables of the binomial.

STUDY QUESTIONS

6–1. What is a probability distribution?

6–2. What is the appropriate distribution to use when the probability that an event will occur is not constant from one trial to another?

6–3. What is the appropriate distribution to use when the probability of an event is a constant?

6–4. What is the relationship between the binomial distribution and the normal distribution?

6–5. Why do Formulas 6–9 and 6–10 for the normal distribution use the standard deviation instead of some other measure of dispersion?

6–6. Formula 6–10 is used to compute the ordinate (Y) of the normal distribution for a given value of $X - \mu$. As the value of $X - \mu$ becomes larger, the value of Y becomes smaller. How large must the value of $X - \mu$ become to give a value of zero for Y? How is this significant?

6–7. When can the Poisson distribution be used to get a good approximation of a binomial probability?

6–8. Distinguish among the terms probability distribution, expected frequency distribution, and observed frequency distribution.

PROBLEMS

6–1. If a product is 5% defective, what is the probability that a sample of 10 from a very large lot will contain no defective items?

6–2. If a product is 10% defective, what is the probability that a sample of 12 will contain no more than 1 defective item?

6–3. If a product is 4% defective, what is the probability that a sample of 10 will contain more than 2 defectives?

6–4. Seven people are available for appointment to a committee of four. How many different committees can be appointed?

6–5. Assume that 70% of the families in a city are homeowners and 30% live in rented homes. If 4 families are drawn from all families in the city in such a manner that each one has the same chance of being selected, what is the probability of getting 4, 3, 2, 1, and 0 families who own homes? (Carry calculations to 4 decimals.)

6–6. A manufacturer believes that 95% of the items turned out by his plant will pass inspection. If this is true, what is the probability that in a sample of 8 items, 1 or fewer will be defective?

6–7. Three urns contain the following numbers of red and green balls:

	Urn 1	Urn 2	Urn 3
Red	4	3	10
Green	6	12	2
Total	10	15	12

Suppose 1 urn is selected at random and 3 balls are drawn at random, without replacement. If 2 of the balls selected are red, compute the probability that the selection was made from Urn 1. *Hint:* Use Bayes' theorem and the hypergeometric probability distribution.

6–8. A random sample of 100 is taken from a lot of 5,000 items. Assume that you want to know the probability of getting 5 or fewer defective units in the sample when the lot has 10% defective items.

(a) Would the hypergeometric be the correct distribution to use in this case? What would be a reason for not using the hypergeometric?

(b) What grounds would you have for assuming that the binomial can be used instead of the hypergeometric?

(c) If it is assumed that the distribution is the binomial, the probability of 5 or fewer defectives is .0576. However, the binomial is much more work than the Poisson or the normal distribution. How logical do you consider it to be to use one of these distributions as an approximation of the binomial?

6–9. A normal distribution has a mean of 600 and a standard deviation of 100. What proportion of the curve is above 650? Above 400? Above 550? Between 625 and 645? Less than 800?

6–10. Compute the probability of getting 6 or fewer defective units in the sample in Problem 6–8:

(a) On the assumption that the distribution is normal.

(b) On the assumption that the distribution follows the Poisson.

(c) Summarize the results and state the relative advantages and disadvantages of using the hypergeometric, the binomial, the Poisson, and the normal distribution.

6–11. A manufacturer claims that his product does not have more than 2% defective items. A random sample of 900 items is taken and tested. If 30 items are defective, do you believe his claim is correct? Show your calculations in arriving at this conclusion.

6–12. A testing station for driver's licenses has determined from past records that one half of the persons who write the test will pass it. If 5 people are writing the test at a given time, what is the probability that 4 of them will pass the test?

6–13. A mimeograph machine fails to print on every sheet. The average number of blank sheets has been computed to be 1% of the sheets run through the machine. What type of distribution describes this situation best? What would be the probability of no blank sheets in a run of 100 sheets through the mimeograph machine?

6–14. Assume that your experience in making interviews indicates that the probability of a family being at home on a given call is .60. An interviewer is instructed to make 4 calls on a home before giving the questionnaire to his supervisor with the report that no one was at home.

(a) How many interviews would you expect to have completed out of 1,000 schedules given to interviewers, assuming that each interviewer made the 4 calls required before reporting no one at home? Also assume that each time a call was made the probability of the family being at home was .60.

(b) Assume that an interviewer called at 5 houses in a given area. What is the probability of finding all 5 families at home? What is the probability of finding 4 out of the 5 families at home? What is the probability of finding fewer than 4 families at home?

6–15. The average number of telephone calls made per five-minute interval from a group of 6 pay telephones was 4.1. The five-minute intervals were all within a two-hour period in the middle of the day not including Saturday and Sunday. What is the probability of 3 or fewer calls being made from these telephones in a 5-minute interval?

6–16. Weather records for 10 stations over a 33-year period show a total of 363 excessive rainstorms in an area of the midwestern United States. An excessive rainstorm was defined as a 10-minute period having half an inch or more of rain. Since there were 10 stations and the record was kept for 33 years, the average number of excessive rainstorms per station-year (λ) equals $\frac{363}{330} = 1.1$.

(a) Using the Poisson, what is the probability that a weather station will not have rainfall totalling one-half inch or more in a 10-minute period during a year?

(b) What is the probability that a weather station will have 1 10-minute period with rainfall of one-half inch or more during a year? Two such cloudbursts? Three or more?

6–17. A manufacturing plant selects a random sample of 800 items drawn from a very large shipment of parts that contains 1% defective items. What is the probability that the sample will contain 12 or more defectives? Use both the normal distribution and the Poisson to answer this question. Which is the better distribution to use in this case? Why?

6–18. A radio repair shop has 10 small radios sitting on a shelf. Four of these radios have defects not visible to the casual observer. If a thief breaks into the shop and carries off half of the radios on the shelf:

(a) What is the probability that all 5 radios stolen are good ones?

(b) What is the probability that 3 of the 5 are defective?

(c) What is the probability that at least 3 of the 5 radios are good ones?

6–19. A consumer panel is made up of 4 men and 6 women. If 4 are selected at random from the panel, compute each of the following probabilities:

(a) All 4 are women.

(b) Half are women.

(c) Three are men.

(d) At least 2 are men.

(e) Women are in the minority.

6–20. A lazy worker spends 20% of his time on the job drinking coffee. If his boss checks up on him at 4 randomly selected times each day, how many times on the average would the worker be caught drinking coffee each day?

6–21. If 4 balanced, 6-sided dice are rolled simultaneously, compute the following probabilities:

(a) Two dice will land with the 6 up.

(b) None will land with a 3 up.

(c) The number on the top side of each die will be the same for all.

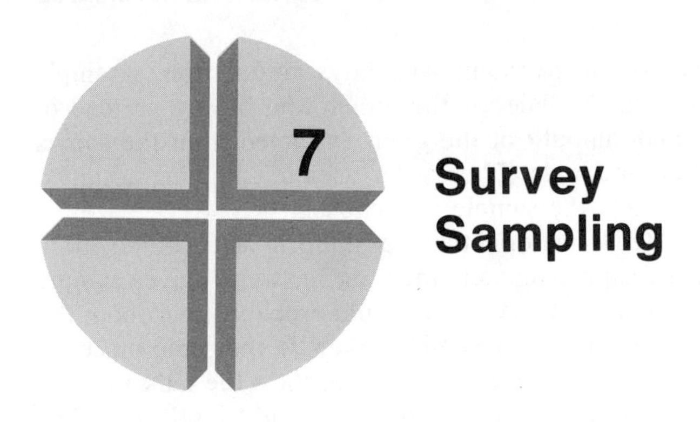

7 Survey Sampling

In the earlier discussion of the collection and analysis of statistical data it was assumed that information was collected from every one of the individual units in the population or universe. However, with the large populations that exist in the business world it would be extremely expensive to collect all the data needed by business if it were necessary to rely entirely on complete enumerations. The ability to make a reasonably accurate estimate from a sample of the information is an extremely valuable technique that is used in a wide range of business activities.

Many reports used in business must be based on a complete record of all transactions. This is particularly true of account information, but many types of internal statistics are also based on complete enumerations. On the other hand, auditors have always made use of sampling techniques, and samples are being used more and more frequently by accountants in collecting many types of data. In collecting external statistical data, samples are widely used because the cost of a complete enumeration is commonly too high to justify the collection of population data. In the past a major use of complete enumerations in compiling external data has been by the government in taking the census, but sampling methods are now being used to such a great extent by the Bureau of the Census that a complete enumeration is rapidly becoming the exception rather than the rule. Many techniques now used in sample surveys by businessmen were perfected by the Bureau of the Census to use instead of the expensive complete enumeration methods.

IMPORTANCE OF SAMPLING

The practice of making an estimate of the characteristics of a universe by examining a sample is a very old procedure, although a scientific approach to the problem is of recent origin. The cook who tastes a spoonful of soup in order to form a conclusion regarding the whole kettleful is using sampling procedures. The cotton merchant who buys cotton from a sample cut from only a few places in the bale, or the miller who buys a carload of wheat on the basis of a small amount of the grain extracted from the car, is putting his trust in sampling methods. The earlier uses of sampling assumed that it made little difference how the sample was chosen.

When sampling was limited to material that was uniform in its composition, any sample chosen would give approximately the same answer. A given sample of seawater taken in the middle of the Atlantic would probably give about the same concentration of salt as any other sample taken at the same place. A given group of ten people, however, would probably not show the same number unemployed as every other group of ten people that could be selected. The selection of a sample from a universe that shows great variation among the constituent elements offers many problems, but it is possible to make satisfactory estimates of the whole from a relatively small sample if the proper methods of selection are used.

The result obtained from analyzing a sample is never assumed to reflect exactly the result that would be secured had the entire universe been examined. However, if the sample is selected in accordance with proper sampling procedures, then it is possible to compute the maximum difference that may be expected to exist between the sample result and the result that would have been secured had the entire universe been analyzed. These computations involve the concept called sampling error, which will be discussed shortly. Sample data so treated are, in most cases, just as useful as a complete enumeration. For many purposes an approximation is all that is needed; therefore, estimates from a sample will serve the same purpose as a complete enumeration of the universe and will ordinarily entail considerably less expense.

A manufacturer of building materials who wants to select a magazine in which to advertise his products might consider as an important factor in his decision the percentage of homeowning subscribers to the magazines. Whether 71, 72, or 73% of the subscribers to a given magazine owned their homes would ordinarily make very little difference in his decision, although it would make a difference if only 30% owned their homes. The degree of approximation that can be tolerated in the data varies with the problem being considered, but it is always important to know the maximum amount of variation to be expected.

It is wasteful to use data that are more precise than needed for the purpose at hand, since the increase in precision increases the cost of the data without making them any more useful. It is doubtful if a very large proportion of external statistical data used by businessmen is worth the cost of making a complete enumeration. As a result, sample surveys are being substituted to an ever-increasing extent. The use of samples has reduced the cost of many data to a point where the businessman can afford them, whereas he would be forced to make decisions without the information if he had to rely on universe data.

The problem of taking a sample that can be depended upon to possess the characteristics of the universe, and the related problem of computing the maximum amount of sampling error that will be present in a given situation, are discussed in this chapter.

SAMPLING AND NONSAMPLING ERROR

Even though a complete enumeration of a universe is made, the information collected is unlikely to be entirely free from error. If business concerns are asked to submit information, it may not be exact; when consumers are surveyed about their buying habits or preferences, they commonly have trouble giving accurate information. This problem of errors in surveys was discussed in Chapter 2; although that discussion was oriented toward errors in complete enumerations, the comments are equally applicable to situations in which only a sample has been examined. It is important to understand that errors which are usually called *nonsampling errors* can occur in any collected data whether the method of collection is a complete enumeration or a sample from the population.

The extent to which sample results deviate from the results that would have been secured from a complete enumeration using the same data collection methods is called the *precision* of an estimate. This deviation is not a result of measurement error, but is a result of examining only a sample of the entire universe, and is called the *sampling error*. The total error in a sample survey, consisting of both the sampling and the nonsampling errors, is referred to as the degree of *accuracy*.

Methods have been developed for measuring sampling error, so even though it is not possible to eliminate error in an estimate, it is possible to determine the maximum degree of error present in an estimate. This permits a businessman to use the data with a full awareness of the degree of inaccuracy that was introduced by the use of a sample. Whenever an estimate from a sample is used, it is extremely important that a measure of the precision of the estimate be taken into account by the user.

SELECTION OF A SAMPLE

When a small portion of the universe is to be studied in order to estimate its characteristics, a great deal of importance attaches to the selection of the sample items. The first reaction of most people is that typical items should be selected if the sample is to be used to represent the universe. However, much experience has demonstrated that it is extremely difficult, if not impossible, for a person to make a selection of typical items. In fact, specifying "typical" items implies that one knows the answer; otherwise, one would not know if a particular item were typical.

Purposive or Judgment Sample

This approach to the problem of choosing a sample involves selecting individual items that are known to be typical with respect to certain characteristics, rather than selecting the items randomly. The assumption is made that if the items are typical in these respects, they will be typical with respect to the characteristics being studied. A sample selected in this manner is called a *purposive* or *judgment sample.* For example, in making a study of the buying habits of farmers in a given state, a sample of farms might be selected in such a manner that the average size of the farms in the sample would be the same as the average for the state; and the distribution of the sample farms among the different types would parallel the distribution throughout the state. Since the sample of farms would be typical with respect to these characteristics, it would be assumed that the answers given to questions on buying habits would also be typical of all farmers in the state, and the sample would be considered a good cross section of the universe.

While it cannot be stated positively that a sample selected as described in the preceding paragraph *will not* possess the characteristics of the universe, there is no way of demonstrating that it *will* be representative. The quality of the results can be evaluated only as a judgment of someone who has expert knowledge of the situation. It is not possible to compute objectively the precision of the estimates of the universe. Though a good deal of business data have been collected by using samples of this kind, the application of the data is limited by the fact that no measure of the precision of the results can justifiably be made.

Probability or Random Sample

When the probability of including each individual item in the sample is known, the precision of a sample result can be computed. If a person makes a decision that determines the selection of an individual item, there is no method of knowing the probability of including that item. Therefore, the sampling error of judgment samples cannot be computed. If the probability of selection is known, the sampling error can be computed; and in this case the sample is called a *probability* or a *random sample.*

The universe from which a random sample is to be drawn will often be a list of people, business establishments, or other names. When it is necessary to draw a random sample from items on a list, each item might be written on a card. In drawing the sample, the cards must be thoroughly mixed before the selection is made and this is often not easy to accomplish. For example, if the universe consists of the personnel file cards for 30,000 employees, there is no practicable method of mixing the cards thoroughly enough to permit a random sample to be drawn.

Undoubtedly the best method of taking a random sample from a list is to make use of *random numbers,* which consist of series of digits from 0 to 9.

Each digit occurs approximately the same number of times, and statisticians have accepted the order as random. (Several pages of random numbers are given in Appendix N.) If the universe consists of nine or less items, one column of numbers is used; if the universe consists of 10 to 99 items, two columns are used, etc. The numbers are read consecutively from the table. If a number is repeated, it is not used a second time. If a number larger than the largest number in the universe is selected, it is ignored. A sample of items selected in this manner will be random; that is, each item in the universe will have the same probability of being selected.

Systematic Sample

If a sample is to consist of 1% of the universe, the first item could be selected at random from the first 100 items, and then every 100th item throughout the remainder of the list. This is a *systematic sample,* which is the equivalent of a simple random sample when the items in the universe are in random order. An example of such a random order is a file of names in alphabetical order, if the items being studied bear no relationship to the names of the individuals. There are some dangers in using a systematic sample, but it is a widely used method of selecting a sample and in general is considered simple and foolproof.

PARAMETER AND STATISTIC

The value of a measure such as an aggregate, a mean, a median, a proportion, or a standard deviation compiled from a universe is called a *parameter.* The value of such a measure computed from a sample is called a *statistic.* The purpose of a sample survey is to estimate a given parameter from the corresponding statistic. A method of estimating a parameter from a statistic is available for the descriptive measures generally used in summarizing statistical data. For many of these measures, the statistic is the estimate of the parameter, although this is not true for all measures in every situation. It is possible to have the following three values for the same measure:

1. The statistic (derived from the sample)
2. The parameter (the universe value)
3. The estimate of the parameter derived from the statistic.

Since only in some cases will the statistic also be the best estimate of the parameter, it is always desirable to distinguish very carefully among these three values.

The general rule for symbols will be to use a Greek letter to represent a parameter and a Roman letter to represent a statistic. In Chapters 3 and 4 the mean and the standard deviation were designated by Greek letters since it was assumed that the measures were computed from complete enumerations.

The values of the X variable that are selected as a sample will be represented by x, and the mean of these samples, by $\bar{x}$. The standard deviation computed from a sample will be designated s, and the variance, s^2. Other symbols will be introduced as needed.

ESTIMATES OF THE MEAN AND THEIR PRECISION

In order to demonstrate how the mean of a random sample may be used to estimate the mean of the universe, and how the precision of the estimate may be computed, a small hypothetical universe will be used as a case study. Assume that the following semimonthly expenditures for utilities comprise the universe, and the problem is to estimate the average expenditure of the ten-family universe from a sample of two families.

Family	Semimonthly Expenditures
1	$ 74.00
2	47.00
3	37.00
4	90.00
5	84.00
6	40.00
7	51.00
8	54.00
9	66.00
10	59.00
Total	$602.00

Distribution of Sample Means

There are 45 different samples of 2 items each that can be drawn from a universe of 10. Formula 6–1 was used to compute n items taken r at a time. The formula used to show the combinations of N items taken n at a time is

$$_{N}C_{n} = \frac{N!}{n!(N-n)!} \qquad _{10}C_{2} = \frac{10!}{2!8!} = 45.$$

When the selection is random, any one of the 45 samples is equally likely to be drawn. The 45 samples are listed in the first column of Table 7–1, and the means of any one of these samples is represented by the symbol $\bar{x}$. Any one of the means in this table can be used as an estimate of the universe mean.

The smallest estimate of the mean expenditure is found in the sample consisting of families 3 and 6, which have an average of $38.50. The largest estimate would be obtained if families 4 and 5, with an average of $87.00, were drawn for a sample. The mean of the 45 sample means is $60.20 ($2,709 ÷ 45), which is the arithmetic mean of the universe ($602.00 ÷ 10).

TABLE 7–1

MEANS OF ALL POSSIBLE SAMPLES OF TWO FROM A UNIVERSE OF TEN FAMILIES

Families in Sample	Samples Values x	Mean $\bar{x}$	$\bar{x}^2$
1 and 2	74 and 47	60.50	3,660.25
1 and 3	74 and 37	55.50	3,080.25
1 and 4	74 and 90	82.00	6,724.00
1 and 5	74 and 84	79.00	6,241.00
1 and 6	74 and 40	57.00	3,249.00
1 and 7	74 and 51	62.50	3,906.25
1 and 8	74 and 54	64.00	4,096.00
1 and 9	74 and 66	70.00	4,900.00
1 and 10	74 and 59	66.50	4,422.25
2 and 3	47 and 37	42.00	1,764.00
2 and 4	47 and 90	68.50	4,692.25
2 and 5	47 and 84	65.50	4,290.25
2 and 6	47 and 40	43.50	1,892.25
2 and 7	47 and 51	49.00	2,401.00
2 and 8	47 and 54	50.50	2,550.25
2 and 9	47 and 66	56.50	3,192.25
2 and 10	47 and 59	53.00	2,809.00
3 and 4	37 and 90	63.50	4,032.25
3 and 5	37 and 84	60.50	3,660.25
3 and 6	37 and 40	38.50	1,482.25
3 and 7	37 and 51	44.00	1,936.00
3 and 8	37 and 54	45.50	2,070.25
3 and 9	37 and 66	51.50	2,652.25
3 and 10	37 and 59	48.00	2,304.00
4 and 5	90 and 84	87.00	7,569.00
4 and 6	90 and 40	65.00	4,225.00
4 and 7	90 and 51	70.50	4,970.25
4 and 8	90 and 54	72.00	5,184.00
4 and 9	90 and 66	78.00	6,084.00
4 and 10	90 and 59	74.50	5,550.25
5 and 6	84 and 40	62.00	3,844.00
5 and 7	84 and 51	67.50	4,556.25
5 and 8	84 and 54	69.00	4,761.00
5 and 9	84 and 66	75.00	5,625.00
5 and 10	84 and 59	71.50	5,112.25
6 and 7	40 and 51	45.50	2,070.25
6 and 8	40 and 54	47.00	2,209.00
6 and 9	40 and 66	53.00	2,809.00
6 and 10	40 and 59	49.50	2,450.25
7 and 8	51 and 54	52.50	2,756.25
7 and 9	51 and 66	58.50	3,422.25
7 and 10	51 and 59	55.00	3,025.00
8 and 9	54 and 66	60.00	3,600.00
8 and 10	54 and 59	56.50	3,192.25
9 and 10	66 and 59	62.50	3,906.25
Total		2,709.00	168,929.00

The mean of all possible sample means will always be exactly equal to the universe mean. This is true for any population no matter what the size of the sample. The reason for this can be seen by checking the 45 samples of 2 items each listed in Table 7–1. Each of the 10 items in the universe is included in 9 samples; thus, each of the items in the universe is given equal weight in computing the mean of the sample means. Each of the items in the universe is also given equal weight in the straightforward computation of the universe mean. Since the same set of values are being equally weighted in two computations, the means resulting from the computation will be the same.

Each of the sample means is an independent estimate of the universe mean, and the average of all the estimates is referred to as the *expected value* of the sample mean. When the expected value of the sample mean equals the universe value, the sample estimates are said to be unbiased. This does not mean that each sample gives a completely *accurate* estimate of the universe mean, for it can be seen from Table 7–1 that many of the sample means differ substantially from the mean of the universe. The method of estimating is unbiased in the sense that if it is used repeatedly, the average of all the estimates will equal the universe mean.

Precision of Estimates

The extent to which an individual estimate may differ from the universe value is the *precision* of the estimate. Some idea of the confidence that may be placed in a single estimate can be obtained by examining the 45 estimates of the mean shown in Table 7–1. The estimates tend to cluster around the universe mean of $60.20, although a few of them depart substantially from this value. A method of summarizing how much the individual means scatter about the universe mean is needed.

The basis for evaluating the precision of an estimate is a measure of dispersion among the sample means. In other words, how much do the 45 sample means vary from the mean of the universe? In Chapter 4 the various measures of dispersion were described and it was pointed out that the best measure of dispersion to use is the standard deviation. This suggests that the standard deviation of the means of the 45 samples be computed as a measure of the dispersion of the sample means. Treating each of the sample means as a value of the variable, the standard deviation can be computed by Formula 4–12.

The standard deviation of the distribution of the means of all possible samples ($\sigma_{\bar{x}}$) is distinguished from the standard deviation of the universe (σ) by the addition of the subscript $\bar{x}$. The term $\sigma_{\bar{x}}$ is called the *standard error of the mean,* and is the measure of precision of a sample. When a standard deviation is calculated from a distribution of sample estimates, such as those shown in Table 7–1, it is called the "standard error." The term "standard deviation" is used when the computation is based on individual measurements.

In Table 7–1 the sum of the squares of the 45 means is 168,929.0000 and the sum of the means is 2,709.00. The standard deviation is computed

$$\sigma^2_{\bar{x}} = \frac{N\Sigma\bar{x}^2 - (\Sigma\bar{x})^2}{N^2}$$

$$\sigma^2_{\bar{x}} = \frac{(45)(168{,}929) - (2{,}709)^2}{45^2} = \frac{7{,}601{,}805 - 7{,}338{,}681}{2{,}025}$$

$$\sigma^2_{\bar{x}} = \frac{263{,}124}{2{,}025} = 129.9378$$

$$\sigma_{\bar{x}} = 11.40.$$

Confidence Intervals

In the preceding section is was stated that the standard error of the mean indicates the degree to which the sample means will vary from the universe mean. Another method of indicating the precision of the estimates from a sample is to determine the probability that a sample mean will fall within a given interval about the universe mean. It can be shown that regardless of the form of the distribution of sample means, the probability of a sample mean exceeding the 3σ interval is less than .11.[1] If the distribution of sample means is normal, the probability is only .0027 that an individual sample mean will differ by more than three times the standard deviation of the sample means.[2] Although the distribution of sample means may not be distributed exactly as the normal distribution, it tends to approach the form of the normal distribution fairly rapidly as the size of the sample is increased. This means that the probability of a sample mean being outside the 3σ interval is nearer to .0027 than to .11. It is generally assumed to be practically certain that a given sample mean will fall within the 3σ interval.

Referring to the example of the 45 sample means in Table 7–1, the standard deviation of this distribution of means (the standard error of the mean) was found to be $11.40. It follows that practically all the sample means should fall within the interval $\mu - 3\ \sigma_{\bar{x}}$ to $\mu + 3\ \sigma_{\bar{x}}$. This interval is $26.00 to $94.40, computed

$$60.20 \pm 3(11.40) = 60.20 \pm 34.20.$$

The largest sample mean in Table 7–1 is $87.00 and the smallest is $38.50. Thus, all the sample means fall well within the 3σ interval.

In the discussion of the normal curve in Chapter 6, it was shown that approximately 68% of the items in a normal distribution will fall within an interval of $\pm 1\sigma$ from the mean of the distribution, that approximately 95% of the items will fall within $\pm 2\sigma$ of the mean, and finally that approximately 99.73% will fall within $\pm 3\sigma$. Thus, if $1\sigma_{\bar{x}}$ is added to μ and $1\sigma_{\bar{x}}$ is subtracted from μ,

[1] This is known as Tchebycheff's inequality. For a fuller discussion, see *Sampling Survey Methods and Theory*, Volume II, Chapter 3, Section 7, Theorem 18, by Morris H. Hansen, William N. Hurwitz, and William A. Madow (New York: John Wiley & Sons, Inc.). 1953.

[2] See discussion of the normal distribution starting on page 131.

that interval should be expected to contain 68% of the sample means. In other words, the probability that any sample mean falls within $\pm 1\sigma_{\bar{x}}$ of μ is .68. If $2\sigma_{\bar{x}}$ is added to μ and $2\sigma_{\bar{x}}$ subtracted, it can be said that the probability of any sample mean falling within this interval is .95. Finally if $3\sigma_{\bar{x}}$ is added to and subtracted from μ, it can be concluded that the probability of any sample mean falling within that interval is .9973. This probability value of .9973 is the equivalent to our phrase "practically certain" used earlier. If one is willing to be less certain, two or possibly even one $\sigma_{\bar{x}}$ can be added to and subtracted from μ. For the introductory discussions in this chapter the "practically certain" or $3\sigma_{\bar{x}}$ criterion will be employed. In Chapter 8 material will be presented which will demonstrate the procedures and rationale for using less stringent criteria.

The standard deviation of the distribution of sample means based on samples of 2 was computed to be $11.40. When the size of the sample is increased, the standard deviation of the distribution of sample means decreases. This is demonstrated in Table 7–2, which gives the values of the standard deviation for sample means based on sample sizes of 2 through 9.

The practical application of this principle is of great importance in applied statistics. If the confidence interval is too large for the purposes for which the data are being collected, this margin of error can be decreased by increasing the size of the sample. As the sample size increases, the estimates will deviate less and less from the universe value being estimated. This fact is described by saying that random samples give *consistent* estimates. This is an important characteristic of estimates from samples, since it permits achieving any desired level of precision by increasing the size of the sample. The amount of error that can be tolerated in an estimate depends on the use that will be made of the estimate. For some purposes it may be desirable that the estimate would vary no more than 2% from the true value, while in other situations a much larger variation might create no problem. It is wasteful to use a degree of precision greater than needed in a given case.

TABLE 7–2

STANDARD DEVIATIONS OF DISTRIBUTIONS OF MEANS

Sample Sizes 2 to 9

Size of Sample	Standard Deviation of the Distribution of Means
2	11.40
3	8.71
4	6.98
5	5.70
6	4.65
7	3.73
8	2.85
9	1.90

Computation of $\sigma_{\bar{x}}$

The previous discussion aimed at demonstrating the significance of measures of sampling error, but the method described cannot be used in a practical case. Taking all possible samples from even a small universe would be absurd, since it would be so much simpler to make a complete enumeration of the universe. The method was used to demonstrate how the dispersion in the sample means serves as a measure of the precision of an estimate made from a sample.

Actually it is not necessary to list all the possible samples in order to compute the standard deviation of the distribution of sample means. If the standard deviation of the universe is known from other sources, the variance and the standard error of the mean can be computed

$$\sigma_{\bar{x}}^2 = \frac{\sigma^2}{n} \cdot \frac{N-n}{N-1} \tag{7-1}$$

$$\sigma_{\bar{x}} = \frac{\sigma}{\sqrt{n}} \sqrt{\frac{N-n}{N-1}}. \tag{7-2}$$

If the standard deviation of the universe is not known, it can be estimated from the sample. This latter procedure will be described in the next section.

When the data on semimonthly expenditures for utilities by 10 families are analyzed, the mean expenditure is found to be \$60.20; the variance, \$292.36; and the standard deviation, \$17.10. The size of the universe is 10, and the size of the sample is 2. The following computations give a value of \$11.40 for the standard error of the mean. This is the same as the value computed on page 157 for the standard deviation of the distribution of sample means.

$$\sigma_{\bar{x}}^2 = \frac{292.36}{2} \cdot \frac{10-2}{10-1} = \frac{292.36}{2} \cdot \frac{8}{9} = 129.94$$

$$\sigma_{\bar{x}} = \sqrt{129.94} = 11.40.$$

Using Formula 7–2, the standard error of the mean is computed

$$\sigma_{\bar{x}} = \frac{17.10}{\sqrt{2}} \sqrt{\frac{10-2}{10-1}} = \frac{17.10}{1.414} \sqrt{\frac{8}{9}} = 11.40.$$

Relative Precision

Instead of computing the average amount of dispersion in the sample means, it may be desirable to express the dispersion as a ratio to or a proportion of the mean that is being estimated. This measure gives a relative measure of dispersion rather than the absolute dispersion of the sample means. The coefficient of variation (V) described in Chapter 4 is used for this purpose and is computed in the same manner as shown in Chapter 4. From the mean expenditure of \$60.20 and the standard error of \$11.40, the coefficient of variation of the distribution of means is computed

$$V_{\bar{x}} = \frac{\sigma_{\bar{x}}}{\mu} 100 \tag{7-3}$$

$$V_{\bar{x}} = \frac{11.40}{60.20} 100 = 19\%.$$

This computation is valid since the standard error of the mean is the standard deviation of the distribution of sample means. The subscript to V is used to identify the value as having been computed for the distribution of sample means. Instead of saying that the sample means will vary on the average \$11.40 from the universe mean, it can be said that the sample means will vary on the average 19% from the universe mean. In many situations the expression in relative form will be more useful. It is possible to express the standard error of any estimate in relative form as well as in absolute units. This is done in the same manner as in the example above by expressing the standard error as a decimal fraction (or percentage) of the value that has been estimated.

MEASURING THE PRECISION OF A SAMPLE ESTIMATE FROM THE SAMPLE

When one knows the standard deviation of the universe from which a sample is drawn, it is a simple matter to compute the standard error of the mean ($\sigma_{\bar{x}}$) by Formulas 7–1 and 7–2. At this point, however, the question inevitably arises as to how one can know the standard deviation of the universe with which he is dealing. If the problem is to estimate the arithmetic mean of the universe, it is certain that the mean is unknown. If this is true, it does not seem likely that the standard deviation of that universe will be known. However, the standard deviation of the universe can be estimated from the same sample that was used to estimate the mean of the universe. It seems reasonable to assume that if the sample will give a satisfactory estimate of the universe mean, it should give an equally good estimate of the universe standard deviation. The mean of a sample is an unbiased estimate of the universe mean, but the standard deviation of the sample is not an unbiased estimate of the universe standard deviation, although with the proper adjustment an unbiased estimate can be derived from the sample.

The universe consisting of the expenditures for utilities by 10 families will be used to present the subject of estimating the variance of a universe from a sample. This discussion deals with the variance, since the standard deviation can be computed from the variance simply by taking the square root. Each of the samples of 2 shown in Table 7–3 may be used to estimate the variance, just as each sample was used to estimate the mean in Table 7–1. Column

TABLE 7–3

VARIANCES OF ALL POSSIBLE SAMPLES OF TWO FROM A UNIVERSE OF TEN FAMILIES, AND MEAN OF SQUARED DEVIATIONS FROM UNIVERSE MEAN

Families in Sample (1)	Sample Values (2)	Mean (3)	s^2 (4)	Squared Deviations from Universe Mean (60.20) (5)	Mean of Squared Deviations from 60.20 (6)
1 and 2	74; 47	60.50	182.25	190.44 and 174.24	182.34
1 and 3	74; 37	55.50	342.25	190.44 and 538.24	364.34
1 and 4	74; 90	82.00	64.00	190.44 and 888.04	539.24
1 and 5	74; 84	79.00	25.00	190.44 and 566.44	378.44
1 and 6	74; 40	57.00	289.00	190.44 and 408.04	299.24
1 and 7	74; 51	62.50	132.25	190.44 and 84.64	137.54
1 and 8	74; 54	64.00	100.00	190.44 and 38.44	114.44
1 and 9	74; 66	70.00	16.00	190.44 and 33.64	112.04
1 and 10	74; 59	66.50	56.25	190.44 and 1.44	95.94
2 and 3	47; 37	42.00	25.00	174.24 and 538.24	356.24
2 and 4	47; 90	68.50	462.25	174.24 and 888.04	531.14
2 and 5	47; 84	65.50	342.25	174.24 and 566.44	370.34
2 and 6	47; 40	43.50	12.25	174.24 and 408.04	291.14
2 and 7	47; 51	49.00	4.00	174.24 and 84.64	129.44
2 and 8	47; 54	50.50	12.25	174.24 and 38.44	106.34
2 and 9	47; 66	56.50	90.25	174.24 and 33.64	103.94
2 and 10	47; 59	53.00	36.00	174.24 and 1.44	87.84
3 and 4	37; 90	63.50	702.25	538.24 and 888.04	713.14
3 and 5	37; 84	60.50	552.25	538.24 and 566.44	552.34
3 and 6	37; 40	38.50	2.25	538.24 and 408.04	473.14
3 and 7	37; 51	44.00	49.00	538.24 and 84.64	311.44
3 and 8	37; 54	45.50	72.25	538.24 and 38.44	288.34
3 and 9	37; 66	51.50	210.25	538.24 and 33.64	285.94
3 and 10	37; 59	48.00	121.00	538.24 and 1.44	269.84
4 and 5	90; 84	87.00	9.00	888.04 and 566.44	727.24
4 and 6	90; 40	65.00	625.00	888.04 and 408.04	648.04
4 and 7	90; 51	70.50	380.25	888.04 and 84.64	486.34
4 and 8	90; 54	72.00	324.00	888.04 and 38.44	463.24
4 and 9	90; 66	78.00	144.00	888.04 and 33.64	460.84
4 and 10	90; 59	74.50	240.25	888.04 and 1.44	444.74
5 and 6	84; 40	62.00	484.00	566.44 and 408.04	487.24
5 and 7	84; 51	67.50	272.25	566.44 and 84.64	325.54
5 and 8	84; 54	69.00	225.00	566.44 and 38.44	302.44
5 and 9	84; 66	75.00	81.00	566.44 and 33.64	300.04
5 and 10	84; 59	71.50	156.25	566.44 and 1.44	283.94
6 and 7	40; 51	45.50	30.25	408.04 and 84.64	246.34
6 and 8	40; 54	47.00	49.00	408.04 and 38.44	223.24
6 and 9	40; 66	53.00	169.00	408.04 and 33.64	220.84
6 and 10	40; 59	49.50	90.25	408.04 and 1.44	204.74
7 and 8	51; 54	52.50	2.25	84.64 and 38.44	61.54
7 and 9	51; 66	58.50	56.25	84.64 and 33.64	59.14
7 and 10	51; 59	55.00	16.00	84.64 and 1.44	43.04
8 and 9	54; 66	60.00	36.00	38.44 and 33.64	36.04
8 and 10	54; 59	56.50	6.25	38.44 and 1.44	19.94
9 and 10	66; 59	62.50	12.25	33.64 and 1.44	17.54
Total		2,709.00	7,309.00		13,156.20

TABLE 7–3 (concluded)

$$\text{Mean of the sample variances in Column 4} = \frac{7{,}309.0}{45} = 162.42$$

$$\text{Mean of the sample variances in Column 6} = \frac{13{,}156.20}{45} = 292.36.$$

4 of Table 7–3 gives the variance for each of the 45 samples that can be drawn from a universe of 10. Each of these estimates of the variance could be used to estimate the variance of the universe, which is known to be 292.36.

The expected value (arithmetic mean) of all possible sample means was shown in Table 7–1 to be equal to the universe mean of $60.20, but it can be seen from Column 4 of Table 7–3 that this is not true of the variance. The mean of the variances estimated from the 45 samples is found to be 162.42, compared to the known variance of 292.36.

The variance of each sample in Column 4 of Table 7–3 was computed from the mean of each sample. In the absence of any knowledge of the value of the universe mean, this is the only possible course. If the universe mean were known, the estimate of the universe variance could be made by computing the deviations of the individual values in the sample from the universe mean. In every case, except when the sample mean happened to coincide with the universe mean, the resulting estimate of the variance of a sample computed from the sample mean would be different from the estimate of the variance that would be obtained if the universe mean had been used.

In Column 5 the deviations were taken from the universe mean instead of from the mean of each sample as was done in computing the value in Column 4. It seems reasonable to assume that the estimate of the variance made by first estimating the mean and then taking deviations of the sample values from this estimated mean would be less accurate than if the deviations had been measured from the correct universe mean. It can be verified that every one of the 45 estimates of the variance in Column 6 is larger than the estimate from the same sample in Column 4. The closer the sample mean is to the universe mean the less the difference between the estimates; if any of the sample means had been exactly the same as the universe mean, the two estimates of the variance would be the same.

The arithmetic mean of the 45 variances in Column 6 is 292.36. The fact that the mean of the variances in Column 4 is considerably smaller (162.42) is to be expected for the following reason. The sum of the squared deviations from the mean of a set of numbers is less than the sum of the squared deviations from any other value. From this it follows that the sum of the squared deviations used in computing the variances in Column 4 were less than those used in Column 6. By being forced to use the mean of the sample in estimating the variance, a biased result is secured; the variance computed from a sample, using the mean of that sample, will always be less than if the deviations were computed from the mean of the universe. It was demonstrated that the mean of the variances in Column 6, which were computed from the universe mean, equals the correct universe variance. It was also demonstrated that the mean of the variances in Column 4, which were computed from the mean of each sample, is less than the correct universe variance.

It would be incorrect to use the biased estimates of the standard deviation, since they would on the average understate the universe variance. However, if it is possible to determine the extent of the bias in these estimates, a correction could be made and the bias removed. An analogous situation would arise if a series of measurements had been made with a measuring tape with one inch of its length cut off. On the average the measurements would be one inch too short, but when this fact was discovered, it would not be necessary to repeat the measuring with an accurate tape. An adjustment could be made in the measurements by simply adding one inch to each of them. In the same manner, the extent of the bias in using the variance of the sample is known, so it is possible to adjust the biased sample results to secure unbiased estimates of the parameter.

It can be demonstrated that for an *infinitely large universe* the variance equals $\frac{n}{n-1}$ times the expected value (mean) of the sample variance. This may be written as

$$\sigma^2 = E\left(s^2 \frac{n}{n-1}\right),$$

which is valid only for an infinite universe or sampling with replacement (see footnote 4). If the variance computed from any given sample is multiplied by $\frac{n}{n-1}$, the result will be an unbiased estimate of the universe variance. This relationship can be written as the following formula for estimating the variance of the universe from the variance computed from a sample, using σ^2 to represent this estimate.

$$\hat{\sigma}^2 = s^2 \frac{n}{n-1}. \qquad \textbf{(7–4)}$$

The estimated variance will be shown by adding a caret (ˆ) to the symbol for the variance of a sample, writing the estimated variance as $\hat{\sigma}^2$. An unbiased estimate of the standard deviation of the universe is the square root of the unbiased estimate of the variance.

$$\hat{\sigma} = \sqrt{s^2 \frac{n}{n-1}}. \qquad \textbf{(7–5)}$$

If it is known that the standard deviation of the sample will be used as an estimate of the universe standard deviation, the correction can be made at the time the standard deviation is computed. If this is done, the formula is

$$\hat{\sigma} = \sqrt{\frac{\Sigma(x - \bar{x})^2}{n - 1}}. \qquad \textbf{(7–6)}$$ [3]

The expected value of the sample variances in Column 4, Table 7–3, was found to be 162.42, compared to the universe variance of 292.36 computed from the finite universe of 10 families. If the mean of the sample variance times $\frac{n}{n-1}$ equaled σ^2, multiplying 162.42 by $\frac{n}{n-1}$ would correct for the bias in the sample variances. However, 162.42 times $\frac{2}{2-1}$ equals 324.84. This is higher than the universe variance of 292.36, which means that in this case the factor $\frac{n}{n-1}$ *overcorrects* for the bias. The amount of overcorrection varies with the size of the universe, becoming insignificant for a large universe. Only for an infinite universe or sampling with replacement[4] does $\frac{n}{n-1}$ times a sample variance give an unbiased estimate of the universe variance. The overcorrection in a finite universe can be allowed for by multiplying the estimate by $\frac{N-1}{N}$, as shown in the following calculations.

$$\sigma^2 = E\left[\left(s^2 \frac{n}{n-1}\right)\left(\frac{N-1}{N}\right)\right]$$

$$= 324.84 \, \frac{10 - 1}{10}$$

$$= 292.36.$$

This is the correct value of the universe variance. It can be seen that $\frac{N-1}{N}$ would have very little effect when the universe is large, and no effect when a universe is infinite.

Substituting $E\left[\left(s^2 \frac{n}{n-1}\right)\left(\frac{N-1}{N}\right)\right]$ for σ^2, Formula 7–1 becomes

$$\sigma_{\bar{x}}^2 = E\left[\left(s^2 \frac{n}{n-1}\right)\left(\frac{N-1}{N}\right)\left(\frac{1}{n}\right)\left(\frac{N-n}{N-1}\right)\right].$$

---

[3] This formula is derived as

$$\hat{\sigma} = \sqrt{s^2 \frac{n}{n-1}} = \sqrt{\frac{\Sigma(x - \bar{x})^2}{n} \frac{n}{n-1}} = \sqrt{\frac{\Sigma(x - \bar{x})^2}{n - 1}}.$$

[4] *Sampling with replacement* is in effect the same as sampling from an infinite universe. If each item drawn in a sample is replaced before the next sample is taken, an infinite number of samples of a given size can be taken from a finite population. Sampling without replacement is the normal method of making sample surveys.

It can be demonstrated that the formula gives the correct value for $\sigma_{\bar{x}}^2$ as

$$\sigma_{\bar{x}}^2 = 162.42 \cdot \frac{2}{2-1} \cdot \frac{10-1}{10} \cdot \frac{1}{2} \cdot \frac{10-2}{10-1} = 292.36 \cdot \frac{1}{2} \cdot \frac{10-2}{10-1}$$

$$= \frac{292.36}{2} \cdot \frac{8}{9}$$

$$= 129.94.$$

This is the value of $\sigma_{\bar{x}}^2$ given by Formula 7.1, page 159.

If one sample value of s^2 is used instead of the expected value, the result will be an unbiased estimate of $\sigma_{\bar{x}}^2$, since the mean of all possible estimates would equal $\hat{\sigma}^2$. The formulas for an estimate of the variance of the mean would be

$$\hat{\sigma}_{\bar{x}}^2 = s^2 \cdot \frac{n}{n-1} \cdot \frac{N-1}{N} \cdot \frac{1}{n} \cdot \frac{N-n}{N-1}.$$

Cancelling $N-1$ from the numerator and the denominator gives

$$\hat{\sigma}_{\bar{x}}^2 = s^2 \cdot \frac{n}{n-1} \cdot \frac{1}{n} \cdot \frac{N-n}{N}.$$

Since $\hat{\sigma}^2 = s^2 \cdot \dfrac{n}{n-1}$, a more compact formula would be

$$\hat{\sigma}_{\bar{x}}^2 = \hat{\sigma}^2 \cdot \frac{1}{n} \cdot \frac{N-n}{N}.$$

The term $\dfrac{N-n}{N}$ reduces to $1-\dfrac{n}{N}$, and letting $f=\dfrac{n}{N}$ (called the sampling fraction) gives the formulas that will be used for estimating the variance and the standard error of the mean from a sample.

$$\hat{\sigma}_{\bar{x}}^2 = \frac{\hat{\sigma}^2}{n}(1-f) \quad \textbf{(7-7)}$$

and

$$\hat{\sigma}_{\bar{x}} = \frac{\hat{\sigma}}{\sqrt{n}}\sqrt{1-f}. \quad \textbf{(7-8)}$$

Formula 7–7 differs from 7–1 in that $\hat{\sigma}^2$ (the universe variance estimated from the sample) is used instead of σ^2 (the universe variance), and $1-f$ is used instead of $\dfrac{N-n}{N-1}$. Both of these terms are known as the *finite multiplier;* the one used depends upon whether the universe variance is known or must be estimated from a sample.

IMPORTANCE OF THE SIZE OF THE UNIVERSE

When persons unfamiliar with sampling first consider the question of the influence of the size of the universe on the precision of sampling results, the intuitive answer seems to be that a sample of a given size, say 1,000, will give a much more precise result when drawn from a universe of 100,000 than when drawn from 800,000. In the first case the sample is one out of every 100 items in the universe, while in the second universe the sample is one out of every 800 items. It appears obvious that the first sample will be more precise than the second. However, if the dispersion in the two universes, as measured by their standard deviations, is the same, the precision of the sample result from a sample of 1,000 will be practically the same for the larger universe as for the smaller.

Formula 7–7 for the standard error of the mean can be used to demonstrate that the statement in the preceding paragraph is true. Assume that 8,295 was the average number of miles driven last year by a sample of 1,000 car owners from a universe of 100,000, and that the standard deviation of the universe was estimated from the sample study to be 3,100 miles. The standard error of the estimate of the average number of miles driven by the 100,000 car owners in the universe would be

$$f = \frac{n}{N} = \frac{1{,}000}{100{,}000} = .01$$

$$\hat{\sigma}_{\bar{x}} = \frac{\hat{\sigma}}{\sqrt{n}} \sqrt{1-f} = \frac{3{,}100}{\sqrt{1{,}000}} \sqrt{1-.01} = 98.03\sqrt{.99} = 98.03 \cdot .995 = 97.5.$$

If the sample of 1,000 had been taken from a universe of 800,000 car owners, the standard error of the sample mean would be

$$f = \frac{n}{N} = \frac{1{,}000}{800{,}000} = .00125$$

$$\hat{\sigma}_{\bar{x}} = 98.03\sqrt{1-.00125} = 98.03\sqrt{.99875} = 98.03 \cdot .999 = 97.9.$$

For all practical purposes the sampling error is the same for the two samples since in each case the second term in the equation approximates unity. Whenever the size of the sample is small in relation to the size of the universe, the finite multiplier will be close to unity and can be ignored in computing the standard error. When the sampling fraction $\frac{n}{N}$ does not exceed 5%, it is customary to ignore the finite multiplier. Sometimes it is ignored when the sampling fraction is as high as 10%, since it results in slightly overstating the standard error. This means that the precision of the estimate from a sample is somewhat greater than shown by the computed standard error.

If the size of the universe is not known, the size of the sampling fraction should be approximated and the term $1-f$ used in the formula. If it is not possible even to approximate the size of N, it should be determined whether or not N is large in relation to n. If it is, it will not cause any serious error in the computation of the confidence interval if $1 - f$ is assumed to be equal to unity.

EXAMPLE OF ESTIMATING A UNIVERSE MEAN FROM A SAMPLE

The following example illustrates the estimation of a universe mean from a sample, and the determination of the precision of this estimate from data supplied by the sample. Table 7–4 gives a frequency distribution of monthly salaries of file clerks in Tennessee. The sample was selected at random from a population of approximately 4,000 file clerks for the purpose of estimating their average straight-time monthly earnings. The mean of the sample is represented by the symbol $\bar{x}$, and the mean of the universe by μ. The unbiased estimate of μ may be written $\hat{\mu}$, to conform to the symbols used for the standard deviation. But since $\bar{x} = \hat{\mu}$, the unbiased estimate of the mean will be represented by $\bar{x}$ in this text. The arithmetic mean of the sample is computed in Table 7–4 to be \$407.14, which is an unbiased estimate of the universe mean.

TABLE 7–4

COMPUTATION OF THE MEAN AND STANDARD DEVIATION FROM A SAMPLE

Monthly Salaries of File Clerks in Tennessee

Average Monthly Earnings (Dollars)	Number f	d'	fd'	$f(d')^2$
250 and under 300	6	−3	−18	54
300 and under 350	41	−2	−82	164
350 and under 400	70	−1	−70	70
400 and under 450	16	0	0	0
450 and under 500	17	1	17	17
500 and under 550	14	2	28	56
500 and under 600	12	3	36	108
600 and under 650	6	4	24	96
Total	182		−65	565

Source: Adapted from U.S. Department of Labor, Regional Report 17 (May, 1972).

$$\bar{x} = A + \frac{\Sigma fd'}{n} i = 425 + \frac{-65}{182} 50 = 425 - 17.8571 = 407.14$$

$$s^2 = i^2 \left[\frac{\Sigma f(d')^2}{n} - \left(\frac{\Sigma fd'}{n}\right)^2\right] = 50^2 \left[\frac{565}{182} - \left(\frac{-65}{182}\right)^2\right] = 2{,}500 \left(3.1044 - \frac{4{,}225}{33{,}124}\right)$$

$$s^2 = (2{,}500)(3.1044 - .1276) = (2{,}500)(2.9768) = 7{,}442.00$$

$$\sigma = \sqrt{s^2 \frac{n}{n-1}} = \sqrt{7{,}442.00 \frac{182}{182-1}} = \sqrt{7{,}483.12} = 86.51$$

The standard deviation computed from the sample is \$86.51, so $s = \$86.51$. The standard deviation of the universe is estimated from the sample standard deviation by Formula 7–5.

$$\hat{\sigma} = \sqrt{s^2 \frac{n}{n-1}} = \sqrt{7{,}442.00 \frac{182}{182-1}} = \sqrt{7{,}483.12} = 86.51.$$

Although the correction for the bias in estimating the standard deviation is very slight, it is wise to use it. In practice it is frequently ignored when the sample is as large as 182, but for small samples the correction is important. Rather than having to decide each time whether or not to use the correction, it is probably easier always to use it.

Using Formula 7–8, the standard error of the mean is estimated to be

$$\hat{\sigma}_{\bar{x}} = \frac{86.51}{\sqrt{182}} \sqrt{1 - \frac{182}{4{,}000}} = \frac{86.51}{13.49} \sqrt{\frac{4{,}000 - 182}{4{,}000}}$$
$$= (6.41)(\sqrt{.9545}) = 6.26.$$

Since the sample of 182 is a very small fraction of the universe, the finite multiplier has very little effect on the standard error of the mean. It is so near unity that it would normally be ignored. When the sampling fraction is as small as in this sample, very little in accuracy is gained by using the finite multiplier. As stated previously, when the sampling fraction is less than .05, it is customary to ignore it.

A measure of relative dispersion is provided by the coefficient of variation computed by Formula 7–9, which substitutes the values derived from the sample for the universe values used in Formula 7–3.

$$\boldsymbol{v_{\bar{x}} = \frac{\hat{\sigma}_{\bar{x}}}{\bar{x}} 100} \qquad \textbf{(7–9)}$$

$$v_{\bar{x}} = \frac{6.26}{407.14} 100 = 1.54\%.$$

The only difference between Formulas 7–3 and 7–9 is that the latter uses estimates from the sample instead of universe values.

The estimate of the mean earnings of file clerks in Tennessee does not have a large amount of error due to the use of a sample. The standard error of \$6.26 gives a measure of the precision of the estimate from the sample. The probability is very high (practically certain) that the mean of the universe would fall within the confidence interval \$407.14 ± 3(6.26), or within the range \$388.36 to \$425.92. It would be appropriate to make the statement that if all the 4,000 file clerks in Tennessee were included in the study, the probability is very small that the average monthly earnings would fall outside the interval \$388.36 to \$425.92.

Another important characteristic of the estimate described above is that it is a *consistent* estimate. In other words, if a more precise estimate is needed, it can be secured by increasing the size of the sample. For example, if the sample just used had been 500, the sampling error would have been considerably less, as shown in the following computations. It is assumed in this case that the increased sample gave the same estimate of the universe standard deviation as the smaller.

$$\hat{\sigma}_{\bar{x}} = \frac{86.51}{\sqrt{500}}\sqrt{1-\frac{500}{4{,}000}} = \frac{86.51}{22.36}\sqrt{1-.125} = 3.869\sqrt{.875}$$

$$\hat{\sigma}_{\bar{x}} = (3.869)(.9354) = 3.619.$$

When the size of the sample is quadrupled, the confidence interval is cut in half, except for the influence of the finite multiplier. This relationship between the size of the confidence interval and the size of the sample shows that increasing the size of the sample becomes more and more costly until the point is reached when the increased precision is not worth the cost.

The example above also illustrates the influence of the finite multiplier. With $n = 182$ the finite multiplier had very little influence on the value of the standard error of the mean. This influence increased considerably when n was increased to 500, since the sampling fraction then became .125 instead of .0455. It should be noted that ignoring the finite multiplier makes the confidence interval appear to be larger than it actually is. In other words, the results will be more precise than one thinks they are.

PRECISION OF A PERCENTAGE

The examples of estimates from samples have all been estimates of the arithmetic mean of the universe, but probably equally important to the business statistician is the estimate of the percentage of units in the universe that possess a given characteristic. The percentage of consumers in a given market that prefer a product, the percentage of subscribers to a magazine that own their homes, and the percentage of voters that will support a given candidate are all examples of parameters that belong in this category.

Standard Error of a Percentage

The universe of 10 families used to illustrate the estimate of the universe mean can be used to illustrate the estimating of the universe percentage from a sample. Of the 10 families in the universe, 4 own a home freezer and 6 do not. If we want to estimate the percentage of families owning home freezers in the universe from a sample of two families, we can draw 45 combinations of 2 from the universe, just as in the example estimating the average semimonthly expenditure for utilities. Assume that families 2, 3, 5, and 8 own a home freezer, and the remaining six do not. A sample of two families will show either zero, 50%, or 100% ownership of a home freezer. All the possible

45 sample percentages are shown in Table 7–5. For 15 of the samples, the percentage of owners is zero; for 24 samples, the percentage of owners is 50; for 6 samples, the percentage of owners is 100. The mean of the 45 percentages is 40%, which is the correct value for the universe (4 out of 10).

TABLE 7–5

PERCENTAGE OF FAMILIES OWNING A HOME FREEZER IN ALL POSSIBLE SAMPLES OF TWO FROM A UNIVERSE OF 10 FAMILIES

Families in Sample (1)	Percentage Owning a Home Freezer (2)
1 and 2 *	50
1 and 3 *	50
1 and 4	0
1 and 5 *	50
1 and 6	0
1 and 7	0
1 and 8 *	50
1 and 9	0
1 and 10	0
2 * and 3 *	100
2 * and 4	50
2 * and 5 *	100
2 * and 6	50
2 * and 7	50
2 * and 8 *	100
2 * and 9	50
2 * and 10	50
3 * and 4	50
3 * and 5 *	100
3 * and 6	50
3 * and 7	50
3 * and 8 *	100
3 * and 9	50
3 * and 10	50
4 and 5 *	50
4 and 6	0
4 and 7	0
4 and 8 *	50
4 and 9	0
4 and 10	0
5 * and 6	50
5 * and 7	50
5 * and 8 *	100
5 * and 9	50
5 * and 10	50
6 and 7	0
6 and 8 *	50
6 and 9	0
6 and 10	0
7 and 8 *	50
7 and 9	0
7 and 10	0
8 * and 9	50
8 * and 10	50
9 and 10	0
Total	1,800
Mean	40

* Family owns a home freezer.

Source: Hypothetical data.

The dispersion in the sample percentages is computed in Table 7–6 in exactly the same manner as the dispersion in the sample means was computed from the data in Table 7–1. The deviation of each percentage from the universe percentage (40) is computed; the deviations are squared, totaled, and divided by N (45); and the square root is taken. The computation uses Formula 4–13 with $X - \mu$ representing the deviation of each percentage from the average percentage for the universe. Since percentages are used, this computation gives the standard deviation of the distributions of the sample percentages and is known as the *standard error of a percentage* (σ_p).

The standard deviation of the percentages resulting from taking all possible samples from the universe measures the scatter in the sample values, and gives an indication of the precision of an estimate made from one sample. The value of the standard deviation computed from the sample percentages is 32.7%. Since approximately 68 out of 100 of the items in a normal distribution fall within $\pm 1\ \sigma$ from the mean, 31 sample percentages should fall within the confidence interval of 40 ± 32.7. A count of sample percentages shows that 24 fall within the confidence interval. In the same manner the $2\ \sigma$ and $3\ \sigma$ confidence intervals can be computed, just as with the distribution of means. In this case all the percentages fall within the $40 \pm 2\ \sigma$ confidence interval.

TABLE 7-6

COMPUTATION OF STANDARD DEVIATION OF ALL POSSIBLE SAMPLE PERCENTAGES

Percentage Owning Home Freezer	No. of Samples f	Deviation from Mean $X - \mu$	$(X - \mu)^2$	$f(X - \mu)^2$
0	15	−40	1,600	24,000
50	24	10	100	2,400
100	6	60	3,600	21,600
Total	45			48,000

Source: Table 7-5.

$$\sigma_p = \sqrt{\frac{48{,}000}{45}} = \sqrt{1{,}066.67} = 32.7$$

It was shown on page 159 that the standard deviation of the distribution of means can be computed without the laborious job of taking all possible samples from the universe, and the same can be done for the distribution of percentages. The formula for the standard deviation of all possible sample percentages from a universe follows. In conformance with the rule that parameters are designated by Greek letters, a universe percentage is represented by pi (π) and a sample percentage by p.

$$\sigma_p = \sqrt{\frac{\pi(1 - \pi)}{n} \frac{N - n}{N - 1}}. \qquad \textbf{(7-10)}$$

When this formula is used, the standard error of the estimate of the percentage of families owning a home freezer is

$$\sigma_p = \sqrt{\frac{40(100 - 40)}{2} \frac{10 - 2}{10 - 1}} = \sqrt{\frac{2{,}400}{2} \frac{8}{9}} = \sqrt{1{,}066.67}$$

$$\sigma_p = 32.7.$$

The value of σ_p is the same as that computed in Table 7–6, the standard deviation of all possible sample percentages. If the value of the parameter π is not known, the value from the sample may be used as an estimate, just as the value of s computed from the sample was used in making an estimate of the standard error of the mean.

Value of π Estimated from a Sample

Assume that a survey is made in a city of 100,000 families to determine the percentage of homes occupied by the owners. In a survey made with a sample of 1,000, it is found that 62% of the families report they own their homes. In the absence of any other information about this percentage, the sample offers the best estimate of the parameter. The value of p from a random sample is an unbiased estimate of the value of π. Using p as an estimate of π, it is possible to estimate the confidence interval from the information provided by the sample. The value of p is substituted for π, and q for $1-\pi$. The finite multiplier becomes $\frac{(N-n)}{N}$ or $1-f$, instead of $\frac{N-n}{N-1}$ used with universe values, giving the following formula for estimating the standard error of a percentage.

$$\hat{\sigma}_p = \sqrt{\frac{pq}{n-1}(1-f)} \qquad \textbf{(7–11)}$$

where

$q = 1 - p.$

The estimate of 62% obtained from the sample above will have a standard error of 1.53, computed

$$\hat{\sigma}_p = \sqrt{\frac{(62)(38)}{1000-1}(1-.01)} = \sqrt{\frac{2{,}356}{999}(.99)} = \sqrt{2.3348} = 1.53.$$

This indicates that if all possible samples of 1,000 were taken from this universe, the parameter would fall within the range 62 ± 1.53 in approximately 68 out of 100 samples. The probabilities associated with two and three times the standard error would be the same as previously discussed for the estimate of the arithmetic mean.

Relative Precision of a Percentage

The standard error of a percentage may be expressed as a relative, in the same manner as the standard error of the mean. In the previous example the value of $\hat{\sigma}_p$ may be expressed as a percentage of p, the estimate of π, by Formula 7–12.

$$\nu_p = \frac{\hat{\sigma}_p}{p} 100 \tag{7-12}$$

$$\nu_p = \frac{1.53}{62} 100 = 2.47\%.$$

The interpretation of the value of ν_p is the same as for the relative precision of the mean.

STANDARD ERROR OF OTHER DESCRIPTIVE MEASURES

The standard error of the mean and the standard error of a percentage have been discussed in considerable detail, but it is desirable to describe briefly the standard error of some of the other descriptive measures. The median, the quartiles, and the standard deviation of a universe may be estimated from a sample and the precision of this estimate computed. It is also possible to estimate an aggregate, such as total retail sales for an area, from a sample and to compute a measure of precision of the estimate.

Median and Quartiles

When the samples are drawn from a normal universe, the standard error of the median and the quartiles may be computed

$$\hat{\sigma}_{Md} = \frac{1.2533\hat{\sigma}}{\sqrt{n}} \sqrt{1-f} \tag{7-13}$$

$$\hat{\sigma}_{Q_1} = \frac{1.3626\hat{\sigma}}{\sqrt{n}} \sqrt{1-f} \tag{7-14}$$

$$\hat{\sigma}_{Q_3} = \frac{1.3626\hat{\sigma}}{\sqrt{n}} \sqrt{1-f}. \tag{7-15}$$

If a large sample is used, the standard error is a reasonably good approximation for a parent universe that is only moderately skewed. It will be noted that the median and the quartiles have larger standard errors than the arithmetic mean.

Since $\hat{\sigma}$ will normally be used, the finite multiplier will be $\sqrt{1-f}$ or $\sqrt{\frac{N-n}{N}}$. If σ is known for the universe, the finite multiplier is $\sqrt{\frac{N-n}{N-1}}$.

Standard Deviation

For large samples the standard error of the standard deviation is computed

$$\hat{\sigma}_s = \sqrt{\frac{m_4 - m_2^2}{4m_2 n}} \tag{7-16}$$

where

m_2 and m_4 = the second and fourth moments of the frequency distribution estimated from a sample.[5]

The formula above is valid for a sample from a nonnormal distribution, provided it is large. However, if the parent distribution from which the sample is taken is normal, Formula 7–16 reduces to

$$\hat{\sigma}_s = \frac{\hat{\sigma}}{\sqrt{2n}}. \tag{7-17}$$

Formula 7–17 assumes a population that is large relative to the size of the sample. For sampling from a finite population, where a large proportion of the population is included in the sample, this formula overstates the sampling error. The formula for the correct sampling error of the standard deviation for a finite population, however, is very complex and is seldom used, in spite of the fact that Formula 7–16 overstates the sampling error in this situation.

Aggregates

An unbiased estimate of an aggregate value in a universe may be obtained by dividing the total for the sample by the sampling fraction. For example, a 1% sample of retail stores in a state is selected at random. There are 99,000 stores in the state, and the sample consists of 990 stores. Sales of the 990 stores total $105,930,000, or an average of $107,000 per store. The estimated total sales would be 100 times this sample total, or $10,593,000,000. Written as an equation, this becomes

$$\Sigma\hat{X} = \frac{\Sigma x}{f} \tag{7-18}$$

$$f = \frac{n}{N}$$

$$\Sigma\hat{X} = \frac{105{,}930{,}000}{.01} = \$10{,}593{,}000{,}000.$$

The estimate above is N times the average sales per store; thus,

$$\Sigma\hat{X} = 107{,}000 \cdot 99{,}000 = \$10{,}593{,}000{,}000.$$

Both computations above give the same answer.

---

[5] In Chapter 4 the moments were computed from a complete enumeration and designated as M_3 and M_4. When the moments are derived from a sample, they will be represented by m_3 and m_4.

The standard error of the mean is estimated from the sample to be \$2,160. The standard error of the total is N times the standard error of the mean, so

$$\hat{\sigma}_{\Sigma X} = N \cdot \hat{\sigma}_{\bar{x}}$$

$$\hat{\sigma}_{\Sigma X} = 99{,}000 \cdot 2{,}160 = \$213{,}840{,}000.$$

It is significant, as the following computations show, that the relative precision is the same for both the mean and the aggregate of the values for which the mean was computed,

$$v_{\bar{x}} = \frac{2{,}160}{107{,}000} 100 = 2\%$$

$$v_{\Sigma X} = \frac{213{,}840{,}000}{10{,}593{,}000{,}000} 100 = 2\%.$$

DETERMINING THE SIZE OF SAMPLE NEEDED

The preceding discussion has dealt entirely with the problem of determining the precision of the estimates of parameters. In actual practice, however, an equally important procedure is to determine how large a sample is needed before a study is made. When a preliminary analysis of a problem at hand indicates that information is needed, one of the first considerations is cost. If there has been any previous experience with collecting data, it will be possible to estimate approximately the cost per interview; but it will also be necessary to know how many interviews will be needed. For example, if it appears that the cost of making an interview and tabulating the results will be approximately \$6 per schedule, and if the precision desired in the study will require a sample of 2,000, the total cost of the study will be approximately \$12,000. Whether management is willing to spend this amount for the information is an important question. Hence, it is extremely important to estimate at least approximately how large a sample is needed before any work is done on the study.

The following three examples demonstrate the steps used in determining the size of the sample to be used.

Example 1

A large dairy was considering the advisability of selling milk in paper cartons through grocery stores; but before going further with their consideration of the idea, the manager wanted to know how well consumers liked this type of container. It was suggested that a small consumer survey might give an approximate answer to the question of how well consumers would like this form of container, and it was decided to explore the cost of making such a survey.

Before any answer could be given to this question, it was necessary to determine (1) how much it would cost per interview to secure the information, and (2) how many interviews would be needed. In making an estimate of the size of the sample needed, the amount of sampling error that could be tolerated in the results had to be established. This decision must always be made by the persons who will make use of the data. The statistician planning the survey generally will discuss the required accuracy with a user of the information in order to find out the degree of precision needed in the data.

The manager of the dairy might decide that, if the survey showed correctly within a range of 6 either way the percentage of people that like paper milk cartons, the data would be satisfactory. This would represent a confidence interval of ± 6, although this statement would not indicate the probability level. If one wanted to be practically certain that the universe percentage would fall within this range, it would be well to use the 3σ level. If this level of probability were used, it follows that $3\sigma_p = 6$ and $\sigma_p = 2$.

Ignoring the correction for the size of the universe, the formula for the standard error of a percentage is

$$\sigma_p = \sqrt{\frac{\pi(1-\pi)}{n}}. \tag{7–19}$$

If the sampling fraction is small, the correction factor can be ignored without having any appreciable effect on the result. If we have an approximation of the value of π, it will be possible to substitute values for π, $1-\pi$, and σ_p, leaving only the value of n unknown. Since the intention is to solve the equation for n, it is more convenient to write

$$n = \frac{\pi(1-\pi)}{\sigma_p^2}. \tag{7–20}$$

If it were believed that the percentage of persons liking paper milk cartons would be approximately 70, this value could be substituted for π, and the calculation made for the size of n needed would be

$$\begin{aligned} \pi &= 70 \\ 100-\pi &= 100-70 = 30 \\ \sigma_p &= 2 \end{aligned} \qquad n = \frac{(70)(30)}{2^2} = \frac{2{,}100}{4} = 525.$$

If the value of π were thought to be approximately 60, the size of the sample needed would be somewhat larger, calculated

$$\begin{aligned} \pi &= 60 \\ 100-\pi &= 100-60 = 40 \\ \sigma_p &= 2 \end{aligned} \qquad n = \frac{(60)(40)}{2^2} = \frac{2{,}400}{4} = 600.$$

If it is impossible to make an approximation of the size of the percentage, it is customary to assign π a value of 50, since this value requires the largest sample for a given degree of reliability. The sample size required for the confidence interval when $\pi = 50$ is computed

$$\pi = 50 \qquad 100 - \pi = 100 - 50 = 50 \qquad \sigma_p = 2 \qquad n = \frac{(50)(50)}{2^2} = \frac{2{,}500}{4} = 625.$$

The sample of 625, required when $\pi = 50$, is the largest sample that would be needed for a confidence interval of ± 6. For any other value of π, the value of $\pi(1 - \pi)$ would be less than 2,500, which means that the sample size needed would be slightly less. The general practice is to use a sample large enough to give the required precision if $\pi = 50$, even though there is reason to believe the value of π will differ considerably from 50. If this results in taking a somewhat larger sample than is needed, it will normally not increase the cost a great deal and will have the advantage of never providing a smaller degree of precision than was intended.

Example 2

If the problem is to estimate the size of the sample needed to determine the mean value of some characteristic, such as the average distance traveled by farmers to buy shopping goods, Formula 7–7 is used, usually after solving for *n*,

$$n = \frac{\sigma^2}{\sigma_{\bar{x}}^2}. \qquad \textbf{(7–21)}[6]$$

Formula 7–21 ignores the correction for the size of the universe since this factor has a very slight effect on the result when the sampling fraction is small.

The problem of estimating the size of sample needed is the same as discussed in the preceding example for a percentage. First, a decision must be made as to the sampling error that can be tolerated; and second, an estimate must be made of the standard deviation in the universe. The latter estimate is not so easy to make for this situation as it is for the percentage. The maximum sample is needed when $\pi = 50\%$, but there is no way of knowing what the maximum standard deviation will be. One approach is to try to estimate the amount of dispersion from a previous study; but if this cannot be done, it may be necessary to make a preliminary study and estimate the size of the standard deviation from these data. A small sample would not give a very accurate estimate of the standard deviation, but it would be better than a guess and would at least give an approximation of the size of the sample that would be needed.

A study was made by a market research organization to learn the distance farmers in the Middle West traveled to buy various kinds of goods. One problem in planning the survey was the determination of the size of sample needed.

[6] It might be argued that $\hat{\sigma}^2$ and $\hat{\sigma}_{\bar{x}}^2$ should be used instead of σ^2 and $\sigma_{\bar{x}}^2$ in Formula 7–21, but it really makes little difference which symbol is used. Sometimes universe values are used, but at other times all that is available is an estimate from a sample.

A number of studies similar to this had been made by market research organizations, and the information in these surveys showed that the average distance traveled was not more than 40 miles and the standard deviation not more than 20 miles. Although these amounts were known to be guesses, it was decided from this information to make a preliminary estimate of the size of the sample needed.

The executives who were to use the data wanted to have a rather small amount of sampling error, and a confidence interval of 1.2 miles was set up as an objective. It was further specified that it be practically certain that the estimate would not differ from the universe value by more than 1.2 miles. Assuming that this was at the 3 σ level, the required standard error of the mean was 0.4 miles, calculated

$$3\,\sigma_{\bar{x}} = 1.2 \text{ miles} \qquad \sigma_{\bar{x}} = 0.4 \text{ miles.}$$

The size of the sample that would be required was then computed

$$n = \frac{\sigma^2}{\sigma_{\bar{x}}^2} = \frac{20^2}{0.4^2} = \frac{400}{.16} = 2{,}500.$$

When the study was made, the standard deviation of the universe was found to be 14.3 miles instead of 20 miles, so the sampling error was less than the specifications required. The sample that was finally secured was 2,467, and the precision of the sample result was

$$\hat{\sigma}_{\bar{x}} = \frac{14.3}{\sqrt{2{,}467}} = \frac{14.3}{49.7} = .29 \text{ miles.}$$

This was somewhat smaller than the 0.4 miles specified for the study. The more accurately the standard deviation of the universe can be predicted, the more accurately the size of sample needed can be determined. If a preliminary study had been made and the size of the universe estimated from a small sample, it usually would have been more accurate than the above example. However, if the decision to make a study depends upon the size of sample needed, there may not be any funds available to take a preliminary sample. In such a case it becomes necessary to make a rough approximation, even though this might be merely a guess.

Example 3

It was stated on page 159 that the confidence interval of the mean might be expressed as a percentage of the mean instead of an absolute amount. If it is specified that with practical certainty the sampling error should not be more than 3% of the universe mean, it would normally be assumed that the confidence interval was three times the coefficient of variation of the mean. Letting the maximum sampling error allowed be represented by E, it can be written that $E = 3V_{\bar{x}}$, and $V_{\bar{x}} = \frac{E}{3}$. $V_{\bar{x}}$ and E are expressed as percentages.

When the confidence interval desired is expressed as a percentage of the value being estimated (in this case the mean), Formula 7–22 gives the same value for n as Formula 7–21, since both the numerator and the denominator were divided by μ^2. Both formulas ignore the finite multiplier.

$$n = \frac{\frac{\sigma^2}{\mu^2}}{\frac{\sigma_{\bar{x}}^2}{\mu^2}} = \frac{V^2}{V_{\bar{x}}^2}.$$

Since

$$V_{\bar{x}} = \frac{E}{3},$$

$$n = \frac{V^2}{\left(\frac{E}{3}\right)^2} = \left(\frac{V}{\frac{E}{3}}\right)^2$$

$$n = \left(\frac{3V}{E}\right)^2. \qquad \textbf{(7–22)}$$

Formula 7–22 is generally easier to use than Formula 7–21, although they both give the same value for n if the same assumptions are used. Using the same facts as in Example 2, assume that it was decided to be practically certain that the sample mean would not vary more than 3% from the universe mean. This is the value of E. It was stated in Example 2 that the mean was about 40 miles and the standard deviation about 20 miles; so $V = \frac{20}{40}\,100 = 50\%$ by Formula 4–15.

Substituting these values in Formula 7–22 gives the same value for n as computed on page 178.

$$n = \left(\frac{3V}{E}\right)^2 = \left(\frac{150}{3}\right)^2 = (50)^2 = 2{,}500.$$

STRATIFIED RANDOM SAMPLE

If it is possible to divide the universe into classes or *strata* that are more homogeneous than the universe as a whole, estimates may be made from a *stratified sample* with greater precision than if the population to be sampled is treated as an undifferentiated whole and simple random sampling is used. Since the sampling error depends on the variation in the universe from which the sample is drawn, estimates are made for each of the strata and then combined by a proper weighting into an estimate for the universe. With more accurate estimates for each of the classes, the estimate of the total is more precise than if the universe had been treated as a whole.

A basic problem in stratified sampling is that considerable knowledge of the structure of the universe is necessary to delineate the strata and to weight the results. For example, if retail stores are being sampled to estimate sales, the efficiency of the estimate will normally be increased if the strata are based on annual sales. But this means that something must be known about the sales volume of the stores in the universe. The largest stores would be set up as one stratum, the smallest stores as another stratum, and as many size groups between the largest and the smallest as seems desirable. In such a stratified sample the stratum consisting of the largest stores would usually contain a small number of stores, but would account for a large proportion of total sales. Conversely, the smallest stores would account for the largest number of stores, but for a small proportion of the total sales.

It would generally be desirable to select a large percentage of the stores in a large-store stratum, even to the extent of including 100% of them in the sample. The stratum with the smallest stores would be represented by a small percentage of the total, and the strata between the extremes with a somewhat larger percentage than for the stratum with the smallest businesses. The allocation of the total sample can be made in many ways, but the optimum allocation takes into account the size of the stratum and the dispersion within the stratum. If the cost of collecting data varies from stratum to stratum, the cost differentials should also be taken into consideration. The simplest method of allocation is in proportion to the size of the strata, but this method loses much of the advantage of stratification and should be used only in special cases. The chief advantage of the method is that it is easier to make estimates and to compute the sampling error than it would be if a disproportionately stratified sample were used.

CLUSTER SAMPLE

In sampling business and human universes, securing a complete list of all the items in the universe is often either impossible or very expensive. Many universes exist in the form of lists, such as telephone directories, city directories, members of trade associations, employees, students in a school, charge customers, and subscribers to a magazine or newspaper. However, for many universes no satisfactory list exists, and compiling one would be too expensive. Sometimes when it is impossible to secure a list, or it is too expensive to compile one for the study, a population may be divided into groups and a sample of these groups drawn to represent the population. A *cluster sample* is defined as a sample in which groups or clusters of individual items serve as sampling units.

In making a consumer survey in a city, the sampling units might be families. It might be intended that the survey consist of a random sample of all families in the city. Assume that a recent listing of families, such as an up-to-date city directory, is not available. It would be possible to make a listing of all the families in the city, and from this list draw a sample of families to be interviewed. However, this approach would be extremely costly. A cheaper method would be to divide a map of the city into a number of small areas consisting of single blocks or even portions of blocks. Each area would be given a number and from this list of areas, a random sample or a systematic sample might be drawn. Each area is a cluster, and clusters rather than the families form the sampling unit.

Since the family represents the unit to be interviewed, the next step is to interview all the families in each area selected, or to choose either a random or a systematic sample of families from each area. Since every part of the city would have a chance of being selected, the sample of families would be random. This type of cluster sample is called an *area sample,* and it can be used in the absence of a complete list of the individual units. If the universe can be broken into geographic units that include all the universe, a random sample of the geographic units can be selected. By selecting a small enough sample of units, it is always possible to list all the individual units in these areas and either interview all the individual units or select a sample of units from the listing. In this manner it is not necessary to list all the units in the universe, but only those that fall in the selected areas.

If the individual units in the sampling areas tend to resemble each other, the sampling error will be greater than would be secured from a simple random sample from all the individual units in the universe. There is a tendency for the individuals in the small areas to be alike. The greater this tendency, the more it increases the sampling error of the results from the sample. Thus, the precision of a cluster sample tends to be lower than for a simple random sample of the same size. If the individual units in the areas show no greater resemblance than for the whole universe, a cluster sample will give as precise an estimate as a simple random sample.

Sampling error can be reduced by increasing the size of the sample, offsetting the cluster sample's loss in precision. In many cases it costs much less to take a considerably larger cluster sample than it would cost to take the smaller simple random sample with the same precision. The cluster sample would save in the listing of the units in the universe, and it would also be possible to make the interviews at a lower cost per interview, since they were grouped into clusters. A simple random sample would be scattered over all portions of the universe, with the result that interview costs would generally be high. The cluster sample, on the other hand, would concentrate the interviews in a limited number of areas, with a resulting saving in cost that often would more than compensate for the larger number of interviews needed. In fact, the saving in the cost of interviews is so great that area samples are used even when a complete list of the individual units in the universe is available.

STUDY QUESTIONS

7-1. Explain why business has found the use of samples so profitable in collecting external statistical data. Give particular attention to the cost of the information in relation to its accuracy.

7-2. Once an individual is chosen at random as part of a probability sample, it is considered wrong to substitute another person if the one drawn is not at home or for any other reason cannot be interviewed. Explain.

7-3. An opinion study is planned among students on the campus to secure their opinion of a compulsory fee of $15 per semester to be used for the support of the student union. At the present time there is no fee for support of the union, and there is a general feeling among the students that money for the support of the union should be provided from some source. With approximately 20,000 students in the school, it is unlikely that anyone would pay for a complete enumeration of all the students.

(a) Describe how a strictly random sample of students could be selected for this study.

(b) How large a sample would you recommend be taken for this study? Explain how you decided upon this size.

7-4. A soft-drink bottling company is considering converting its operation to no-return bottles because the sales manager believes this would improve the company's position with soft-drink consumers. None of the company officials knows what their customers prefer so it is decided to conduct an opinion survey to attempt to find out which type of bottle their customers prefer and how strong this preference is.

Describe a correct method of taking a random sample of the people who buy the company's product. It is widely sold throughout the city in all types of retail outlets that sell soft drinks by the carton.

7-5. Describe how you would select a random sample for a sample survey of tire dealers in your state to inquire about their customers' acceptance of radial tires.

7-6. A random sample of 1,000 passengers is taken at stated intervals during the day at a large city airport, and the passengers are asked whether they are changing planes at this airport or whether they are beginning or ending their flight there.

(a) How accurate would you expect the answers to the 1,000 interviews to be in comparison with the answers you would get if you asked all the passengers the same question?

(b) How would you take a random sample of all passengers leaving and arriving at the airport?

7-7. Assume that you have a random sample of retail stores in a state that reports retail sales each month. An estimate of total retail sales for the state is also made from the sample each month. The confidence interval is not too large and the information is very useful to businessmen.

You are asked to use your sample of stores to make estimates of retail sales in each county since you have stores in the sample from every county in the state. You are not expected to secure any larger sample; you are asked merely to tabulate the sample results by counties and use the sample for each county to estimate retail sales in the county. It is argued that if the sample gives a good enough estimate of retail sales for the state, it should give equally good estimates for the individual counties, since the sum of the county estimates would be the same as the estimate for the state.

What answer would you give to this request?

7-8. A manufacturer has designed a completely new package for his product and after a year of using this new package he is considering making a consumer survey to find out if his customers like it. It is obvious that a sample should be used, but the question of cost immediately arises. You estimate that it will cost $2.35 to make each interview and to tabulate its results.

Explain how you would go about estimating the total cost of conducting the survey.

7–9. Assume that you are able to divide the universe into ten strata with all the items in each stratum exactly the same. How large a sample would you need to make an estimate of the parameter? How good would this estimate be?

7–10. Under what circumstances does it pay to use a stratified sample instead of a simple unrestricted sample?

7–11. Assume that you believe a cluster sample of 1,500 items would give the same degree of precision as a simple unrestricted sample of 1,000. How would you decide which type of sample to use?

7–12. A group of businessmen planning to organize a bank in a new shopping center wanted information indicating whether a bank was needed in the area. Interviews concerning the need for a new bank were made among residents in an area of 23 square miles. The area was further divided into 4 approximately equal subareas. A quota of 10 interviews (5 men and 5 women) was established for each subarea. No more than 2 interviews were conducted in any block, and only one in a household. No instructions were given as to how the individuals were to be selected.

Do you think this was a good sample design? Discuss.

7–13. A research organization is making a study of the consumption of dairy products in a city. It has been decided that an inexpensive method of getting questionnaires filled out is to distribute them through the public schools. The superintendent of schools has agreed to distribute the questionnaires to all sixth-grade students in the city schools. If a report is secured from each student in the sixth grade in the city, do you consider this a good sample from which to estimate the consumption of dairy products in the city? Discuss.

7–14. A retail clothing store manager wanted to know what her customers think of the services rendered by the store. She mailed a questionnaire to all the charge accounts maintained by the store. The returns from this questionnaire represented 24% of the total number of charge accounts. Would you consider these questionnaires a random sample of the customers of the store?

7–15. Is it possible under any circumstances for a sample survey to give a more accurate result than can be secured from a complete enumeration? Discuss.

7–16. In many of the opinion polls, the size of the sample used is not published. What justification can you give for this practice? What objections can be raised against this practice?

7–17. A sample survey is designed to take a sample that is 0.5% of the universe. The question is raised as to whether such a small fraction of the universe can give a result that is sufficiently accurate. What answer would you give to this question?

7–18. What is the important difference between the methods of taking a judgment sample and a probability (random) sample? Explain why the probability sample is considered better than the judgment sample.

PROBLEMS

7–1. A market research study was undertaken to determine how much was spent in a particular year by automobile owners in a certain city for gasoline and oil. A sample of 1,000 automobile owners was taken at random. The average expenditure of this sample was $300. The standard deviation of the 1,000 amounts reported was $90.

(a) What would you estimate to be the average expenditure for gasoline and oil of all automobile owners in this city?

(b) Compute the confidence interval of this estimate and explain what it means.

(c) How would you select a random sample of all automobile owners in the city in order to make the survey described?

7-2. Assume that you want a quick estimate of the average age of the 10,000 employees in the field offices of a company. You select 400 cards at random from the personnel file and compute the average age of these 400 employees to be 34.8 years. The standard deviation computed from the 400 ages on the personnel cards is 10.8 years.

(a) Make an estimate of the average age of the 10,000 employees and compute the confidence interval of this estimate. Explain what these two measures mean.

(b) Describe how you would take a random sample of the 10,000 cards in the personnel file.

7-3. A state employment commission has reports on employment of 300,000 firms in the state during the first quarter of the year. A random sample of 1,000 of these reports is taken, and the number of employees for each firm is secured from the report. The total number of employees reported by the 1,000 firms is 28,120. The sum of the squares of these 1,000 numbers is 984,000. (In other words, $\Sigma x = 28{,}120$ and $\Sigma x^2 = 984{,}000$.)

(a) What would you estimate to be the number of employees per firm in the state?

(b) Compute the confidence interval of this estimate and explain what it means.

7-4. A manufacturing company employing 20,000 people has a progressive employee relations program. It was decided to make a survey to determine the state of employee morale and to secure information that would be used in future labor relations efforts.

A random sample of 400 employees was interviewed at home. One question asked was how much the company contributed to the retirement plan and how much the employees contributed. Thirty percent of those interviewed did not know.

(a) If all the 20,000 employees had been interviewed, what would you estimate to be the percentage that would reply that they did not know the answer to the retirement plan question?

(b) Do you think the sample is accurate enough for the use of management in deciding whether or not an educational program should be initiated to inform the employees about the company's retirement plan? (The company paid all of the contribution to the retirement plan.) Before answering, it would be wise to compute the confidence interval of the estimate secured from the sample.

7-5. Assume that you are making a survey to ask a number of questions about consumer brand preferences. All your tabulations will be in the form of percentages (such as the percentage of people who prefer your brand, and the percentage of people who do not use your product). You are instructed to secure a confidence interval of not less than ± 2.7 for each percentage. (It is satisfactory if the confidence interval is less than ± 2.7 for some percentages, but it must not be more.)

Compute the size of the sample that you would need to be certain that the confidence interval did not exceed ± 2.7.

7-6. Assume that you are planning a study of family incomes in a city and you have reached the point where you must make an estimate of the cost of the study. This requires an approximation of the number of interviews to be taken, which will vary with the accuracy required in the estimate of family incomes. It is decided that the estimates must not vary more than \$270 from the average income that would have been found if a complete study were made of all families in the city.

Other income studies suggest that the average income per family might be somewhere in the neighborhood of $7,500 and that the standard deviation might be approximately $3,000. This information is not much better than a guess, but it is the best information that you have on which to base your computation of the size of sample needed.

(a) Compute the size of the sample that you would recommend, taking into consideration the information given.

When you tabulate your data, you find the mean income of your sample is $6,850 and the standard deviation of the sample is $3,400.

(b) Compute the confidence interval you have secured in your study. If your confidence interval is smaller than the $270 specified, your study is complete.

(c) If the confidence interval is larger than $270, what should you do?

7–7. The study described in Problem 7–6 was made in a city of 700,000. Later a survey was wanted in a city of 5 million, and it became necessary to compute the size of the sample needed in the second survey. There was good reason to believe that the dispersion in incomes for the larger city would be essentially the same as the first city studied.

What size sample would you recommend for the city of 5 million, assuming that the same degree of reliability was wanted as in the first survey?

7–8. Assume that you are planning a sample survey to estimate the travel expenditures of families in a large city during the coming year. Previous studies indicate that the coefficient of variation of these expenditures is approximately 30%. You wish the estimate of average expenditures to be practically certain to fall within 5% of the value that would be obtained from a complete enumeration. Estimate the size of the sample that would be needed to obtain this degree of precision.

7–9. A trade association plans to estimate the amount of money that its members will spend on advertising during the next year. It has been determined from previous studies that these expenditures will have a standard deviation of approximately $250,000. The secretary of the association states that he wants to be certain that the estimate from the sample will not exceed the true amount by more than $10,000. Estimate the size of the sample that will be needed to make this estimate of expenditures.

7–10. A sample survey in Austin, Texas, in 1963 showed that 8% of the television sets could pick up ultrahigh frequency broadcasts.

(a) What is the confidence interval of this estimate if it is based on a sample of 100? A sample of 400? A sample of 1,000?

(b) Another sample survey made in Austin, Texas, in 1973 showed that 82% of the television sets could pick up ultra-high frequency broadcasts. What is the confidence interval of this estimate if it is based on a sample of 100? A sample of 400? A sample of 1,000?

7–11. The following four samples are drawn from infinite universes for which the standard deviations are known. For each sample compute the confidence interval for the universe mean (μ) at the probability level called for.

	Sample Mean	Sample Size	Standard Deviation	Probability Level
(a)	14	25	2.0	2.33
(b)	206	100	4.8	2.58
(c)	75.8	16	16	3.00
(d)	53	225	12	1.96

7–12. If in Problem 7–11 the universes had been small and finite as shown below, how much difference would this make in each confidence interval?

	Universe Size
(a)	50
(b)	100
(c)	1,600
(d)	4,500

7–13. Assume that all possible samples of $n = 2$ are drawn from the following universe of $N = 4$. There will be 6 samples.

Universe Values
16
8
12
4

(a) Compute the universe mean and universe standard deviation.

(b) Compute the standard error of the mean by using the sample means of the 6 samples of 2 items each.

(c) Compute the standard error of the mean using the universe standard deviation and Formula 7–2. Is this answer the same as (b) above?

(d) Demonstrate that the expected value of the sample mean (the mean of the sample means) is the same as the mean of the universe.

(e) Demonstrate that the expected value of the sample standard deviation (s) is less than the standard deviation of the universe.

(f) If all possible samples of size three had been drawn from this universe, what effect would this have on the size of the standard error of the mean?

7–14. The registrar of a large university wishes to estimate the proportion of students registered for the regular term who will go to summer school. If he uses a probability level of $3\sigma_p$, how large a sample would be needed to estimate the universe percentage within $\pm 6\%$? Assume that 60% of the student body went to summer school the previous summer.

7–15. An accounting firm plans to use a random sample to audit the accounts payable of a client. A previous audit showed a standard deviation in the size of these accounts to be \$50. What size sample would be required to produce a confidence interval no wider than \$4.00 with a 95% probability of containing the true universe mean? What would be the effect on sample size if the probability requirement was reduced to 90%? What would be the effect if the probability requirement is set at 95% and the confidence interval reduced to $\pm$ \$2.00?

7–16. Compute measures of relative precision for the four examples in Problem 7–11 using Formula 7–3. Which sample is the most accurate when measured by the coefficient of variation?

7–17. A large retailer issues maintenance contracts for the refrigerators it sells. A random sample is used to determine the frequency of service calls requiring new parts. A random sample of 100 calls is drawn from a total of 980 calls recorded in one month.

Forty-five of the sampled calls required new parts. Compute the 95% confidence interval estimate of the universe percentage.

7–18. Problem 2–3 gives the weights of 84 cans of tomatoes. Since these 84 cans represented a random sample from production of an automatic filling machine, the mean of the 84 weights computed in Problem 3–22 is an unbiased estimate of the mean weight of all cans filled by this automatic machine while it is operating in its present state of adjustment. From the frequency distribution constructed in Problem 2–3, compute the standard error of the mean. Explain what this measure shows about the reliability of the sample mean as an estimate of the universe mean. Explain the significance of this measure.

7–19. Assume that the 210 employees in Problem 2–2 represent the universe of all employees of a governmental department and draw a random sample of 50 from this universe. (Use the table of random numbers to select this sample.)

7–20. The table below gives the fineness of 30 random samples of cotton from central Texas, measured by the microgram weight of one inch of fiber.

4.0	3.4	2.5
3.0	4.2	2.9
3.2	4.4	2.8
3.7	4.1	2.8
3.8	4.5	4.5
3.4	4.7	3.7
3.6	4.8	4.6
4.2	3.5	4.5
4.4	4.4	4.7
3.4	4.3	4.2

Source: Hypothetical data.

(a) From these data estimate the mean fineness of cotton from this area of Texas in micrograms.

(b) Compute the standard error and the coefficient of variation for this estimate.

7–21. Table 2–1, page 22, gives a universe of 125 lots of cotton yarn. This universe was analyzed in Chapters 3 and 4, and all the measures normally used to describe a frequency distribution were computed. This problem is to demonstrate how a sample may be used to estimate the various measures that have been computed from this universe.

(a) Using random numbers, select a sample of 10 from the universe in Table 2–1.

(b) Estimate the mean of the universe and compare it with the mean given in Chapter 3.

(c) Compute the standard error of the mean ($\sigma_{\bar{x}}$) for the sample of 10, using the value of the universe standard deviation computed in Chapter 4.

(d) Does the estimate of the mean you made in (b) fall within the 95% confidence interval? Would you reasonably expect that it would?

7–22. This problem continues the use of the sample selected in Problem 7–21.

(a) Estimate the standard deviation of the universe in Table 2–1, and compare it with the standard deviation computed in Chapter 4.

(b) Compute the standard error of the standard deviation for a sample of ten, using the value of the standard deviation of the universe given in Chapter 4.

(c) Does the estimate of the standard deviation you made in (a) fall within the 95% confidence interval? Would you reasonably expect that it would?

7–23. This problem, which is to be worked in cooperation with other members of the class, continues the use of the sample selected in Problem 7–21.

(a) Divide the class into groups of 4 and combine the answers of each group into a mean based on a sample of 40.

(b) Estimate the mean for the sample of 40 and compute its standard error and the 95% confidence interval.

(c) How does the standard error of this mean compare with the standard error of the mean estimated from a sample of ten?

(d) Does the estimate of the mean from a sample of 40 fall within the confidence interval? Would you reasonably expect that it would?

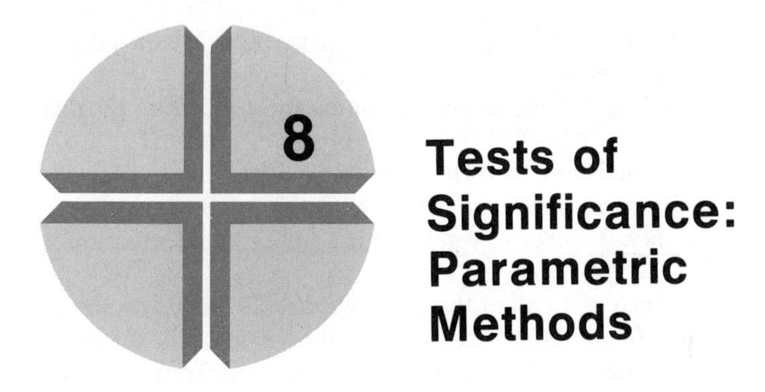

8 Tests of Significance: Parametric Methods

In Chapter 7 sample statistics were used to estimate certain characteristics of the universe from which the sample was drawn. It was pointed out that estimates of population parameters are subject to error, but that limits may be determined within which, given a degree of confidence, the universe values may be expected to fall.

In this chapter and in Chapter 9, sample statistics will be used to test hypotheses that will be made about certain universe parameters. The tests of significance introduced in this chapter are generally known as *parametric tests* or *distribution tests*. The name comes from the assumption that the nature of the population distribution from which the sample is drawn is known.

Since a properly designed sample study provides the confidence interval of the estimate, there is enough information available to permit drawing conclusions even though the information is only approximate. The following sections illustrate how this can be done.

DIFFERENCE BETWEEN UNIVERSE MEAN AND SAMPLE MEAN, UNIVERSE STANDARD DEVIATION KNOWN

The advertising department of a farm journal publisher believed that the farmers who subscribed to one of its papers had a higher average income than all farmers in the area. To investigate this belief the circulation department took a random sample of 3,600 names from the circulation list in one state, and the Bureau of the Census made a special tabulation of the census schedules for these farmers. (The Bureau of the Census will make special tabulations when it can be done without revealing any confidential information from the individual schedules.) According to the census report, the average value of farm products sold per farm was $5,509 for the 3,600 subscribers, and $5,374 for all farms in the state. This seems to indicate that the gross income of the subscribers to the farm paper was higher than the income for all farmers in the state; but the average for the subscribers to the farm paper was a sample and, therefore, only an approximation.

The problem presented by this situation illustrates the procedure of testing hypotheses by setting the hypotheses against observed data. If the observed facts are clearly inconsistent with the hypothesis, then the hypothesis must be rejected. If the facts are not inconsistent with the hypothesis, it may be given tentative acceptance, subject to rejection if conflicting information is secured.

The hypothesis to be tested is called the *null hypothesis*. This hypothesis asserts that there is *no* significant difference between the value of the universe parameter being tested and the value of the statistic computed from a sample drawn from that universe. The analyst may believe there is a difference, but hypothesizes no difference and then ascertains whether this null hypothesis can be supported. This is clarified in the following discussion, which continues the example begun earlier.

Type I and Type II Errors

The null hypothesis to be tested in the farm income example is: The mean value of farm products sold by subscribers to the farm paper is the same as the mean value of farm products sold by all farmers. Or, the hypothesis might be stated that the real difference between the universe mean of $5,374 and the sample mean of $5,509 is zero. The sample mean does not give the information needed to determine the validity of a stated hypothesis with absolute certainty, for there is always the risk of making one of two kinds of errors. The null hypothesis as stated may actually be true, but there is a possibility that it will be decided as not true. Rejecting as false a hypothesis that is in fact true is called a *Type I error*. On the other hand, it is possible that the stated hypothesis is not true but will be accepted as true. This is a *Type II error*.

Level of Significance (α)

As long as sample data are used, it is impossible to be absolutely certain that the hypothesis being accepted is true. When using sample data, it is important to state a probability that a mistake has been made. If the probability of making a Type I error is small enough, the null hypothesis may be rejected. The probability of making a Type I error is the *level of significance* and is designated as alpha (α). Frequent use is made of 5% and 1% values of α, although other levels of significance are also used. How much risk one wants to take in a given problem of rejecting a true hypothesis determines the value of α that will be chosen.

In the farm journal example the standard deviation of the universe was $2,460 and will be represented by σ. It has already been demonstrated that the means of all possible samples of size n from a universe will form approximately a normal distribution, except when n is very small. The distribution of means will have an average value that is the same as the mean of the universe. Since in this case the mean of the universe was known to be \$5,374, the mean of the distribution of all sample means would be \$5,374 also. The standard deviation of the distribution of sample means may be computed

$$\sigma_{\bar{x}} = \frac{\sigma}{\sqrt{n}} = \frac{\$2{,}460}{\sqrt{3{,}600}} = \$41.$$

The difference between the universe mean (μ) and the sample mean ($\bar{x}$) is expressed in units of the standard error of the mean.

$$z = \frac{\bar{x} - \mu}{\sigma_{\bar{x}}} \tag{8-1}$$

$$z = \frac{5{,}509 - 5{,}374}{41} = 3.29.$$

A complete discussion of the normal distribution z is found on page 132 of Chapter 6. The question to be answered is whether a difference of this size can be explained as a variation due to sampling, or whether there is a *significant* difference between the universe mean and the mean of the sample. Significance is used in the sense that some factor has produced the observed differences, other than the variations due to random fluctuations present in the drawing of successive samples from the same universe. This decision is made by comparing the computed value of z (3.29) with the value of z that cuts off a fraction of the normal curve equal to the value of the level of significance (α) selected.

If α is set equal to .05, we must locate the value of z that cuts off .025 of the normal curve at each end. Since the curve is symmetrical, this value will be at the same point on each side of the curve and will give a total of .05 of the area of the curve. The values in Appendix H represent the fraction of the normal curve contained on one side of the symmetrical curve between the maximum ordinate and the value of z. Thus, it is necessary to find the z value for .4750 in the table. This will leave .025 beyond this value of z. The same fraction of the curve is found at the other tail of the curve, making a total of .05 of the curve beyond the 1.96 value of z.

Since the computed value of z is 3.29, the decision is made that the difference between the sample mean and the universe mean is significant. The probability of securing this large a deviation from the universe mean is so small that there is little risk in rejecting the hypothesis that the difference between the universe mean and the sample mean is zero. Approximately 8 out of 10,000 samples of 3,600 would fall this far or farther from the mean, so the probability that this particular sample is a random sample from the universe of all farms is so small that it can be ignored. Since it was decided to reject the null hypothesis if the probability of rejecting a true hypothesis was no more than 5 out of 100, this hypothesis should be rejected. This is another way of saying that the farmers who subscribe to the farm paper do not have the same average income as all farmers in the state.

The level of significance to use should be chosen before an analysis is made. There is always danger that the value of z and its probability will influence the decision as to the level of significance to be used to reject the hypothesis. For example, if one had hoped that the above analysis would show a significant difference for the farm paper and the value of z turned out to be 1.86, it might be a temptation to set $\alpha = .10$. At this value of α the table value of z would be 1.64; hence, the hypothesis could be rejected at this level. The decision as to the level of significance to use should be completely independent of the results of the analysis, and the best way to insure this independence is to set the level of significance before computing the value of z.

The preceding computations may be summarized in a few simple steps:

1. Decide on the level of significance (α), which is the probability of committing a Type I error.
2. Select a value of z that cuts off a fraction of the normal curve equal to this level of significance. The most commonly used values of z are 1.96 and 2.58, which represent $\alpha = .05$ and $\alpha = .01$ respectively. However, $z = 3$, representing $\alpha = .0027$, is also used.
3. Compute z for the sample. This is done by finding the difference between the universe mean and the sample mean, divided by the standard error of the mean for this size sample.
4. If this value of z is equal to or larger than the value of z for the level of significance chosen, the hypothesis is rejected.
5. If the value of z is less than that for the level of significance chosen, the hypothesis is *not* rejected. It is worth noting that this smaller value of z does not *prove* the hypothesis is correct; it simply *fails to disprove it.*

Decision Criteria

Figure 8–1 refers to the farm income example and shows the distribution of the means of samples of 3,600 drawn from a universe with a mean of \$5,374 and a standard deviation of \$2,460. The mean of this distribution of sample means is the same as the mean of the universe from which the samples were taken. The standard deviation of the distribution of sample means is \$41, computed on page 191. The X scale on the figure, shown in standard deviation units and also in dollars, runs from \$5,230.50 to \$5,517.50, or from $-3.5\sigma_{\bar{x}}$ to $+3.5\sigma_{\bar{x}}$.

The shaded portions of the distribution represent the proportion of the sample means that deviates more than $1.96\sigma_{\bar{x}}$ from the mean of the universe. The table of Areas of the Normal Curve in Appendix H shows that .025 of the area falls above $+1.96\sigma_{\bar{x}}$ and the same fraction of the area falls below $-1.96\sigma_{\bar{x}}$. Since the distribution of sample means is approximately normal, .05 of the means fall in the two tails of the distribution cut off by the ordinates at $\pm 1.96\sigma_{\bar{x}}$. The sample mean of \$5,509 falls in the portion of the curve above

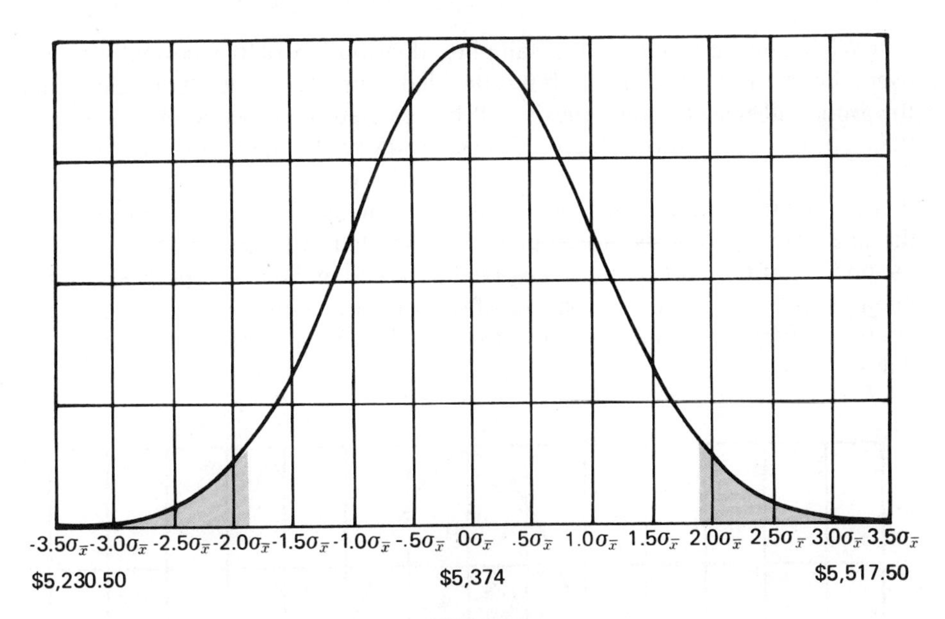

FIGURE 8–1

TWO-TAIL TEST

$+1.96\sigma_{\bar{x}}$, and the hypothesis that it was a sample from the universe of all farmers must be rejected. This is called a *two-tail test,* since the combined areas of both tails of the normal curve are used.

It is very important to note that the standard deviation of the distribution of sample means is $\sigma_{\bar{x}}$ or \$41, and not σ (\$2,460), which is the standard deviation of the universe of values of farm products sold.

It helps in interpreting the test of significance if the null hypothesis to be tested is stated specifically and identified as H_o, and if the alternative hypothesis, H_a, is stated so the consequences of rejecting the null hypothesis will be clearly understood. Also, the criterion on which H_o will be rejected should be clearly stated.

Null hypothesis *(H_o)*: *The mean value of farm products sold by subscribers to the farm paper equals the mean value for all farmers in the state.* ($\mu = \$5{,}374$)

Alternative hypothesis *(H_a)*: $\mu \neq \$5{,}374$

Criterion for decision: Reject H_o and accept H_a if $z > +1.96$ or $z < -1.96$.

The criterion gives $\alpha = .05$. If a value of $\alpha = .01$ is desired, the criterion will be 2.58 instead of 1.96. Since the value of z is 3.29 in this example, the null hypothesis would be rejected for either value of α. In other words, this information indicates that the incomes of subscribers to the farm paper are *not equal* to the average of all farmers in the state.

When only one tail of the sampling distribution of the normal curve is used, the test is a *one-tail test.* If it is decided to use the 5% level of significance for a one-tail test, the hypothesis will be rejected if the value of z is *greater* than $+1.645$, since this value of z cuts off .05 at the upper end of the curve. Since the value of z was computed previously to be 3.29, the hypothesis that the mean income of subscribers is the same as all farmers is rejected and the alternate hypothesis is accepted. The shaded area in Figure 8–2 shows the portion of the normal curve involved in rejecting the null hypothesis for the one-tail test. The same area on the left side of the figure would be used if the alternate hypothesis stated that subscribers' average income was *less* than that of all farmers.

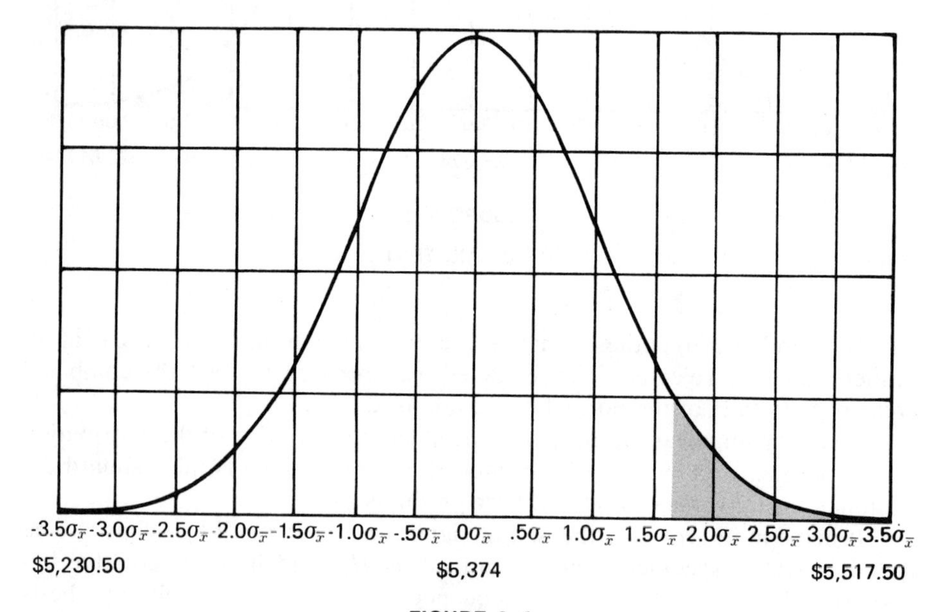

FIGURE 8–2

ONE-TAIL TEST

The one-tail test may be summarized in the same manner as the two-tail test.

Null hypothesis (H_o): *The mean value of farm products sold by subscribers to the farm paper equals the mean value for all farmers in the state.* ($\mu = \$5,374$)

Alternative hypothesis (H_a): $\mu > \$5,374$

Criterion for decision: Reject H_o and accept H_a if $z > +1.645$.

This criterion gives $\alpha = .05$. If a value of $\alpha = .01$ is desired, the criterion will be 2.33 instead of 1.645.

DIFFERENCE BETWEEN UNIVERSE PROPORTION AND SAMPLE PROPORTION

A common test of significance used in making business decisions is to compare the value of a proportion derived from a sample with the value of a known universe proportion.[1] The exact test of significance involves the evaluation of the probabilities of obtaining certain values of the binomial distribution. To avoid lengthy calculations, the normal distribution is used as an approximation of the binomial. For small samples the normal distribution gives a good approximation when the proportion is near .50; but as the proportion departs from .50, a larger sample is needed to give accurate results. The table of Areas of the Normal Curve in Appendix H makes it possible to test the significance of the difference between proportions with a small amount of calculation.

Assume that a metal-stamping machine, when properly set and adjusted, will turn out an average of .05 defective products. Management is satisfied with this proportion of defective units, but wants to be certain that the equipment is doing this well at all times. Therefore, it has been decided that whenever there is convincing evidence that the machine is turning out an average of more than .05 defective products, production will be stopped and the machine adjusted.

Inspection of a lot of 400 items showed 32 defectives. Since this represents .08 defective items, the decision must be made as to whether or not the output of the machine changed from .05 defective. If we were dealing with data from a complete enumeration, there would be no question that .08 is larger than the desired .05. But in this case sampling error is expected in the data, and it becomes necessary to test the sample proportion for a significant difference from the .05 universe value. This involves testing the hypothesis that the difference between the sample proportion .08 and the assumed universe proportion .05 is zero.

The procedure is basically the same as testing the difference between the universe mean and a sample mean. The value of z is given by the formula

$$z = \frac{p - \pi}{\sigma_p} \tag{8-2}$$

where

π = the assumed universe proportion
p = the sample proportion.

The standard error of a proportion was computed in Chapter 7 by Formula 7–8.

$$\sigma_p = \sqrt{\frac{\pi(1-\pi)}{n}\,\frac{N-n}{N-1}}.$$

[1] In Chapter 7 the proportion was expressed as a percentage, but in this chapter and in Chapter 9 the proportion will be expressed as a decimal fraction. This involves nothing more than moving the decimal point two places to the left.

It was also pointed out that when the sampling fraction is small, the formula can be written

$$\sigma_p = \sqrt{\frac{\pi(1-\pi)}{n}}.$$

It is safe to assume in this case that the size of the sample is very small in relation to the universe since the total number of articles that can be made on the machine may be considered the universe. With a small sampling fraction, the correction for the size of the universe may be considered to be equal to unity.

The standard error of the proportion and the value of z are computed

$$\sigma_p = \sqrt{\frac{(.05)(.95)}{400}} = .0109$$

$$z = \frac{.08 - .05}{.0109} = 2.75.$$

Since we are interested in determining the probability of securing a proportion defective greater than .05, a one-tail test should be applied. If we use the 5% level of significance, as in the previous example, the null hypothesis will be rejected if the value of z is greater than 1.645. Since the value of z was computed to be 2.75, the hypothesis that the proportion defective has not changed from the .05 is rejected. In other words, something apparently has gone wrong with the process and corrective measures should be taken before any more items are produced.

This test may be summarized

Null hypothesis (H_o): *The machine is turning out material that has a proportion defective of .05.* $(\pi = .05)$

Alternative hypothesis (H_a): $\pi > .05$

Criterion for decision: Reject H_o and accept H_a if $z > +1.645$

This criterion gives $\alpha = .05$.

DIFFERENCE BETWEEN UNIVERSE MEAN AND SAMPLE MEAN, UNIVERSE STANDARD DEVIATION NOT KNOWN

In the example discussed on pages 189 to 194, the standard deviation of the universe was known. There are many situations, however, in which it is not known. In Chapter 7 the standard deviation of the sample was used to estimate the universe standard deviation in computing the confidence interval of an estimate. This same device can be used in the testing of hypotheses when the universe standard deviation is not known.

Estimating the Universe Standard Deviation

Assume that a textile mill wants to test the fiber strength of a certain lot of cotton. The Pressley strength index—thousands of pounds of pressure required to break the equivalent of one square inch of fiber—is used to measure the strength of the fiber. A sample of 41 pieces of fiber is drawn. The sample has a mean strength of 85.78 pounds and a standard deviation of 15.49 pounds. From these sample data it must be determined whether or not the fiber strength of this lot of cotton differs significantly from 90 pounds, the desired strength. The only information known about the variation in the strength of the fiber is the standard deviation of the sample; the standard deviation of the universe is unknown.

The standard deviation of the sample can be used to estimate the standard deviation of the universe, but the estimate of the standard deviation is only approximate since it contains sampling error, which is inherent in any estimate of a parameter from a statistic. The distribution of the value of z computed from $\frac{\bar{x}-\mu}{\sigma_{\bar{x}}}$ is distributed normally, and in the test of significance described previously, the table of Areas of the Normal Curve in Appendix H was used to interpret the values of z. However, if $\hat{\sigma}$, the estimated value of the parameter σ, is substituted for σ to compute $\hat{\sigma}_{\bar{x}}$ instead of $\sigma_{\bar{x}}$, the resulting distribution of the values $\frac{\bar{x}-\mu}{\hat{\sigma}_{\bar{x}}}$ is not normal, but takes the form of the t distribution described below.

The *t* Distribution

The *t distribution* may be used when a sample has been drawn from a normal parent population. The t distribution is symmetrical but slightly flatter than the normal distribution. In addition, it differs for each different number of degrees of freedom.[2] When the number of degrees of freedom is very small, the variation from the normal distribution is fairly marked; but as the degrees of freedom increase, the t distribution resembles the normal distribution more and more. When there are as many as 100 degrees of freedom, there is very little difference; and as the degrees of freedom approach infinity, the t distribution approaches the normal distribution.

In the type of test discussed in this section, degrees of freedom are defined as $n - 1$. Degrees of freedom may be defined in other ways for other types of problems.

Instead of computing t from the estimate of the universe standard deviation, it is possible to compute the value of t directly from the sample standard deviation.

[2] Since sample data were used in estimating the standard error used in computing the value of t, degrees of freedom must be used instead of n in the interpretation of the results.

$$t = \frac{\bar{x} - \mu}{\frac{s}{\sqrt{n-1}}}. \qquad \textbf{(8-3)}$$

For the data on the strength of cotton fiber, the value of t is

$$t = \frac{85.78 - 90.00}{\frac{15.49}{\sqrt{41-1}}} = \frac{-4.22}{2.45} = -1.72.$$

Certain values of the t distribution are given in Appendix K. The interpretation of the value of t is the same as the interpretation of z in the previous example. However, the table of t values is not quite so simple to use as the table to Areas of the Normal Curve in Appendix H, because there is a different distribution for each number of degrees of freedom. In the table of t values, degrees of freedom are used instead of n, which is used in the table of z values. When dealing with a standard deviation, one degree of freedom is lost in estimating the universe standard deviation from the sample, as explained in Chapter 7.

Use of the table of t values first requires finding the line for the correct number of degrees of freedom and then reading across the table to the column giving the desired level of significance. The table gives the values of t for the two-tail test associated with the probabilities in the captions.[3] Assume that it has been decided to test at the .05 level ($\alpha = .05$) the hypothesis stated earlier relating to fiber strength. The value of t for 40 degrees of freedom at the .05 level of significance is found to be 2.021. Since the computed value of t is less than 2.021, we cannot reject the hypothesis. It will be observed that a somewhat larger value of t is required to reject a hypothesis at a given level of significance than when the value of z is being used. For example, the value of z for the .05 level of significance is only 1.96 for a two-tail test, compared with the t value of 2.021 for 40 degrees of freedom. However, the larger the size of the sample used, the closer the value of t needed approaches the value of z. The last line in the table of t values represents the values for the normal curve.

This test may be summarized

Null hypothesis (H_o): *The mean strength of a lot of cotton fiber is 90 on the Pressley index.* ($\mu = 90$ pounds)
Alternative hypothesis (H_a): $\mu \neq 90$ pounds
Criterion for decision: Reject H_o if $t > +2.021$ or $t < -2.021$
This criterion gives $\alpha = .05$.

[3] The table of Areas of the Normal Curve (Appendix H) gives area values for one tail of the curve. To make a two-tail test, it is necessary to double the area values given in the table. The difference between the values of areas of the normal curve and the values of the t distribution should be noted carefully.

DIFFERENCE BETWEEN TWO SAMPLE MEANS

In experimental work and in quality control, the means of two samples are frequently tested to determine whether there is a significant difference between them. For example, two lots of cotton were tested for fineness, using five samples taken at random from each lot. One lot came from the Lower Rio Grande Valley and the other from the High Plains of Texas. It was desired to determine whether there was a significant difference in the fineness of the two lots. The samples from the two lots are given in Table 8–1 in micronaire readings, that is, the weight of one inch of fiber in micrograms.

The mean fineness of the cotton from the Lower Rio Grande Valley ($\bar{x}_1$) is 4.3 micrograms and that of the High Plains cotton ($\bar{x}_2$) is 4.6 micrograms. The problem is to decide on the basis of these samples whether the two lots of cotton differ significantly in fineness. If the two universes are normal and have the same variance (or standard deviation), it is appropriate to use the *t* distribution to test for a significant difference. Even if the universe departs moderately from normal, the *t* test will give fairly satisfactory results. However, the universes must have a common variance; if they do not, no exact test can be applied.

Assume that all possible pairs of two samples are taken from a normal universe and the difference between the means of each pair of samples is computed. If this assumption is carried out, it will generate a distribution of differences, the mean of which is zero. The standard deviation of the distribution of differences is

$$\hat{\sigma}_{\bar{x}_1 - \bar{x}_2} = \hat{\sigma} \sqrt{\frac{1}{n_1} + \frac{1}{n_2}}. \tag{8-4}$$

TABLE 8–1

FINENESS OF TWO SAMPLES OF TEXAS COTTON

(Micrograms)

Lower Rio Grande Valley		High Plains	
x_1	Computations	x_2	Computations
4.5	$\Sigma x_1 = 21.5$	4.7	$\Sigma x_2 = 23.0$
3.7	$\Sigma x_1^2 = 93.07$	5.1	$\Sigma x_2^2 = 106.44$
4.6	$\bar{x}_1 = \frac{21.5}{5} = 4.3$	4.8	$\bar{x}_2 = \frac{23.0}{5} = 4.6$
4.1	$s_1^2 = \frac{93.07}{5} - \left(\frac{21.5}{5}\right)^2$	4.3	$s_2^2 = \frac{106.44}{5} - \left(\frac{23.0}{5}\right)^2$
4.6	$s_1^2 = .124$	4.1	$s_2^2 = .128$

Source: Cotton Economic Research, The University of Texas at Austin.

The value of the t distribution for any two sample means is computed

$$t = \frac{\bar{x}_1 - \bar{x}_2}{\hat{\sigma}_{\bar{x}_1 - \bar{x}_2}}. \tag{8-5}$$

The value of $\hat{\sigma}$ is the estimated standard deviation of the universe from which the pairs of samples were drawn. The problem of finding an estimate of the standard deviation of the universe differs somewhat from the previous example of the computation of the value of t. Since it is assumed that the two samples were drawn from the same universe, both samples may be considered as an estimate of the universe standard deviation and may thus be combined to give a better estimate than if only one of the samples was used. This is accomplished by first computing for each sample the sum of the squared deviations from the mean of the sample. The two sums of squares are then added and the variance is computed from this total. Instead of dividing by the total number of items in the two samples, however, allowance must be made for the loss of one degree of freedom in the computation of each of the means. This results in the total number of degrees of freedom being two less than the total of n_1 and n_2, as shown in Formula 8–6. The resulting value of $\hat{\sigma}$ is the best estimate that can be made of the common variance of the two samples, and this value is used in Formula 8–4 to compute the value of $\hat{\sigma}_{\bar{x}_1 - \bar{x}_2}$.

$$\hat{\sigma} = \sqrt{\frac{\Sigma(x_1 - \bar{x}_1)^2 + \Sigma(x_2 - \bar{x}_2)^2}{n_1 + n_2 - 2}}. \tag{8-6}$$

Since $s_1^2 = \dfrac{\Sigma(x_1 - \bar{x}_1)^2}{n_1}$ by Formula 4–6, the value of $\Sigma(x_1 - \bar{x}_1)^2$ is

$$\Sigma(x_1 - \bar{x}_1)^2 = s_1^2 n_1.$$

The computation of the value of t for the data on fineness of cotton, given in Table 8–1, is shown below.

$$s_1^2 = .124 \qquad \Sigma(x_1 - \bar{x}_1)^2 = (.124)(5) = .62$$

$$s_2^2 = .128 \qquad \Sigma(x_2 - \bar{x}_2)^2 = (.128)(5) = .64$$

$$\hat{\sigma} = \sqrt{\frac{\Sigma(x_1 - \bar{x}_1)^2 + \Sigma(x_2 - \bar{x}_2)^2}{n_1 + n_2 - 2}} = \sqrt{\frac{.62 + .64}{5 + 5 - 2}} = .3969$$

$$\hat{\sigma}_{\bar{x}_1 - \bar{x}_2} = \hat{\sigma}\sqrt{\frac{1}{n_1} + \frac{1}{n_2}} = .3969\sqrt{\frac{1}{5} + \frac{1}{5}}$$

$$= .3969\sqrt{.4} = .251$$

$$t = \frac{\bar{x}_1 - \bar{x}_2}{\hat{\sigma}_{\bar{x}_1 - \bar{x}_2}} = \frac{4.3 - 4.6}{.251} = -1.2.$$

Since the value of t at the .01 level of significance with 8 degrees of freedom is 3.355, we conclude that the difference is not significant at the .01 level. This means that there is not a real difference between the fineness of the two lots of cotton. In other words, the difference can be explained logically by sampling variation.

It should be noted that for this test, degrees of freedom have been defined as the sum of the two sample sizes minus 2, $n_1 + n_2 - 2$.

This test may be summarized

Null hypothesis (H_0): *The two lots of cotton do not differ significantly in fineness; in other words, they are samples from two universes with the same fineness.* ($\mu_1 = \mu_2$)

Alternative hypothesis (H_a): $\mu_1 \neq \mu_2$

Criterion for decision: Reject H_0 and accept H_a if $t > +3.355$ or $t < -3.355$

This criterion gives $\alpha = .01$ and degrees of freedom $= 8$.

It should be emphasized that the choice between using the z test and the t test depends upon the knowledge one possesses about the distribution of the universe. If we know the value of the standard deviation of the universe, a z test is appropriate. If we must estimate the standard deviation from a sample, the appropriate probability distribution is the t distribution.

DIFFERENCE BETWEEN TWO SAMPLE PROPORTIONS

The comparison of two sample proportions or percentages to determine if there is a significant difference is similar in concept to the comparison of the arithmetic means of two samples. The probabilities associated with the normal curve may be used as an approximate test of the difference between two proportions. When the samples are reasonably large, this approximation is sufficiently accurate and is almost universally used.

Assume that one trimming machine turns out 25 defective items in a lot of 400 and that another machine turns out 42 defectives in a lot of 600. We want to know whether there is a significant difference in the percentage of defective items turned out by the two machines. Since nothing is known about what to expect from the machines, the decision must be based on the evidence furnished by the two samples.

We may test the hypothesis that the two machines produce the same proportion of defective items, which means that the difference between the two proportions is zero. The proportion of defective items produced by the first machine is $p_1 = \frac{25}{400} = .0625$. The proportion of defectives produced by the second machine is $p_2 = \frac{42}{600} = .07$. Since the hypothesis states that the two machines are the same with respect to the proportion of defective items produced, we may use the two samples to compute the estimate of the proportion defective ($\bar{p}$), which is a weighted average of the two samples:

$$\bar{p} = \frac{25 + 42}{400 + 600} = .067 \qquad \bar{q} = 1 - \bar{p} = 1 - .067 = .933.$$

The standard error of the difference between two proportions is computed

$$s_{p_1 - p_2} = \sqrt{\frac{\bar{p}\bar{q}}{n_1} + \frac{\bar{p}\bar{q}}{n_2}}. \tag{8-7}$$

For the two machines this value is

$$s_{p_1 - p_2} = \sqrt{\frac{(.067)(.933)}{400} + \frac{(.067)(.933)}{600}} = .0161.$$

To test the hypothesis that the difference between the two proportions is zero requires the computation of the value of z. If it is decided to test the hypothesis at the .01 level of significance, the hypothesis will be rejected if the computed value of z is 2.58 or greater. The computation of z is

$$z = \frac{p_1 - p_2}{s_{p_1 - p_2}} \tag{8-8}$$

$$z = \frac{.0625 - .07}{.016} = -.47.$$

Since the value of z does not exceed the .01 level, one is not justified in rejecting the hypothesis that the difference between the two proportions is zero.

This test may be summarized

Null hypothesis (H_o): *The difference between the proportion defective turned out by the two machines is zero.* $(\pi_1 = \pi_2)$

Alternative hypothesis (H_a): $\pi_1 \neq \pi_2$

Criterion for decision: Reject H_o and accept H_a if $z > +2.58$ or $z < -2.58$

This criterion gives $\alpha = .01$.

ANALYSIS OF VARIANCE

In testing the difference between two sample means, it was assumed on page 199 that both samples were drawn from the same universe. If this assumption is correct, the two sample variances computed in Table 8–1, page 199, can be used to make two estimates of the variance of the universe from which the two samples were presumably drawn. Using the values of s_1^2 and s_2^2 computed in Table 8–1, the estimates of the variance are

$$s_1^2 = .124$$

$$\hat{\sigma}_1^2 = s_1^2 \left(\frac{n_1}{n_1 - 1}\right) = .124 \frac{5}{5 - 1} = .1550$$

$$s_2^2 = .128$$

$$\hat{\sigma}_2^2 = .128 \frac{5}{5-1} = .1600.$$

Comparison of Variances

The estimated variance of the population based on the sample of cotton from the High Plains of Texas is greater than that from the Lower Rio Grande Valley of Texas. But since both of the estimates are based on sample data, the difference must be tested for significance. The ratio of the larger variance to the smaller is computed

$$\frac{\text{Larger variance (High Plains)}}{\text{Smaller variance (Lower Rio Grande Valley)}} = \frac{.1600}{.1550} = 1.03.$$

The cotton from the High Plains shows the greater variation, but there is no way to determine by looking at the ratio whether this variation is a random variation due to sampling alone or whether it is great enough to be attributed to a specific cause. If successive pairs of samples are drawn from a given universe and the variances of both samples computed, the ratio of the larger variance to the smaller will vary from sample to sample, even though the samples are from the same universe. The distribution of these ratios will form the *F distribution.*

The *F* Distribution

The F distribution starts out at zero and rises to a peak at the value equal to $\frac{n_2(n_1 - 2)}{n_1(n_2 + 2)}$ and then falls again to zero as F increases without limit. The mean of this distribution is $\frac{n_2}{n_2 - 2}\cdot$ The shape of the distribution varies with n_1 and n_2, but as n_1 and n_2 become larger the distribution tends to become symmetrical. As n_1 and n_2 increase without limit, the F distribution approaches the normal distribution. If one of the n's increases without limit while the other n remains small, the F distribution approaches the χ distribution. If $n_1 = 1$ and n_2 increases without limit, the term $\sqrt{F}$ approaches the t distribution, so $F = t^2$. In other words, the normal distribution, χ^2, and t are special cases of the F distribution, which lends itself to a large number of applications in the analysis of sample data. One use is the problem of testing two variances to determine the probability that they are random samples from the same universe.

An abridged table of the F distribution is given in Appendix M. The complete F distribution for all the combinations of sample sizes would be a very large table, so published tables are limited to certain points on the distribution, and these values are given for only a limited number of samples. The values of F are always given for the number of degrees of freedom ($d.f._1$ and $d.f._2$) rather than the total number in the sample.

The values of the upper tail of the F distribution give the values of F that are exceeded by 5% (in roman type) and 1% (in boldface type) of the distribution. If the value of F computed from the two samples is larger than the .01 value, it is concluded that the probability of occurrence of a ratio this large is .01. In other words, in this case the value of α is .01 and the null hypothesis that $\sigma_1^2 = \sigma_2^2$ is rejected.

The value of F computed from the two variances above is 1.03, which is substantially below both the .01 and the .05 levels of F for 4 degrees of freedom for each variance. For these two samples to have shown a significant difference at the .01 level, the ratio of the two variances would have had to exceed 15.98. For a significant difference to be shown at the .05 level, the value of F required would be 6.39. This test does not reject the null hypothesis that the two variances are equal, so the cotton from the Lower Rio Grande Valley and from the High Plains should be considered to have the same degree of variation until the time that further evidence gives a different indication.

This test may be summarized

Null hypothesis (H_o): $\sigma_1^2 = \sigma_2^2$
Alternative hypothesis (H_a): $\sigma_1^2 \neq \sigma_2^2$
Criterion for decision: Reject H_o if $F_{.01} > 15.98$
This criterion gives degrees of freedom = 4 for each variance.

If one thinks of F as a ratio of two variances,

$$F = \frac{\hat{\sigma}_2^2}{\hat{\sigma}_1^2} = \frac{\dfrac{\Sigma(x_2 - \bar{x}_2)^2}{n_2 - 1}}{\dfrac{\Sigma(x_1 - \bar{x}_1)^2}{n_1 - 1}},$$

then the degrees of freedom for the greater mean square (greater variance) is $n_2 - 1 = 5 - 1 = 4$. The degrees of freedom for the smaller mean square (smaller variance) is $n_1 - 1 = 5 - 1$, which is also 4. If the two samples had been of different sizes, the two sets of degrees of freedom would have been different.

Testing the Difference Between Three or More Means

On page 199 a method was given for testing whether or not two sample means differ significantly. This method can be used for two means, but if the means of more than two groups are used, another approach is necessary. The data in Table 8–2 will be used to illustrate the method of analysis of variance, using the F distribution for testing more than two sample means for a significant difference. Table 8–1 gives the results of testing two lots of cotton for fineness. One lot came from the Lower Rio Grande Valley and the other from the High Plains. Table 8–2 contains the same information as Table 8–1, plus data on fineness for two additional lots of cotton, one from East Texas and

TABLE 8–2

FINENESS OF FOUR SAMPLES OF TEXAS COTTON FROM FOUR REGIONS OF TEXAS

(Micrograms)

Region	Individual Items					Σx	$\bar{x}$	Σx^2
Lower Rio Grande Valley	4.5	3.7	4.6	4.1	4.6	21.5	4.30	93.07
High Plains	4.7	5.1	4.8	4.3	4.1	23.0	4.60	106.44
East Texas	3.9	4.2	4.3	3.8	3.4	19.6	3.92	77.34
Coastal Plains	5.4	3.3	4.0	3.8	4.6	21.1	4.22	91.65
Total						85.2	17.04	368.50

$$\bar{\bar{x}} = \frac{17.04}{4} = 4.26$$

the other from the Texas Coastal Plains. Each of these four lots of cotton is considered a random sample from a different area in Texas, and the conditions under which cotton is grown differ substantially from region to region. The problem is to determine whether or not the four samples differ significantly with respect to fineness, or whether the four samples can be assumed to differ no more than if they had all been taken from the same universe. This problem is the same as testing the hypothesis that two samples came from the same universe; the only difference in the present example is the fact that four sample means are involved instead of two.

It would be possible to take each pair of locations separately and test the difference between the two means by using the t distribution, but this is not quite the same as testing whether the four means differ significantly among themselves. A method of making this test is to determine whether the variation in fineness is significantly greater between the four means of the regions than the variation that exists within the individual classes representing different regions of Texas.

The null hypothesis to be tested is that the four samples are random samples from the same universe. If the probability of this occurrence is very small, the null hypothesis will be rejected and the variation among the means will not be considered significant.

The variation between the four means is computed by finding the deviation of the mean of each region from the grand mean, that is, the mean of all the cotton in the study regardless of where it was grown. The deviation of each class mean from the grand mean is squared and the squared deviation multiplied by the number in the group. These calculations are made for each region, and the totals for the four regions are added. This total for the four groups measures the variation in fineness that occurred between the four cotton-growing regions.

The computation of the total squared deviations of the means is shown in Table 8–3. The mean of all 20 items is 4.26 micrograms, and the mean

TABLE 8–3

VARIATION OF FINENESS OF TEXAS COTTON BETWEEN FOUR REGIONS OF TEXAS

Computation of the Sum of the Squared Deviations of Group Means from Grand Mean

Region	n	$\bar{x}$	$\bar{x} - \bar{\bar{x}}$	$(\bar{x} - \bar{\bar{x}})^2$	$n(\bar{x} - \bar{\bar{x}})^2$
Lower Rio Grande Valley	5	4.30	.04	.0016	.0080
High Plains	5	4.60	.34	.1156	.5780
East Texas	5	3.92	−.34	.1156	.5780
Coastal Plains	5	4.22	−.04	.0016	.0080
Total	20				1.1720

$\bar{\bar{x}} = 4.26$

of the 5 batches of cotton from the Lower Rio Grande Valley is 4.30 micrograms. The deviation, .04 micrograms, is squared and multiplied by 5, the size of the sample, giving .0080, which is entered on the first line of the last column. This calculation is made for each of the regions, and the total of the last column, the squared deviations between the class means, is found to be 1.1720. This total represents the portion of the total variance that is associated with the regions, and may be referred to as deviations between regions.

The variation within each of the four regions must next be computed. This will involve subtracting the mean of the items in the region from each of the individual items, squaring these differences, and summing the five squared deviations, just as was done in Table 4–3 in computing the standard deviation. The term being computed for each district is represented by $\Sigma(x - \bar{x})^2$, but it was shown by Formula 4–8 that this term can be computed

$$\Sigma(x - \bar{x})^2 = \Sigma x^2 - \frac{(\Sigma x)^2}{n}. \qquad \textbf{(8–9)}[4]$$

Using the sum of the squared items, Σx^2, and the sum of the items, Σx, from Table 8–2, the sum of the squared deviations from the mean of the data for the Lower Rio Grande Valley is computed

$$\Sigma(x - \bar{x})^2 = 93.07 - \frac{(21.5)^2}{5} = 93.07 - 92.45 = .62.$$

Similar computations are made for each of the regions, and the results are entered in the last column of Table 8–4. The total of the four sums of the squared deviations is 4.376 and represents the total deviation within the four classes. This variation is in no way related to any variation in the fineness of cotton resulting from the region in which it was grown. The deviation of each sample is computed from the mean of the items in the region, so the variation is the result of factors other than those related to the growing region.

[4] Refer to page 86, Formula 4–8, and substitute n for N, x for X, and $\bar{x}$ for μ.

TABLE 8–4

VARIATION OF FINENESS OF TEXAS COTTON WITHIN FOUR REGIONS OF TEXAS

Computation of Squared Deviations from Mean of Each Class

Region	n	Σx^2	Σx	$(\Sigma x)^2$	$\frac{(\Sigma x)^2}{n}$	$\Sigma x^2 - \frac{(\Sigma x)^2}{n}$
Lower Rio Grande Valley	5	93.07	21.5	462.25	92.450	.620
High Plains	5	106.44	23.0	529.00	105.800	.640
East Texas	5	77.34	19.6	384.16	76.832	.508
Coastal Plains	5	91.65	21.1	445.21	89.042	2.608
Total	20					4.376

The calculations made have produced two totals, one that represents the variation in fineness of cotton within the four regions, and another that represents the variation in fineness between these regions. The sum of the variations from these two sources is the total variation in the fineness of cotton produced in the area made up of the four regions. In other words, the batches of cotton may be treated as a sample from the area consisting of the four regions, and the total variation in these 20 items computed. The total of the squared deviations from the mean of the 20 observations is the same as the sum of squared deviations between classes plus the sum of the squared deviations within classes.

In Table 8–2 the sum of all the squared items is 368.50, and the sum of all the items is 85.2. The computation of the sum of the squared deviations from the mean, using Formula 8–9, is

$$\Sigma(x - \bar{x})^2 = 368.5 - \frac{(85.2)^2}{20} = 368.5 - \frac{7{,}259.04}{20}$$

$$\Sigma(x - \bar{x})^2 = 368.5 - 362.952 = 5.548.$$

The fact that the sum of the squared deviations within the regions and the sum of the squared deviations between the regions equals the total squared deviations computed directly from the original 20 observations proves the accuracy of the calculations. These figures are entered in the second column of Table 8–5, where the final computations in the analysis of variance are made.

The sum of the squared deviations within groups was computed from four groups of five items each, while the total squared deviations of the means was computed from only four means. To adjust for the different number of items considered in the two groups, it is necessary to compute the mean of the squared deviations both between groups and within groups. In Chapter 4 the mean of the squared deviations is designated σ^2, or the variance, and either term may be used. For a sample variance, s^2 is the symbol used.

In determining the number to be used in computing the variance, attention must be given to the number of degrees of freedom (*d.f.*) since one is dealing with sample data.

TABLE 8–5

ANALYSIS OF VARIANCE OF THE FINENESS OF TEXAS COTTON FROM FOUR REGIONS IN TEXAS

Source of Variation	Sum of Squared Deviations	Degrees of Freedom	Variance
Between regions	1.1720	3	.3907
Within regions	4.3760	16	.2735
Total	5.5480	19	

$$F = \frac{.3907}{.2735} = 1.4285 \qquad d.f._1 = 3 \qquad F_{.01} = 5.29$$

$$d.f._2 = 16 \qquad F_{.05} = 3.24$$

When making estimates of variance from a sample, one degree of freedom is lost, so the degrees of freedom equal $n - 1$. When computing the variance of a group with $n = 5$, there are 4 degrees of freedom. Each of the four groups has 1 degree of freedom less than the number in the sample, so a total of 16 degrees of freedom is associated with the total of the squared deviations, 4.376, obtained from Table 8–4. This number is entered on the second line of the first column in Table 8–5.

The total of the squared deviations of the group means from the mean of all the items involves 3 degrees of freedom since there are 4 group means. This number of degrees of freedom is entered on the first line of the second column in Table 8–5. The sum of the degrees of freedom within groups and between groups is 19. This is the correct number of degrees of freedom for the variance of the 20 items treated as 1 sample, thus checking the computation of the various degrees of freedom occurring in this problem. The average of the squared deviations within groups, involving 16 degrees of freedom, and the average of the squared deviations between groups, involving 3 degrees of freedom, are entered in the last column of Table 8–5. Since we are not interested in the total variance of the 20 items, this value was not computed.

In this problem the variance between groups is larger than the variance within groups, which suggests that in the universe the variance between groups may be greater than within the groups. Since the difference may be only sampling error, the ratio of the larger variance to the smaller is tested for significance for the same reason that the two means were tested for a significant difference on pages 199 to 201. The ratio of the variance between groups to the variance within groups is

$$F = \frac{.3907}{.2735} = 1.4285.$$

If the variance between groups and the variance within groups were the same, this ratio would equal 1. If the ratio is less than 1, it suggests that the variation between regions is *less* than within regions. However, when the ratio computed from a sample is greater than unity, as it is in this example, further testing will be necessary to decide whether the ratio of 1.43 is far enough above 1.00 to conclude that something other than sampling variation accounts for the difference. In the example above, the variance between groups has 3 degrees of freedom and the variance within groups has 16 degrees of freedom. In Appendix M the column for the number of degrees of freedom for the larger variance, in this case 3, is located and the row for 16 degrees of freedom (the smaller variance) is read in the stubs. At the intersection of this column and this row the value of F that cuts off the upper 5% of the distribution is given in roman type and the value of F that cuts off the upper 1% of the distribution is in boldface type. In this example the value of F at the 1% level ($F_{.01}$) is 5.29 and the value at the 5% level ($F_{.05}$) is 3.24. If we had decided that the hypothesis would be rejected if the probability of making a Type I error was .05, the decision to reject the hypothesis would depend upon the computed value of F being larger than 3.24. If the decision was made to reject the hypothesis if the probability of making a Type I error was .01, the hypothesis would be rejected if the value of F was greater than 5.29. Since the computed value of F was 1.43, the null hypothesis that the true ratio between the variances is 1.00 would not be rejected at either the .05 or the .01 level. With only these two values of the F distribution given, only values of α of .05 or .01 can be used. The problem of choosing a value of α for rejection of the null hypothesis has been discussed earlier in this chapter, and that discussion applies to this situation.

This test may be summarized

Null hypothesis *(H_o)*: *The fineness of the cotton from the four regions is the same; or, the variation between means equals the variance within regions.*

Alternative hypothesis (H_a): The fineness of cotton differs significantly with the regions.

Criterion for decision: Reject H_o if $F_{.01} > 5.29$

This criterion gives degrees of freedom = 3 for the variance between groups, and degrees of freedom = 16 for the variance within groups.

The interpretation of the results of this analysis is valid only if the distributions from which the within-group variance is estimated are normally distributed universes. Some departure from normal does not introduce serious error into the analysis, but extremely skewed distributions should be avoided. It is assumed that the several groups used to compute the variance within classes are homogeneous with respect to their variance. Extreme heterogeneity will seriously distort the test of significance, but small departures from a common variance will not affect it.

It should be noted that the above test is one-tail, since the null hypothesis is to be rejected if the computed value of F falls in the upper tail of the distribution. Problems of testing for a significant difference in the variance between groups and the variance within groups use a one-tail test. Since a two-tail test is not commonly used, percentages on the tail of the F distribution at the lower end are usually not given in tables. If it is needed, however, the F value which has 5% of the curve below it is the reciprocal of the 5% value found in Appendix M by using this table with the degrees of freedom reversed. In other words, 5% of the F distribution used in the example above falls below the F value of $\frac{1}{8.69}$ or .12. The two-tail test would use the probability of 90% that the value of F would fall between .12 and 3.24. The value of F at the 1% level in the lower tail of the distribution is computed in the same way.

Testing the Difference Between Two Means

The use of the F distribution to test for significant variations between groups is similar to the problem for which the t distribution was used. In fact, the t distribution is a special case of the more general F distribution. When the number of degrees of freedom between groups equals 1, $\sqrt{F} = t$ and the F distribution equals t^2 for the same number of degrees of freedom used in computing the variance within groups.

The difference between two means tested by the t distribution in the illustration on pages 200 to 201 can also be tested by the F distribution. Using the values of Σx^2 and Σx for the Lower Rio Grande Valley and the High Plains regions of Texas in Table 8–4, the sum of the squared deviations within groups is found to be 1.26.

Lower Rio Grande Valley:

$$\Sigma(x - \bar{x})^2 = 93.07 - \frac{(21.5)^2}{5} = 93.07 - 92.54 = .620$$

High Plains:

$$\Sigma(x - \bar{x})^2 = 106.44 - \frac{(23.0)^2}{5} = 106.44 - 105.80 = .640$$

$$\text{Total squared deviations} = 1.260$$

The means of the two groups are 4.30 and 4.60, and the mean of the two groups combined is 4.45. This value has been designated $\bar{\bar{x}}$. The squared deviations of the two group means from the grand mean, weighted by the number of items in each group, is computed as shown below and is found to be .225.

Lower Rio Grande Valley: $n(\bar{x} - \bar{\bar{x}}) = 5(4.30 - 4.45)^2 = 5(.0225) = .1125$

High Plains: $n(\bar{x} - \bar{\bar{x}}) = 5(4.60 - 4.45)^2 = 5(.0225) = .1125$

Total squared deviations = .2250

The calculation of the two variances is shown in Table 8–6, together with the number of degrees of freedom. The value of F is computed from the variances as

$$F = \frac{\text{Variance between groups}}{\text{Variance within groups}} = \frac{.225}{.1575} = 1.43.$$

From Appendix M the values of F are found to be

$$F_{.05} = 5.32 \qquad d.f._1 = 1$$

$$F_{.01} = 11.26 \qquad d.f._2 = 8$$

TABLE 8-6

ANALYSIS OF VARIANCE OF FINENESS OF COTTON FROM TWO REGIONS IN TEXAS

Source of Variation	Sum of Squared Deviations	Degrees of Freedom	Variance
Between regions	.2250	1	.2250
Within regions	1.2600	8	.1575
Total	1.4850	9	

The hypothesis to be tested is

Null hypothesis (H_o): *Variance between groups = variance within groups.*

Alternative hypothesis (H_a): Variance between groups is greater than within groups.

Criterion for decision: Reject H_o if $F_{.01} > 11.26$

This criterion gives degrees of freedom = 1 for the variance between groups and degrees of freedom = 8 for the variance within groups.

Since the value of F is 1.43, the null hypothesis is not rejected and it appears that the difference between the means of the groups is not large enough to warrant the conclusion that such variation between groups is significant.

This conclusion, reached from applying the F test to the data, is the same as reached when the data were tested by the t distribution.

STUDY QUESTIONS

8-1. You are inspecting a certain lot of a product to determine whether it meets the quality that has been specified. What type of error have you made if you decide that the lot is acceptable when in fact it is below the quality you should accept? What undesirable consequences might follow from making this type of error?

8-2. What type of error have you made if you reject a lot as unsatisfactory when in fact it meets the specified standards? What undesirable consequences might likely follow from making this type of error?

8-3. Is it possible to make a Type I error and a Type II error on the same decision? Explain.

8-4. How do you determine the probability of making a Type I error that you are willing to risk in a given decision?

8–5. Is there any difference between saying that a hypothesis has been accepted and saying that it has not been rejected? Explain.

8–6. What is the difference between a one-tail test and a two-tail test?

8–7. Why is it unwise to trust one's intuition in deciding whether or not a sample statistic (such as a proportion) shows a significant deviation from the parameter?

8–8. Under what circumstances should the t distribution be used instead of the normal distribution?

8–9. Under what circumstances may the normal distribution be used as an approximation of the t distribution?

8–10. What are the practical advantages of using the normal distribution as an approximation of the binomial distribution?

8–11. Why is it necessary to test the difference between two sample means or two sample proportions for significance?

8–12. Why is the correction for the size of the universe commonly ignored in making tests of significance?

PROBLEMS

8–1. If the following observations are known to be a random sample from a universe with a standard deviation of 16, can you reject the hypothesis that the universe mean is 900? Use a one-tail test.

925	905	985	920
910	915	900	910
890	904	887	904

8–2. A production line is producing a steel part with a diameter of 1.240 inches. The standard deviation of the parts produced by this process is .0575 inches. A random sample of 5 is taken from current production and the mean diameter is found to be 1.359 inches. Does this mean that something has gone wrong with the process, or is the variation due to sampling? Explain.

8–3. A foundry is producing castings with an average weight of 115.4 pounds and a standard deviation of 12.45 pounds. A sample of 16 is taken from current production and found to have an average weight of 112.6 pounds. It is important that the castings weigh 115.4 pounds or more. On the basis of the sample, would you accept the lot as being 115.4 pounds or more? Explain how you would make the decision, based on the sample information.

8–4. In a certain city 65% of the families own their homes. A survey made among the subscribers to a magazine asked whether or not the family owned the home in which they were living. A random sample of 1,200 homes was taken and 62% of the subscribers reported that they owned their homes.

Is it correct to conclude that the proportion of subscribers who own their homes is significantly different from the percentage in the population of the city? Show your calculations and state your reasoning.

8–5. A life insurance salesman with a great deal of experience in estate planning believes that the peak earning age of businessmen is reached at age 52. A random sample of 50 retired businessmen was studied and the age at which they received their largest salaries determined. This average was 57 years; the standard deviation of the sample was 9 years. What does this information indicate about the insurance salesman's belief that the peak earning age is 52? Explain.

8–6. Two methods of training in a shop have been developed and are being tested. Two random samples of 17 employees each were selected for the test. The group using method A reached the accepted level of output in an average of 75 minutes of training and practice. The standard deviation of this sample was 22 minutes. The group using method B reached the same level of output in an average of 69 minutes. The standard deviation of this sample was 20 minutes. Was there a significant difference between the length of time required by the two methods? Explain.

8–7. If in Problem 8–1 the universe standard deviation is not known, can you reject the hypothesis that the universe mean is 895?

8–8. Two paints are tested by a testing company for a hardware chain to determine if one is superior to the other on coverage per gallon. The results of the tests are shown below.

	Amazon Paint	Coverup Paint
Number of tests	36	36
Mean coverage (square feet)	223	198
Standard deviation (square feet)	14	24

Is the Amazon paint significantly better than the Coverup paint? Use the t distribution and a one-tail test.

8–9. A machine normally produces parts with only 2% of the parts being defective. Periodically a sample is tested to be sure there is no change in the quality of the product. If a sample of 121 items is tested and 4% of them are defective, do you conclude that the machine is still operating with an output that is only 2% defective. Show your calculations.

8–10. The following data are for two electrical products that were tested for the length of time they would operate before failure.

	Product A	Product B
Number of units tested	81	64
Mean life (hours)	1,216	1,190
Standard deviation (hours)	36	40

It appears that product A has a longer average life than product B. Is the difference significant? Show your computations and explain your reasoning. Use the t distribution.

8–11. If a random sample has a mean of 22, how large would the sample have to be to reject a hypothesis that the universe mean is 23? Assume a universe standard deviation of 6, $\alpha = 0.05$, and a one-tail test.

8–12. Using the data shown, test the following null hypotheses at a level of significance of 0.05.

Null Hypothesis	Sample Mean	Sample Size	Universe Standard Deviation	Type of Test
a. $\mu = 16$	15.5	25	2.0	one-tail
b. $\mu = 207$	206.0	100	6.2	two-tail
c. $\mu = 78$	75.8	14	16.0	two-tail
d. $\mu = 50$	56.0	225	12.0	one-tail

8–13. Problem 8–10 gives data on tests made of the length of life of two electrical products. In using the *t* distribution to test the difference between the two means it was assumed that the two universes were the same with respect to both dispersion as measured by the standard deviations and the average length of life as measured by the means. The *t* distribution was used to test the hypothesis that the means were the same, but no test was made regarding the standard deviations. Use the *F* distribution to test these data to determine whether or not there is a significant difference between the two standard deviations.

8–14. Problem 8–8 gives data on tests made of the coverage of two brands of paint. In using the *t* distribution to test the difference between two means it was assumed that the two universes were the same with respect to both the dispersion as measured by the standard deviations and the average number of square feet covered by the two paints. The *t* distribution was used to test the hypothesis that the mean coverage was the same for both paints, but no test was made regarding the standard deviations. Use the *F* distribution to test these data to determine whether there is a significant difference between the two standard deviations.

8–15. Test the data in Problem 8–8 by using the *F* distribution to determine whether or not there is a significant difference between the square feet of surface covered by the two paints used. Compare your results with those obtained by using the *t* distribution to make the same test in Problem 8–8.

8–16. It has been claimed that 40% of all college freshmen who enter during the summer term drop out of school before completing 30 hours of work. Test this claim against the alternative that it is less than 40% if a random sample of 480 students shows 173 dropouts. Use a level of significance of 0.05.

8–17. A company has a machine which is designed to produce parts with no more than 12% defective. If an operator draws a random sample of 60 parts each hour to determine whether or not the machine is running properly, how many defective parts would he need to find in a sample to cause him to shut down the machine because it would be producing too high a proportion of defects? Use a level of significance of 0.10.

8–18. An operator in a feed mill is told to fill sacks with an average of 60 lbs. of feed per sack. A random sample of his work is shown in the following table.

Sack Number	Weight (lbs.)
1	62
2	56
3	60
4	55
5	57

Test at a level of significance of 0.05 the hypothesis that he is doing the job correctly against the alternate hypothesis that he is not.

8–19. A machine is built to produce lead weights for fishing tackle with an average weight of 3 ounces. A random sample of 50 weights produced on the first day of operation has a mean of 3.1 ounces and a standard deviation of 0.2 ounce. Test at a level of significance of 0.05 the hypothesis that $\mu = 3$ against the alternate hypothesis that $\mu > 3$.

8–20. A manufacturing process is designed to bore a hole 1.534 inches from the edge of a piece of metal. The variation in the process, measured by the standard deviation, is .009 inches. A sample of 4 pieces taken at random from the assembly line is found to have the following measurements.

Sample Number	Inches from Edge
1	1.576
2	1.558
3	1.546
4	1.567

Can you conclude that the distance between the hole and the edge at the time this sample was taken differs significantly from the universe value of 1.534 inches? Show your computations and state your reasoning.

8–21. At a later date another random sample of 4 pieces was taken from the assembly line described in Problem 8–20. This sample was found to have the following measurements.

Sample Number	Inches from Edge
1	1.559
2	1.529
3	1.535
4	1.549

Do you conclude that the distance between the hole and the edge differs at this time significantly from the universe value of 1.534 inches? Show your computations and state your reasoning.

8–22. An oil company is interested in attracting women drivers to its stations. An architect designs and builds a new station on a busy street close to one of the old stations. During a randomly selected period of time, records are kept on both stations with the following results.

	Old Station 1	New Station 2
Number of male drivers	416	330
Number of female drivers	234	470
Total	650	800

Test at a level of significance of 0.05 the hypothesis that $\pi_1 = \pi_2$ against the alternate hypothesis that $\pi_1 < \pi_2$, where π is the percentage of female drivers.

8–23. A manufacturing process is turning out cotton yarn. It is specified that the strength of the yarn be such that on the average not less than 99.8 pounds of pressure be required to break a skein of yarn. It has been determined that the process being used is capable of producing yarn with an average strength of 99.8 pounds and a standard deviation of 12.4 pounds. A sample of 4 skeins is taken at random from a large lot of yarn and tested for strength with the following results, expressed in pounds of pressure required to break a skein.

Test Number	Pounds of Pressure Required
1	106.8
2	78.7
3	76.7
4	85.8

On the basis of this sample you are required to make a decision as to whether or not the yarn being produced meets the standard of strength specified (99.8 pounds of pressure to break a skein). Show all your computations and state your reasoning.

8–24. At a later date another sample of 4 skeins of yarn was taken at random from the process described in Problem 8–23. These 4 skeins were tested for strength with the following results, expressed in pounds of pressure required to break a skein.

Test Number	Pounds of Pressure Required
1	74.2
2	97.6
3	86.3
4	62.6

On the basis of this sample make the same decision you were required to make in Problem 8–23 as to whether or not the yarn being produced meets the standard of strength specified (99.8 pounds of pressure to break a skein). Show all your computations and state your reasoning.

8–25. Assume that in Problem 8–19, a random sample of 18 weights taken on the second day of operation has a mean of 2.9 ounces and a standard deviation of 0.15 ounce. Test at a level of significance of 0.05 the hypothesis that $\mu_1 = \mu_2$ against the alternate hypothesis that $\mu_1 > \mu_2$.

8–26. Ten employees are selected at random from each of two departments of a company and their weekly earnings are shown in the following table.

Number	Billing Department	Accounting Department
1	$175	$200
2	220	330
3	196	150
4	225	310
5	345	600
6	275	175
7	210	350
8	224	331
9	300	420
10	320	210

Can you state on the basis of the above information that employees in the accounting department are paid significantly more than those in the billing department?

8–27. A textile mill buying cotton specifies that the upper half mean length (U.H.M.) should be 1.03 inches. (Upper half mean length is the mean length in inches of the longest half of the fiber.) Ten tests on samples of cotton taken at random from a lot gave the following upper half mean length in inches:

Test Number	U.H.M. (Inches)	Test Number	U.H.M. (Inches)
1	1.05	6	1.05
2	1.08	7	1.10
3	1.01	8	1.10
4	1.05	9	1.11
5	1.08	10	1.06

Does the upper half mean length for these samples of cotton differ significantly from 1.03 inches? Show all your calculations and state your reasoning.

8–28. Another lot of cotton purchased by the mill in Problem 8–27 gave the following upper half mean length for 10 samples of cotton taken at random.

Test Number	U.H.M. (Inches)	Test Number	U.H.M. (Inches)
1	.87	6	.98
2	1.09	7	1.21
3	.90	8	.89
4	.97	9	.98
5	.80	10	1.02

Does the upper half mean length for these samples of cotton differ significantly from 1.03 inches? Show all your calculations and state your reasoning.

8–29. Two lots of cotton yarn were purchased from different mills, and it was desired to test whether or not there was a significant difference in the strength of the two lots. A sample of 10 skeins from each lot was tested for the pounds of pressure required to break a skein of yarn.

Test Number	Lot 1	Lot 2
1	87	93
2	85	88
3	92	85
4	87	77
5	114	86
6	104	89
7	96	83
8	84	80
9	111	68
10	91	79

Was there a significant difference between the strength of the yarn in the two lots? Show all your calculations and state your reasoning. Use the t distribution, and retain your solution for use in Problem 8–32.

8–30. A manufacturer of cake mixes designed a new package for his product and ran extensive tests in different markets to determine the acceptance of the new package by consumers at various income levels. A typical test that was used is described in the following paragraph.

Two stores were selected, one in a high-income neighborhood and another in a low-income neighborhood. A display of cake mixes was arranged, consisting of packages of the old design and the new. At the end of the test period, 320 packages had been sold in store 1, of which 192 were the new package and 128 the old. In store 2, 400 packages were sold, 252 of the new package and 148 of the old. In other words, 60% of the sales in store 1 were the new package and 63% of the sales in store 2 were the new package.

Test whether there is a significant difference between the percentage of new packages sold in the two stores. Show your calculations and state your reasoning.

8–31. Two random samples show the following results.

	Sample 1	Sample 2
Sample size	12	25
$\Sigma(x - \bar{x})^2$	25	84

Use an F test and a level of significance of 0.05 to test the hypothesis that the sample came from universes with the same variances against the alternate hypothesis that one sample comes from a universe with a larger variance than the other.

8–32. In Problem 8–29 two lots of cotton yarn were tested to determine whether or not there was a significant difference between the strength of the yarn in the two lots using the t distribution. Make this same test using the F distribution. Compare your results with those obtained by using the t distribution.

8–33. The following table gives the results of tests made of the strength of the lead in four pencils manufactured by a certain company. Four tests were made of the lead in each pencil. Do these four pencils differ significantly from each other with respect to the strength of the lead? Show your calculations.

	Pencil 1	Pencil 2	Pencil 3	Pencil 4
	1.81	1.68	1.66	1.76
	1.78	1.33	2.03	1.85
	1.72	1.89	1.82	1.93
	1.57	1.54	1.97	1.62
Σx	6.88	6.44	7.48	7.16
Σx^2	11.8678	10.5350	14.0698	12.8694

8–34. A manufacturing plant collected the following data on the strength of spot welds made on five machines. The strength of each weld was measured by testing to destruction in a shear testing machine. The strength of each weld was measured in pounds and the total pounds in the following table (Σx) represent the sum of all the tests made on a given machine. The average shear strength was computed by dividing the total pounds by the number of tests to give the mean shear strength in pounds. The column Σx^2 was computed by summing the squares of the individual measurements for the tests for each machine. These values of Σx^2 are given to save time in working this problem. Use the F distribution to determine whether or not there is a significant difference between the mean shear strength of the welds from the different machines.

Machine	Number of Tests	Average Shear Strength (Pounds)	Σx	Σx^2
1	125	710	88,750	63,412,225
2	115	670	77,050	52,299,115
3	126	791	99,666	79,423,864
4	110	801	88,110	70,931,040
5	120	604	72,480	44,491,520

9 Tests of Significance: Nonparametric Methods

Chapter 8 introduced the concept of tests of significance and demonstrated tests of means and proportions for one and two samples. In this chapter *nonparametric,* or *distribution-free, methods* of tests of significance are discussed. This special group of tests has been used with great success for a number of years by social scientists and has attracted the interest of business researchers in more recent years.

LEVELS OF MEASUREMENT

In statistics, *measurement* is the assignment of numbers to attributes of objects or observations. The level of measurement is a function of the rules used to assign numbers and is an important factor in determining what type of statistical analysis can be appropriately applied to the data. Before any attempt is made to distinguish between parametric and nonparametric tests, the various levels of measurement of statistical data used in tests of significance must be discussed.

The Nominal Scale

The lowest or weakest level of measurement is the use of numbers to classify observations into mutually exclusive groups or classes. These observations are known as *nominal data.* For example, employees might be classified by sex (the number 1 might be used to designate a female and a 2 might be used to designate a male), or parts produced by a machine might be classified as effective or defective, using numbers to designate these classifications.

Numbers on the doors of rooms in a building are another example of numbers used as labels to put objects into classifications. The first digit of the room number might represent the floor in the building and the last digit might tell whether the room is on the north or the south side of the building.

The Ordinal Scale

When observations are ranked so that each category is distinct and stands in some definite relationship to each of the other categories, the data are *ordinal data*. For example, if people are asked to rate a product as better than most, average, or poorer than most, the categories are ordinal in nature. It is known that the first category is better than the second, which is still better than the third. Each group is "greater than" or "less than" every other group.

If 10 employees are ranked by their supervisor from 1 to 10 in order of preference for merit raises, the numbers 1 through 10 are ordinal data. It cannot be assumed that the difference between employees 1 and 2 is the same as the difference between employees 2 and 3. It is known only that 1 is better than 2, who is better than 3.

The Interval Scale

When the exact distance between any two numbers on the scale are known and when the data meet all of the other requirements of ordinal data, they can be measured as *interval data*. The unit of measure and the origin of an interval scale are arbitrary. The classic examples are the two measures of temperature, centigrade and Fahrenheit. The zero point for each scale is a different temperature and the unit of measure is different for each, but there is a fixed relationship between the two scales which is expressed by the formula

$$\text{Fahrenheit} = \frac{9}{5}\,\text{centigrade} + 32.$$

The Ratio Scale

When measurements having all the characteristics of the interval scale also have a true zero point, they have attained the highest level of measurement and are called *ratio data*. The ratio scale derives its name from the fact that the ratio of any two values is independent of the units in which they are measured. For example, the weight of an object may be measured in grams, ounces, stones, or any one of several other systems of measure. The origin of each system is the same and is zero weight. One can say that 20 ounces is twice as heavy as 10 ounces and that 30 grams is twice as heavy as 15 grams, $20/10 = 30/15 = 2$. On the other hand, it would not be true to say that 20° Fahrenheit is twice as warm as 10°, as the scale used to measure temperature is an interval scale and does not have a true zero.

PARAMETRIC AND NONPARAMETRIC STATISTICAL TESTS

A *parametric statistical test* is a test whose model makes certain assumptions about the parameters of the population from which the sample was drawn. For example, when the researcher uses a t test he must be able to assume that:

1. The observations in the sample are independent.
2. The observations are drawn from a normally distributed population.
3. The populations have the same (or known) variances.
4. The variables involved were measured on at least an interval scale.

A *nonparametric statistical test* is a test whose model does not specify the parameters of the population from which the sample was drawn. Nonparametric tests do not require a level of measurement as high as that for parametric tests. In fact, most nonparametric tests are designed to require only nominal or ordinal data, although there are some nonparametric tests that can be applied to interval data that do not meet the other requirements for a parametric test.

The growing interest in nonparametric methods among business researchers is due to the fact that these methods have certain advantages over the more exacting parametric methods:

1. Nonparametric tests are often much easier to understand and to use than parametric tests.
2. Nonparametric tests may frequently be used with very small samples.
3. When data have only a nominal or ordinal level of measurement, there are no appropriate parametric tests available that can be used.
4. Nonparametric tests require no assumptions about the nature of the population from which the sample is drawn.

CHI-SQUARE DISTRIBUTION

The chi-square distribution is used for many nonparametric tests. Chi-square tests of independence commonly involve the use of nominal data, and chi-square tests of goodness of fit usually involve data with higher levels of measurement. In Chapter 8 tests of significance on sample proportions were discussed as if they were parametric tests because the samples were large, and it could be assumed that the theoretical sampling distribution of proportions or the difference between proportions was approximately normal. As a matter of fact, the level of measurement was nominal and the tests of differences between two sample proportions could have been made using the chi-square distribution.

In Chapter 8 the problem of testing for a significant difference between two sample proportions treated data classified either as defective or effective and according to the machine on which the items were produced. A chi-square test can be made to see whether or not the two bases of classification are independent.

Test of Independence

The simplest form of the test for independence is found when there are only two groups within each basis of classification, giving four subgroups. Such a classification is designated as a *2 × 2 contingency table*. It is possible, however, to have more than two groups in each basis of classification. In this situation a cross-classification table is set up with the groups in the two bases of classification in the stubs and the captions. The basis in the rows is designated as r and the one in the column as c. The whole table is called the *$r \times c$ contingency table*. Since the procedures for these two situations differ slightly, each will be discussed.

The 2 × 2 Contingency Table. Table 9–1 gives the results of the inspection of two lots of output from two different machines. Of the 400 items produced on Machine 1, 25 were defective and 375 were effective. The 600 items in the lot produced on Machine 2 were classified as 42 defective and 558 effective. In other words, the 1,000 items were classified first according to the machine on which they were made, and then the items produced by each machine were classified as defective or effective.

The hypothesis to be tested is that the two bases of classification are independent; that is, the number of defective items is not related to the particular machine on which they were made, since one machine is no more likely to turn out a defective item than the other. If this hypothesis is true, the variations of the observed values from the expected values may be attributed to sampling fluctuations. The value to be tested by the chi-square distribution is computed by: (1) finding the difference between each observed value and the corresponding expected value, (2) expressing the square of the difference as a fraction of the expected value, and (3) summing. If the observed value is designated as f_o and the expected value as f_e, these operations may be written as $\sum\left[\frac{(f_o - f_e)^2}{f_e}\right]$. The distribution of this term computed from successive samples is approximately the chi-square distribution. Therefore, the chi-square distribution may be used to test whether any given value computed as just described may reasonably be expected to be zero. Using the .01 level of significance, the hypothesis that the difference between expected values and observed values

TABLE 9–1

CLASSIFICATION OF THE OUTPUT OF TWO MACHINES AS DEFECTIVE AND EFFECTIVE

Machine Number	Defective	Effective	Total
1	25	375	400
2	42	558	600
Total	67	933	1,000

is zero will be rejected if the computed value is larger than the value in the chi-square table at probability = .01. The application of this test is illustrated with the data in Table 9–1.

It can be seen from Table 9–1 that 67 out of 1,000, or 6.7%, of the items were defective and 933 out of 1,000, or 93.3%, were effective. If the number defective is independent of the machine on which they were made, 6.7% of the 400 items made on Machine 1 should be defective. These numbers are computed and entered in Table 9–2 as 26.8 defective and 373.2 effective items from Machine 1. The same percentages are also applied to the 600 items from Machine 2 and entered in the second row of the table. The totals for the columns and the rows are the same as those in Table 9–1, and the difference between the observed and the expected values in the individual cells of the tables represents the difference between the observed values and the expected values set up by the hypothesis.

In Table 9–3 (page 224) the observed frequency in each cell is represented by f_o, and the expected frequency, by f_e. The difference between f_o and f_e is shown in Column 4, the square of the difference is given in Column 5, and the ratio of the values in Column 5 to the value of f_e is given in Column 6. The sum of the values in Column 6 represents the value of chi-square for this classification. These calculations are summarized in Formula 9–1.

$$\chi^2 = \sum\left[\frac{(f_o - f_e)^2}{f_e}\right]. \qquad \textbf{(9–1)}$$

The value of chi-square varies with the number of degrees of freedom in a given situation. In this problem there are four cells in the classification, but there is only one degree of freedom. The number of frequencies entered in any one cell could be assigned any value, but once an amount has been entered in a cell, the other three frequencies are determined by the totals of the rows and the columns of Table 9–1. The value of chi-square at the .01 level of significance is found in Appendix L to be 6.635. Since the computed value of chi-square was found to be .2160, the conclusion is that there is no significant difference between the percentage of defective items produced on the two

TABLE 9–2

EXPECTED DISTRIBUTION CLASSIFIED AS DEFECTIVE AND EFFECTIVE FOR THE OUTPUT OF TWO MACHINES, ASSUMING BASES OF CLASSIFICATION TO BE INDEPENDENT

Machine Number	Defective	Effective	Total
1	26.8	373.2	400.0
2	40.2	559.8	600.0
Total	67.0	933.0	1,000.0

TABLE 9-3

COMPUTATION OF CHI-SQUARE

Output of Two Machines Classified as Defective and Effective

Cell (1)	Observed Frequencies f_o (2)	Expected Frequencies f_e (3)	$f_o - f_e$ (4)	$(f_o - f_e)^2$ (5)	$\frac{(f_o - f_e)^2}{f_e}$ (6)
Machine 1 defective	25	26.8	−1.8	3.24	.1209
Machine 1 effective	375	373.2	1.8	3.24	.0087
Machine 2 defective	42	40.2	1.8	3.24	.0806
Machine 2 effective	558	559.8	−1.8	3.24	.0058
Total	1,000	1,000.0	0		.2160

$$\chi^2 = \sum\left[\frac{(f_o - f_e)^2}{f_e}\right]$$

$$\chi^2 = .2160.$$

machines. This is the same conclusion reached by the use of the test for the difference between two sample proportions.

When many tests are to be performed, Formula 9–2 may be used for the 2×2 contingency table. The diagram below the formula identifies the cells of Table 9–1 by letters.

$$\chi^2 = \frac{(ad - bc)^2 n}{(a + b)(c + d)(a + c)(b + d)}. \qquad \textbf{(9–2)}$$

a	b	$a + b$
c	d	$c + d$
$a + c$	$b + d$	n

When Formula 9–2 is used to compute χ^2, there is no requirement to compute the expected frequencies as was demonstrated previously.

The following computations show that Formula 9–2 gives the same value of χ^2 as Formula 9–1.

$$\chi^2 = \frac{[(25)(558) - (375)(42)]^2 1{,}000}{(400)(600)(67)(933)} = \frac{(3{,}240{,}000)(1{,}000)}{15{,}002{,}640{,}000}$$

$$\chi^2 = \frac{3{,}240{,}000{,}000}{15{,}002{,}640{,}000} = .2160.$$

This test may be summarized

Null hypothesis (H_o): *The two bases of classification are independent; that is, the difference between the proportion defective turned out by the two machines is zero.* ($\pi_1 = \pi_2$)

Alternative hypothesis (H_a): $\pi_1 \neq \pi_2$

Criterion for decision: Reject H_o and accept H_a if $\chi^2 > 6.635$

This criterion gives $\alpha = .01$ and degrees of freedom = 1.

The r × c Contingency Table. The example just described involved two bases of classification and only two groups within each basis. Any number of rows (designated as *r*) and any number of columns (designated as *c*) may be analyzed in a similar manner. It is also possible to test more than two bases of classification for independence, and the methods may be applied to both qualitative and quantitative classifications.

The following example uses a classification of accidents shown in Table 9–4 by departments and by days of the week covering a two-year period. The question is whether or not the accident pattern for the different departments is independent of the days of the week. The method is basically the same as the analysis of a 2 × 2 contingency table. The first step is to compute the expected frequency in each cell, assuming that the accidents in each department are distributed throughout the week in the same proportion as the total accidents in the four departments. These computations give the expected values in Table 9–5. For example, 70 out of 393 of the accidents occurred on Monday; thus, it is assumed that this proportion of the accidents in department D would occur on Monday. The expected frequency of this cell is

$$f_e = \frac{70}{393}\,132 = 23.5.$$

This computation is repeated for each cell in the same manner, using the marginal frequencies.

TABLE 9–4

ACCIDENTS IN 1972 AND 1973
CLASSIFIED BY DEPARTMENTS AND BY DAYS OF THE WEEK

f_o

Dept.	Mon.	Tue.	Wed.	Thurs.	Fri.	Total
D	21	13	22	30	46	132
E	11	17	12	17	12	69
F	10	10	12	12	10	54
G	28	26	25	28	31	138
Total	70	66	71	87	99	393

Source: Hypothetical data.

TABLE 9–5

EXPECTED FREQUENCIES OF ACCIDENTS CLASSIFIED BY DEPARTMENTS AND BY DAYS OF THE WEEK ON THE ASSUMPTION OF INDEPENDENCE

f_e

Dept.	Mon.	Tue.	Wed.	Thurs.	Fri.	Total
D	23.5	22.2	23.8	29.2	33.3	132.0
E	12.3	11.5*	12.5	15.3	17.4	69.0
F	9.6	9.1	9.8	12.0	13.5*	54.0
G	24.6	23.2	24.9	30.5	34.8	138.0
Total	70.0	66.0	71.0	87.0	99.0	393.0

* These expected frequencies have been changed to force the total to read 393.0.

Source: Table 9–4.

With the observed and expected frequencies in Tables 9–4 and 9–5, it would be possible to compute $(f_o - f_e)^2$ and to derive $\frac{(f_o - f_e)^2}{f_e}$ for substitution in Formula 9–1. However, a shorter computation formula is derived

$$\chi^2 = \sum \left[\frac{(f_o - f_e)^2}{f_e}\right] = \sum \left[\frac{f_o^2 - 2f_o \cdot f_e + f_e^2}{f_e}\right] = \sum \frac{f_o^2}{f_e} - 2\sum \frac{f_o \cdot f_e}{f_e} + \sum \frac{f_e^2}{f_e}$$

$$\chi^2 = \sum \frac{f_o^2}{f_e} - 2\Sigma f_o + \Sigma f_e.$$

But $\Sigma f_o = \Sigma f_e = n$,

therefore

$$\chi^2 = \sum \frac{f_o^2}{f_e} - n. \qquad \textbf{(9–3)}$$

Formula 9–3 requires the computation of the value $\sum \frac{f_o^2}{f_e}$ instead of $\sum \frac{(f_o - f_e)^2}{f_e}$ for each cell. For example, the value of $\sum \frac{f_o^2}{f_e}$ for the upper left-hand cell is

$$\sum \frac{f_o^2}{f_e} = \frac{21^2}{23.5} = \frac{441}{23.5} = 18.766.$$

This computation is repeated for each cell and the total is found by adding the individual entries in Table 9–6. Formula 9–3 gives a value of chi-square of 16.668.

The interpretation of this value requires the value of chi-square for 12 degrees of freedom. The number of degrees of freedom is the product of $(r-1)(c-1)$. In the 2×2 contingency table, this gave 1 degree of freedom by the following calculation: $d.f. = (2-1)(2-1) = 1$. In the present analysis there

TABLE 9–6

COMPUTATION OF $\sum \frac{f_o^2}{f_e}$

Dept.	Mon.	Tue.	Wed.	Thurs.	Fri.	Total
D	18.766	7.613	20.336	30.822	63.544	141.081
E	9.837	25.130	11.520	18.889	8.276	73.652
F	10.417	10.989	14.694	12.000	7.407	55.507
G	31.870	29.138	25.100	25.705	27.615	139.428
Total	70.890	72.870	71.650	87.416	106.842	409.668

Source: Table 9–4.

$$\chi^2 = \sum \frac{f_o^2}{f_e} - n$$

$$\chi^2 = 409.668 - 393 = 16.668.$$

are 4 rows ($r = 4$) and 5 columns ($c = 5$). Thus, there are 12 degrees of freedom, computed as follows: $d.f. = (4 - 1)(5 - 1) = 12$. The value of chi-square for 12 degrees of freedom is 26.217 at the .01 level of significance. Since the value just computed is less than 26.217, the hypothesis of independence cannot be rejected. In other words, on the basis of this information, the distribution of accidents among the departments is independent of the day of the week.

This test may be summarized

Null hypothesis (H_0): *The accidents in the different departments are independent of the days of the week.*

Alternative hypothesis (H_a): Accidents in different departments are related to the days of the week.

Criterion for decision: Reject H_0 and accept H_a if $\chi^2 > 26.217$

This criterion gives $\alpha = .01$ and degrees of freedom = 12.

Tests of Goodness of Fit

It is often useful to test the hypothesis that a sample comes from a universe that has a given distribution. One might wish to know, for example, whether a sample of telephone calls arriving at a switchboard comes from a Poisson distribution or whether a sample of times required for a bus to run between two cities comes from a normal distribution. Such tests can be made using the chi-square distribution.

Fitting a Uniform Distribution. A uniform distribution is one in which each possible outcome has the same chance of selection. An example is shown in Figure 9–1.

A test which determines whether an observed distribution comes from a universe which is a uniform distribution is demonstrated by the following

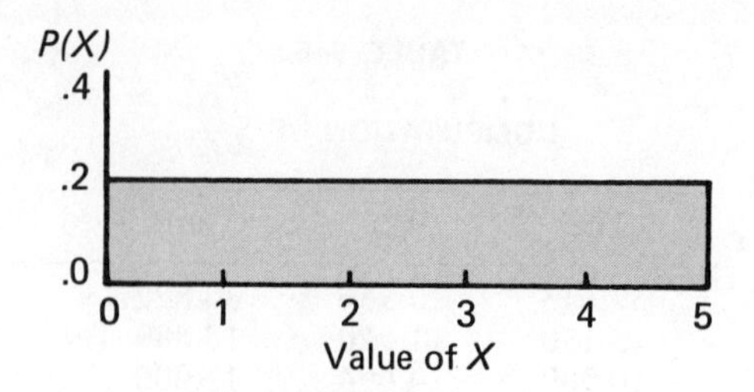

FIGURE 9–1

UNIFORM PROBABILITY DISTRIBUTION OF THE CONTINUOUS VARIABLE *X*

example. An automobile dealer wishes to determine whether his sales of station wagons vary significantly from one season of the year to another. He must make his decision based on a sample of information gathered for one year. Sales are shown in Table 9–7.

The computation of chi-square is shown in Table 9–8. The same procedure used in the previous examples is followed here, except that in goodness of fit tests the expected frequency must be derived from the distribution being fit. In this case a uniform distribution is being fit which implies that the total sales of 96 units would be uniformly distributed over the four seasons. The values

TABLE 9–7

SALES OF STATION WAGONS IN 1973

Season	Units Sold
Winter	15
Spring	35
Summer	26
Fall	20
Total	96

Source: Hypothetical data.

TABLE 9–8

COMPUTATION OF CHI-SQUARE

Sales of Station Wagons Shown by Season

Season	f_o	f_e	$f_o - f_e$	$(f_o - f_e)^2$	$\frac{(f_o - f_e)^2}{f_e}$
Winter	15	24	−9	81	3.375
Spring	35	24	11	121	5.042
Summer	26	24	2	4	0.167
Fall	20	24	−4	16	0.667
Total	96	96	0		9.251

Source: Table 9–7.

of f_e would be determined by dividing the total, 96, by 4, the number of seasons. Once this is completed the computation can proceed.

This may be summarized

Null hypothesis (H_o): *The sample comes from a universe which is a uniform distribution. (Seasons are equally good sales times.)*

Alternate hypothesis (H_a): The universe is not a uniform distribution. (Some seasons are better for sales than others.)

Criterion for decision: Reject H_o and accept H_a if $\chi^2 > 7.815$.

This criterion is based on $\alpha = .05$ and degrees of freedom $= 3$ (the number of categories or seasons minus one).

Since the computed value of χ^2 (9.251) is greater than the critical value (7.815), the dealer should reject H_o for this test. His sample data indicate that the seasons are different.

Fitting the Normal Distribution. The first step in testing the hypothesis of normality is to fit the normal curve to a sample distribution. The hypothesis to be tested is that the universe from which the sample was selected is normal and the deviations of the sample distribution from normal represent simple sampling error. The term $\sum\left[\frac{(f_o - f_e)^2}{f_e}\right]$ can be computed and the table of values of chi-square used to determine the probability that this large a deviation from zero would result if the universe from which the sample was taken were in fact normal.

Fitting the normal distribution to an observed distribution consists of finding the normal distribution that has the same mean, standard deviation, and number of frequencies as the sample distribution. This is more complex than finding the expected frequencies in Table 9–8, but the principle is the same. In Table 9–8 the expected frequencies have the same total as the observed values, and they were secured by dividing the total by 4, the number of seasons. The assumption in the null hypothesis was that all the seasons were alike. In fitting a normal curve the mean, the standard deviation, and the total frequency of the expected distribution are the same as those of the observed distribution.

This method is illustrated in the frequency distribution of average monthly salaries of file clerks in Tennessee given in Table 7–4. The estimates of the mean and the standard deviation of the universe calculated in Table 7–4 were found to be

$$\bar{x} = \$407.14 \qquad \hat{\sigma} = \$86.51.$$

The values of β_1 and β_2 for this distribution, computed from Formulas 4–19 and 4–23, are

$$\beta_1 = .858 \qquad \beta_2 = 2.963.$$

The β_2 is reasonably close to 3, but the β_1 is not very close to zero, which makes it doubtful that the distribution is normal.

The hypothesis that the variation from the normal curve is not greater than might be expected to result from sampling should be tested. The first step is to compute the normal distribution that would have a mean of \$407.14, a standard deviation of \$86.51, and total frequencies of 182. The frequencies of the normal distribution of earnings will represent the expected frequencies (f_e), while the frequencies in Table 7–3 will represent the observed frequencies (f_o).

There are two methods commonly used in fitting a normal curve to an observed frequency distribution. One method makes use of the table in Appendix G, Ordinates of the Normal Distribution, and the other uses the table of Areas of the Normal Curve given in Appendix H. The second method is illustrated in the following discussion.

The total number of frequencies in the distribution of earnings of file clerks is the total number of frequencies in the normal distribution and, therefore, represents the total area under the curve. The frequency of each class in the expected distribution is a proportion of the normal curve. This proportion is derived from the table of areas as described in the following discussion and shown in Table 9–9.

Column 1 in Table 9–9 gives the real lower limits of the classes in the frequency distribution. Following the principles set forth in Chapter 2, the real lower limit of the class "250 and under 300" is 250. The real lower limit of the second class is 300, etc. Column 2 gives the values of the lower class limit minus the mean of the distribution, letting x represent the value of the lower class limit. The z values in Column 3 express the deviations $(x - \bar{x})$ in units of the estimated standard deviation, $\hat{\sigma}$. Each of the $(x - \bar{x})$ values in Column 2 is divided by the value of $\hat{\sigma}$ and the resulting values of $\frac{(x - \bar{x})}{\hat{\sigma}}$ are

TABLE 9–9

FITTING A NORMAL CURVE TO A FREQUENCY DISTRIBUTION

Average Monthly Salaries of File Clerks in the State of Tennessee, June, 1971

Real Class Limits x (1)	$x - \bar{x}$ (2)	$\frac{x - \bar{x}}{\hat{\sigma}}$ (3)	Proportion of Area Between Y_o and x (4)	Expected Frequencies Between Y_o and x (5)
250	−157.14	−1.82	.46562	84.74
300	−107.14	−1.24	.39251	71.44
350	− 57.14	−0.66	.24537	44.66
400	− 7.14	−0.08	.03188	5.80
450	42.86	0.50	.19146	34.85
500	92.86	1.07	.35769	65.01
550	142.86	1.65	.45053	82.00
600	192.86	2.23	.48713	88.66
650	242.86	2.81	.49752	90.55

entered in Column 3. For example, the first item in Column 3 is computed as follows and the remaining values are computed in the same manner.

$$z = \frac{x - \bar{x}}{\hat{\sigma}} = \frac{250 - 407.14}{86.51} = \frac{-157.14}{85.61} = -1.82.$$

The proportion of the area between the maximum ordinate (Y_o) and the values of z are taken from the table of Areas of the Normal Curve in Appendix H. The values of $\frac{(x - \bar{x})}{\hat{\sigma}}$ are used instead of $\frac{(X - \mu)}{\sigma}$, since the true value of the standard deviation of the universe is not known and the values of $\frac{(X - \mu)}{\sigma}$ cannot be computed. But the estimate of z from $\frac{(x - \bar{x})}{\hat{\sigma}}$ is accurate enough to use for this purpose. Using the value of $z = -1.82$, Appendix H shows that the proportion .46562 of the normal curve lies between the maximum ordinate and -1.82. Likewise, the proportion of the curve that lies between the maximum ordinate and -1.24 is found to be .39251. Each of the values in Column 4 is found in the same manner.

The entries in Column 5 represent the number of frequencies between the maximum ordinate Y_o and the given values of x for a normal distribution with a total frequency of 182. The first frequency in Column 5 is computed by multiplying 182 by .46562. The values in the remainder of the column are computed by multiplying each proportion by 182, the total number of frequencies in the distribution.

From the cumulative frequencies between the maximum ordinate and the lower class limits, it is possible to compute the expected frequencies of each of the classes. These expected frequencies are entered in Column 3 of Table 9–10. The first class consists of earnings under \$250, although there were no earnings under \$250 in the observed frequency distribution. Since the normal distribution is a theoretical distribution that extends to plus and minus infinity, a portion of the area under the normal curve extends beyond the value of \$250, which marks the limit of the observed distribution. According to Table 9–9, the number of frequencies falling between the limit of \$250 and the maximum ordinate is 84.74. Since one half of the total frequencies, or 91, fall below the maximum ordinate, the frequency below the \$250 limit equals 6.26 ($91 - 84.74 = 6.26$). In an actual distribution it is impossible to have a fractional frequency, but when dealing with a theoretical distribution there is no reason for not recording a frequency of 6.26. All the expected frequencies are carried out to two decimals in order to give a more accurate measure of the relative frequencies.

The expected frequency for the class with the limits 250 and under 300 is found by taking the difference between the cumulative frequency at 250 and at 300 ($84.74 - 71.44 = 13.30$). Each of the other frequencies is computed in the same way, except for the class in which the maximum ordinate falls. The frequency between the maximum ordinate and 400 is 5.80, which is the

TABLE 9–10

OBSERVED AND EXPECTED FREQUENCY DISTRIBUTIONS OF AVERAGE MONTHLY SALARIES OF FILE CLERKS IN THE STATE OF TENNESSEE, JUNE, 1971

Earnings (Dollars) (1)	Observed Frequencies (2)	Expected Frequencies (3)
Under 250		6.26
250 and Under 300	6	13.30
300 and Under 350	41	26.78
350 and Under 400	70	38.86
400 and Under 450	16	40.65
450 and Under 500	17	30.16
500 and Under 550	14	16.99
550 and Under 600	12	6.66
600 and Under 650	6	1.89
650 and Over		0.45
Total	182	182.00

portion of the total frequency of the class that falls below the maximum ordinate. The frequency between the maximum ordinate and 450 is the portion of the total frequency in the class that is above the maximum ordinate or 34.85. The sum of these two frequencies is the total frequency of the class in which the maximum ordinate falls, or 40.65.

The frequency of the open-end class 650 and over is computed in the same manner as for the class under 250. The cumulative frequency between the maximum ordinate and 650 is 90.55. The frequency above 650 is computed by subtracting 90.55 from 91 (one half the total number of frequencies) and is found to be .45.

Table 9–10 gives the observed frequencies in addition to the expected frequencies. Figure 9–2 presents a graphic comparison of the goodness of fit of the normal curve. The highest point of the normal curve (the maximum ordinate) is at the mean of the distribution. Since the mean is above the midpoint of the class, 400 and under 450, the ordinate at the mean is higher than the ordinate at the midpoint of the class. The maximum ordinate is computed by Formula 6–9.

$$Y_o = \frac{N}{\hat{\sigma}} .39894 = \frac{182}{1.7302} .39894 = 41.9646.$$

The value of $\hat{\sigma}$ is used instead of σ since the estimate of the universe standard deviation is all that is available.

Computing the Value of χ^2. The observed and expected frequencies from Table 9–10 are used as the values of f_o and f_e in Table 9–11. Formula 9-3 is the most convenient one to use and gives the value of 254.9873 for χ^2.

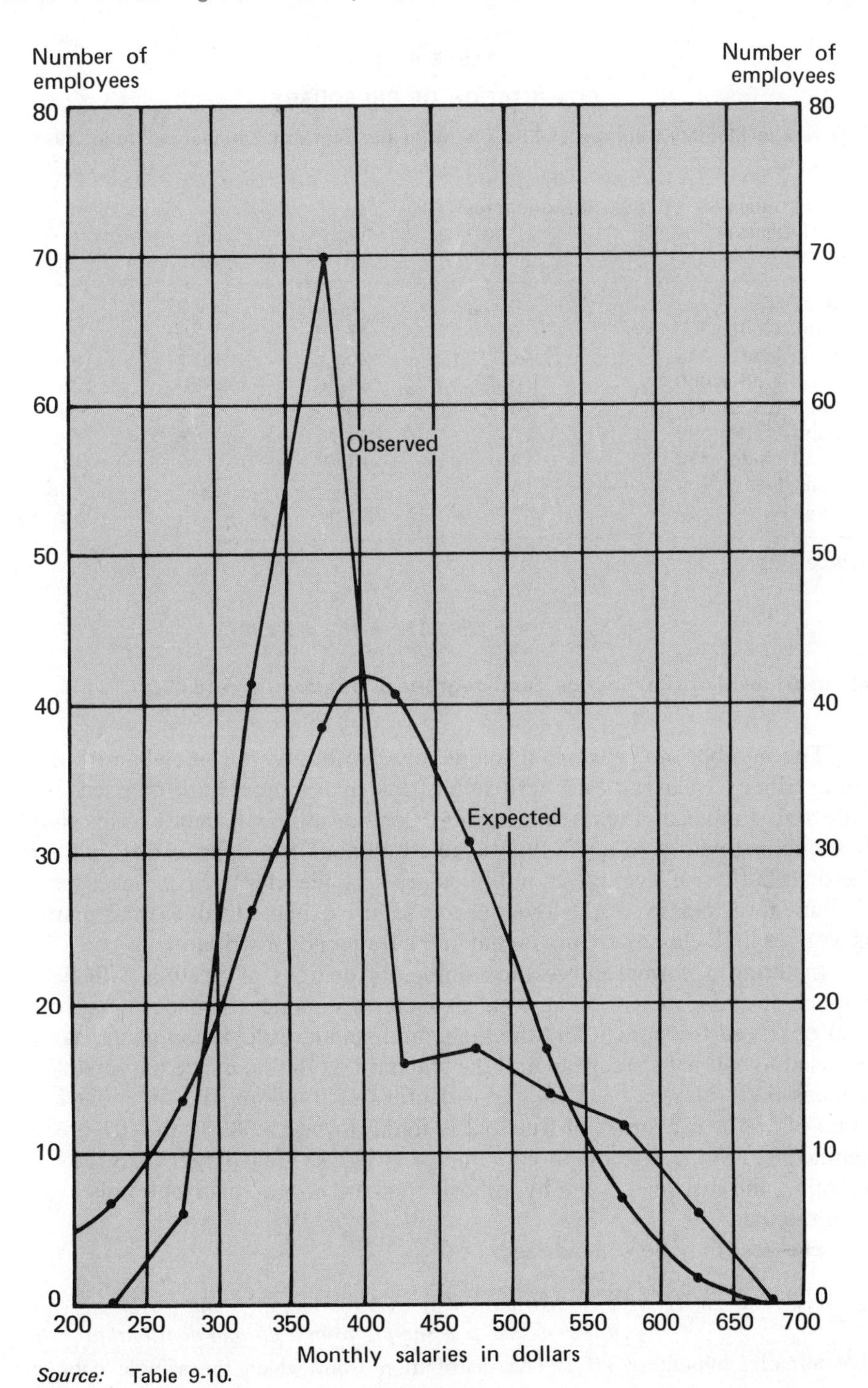

Source: Table 9-10.

FIGURE 9–2

AVERAGE MONTHLY SALARIES OF FILE CLERKS IN TENNESSEE, JUNE, 1971

TABLE 9–11

COMPUTATION OF CHI-SQUARE

Average Monthly Salaries of File Clerks in the State of Tennessee, June, 1971

Earnings (Dollars) (1)	Observed Frequencies f_o (2)	f_e (3)	f_o^2 (4)	$\frac{f_o^2}{f_e}$ (5)
Under 250	0	6.26	0	0
250 and Under 300	6	13.30	36	2.7068
300 and Under 350	41	26.78	1,681	62.7707
350 and Under 400	70	38.86	4,900	126.0937
400 and Under 450	16	40.65	256	6.2977
450 and Under 500	17	30.16	289	9.5822
500 and Under 550	14	16.99	196	11.5362
550 and Over	18	9.00	324	36.0000
Total	182	182.00		254.9873

Source: Table 9–10.

$$\chi^2 = \sum \frac{f_o^2}{f_e} - n = 254.9873 - 182 = 72.9873$$

At the .01 level of significance, for 5 degrees of freedom, $\chi^2 = 15.086$.

The number of classes in a frequency distribution is somewhat arbitrary, but in using χ^2, classes with very small frequencies should not be used. If the expected number in a class is less than 5, one or more adjacent classes should be combined with it to make the expected number 5 or more. Thus, in testing the distribution of average monthly salaries of file clerks, it is necessary to combine three classes at the lower end to secure expected values of as many as 5. As a result, 8 classes are presented in the frequency distribution.

In fitting a normal curve the number of degrees of freedom will be the number of classes less 3. The total expected frequency is made equal to the total observed frequency, and the mean and standard deviation of the sample are used to estimate the mean and the standard deviation of the universe. This reduces the 8 classes by 3 to give 5 degrees of freedom. In Appendix L the value of χ^2 for 5 degrees of freedom is found to be 15.086 at the .01 level of significance. Since the computed value of χ^2 (72.9873) is larger than the tabular value, the test rejects the hypothesis that the normal distribution is a good fit to the data.

This test may be summarized

Null hypothesis (H_o): *The distribution of average monthly salaries of file clerks in Tennessee is a sample from a normal distribution.*

Alternative hypothesis (H_a): The distribution from which the sample data was selected is not normal.

Criterion for decision: Reject H_o and accept H_a if $\chi^2 > 15.086$

This criterion gives $\alpha = .01$ and degrees of freedom $= 5$.

RUNS TEST FOR RANDOMNESS

In the chapters on sampling and tests of significance (parametric methods) it was assumed that the samples used were pure random samples. When the business researcher uses sample data gathered by someone else, he will always want to check the method used to draw the sample to assure the correctness of the sampling procedure.

There is a nonparametric test which can be used to test sample observations for randomness if the order of their selection is known. The test is based on the number of runs observed in the sample as compared with the number of runs that might result under random conditions. A *run* is defined as a series of identical observations which are preceded and followed by different observations or by none at all.

For example, suppose a coin is flipped 10 times and the order of heads and tails is

$$\underset{1}{\underline{\text{HH}}}\ \underset{2}{\underline{\text{T}}}\ \underset{3}{\underline{\text{H}}}\ \underset{4}{\underline{\text{TTT}}}\ \underset{5}{\underline{\text{HHH}}}.$$

The series has 5 runs made up of 2 heads, 1 tail, 1 head, 3 tails, and 3 heads. The runs are underlined and numbered for emphasis. It is customary to designate the number of runs with the letter, R. In this case $R = 5$.

The average (expected) number of runs and the standard deviation of the number of runs may be computed using Formulas 9–4 and 9–5.

$$\mu_R = \frac{2n_1n_2}{n_1 + n_2} + 1 \tag{9–4}$$

$$\sigma_R = \sqrt{\frac{2n_1n_2(2n_1n_2 - n_1 - n_2)}{(n_1 + n_2)^2(n_1 + n_2 - 1)}} \tag{9–5}$$

where

$n_1 =$ number of occurrences of outcome 1
$n_2 =$ number of occurrences of outcome 2.

If either n_1 or n_2 is greater than 20, the theoretical sampling distribution of R is approximately normal and

$$z = \frac{R - \mu_R}{\sigma_R}. \tag{9–6}$$

The test will always be two-tail. When both n_1 and n_2 are equal to or less than 20, special tables are needed to interpret the sample results. Such tables are usually found in books which specialize in nonparametric methods and are not given here. As long as n_1 or n_2 is greater than 20, the table of Areas of the Normal Curve in Appendix H may be used.

The use of a runs test may be shown with this example. Random numbers are used to select a sample of 34 insurance policies from the files of a large insurance firm. The sample is to be used to estimate the proportion of females holding policies. In order to test the sample for randomness, the sex of the insured is noted in the order in which the policies are drawn. The results are shown below.

F MMM F M FF MMMM FF MMM F M FFF MM F MMM FF MMMM

The test may be designed

Null hypothesis (H_o): *The order of males and females is random.*
Alternate hypothesis (H_a): The order of males and females is not random.
Criterion for decision: Reject H_o and accept H_a if $z < -1.96$ or $z > +1.96$
This criterion gives $\alpha = .05$.

The runs are underlined and there are 16 of them, so $R = 16$. The value of n_1, which represents the number of females, is 13. The value of n_2, which represents the number of males, is 21.

The expected number of runs and the standard deviation are computed using Formulas 9–4 and 9–5.

$$\mu_R = \frac{2n_1n_2}{n_1 + n_2} + 1 = \frac{2(13)(21)}{13 + 21} + 1 = 17.06.$$

$$\sigma_R = \sqrt{\frac{2n_1n_2\,(2n_1n_2 - n_1 - n_2)}{(n_1 + n_2)^2(n_1 + n_2 - 1)}} = \sqrt{\frac{2(273)(546 - 13 - 21)}{(34^2)(33)}}$$

$$= \sqrt{7.33} = 2.71$$

$$z = \frac{R - \mu_R}{\sigma_R} = \frac{16 - 17.06}{2.71} = -0.39.$$

Since z $(-0.39) > z_\alpha$ (-1.96), the null hypothesis cannot be rejected, and it can be assumed that the sample is random.

THE SIGN TEST

The *sign test* is used when the experimenter wishes to show that two conditions are different without making any assumptions about the form of the distribution of differences, as would be necessary in using a t test. The sign test can be used only when the data consists of two observations on each of several entities. With such data it is then possible to rank the observations with respect to the members of each pair. If the first member of the pair is greater than the second, the observation is recorded as a plus. If the second is larger than the first, the observation is recorded as a minus. If the two observations are equal, that pair is dropped from the sample. The sign test gets its name from the fact that + and − signs are used as data rather than the original observations.

The method used to make the sign test is a simple one. Let x_i and y_i represent two conditions or treatments, and for each value of i, $d_i = x_i - y_i$. If d_i is > 0, record a plus, and if $d_i < 0$, record a minus. If $d_i = 0$, discard the pair of observations.

Under the assumption that there would be approximately the same number of +'s and −'s if the two conditions are the same, the null hypothesis can be stated

$$H_o\colon P(x_i < y) = P(x_i > y_i) = \frac{1}{2}.$$

If S is used to designate the number of times that the less frequent sign occurs in the series, then S has a binomial distribution with $\pi = 1/2$ and $n =$ the number of values of d_i.

Suppose a grocery store manager wishes to compare prices in his store, designated Store A, with those of a competitor who runs Store B. The manager selects 21 products at random from his store and records the prices. He prices the same items in his competitor's store. The results are shown in Table 9–12.

TABLE 9–12

COMPUTATION OF A SIGN TEST

Product Number	Price		Sign of d_i
	Store A	Store B	
1	$1.27	$1.19	+
2	.36	.33	+
3	.19	.21	−
4	.27	.25	+
5	.18	.18	0
6	.59	.44	+
7	1.03	.97	+
8	2.46	2.99	−
9	.26	.23	+
10	.53	.57	−
11	.44	.40	+
12	.98	.95	+
13	.13	.12	+
14	.67	.65	+
15	.83	.81	+
16	.69	.59	+
17	.45	.37	+
18	3.37	3.89	−
19	2.11	2.45	−
20	.85	.84	+
21	.33	.30	+

Source: Hypothetical data.

It is clear from studying the price information that in a majority of the cases the prices for Store A are higher than the prices for Store B, but there are exceptions. The problem is to determine if the prices of Store A are significantly higher or if the difference results from the fact that a small sample has been used to represent price levels.

Since there are 5 minuses, $S = 5$. Product 5 has the same price in both stores so that observation is eliminated and $n = 20$.

The test may be stated

Null hypothesis (H_o): $P(x_i < y_i) = P(x_i > y_i) = 1/2$
Price levels in two stores are the same.

Alternate hypothesis (H_a): $P(x_i > y_i) > 1/2$
Price levels in store A are higher than store B.

Appendix F shows that $P(r \leqslant 5 \mid \pi = .5 \text{ and } n = 20) = .0000 + .0000 + .0002 + .0011 + .0046 + .0148 = .0207$. Since $.0207 < .05$, the null hypothesis would be rejected at the .05 level of significance.

When n is small and tables of the binomial distribution are not available, it is possible to compute a rejection value called "K" with the formula

$$K = \frac{n-1}{2} - (.98)\sqrt{n+1}. \tag{9–7}$$

If $S \leqslant K$, reject H_o.

In the example above, $K = \frac{20-1}{2} - (.98)\sqrt{20+1} = 5.01$. Since $S(5) < K(5.01)$, reject H_o.

If $n > 25$, a normal approximation of the binomial is possible using the formula

$$z = \frac{S + .5 - n\pi}{\sqrt{n\pi(1-\pi)}}. \tag{9–8}$$

While the sample in this example is too small to justify using Formula 9–8, it is applied nevertheless for illustration.

$$z = \frac{5 + .5 - 20(.5)}{\sqrt{20(.5)(.5)}} = \frac{-4.5}{\sqrt{5}} = -2.01.$$

Since $-2.01 < -1.64$, it is possible to reject H_o at the .05 level of significance.

The data in this problem could also have been tested for significance by using a chi-square test of goodness of fit to a uniform distribution with the two observed frequencies being 15 and 5 and the two expected frequencies being 10 and 10. The result of this test would also be to reject the null hypothesis.

THE KRUSKAL-WALLACE *H* TEST FOR ANALYSIS OF VARIANCE

In Chapter 8 the technique of analysis of variance was used to test for significant differences between two or more sample means. The assumptions are that the variables were measured on at least an interval scale and came from normal distributions having the same variance.

The *Kruskal-Wallace one-way analysis of variance for ranked data test,* commonly called the *H test,* makes it possible to test the hypothesis that the samples come from the same population or from populations with the same averages when the data are ordinal. No assumptions need to be made about the distribution of the populations from which the samples were drawn, except that they are continuous.

The samples to be compared are arranged in columns and the letter "c" is used to represent the number of columns or the number of samples. If n_j represents the number of observations in the jth sample, then $n_1 + n_2 + \ldots + n_j + \ldots + n_c = n$, which is the total number of observations in all of the samples. Each of the n observations is converted to a rank, the rank of 1 going to the highest value (if ranked in descending order), the rank of 2 going to the second highest value, and so on. The smallest value will have the rank n.

The statistic to be computed for this test is called H and can be computed using Formula 9–9.

$$H = \frac{12}{n(n+1)} \sum_{j=1}^{c} \frac{R_j^2}{n_j} - 3(n+1) \qquad \textbf{(9–9)}$$

where

c = number of samples (columns)

n_j = number of observations in the jth sample (column)

$n = \sum_{j=1}^{c} n_j$ is the total number of observations in all of the samples

R_j = sum of the ranks in the jth sample (column).

The statistic H is distributed as chi-square with $c - 1$ degrees of freedom when each of the samples contains at least 6 observations. When any sample is less than 6, special tables are needed to interpret H.

Computation with No Tied Observations

The data and computations in Table 9–13 are used to illustrate the Kruskal-Wallace one-way analysis of variance test. Suppose that a sales aptitude test is given to three random samples drawn from three occupational

TABLE 9–13

COMPUTATION OF KRUSKAL-WALLACE *H* TEST

Engineers		Salesmen		Administrators	
Test Score	Rank	Test Score	Rank	Test Score	Rank
70	14	97	1	65	16
92	4	79	12	83	10
67	15	91	5	42	21
88	7	95	2	61	18
64	17	85	9	50	20
55	19	90	6	81	11
86	8	94	3		
		75	13		
$R_1 = 84$		$R_2 = 51$		$R_3 = 96$	
$R_1^2 = 7{,}056$		$R_2^2 = 2{,}601$		$R_3^2 = 9{,}216$	
$n_1 = 7$		$n_2 = 8$		$n_3 = 6$	
$\frac{R_1^2}{n_1} = 1{,}008$		$\frac{R_2^2}{n_2} = 325.1$		$\frac{R_3^2}{n_3} = 1{,}536$	

Source: Hypothetical data.

$$\sum_{j=1}^{c} \frac{R_j^2}{n_j} = 1{,}008 + 325.1 + 1{,}536 = 2{,}869.1$$

$$n = 7 + 8 + 6 = 21.$$

groups. One must determine whether there is a difference between the performances on this test of the three occupational groups. The null hypothesis would be that there is no difference in the average scores of engineers, salesmen, and administrators.

The statistic H can now be computed using Formula 9–9. The computations are shown in Table 9–13.

$$H = \frac{12}{n(n+1)} \sum_{j=1}^{c} \frac{R_j^2}{n_j} - 3(n+1) = \frac{12}{21(22)}(2{,}869.1) - 3(22) = 8.52.$$

The value of 8.52 should be compared to the value of chi-square with degrees of freedom equal to $c - 1$ or 2. If the level of significance is set at 0.05, the critical value of chi-square taken from Appendix L is 5.991. Since H is greater than 5.991, the null hypothesis would be rejected. One may conclude that the three occupational groups from which the samples were drawn do not have the same average grades on the sales aptitude test.

Tied Observations

When two or more scores are tied, each score is given the arithmetic mean of the ranks for which it is tied. Suppose in the previous example there were two 70-point scores and two 90-point scores. The two 70's tie for the ranks 14 and 15 and are both given the rank 14.5. The two 90's tie for the ranks 1 and 2 and are both given the rank 1.5. See the circled figures in Table 9–14.

To correct for the effects of ties, H is computed as before using Formula 9–9, and that value is then divided by another value which is a correction factor and is computed

$$1 - \frac{\Sigma T}{n^3 - n}$$

where

$T = t^3 - t$

$t =$ the number of tied observations in a tied group of scores.

For the example shown in Table 9–14 there are two groups of ties with two observations in each group. If T_1 represents the first set of ties, and T_2 represents the second, then

$$T_1 = 2^3 - 2 = 8 - 2 = 6$$

$$T_2 = 2^3 - 2 = 8 - 2 = 6.$$

TABLE 9–14

COMPUTATION OF KRUSKAL-WALLACE *H* WITH TIED OBSERVATIONS

Engineers		Salesmen		Administrators	
Test Score	Rank	Test Score	Rank	Test Score	Rank
(70)	(14.5)	(95)	(1.5)	65	16
92	4	79	12	83	10
(70)	(14.5)	91	5	42	21
88	7	(95)	(1.5)	61	18
64	17	85	9	50	20
55	19	90	6	81	11
86	8	94	3		
		75	13		
	$R_1 = 84$		$R_2 = 51$		$R_3 = 96$

Source: Hypothetical data.

The correction factor is computed

$$1 - \frac{T_1 + T_2}{n^3 - n} = 1 - \frac{6 + 6}{21^3 - 21} = 1 - \frac{12}{9{,}240} = .9987.$$

The corrected value of H is the old value of H which was not affected by the ties (which occurred in each case in the same sample) divided by the correction factor.

$$H = \frac{8.52}{.9987} = 8.53.$$

Comparing this computed value of H to the critical chi-square value of 5.991, one would reject the null hypothesis.

STUDY QUESTIONS

9–1. What is a nonparametric method?

9–2. Distinguish between the following levels of measurement:

a. Nominal.

b. Ordinal.

c. Interval.

d. Ratio.

9–3. What are the assumptions which must be met in order to use the t distribution in connection with a test of hypothesis?

9–4. What are the advantages of using nonparametric rather than parametric tests of significance?

9–5. Explain how chi-square may be used to test whether or not two bases of classification are independent of each other.

9–6. What level of measurement is necessary for data to be tested using the chi-square distribution?

9–7. Why is it desirable to use chi-square to test whether the normal curve is a good fit for an actual frequency distribution?

9–8. Give an example of a situation where it would be appropriate to use a chi-square test of goodness of fit to a uniform distribution.

9–9. What is a run?

9–10. What is the purpose of a runs test? What level of measurement is required for the data used in a runs test?

9–11. What are the data used in a sign test? Could a t test be used on the same data?

9–12. What is the underlying probability distribution used in connection with the sign test? Under what conditions can the normal distribution be used?

9–13. What level of measurement is necessary for data to be tested with a Kruskal-Wallace H test? Could the same data be tested using the F distribution?

9–14. In the Kruskal-Wallace H test how large should the samples be to interpret the results using the chi-square distribution?

PROBLEMS

9–1. In one week four salesmen made the sales shown below.

Salesman	Number of Sales
Friedhof	50
Watson	96
Rodriguez	38
Lanham	44

Use a chi-square test of goodness of fit to a uniform distribution to test the hypothesis that the four are equally good salesmen.

9–2. A normal curve is fitted to a frequency distribution of 525 test scores on a placement test given by a personnel office. The frequency distribution has 16 classes. The value of $\sum\left[\frac{(f_o - f_e)^2}{f_e}\right]$ is computed to be 31.426. Test this distribution of test scores for normality.

9–3. The following distribution shows food costs as a percentage of total costs for a random sample of 160 restaurants.

Food Costs as Percentage of Total Costs	Number of Restaurants
40 and under 45	7
45 and under 50	27
50 and under 55	40
55 and under 60	43
60 and under 65	38
65 and under 70	5
Total	160

Use a chi-square test and a level of significance of 0.10 to test the hypothesis that the universe distribution is normal. *Hint:* Compute a normal distribution with the same mean, standard deviation, and total as the sample distribution.

9–4. The following tabulation gives the number of defective and effective parts produced by each of three shifts in a plant. Test whether the production of defective parts is independent of the shift on which they were produced.

	Number Defective	Number Effective	Total
Day shift	28	560	588
Evening shift	30	473	503
Night shift	25	327	352
Total	83	1,360	1,443

9–5. A die is rolled 120 times and a count is made of the number of times each side appears on top. Use a chi-square test of goodness of fit to a uniform distribution to test the hypothesis that the die is fair. The table is shown at the top of page 244.

Side	Number of times each value appears on top
1	22
2	23
3	15
4	17
5	25
6	18
Total	120

9–6. A study is made in a large insurance company to determine if female employees with small children are absent more often than other female employees. A random sample of the absence records for 400 women workers is shown in the following table.

	Often Absent	Seldom Absent	Total
Small children	75	100	175
No small children	75	150	225
Total	150	250	400

Use a chi-square test and a level of significance of 0.05 to determine if the universe proportion of "often absent" is higher for female employees with small children.

9–7. The following tabulation gives the number of defective and effective parts produced by each of three shifts in a plant. Test whether the production of defective parts is independent of the shift on which they were produced.

	Number Defective	Number Effective	Total
Day shift	26	560	586
Evening shift	30	473	503
Night shift	30	327	357
Total	86	1,360	1,446

9–8. A true-false quiz shows the following pattern of answers. Test the hypothesis that the arrangement of T and F answers is random. Use a runs test.

T T F T T T F F F T T F T F F T F F F F

T T T F F F F T F F T F T F F F F F F F

9–9. In a research project carried out by the faculty of a department of office administration, a survey was made among office employees. One question asked was whether the employee preferred a standard electric or an IBM Selectric typewriter. Of the 411 employees who used a typewriter in their work, 61% preferred the standard electric and 39% the Selectric typewriter; 51% of the men and 62% of the women preferred the standard electric.

	Number of Men	Number of Women	Total
Standard electric	25	226	251
IBM Selectric	24	136	160
Total	49	362	411

Test whether there is a significant difference between the percentage of men and women expressing a preference for a standard electric typewriter. Show your calculations and state your reasoning.

9–10. Refer to Problem 9–9 and use the chi-square distribution to test whether there is a significant difference between the preference of men and women for the standard electric typewriter.

9–11. A large company considering the adoption of a new retirement plan made a sample survey among its employees to find out their reactions to three plans under consideration. Employees were classified as shop employees, office employees, supervisors, and executives. The number in each class preferring each of the plans considered is given in the tabulation below. Test whether the plan preferred is independent of the classification by type of employee.

	Plan A	Plan B	Plan C	Total
Shop employees	361	273	136	770
Office employees	103	79	53	235
Supervisors	16	21	24	61
Executives	6	8	11	25
Total	486	381	224	1,091

9–12. Each value in a time series is compared to the median value for the series. If the value is above the median, it is shown as *A*. If the value is below the median, it is shown as *B*. Use a runs test to determine whether the order of *A*'s and *B*'s is random.

BBBBBBBBBABBBAAABBAAAAAAAA

9–13. A random sample of 328 automobile drivers is selected from a universe of drivers known to have had accidents during the past 5 years with damage of \$500 or more. The following table shows the incidence of injury and use of seat belts in those accidents.

	Seat Belts Used	No Belts Used
Serious injury reported	30	70
No serious injury reported	134	94
Total	164	164

Use a chi-square test with a level of significance of 0.05 to test the hypothesis that the proportion of serious injuries is independent of the use of seat belts.

9–14. A random sample of 150 college professors was each asked to express an opinion as to whether research, teaching, or total performance is the most important basis for academic promotion. The survey results are shown in the following table.

	Teaching Field			
	Sciences	Professional	Arts	Total
Research	30	15	15	60
Teaching	10	20	20	50
Total performance	10	20	10	40
Total	50	55	45	150

Use a chi-square test with a level of significance of 0.05 to test the hypothesis that the universe distribution of proportion of opinion is the same for all faculty groups.

9–15. Thirty campers attending a one-week summer camp are weighed at the beginning and the end of camp. If a camper gains weight during the week, a + sign is recorded. If he loses weight, a − sign is recorded. Use the following results to test the hypothesis that campers are just as likely to gain as they are to lose weight during a week of camp.

+ + + + − − + − + + + − + + + + − + + + + + − + + + − + + +

9–16. Small batches of raw materials are tested for strength before and after applying a heat treatment which is intended to improve this characteristic. Use the following results and a sign test to test the hypothesis that the heat treatment does not improve the strength.

Test Number	Result	Test Number	Result
1	better	13	worse
2	better	14	better
3	better	15	better
4	worse	16	worse
5	worse	17	worse
6	better	18	better
7	better	19	better
8	better	20	better
9	worse	21	better
10	better	22	worse
11	better	23	better
12	better	24	worse

9–17. Samples of three brands of heat lamps are tested to determine their average burning life (in thousands of hours). Use a Kruskal-Wallace H test to test the hypothesis that the samples come from universes with the same average burning life.

Brand A	Brand B	Brand C
14.7	22.4	17.2
19.2	25.9	18.9
20.6	23.1	20.4
15.1	20.5	20.1
13.4	24.4	19.9
16.8	23.6	19.8

9–18. A machine is designed to fill sacks of fertilizer with just over 60 pounds each. Random samples of sacks filled on two different days show the following weights above 60 pounds (in ounces).

First Day	Second Day
1.6	.2
.5	.1
.3	.4
.7	.6
1.1	.4
1.2	.3
1.8	
.8	

Use a Kruskal-Wallace H test to test the hypothesis that the machine was putting the same average amount of fertilizer in sacks on both days.

10 Statistical Quality Control

Since World War II the use of sampling in the control of the quality of manufactured products has become standard practice in industry. Sampling has been used in the inspection of products for a long time but it has been only since the development of statistical quality control that widespread use of sampling techniques has been undertaken. The first application of sampling theory to the problems of quality control was made by Walter A. Shewhart of the Bell Telephone Laboratories in the 1920's. During the decade of the 30's the rate of adoption of these new methods was slow, with more emphasis being placed on it in Great Britain than in the United States. But when the rearmament drive started in 1940, interest immediately increased in the application of sampling methods because of the significant saving in manpower required for inspection. Government procurement agencies began to apply the new methods to their procedures and undertook widespread training programs to acquaint industry with the advantage of sampling methods over the traditional methods of complete inspection of all items. The insistence of the federal government on the use of the new sampling methods by suppliers spread the knowledge of the new techniques with unusual speed. At the end of the war there was a rush by manufacturers to apply these methods to their own manufacturing processes. The American Society for Quality Control, organized in 1946, has taken a leading role in promoting the use of statistical quality control methods in the United States and in spreading the understanding of these methods to industrial concerns in other countries.

The term *quality control* has sometimes been used to describe the field of activity represented by the application of statistical methods to the problems of manufacturing, but there is valid objection to using such a general term to apply to one specific method of controlling quality. The term that best describes the use of statistical methods in controlling the quality of manufactured products is *statistical quality control,* which is used in this chapter.[1]

[1] Although statistical quality control is an application of statistical methods, the symbols used in the field of quality control differ considerably from the practice followed in other fields of statistics. In this chapter, however, the symbols are the same as those used in previous chapters.

VARIABILITY IN MANUFACTURED PRODUCT

Although approximate measurements may not show that a difference exists, specifications recognize the fact that no two items are ever exactly alike by giving a tolerance range within which measurements must fall. Individual items falling outside this range are not acceptable; if they are final products they will not perform satisfactorily, or if they are component parts, they will not fit properly when assembled with other parts. The basic principle of mass production is that the individual parts are all near enough the same size to be interchangeable.

Because manufacturing processes cannot produce any two items that are exactly alike, there is a need to locate and segregate defective products through quality checks. A simple and direct method of insuring that all parts meet manufacturing specifications is to inspect each item and reject any that fail to fall within the limits specified. The rejected items may be reworked to make them conform to the specifications or, if this is impossible, they must be scrapped. Inspection of each item produced, a method which is probably as old as manufacturing, will give control over the quality of manufactured products by eliminating most of the defective items. However, for most quality checks the cost of examining every piece would be astronomical. In some cases quality checks may even destroy the item. Such tests have cost practicality only when they are performed on a sample of the universe. Therefore, sampling of the items is employed, and entire lots of parts are deemed acceptable or unacceptable on the basis of sample results.

The use of statistical methods has proven to be so much more efficient in controlling quality that statistical quality control is one of the very significant uses of statistical methods in business management. Its proper use requires such a detailed knowledge of manufacturing as well as statistical methods that it is now recognized as a specialized professional activity. This chapter merely introduces the subject by describing a few basic applications.

The variations in a given characteristic of a manufactured product may be grouped into two classes on the basis of the causes of the variations—assignable causes and random causes.

Assignable Variations

Assignable variations comprise those that result from specific causes which can be identified. Variations in the product due to mistakes of inexperienced workmen, worn tools, machines in need of adjustment, and defective raw materials are examples of this class. Since they represent a relatively large variation in the product, their cause should be identified and removed. For example, a machine that needs adjustment should be located and the adjustment corrected, or an employee who is performing an operation incorrectly should be given further instruction.

Random Variations

Random variations may result from a random combination of circumstances that cause slight differences in the individual units produced, differences that individually have so little effect on the result that it is impractical to try to locate them or to trace their effects. The random variations that result from numerous minor causes may be considered simply as characteristics of the manufacturing process. Even though the same machines, materials, labor, and manufacturing techniques are used, some variations will occur in the product. However, it is not worth the cost of trying to find the reason for each of these variations since they are in reality the result of chance.

When it has been established that a given degree of random variation is inherent in a process, a decision must be made as to whether or not this variation is greater than can be tolerated. If the variations are greater than can be permitted, a change in the process should be made to make it possible to turn out an acceptable product. This may mean purchasing better machines, hiring more-skilled labor, providing labor with better training, or buying better raw materials.

When it has been determined that the degree of random variations in the product can be tolerated, the problem becomes one of preventing any assignable cause from introducing variations. If variations due to an assignable cause appear, it is important to locate and remove the cause. Since the fundamental problem is distinguishing between the random variations and the variations that can be attributed to a specific cause, the inspector who has the responsibility for passing on the acceptability of the product needs some guide that will enable him to distinguish between causes.

The techniques used in statistical quality control may be broken down into two major classes: (1) control charts, which may be used on a continuous basis to check a process; and (2) acceptance sampling, which may be used at the end of a process. These techniques are discussed in detail in the following sections.

CONTROL CHARTS

The earliest work in developing the methods of statistical quality control was devoted to devising a method for determining whether or not the variation that occurred in the output of a process was greater than the random variation inherent in the process. As long as variations in the product do not exceed the limits set up, the process is considered to be in control and production is permitted to continue. But when the variations exceed these limits, it becomes necessary to find out what happened to the process to cause such large variations from the desired characteristics. In other words, variations in a given characteristic of a manufactured product may be grouped into two categories on the basis of the cause of the variations – variations for which a specific cause can be assigned and variations resulting from random causes that are so numerous that it is not feasible to try to isolate them individually.

One of the techniques used in statistical quality control to determine the cause of variations is the control chart. A *control chart* is a device used to make a large number of tests of significance in a systematic manner. It may be used to monitor production on a continuous basis to detect when a process is not performing in an acceptable manner. Since it is important to discover promptly that a process has gone out of control, a schedule is set up to test the hypothesis that the process is performing satisfactorily. The control chart is an efficient device for making these numerous tests and giving a warning when a hypothesis should not be accepted.

$\bar{x}$ Chart[2]

The $\bar{x}$ *chart* is constructed to show the fluctuations of the means of samples about the mean of the process and can be used to determine whether or not the fluctuations are due to random causes or to an assignable cause. The data given in Table 10–1 are used to illustrate the construction and interpretation of the $\bar{x}$ control chart. A critical dimension was measured to the nearest ten thousandth of an inch. The first item measured .5025. Then .5000 inches was subtracted from .5025, and the decimal was moved four places to the right (.5025 − .5000 = .0025, or 25). Thus, the first measurement is 25.

It is not necessary to convert the data as described above, but since it results in considerably less clerical work, it is standard practice in many types of statistical work. The amount to subtract from the individual observations (.5000 in this case) was chosen because it seemed to effect the greatest saving in calculations. This is true because, by subtraction, each individual value was converted from a four-digit number to a two-digit number, with an obvious reduction in clerical work. Expressing the dimensions in .0001 inches had the additional advantage of eliminating the need to work with decimals. It is, of course, necessary to show exactly what operation was performed on the data; the subtitle in Table 10–1 is used to explain the conversion in this example.

The measurements in Table 10–1 are given for 20 subgroups, each containing 4 items. At intervals of 30 minutes, 4 items were taken from the assembly line for inspection and the measurements of the critical dimension were entered in the table. The subgroup numbers represent the order in which the items were produced. If we assume that there were no assignable causes of variation in the dimension present while these items were being produced, the mean of each of the 20 samples of 4 is an unbiased estimate of the universe mean, although a sample of 4 would be expected to produce an estimate with a wide confidence interval. The mean of the 20 sample means would, however, be a more reliable estimate of the universe mean than any of the individual sample means. The mean of the sample means is represented by $\bar{\bar{x}}$ and is used as an estimate of the mean of the universe. A method of checking on the assumption that no assignable causes of variation in the dimension were present during the period in which the samples were

[2] Texts on statistical quality control use $\bar{X}$ to represent a sample mean, but in this chapter $\bar{x}$ will be used in order to be consistent with previous chapters.

TABLE 10–1

MEASUREMENT OF A CRITICAL DIMENSION OF 20 SAMPLES OF 4 ITEMS

Unit: .0001 Inches in Excess of .5000 Inches

Subgroup Number	Measurement of Individual Items	Σx	$\bar{x}$	Σx^2	s^2	s	R
1	25 14 19 18	76	19.00	1,506	15.5000	3.9	11
2	22 16 20 19	77	19.25	1,501	4.6875	2.2	6
3	24 12 15 24	75	18.75	1,521	28.6875	5.4	12
4	18 17 23 21	79	19.75	1,583	5.6875	2.4	6
5	26 19 16 21	82	20.50	1,734	13.2500	3.6	10
6	18 17 16 15	66	16.50	1,094	1.2500	1.1	3
7	19 22 15 14	70	17.50	1,266	10.2500	3.2	8
8	18 20 21 23	82	20.50	1,694	3.2500	1.8	5
9	17 21 20 17	75	18.75	1,419	3.1875	1.8	4
10	20 16 22 17	75	18.75	1,429	5.6875	2.4	6
11	18 19 21 20	78	19.50	1,526	1.2500	1.1	3
12	19 13 20 18	70	17.50	1,254	7.2500	2.7	7
13	19 22 21 18	80	20.00	1,610	2.5000	1.6	4
14	21 16 17 19	73	18.25	1,347	3.6875	1.9	5
15	15 23 15 16	69	17.25	1,235	11.1875	3.3	8
16	20 22 20 19	81	20.25	1,645	1.1875	1.1	3
17	17 23 18 19	77	19.25	1,503	5.1875	2.3	6
18	22 19 21 18	80	20.00	1,610	2.5000	1.6	4
19	17 25 24 20	86	21.50	1,890	10.2500	3.2	8
20	18 16 14 21	69	17.25	1,217	6.6875	2.6	7
Total		1,520	380.00	29,584	143.1250	49.2	126

Source: Confidential.

taken will be explained later. Any sample that is known to have been taken when an assignable cause was present should be eliminated from the calculation of $\bar{\bar{x}}$. The values of the 20 sample means are given in Table 10–1, and from these means the value of $\bar{\bar{x}}$ was found to be 19.0 $\left(\frac{\Sigma\bar{x}}{20} = \frac{380.00}{20} = 19.0\right)$.

Table 10–1 also gives the values of s^2 and s for each of the samples. It was explained in Chapter 7 that the standard deviation and the variance of a small sample are biased estimates of the universe values. The bias may be removed by multiplying the sample variance (s^2) by the ratio $\frac{n}{n-1}$, after which the average variance can be computed. Since the same correction is applied to each of the variances being averaged, it is less work to average the uncorrected variances and then correct the average. The average variance is found by dividing the sum of the 20 variances (143.1250) by 20.

$$\bar{s}^2 = \frac{143.1250}{20} = 7.15625.$$

Using the value of $\bar{s}^2$ as the variance (s^2) in Formula 7–5, page 163, the estimate of the universe standard deviation $(\hat{\sigma})$ is computed

$$\hat{\sigma} = \sqrt{\bar{s}^2 \frac{n}{n-1}} = \sqrt{(7.15625)\left(\frac{4}{4-1}\right)} = \sqrt{9.5417} = 3.09.$$

If the mean of the universe, in this case all the production of this particular item, is assumed to be .5019 inches (using only the last two digits, it becomes 19), the problem is to determine within what range the means of samples of four will fluctuate. The standard error of the mean was computed in Chapter 7 by the equation (ignoring the correction for a finite universe)

$$\hat{\sigma}_{\bar{x}} = \frac{\hat{\sigma}}{\sqrt{n}}.$$

For a sample of four items and a universe standard deviation of 3.09, the standard error of the mean is computed

$$\hat{\sigma}_{\bar{x}} = \frac{3.09}{\sqrt{4}} = 1.545.$$

It is almost universal practice in statistical quality control work in the United States to use a confidence interval of three standard errors. The probability of a sample falling outside the 3σ confidence interval is so small that it is considered practically certain that the mean of a random sample will fall within the limits of $\pm 3\sigma_{\bar{x}}$ from the mean of the universe. Since the standard error of the mean was computed from an estimate of the standard deviation of the universe, it is represented by $\hat{\sigma}_{\bar{x}}$. It is assumed that a sample of four items will fall below $\bar{\bar{x}} + 3\hat{\sigma}_{\bar{x}}$, the upper control limit (UCL), but above $\bar{\bar{x}} - 3\hat{\sigma}_{\bar{x}}$, the lower control limit (LCL).

$$\text{UCL} = \bar{\bar{x}} + 3\hat{\sigma}_{\bar{x}} = 19.0 + (3)\,(1.545) = 23.635$$

$$\text{LCL} = \bar{\bar{x}} - 3\hat{\sigma}_{\bar{x}} = 19.0 - (3)\,(1.545) = 14.365.$$

Since the samples were taken in chronological order, it is possible to set up a time series chart showing the estimate of the universe mean, the upper and lower limits of the confidence interval, and the mean of each of the samples. The chart will show graphically whether or not any of the sample means fall outside the confidence interval. Figure 10–1 is a control chart for $\bar{x}$ and is used to determine each time a sample mean is plotted whether the variation of this sample mean from the universe mean is to be ascribed to random causes or assignable causes. All the 20 sample means in Figure 10–1 fall within the control limits, which indicates that the process is in control, and that only random variations were present in the production at the time the samples were taken. If a sample mean had fallen outside the control limits,

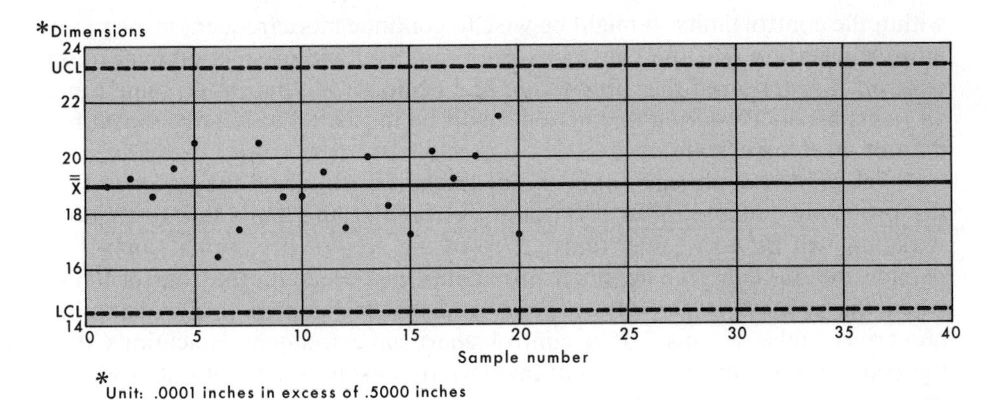

Source: Table 10-1.

FIGURE 10–1

CONTROL CHART FOR MEANS

it would indicate that such a large deviation from the universe mean was the result of something more than the forces causing random variations. The very small probability that a sample mean will deviate more than 3 σ's from the universe mean justifies drawing the above conclusion. It is considered that such an extreme variation from the universe mean can be explained more logically by some assignable cause than by assuming that it was merely due to random variations. It would not be correct to say that it is *certain* that such a deviation is not due to random variations, but the probability that the deviation would be due to a random variation is very small. It is much more reasonable to conclude that some nonrandom cause has influenced production, and a search should be made for this cause of variation. When an assignable cause is believed to be present, the process is said to be *out of control.* As long as only random variations are present, the process is *in control.*

If the mean of any of the subgroups plotted in Figure 10–1 had been either above or below the control limits, the $\bar{\bar{x}}$ and $\bar{s}$ should be recomputed without including the data for the subgroup that was out of control. The control chart should be revised on the basis of the new value for $\bar{\bar{x}}$ and the new upper and lower control limits. There is a possibility that the new control limits will show another subgroup out of control; if so, the process is repeated until all the sample means fall within the control limits.

After the control limits have been established, the chart may be used to check on the manufacturing process during subsequent production. The chart may be extended as shown in Figure 10–1, with a sample of four items taken at 30-minute intervals, and the average for the sample plotted on the chart. If the plotted mean falls outside the control limits, it indicates that an assignable cause is present; and when this occurs, the usual procedure is to start an immediate search for the cause of the variation. When the cause has been found and corrected, the average of the next sample of four items should fall within the control limits. It might be wise to continue more frequent inspections until it is certain that the cause has been removed and production is again in control. If it appeared that production had been out of control for some time, all items produced during this period might be inspected to remove those that did not meet specifications.

The control limits are set at three standard errors from the mean so that the probability of the mean of a sample exceeding the limits due to random variation will be very small (only .27% of the area of the normal curve lies outside the 3σ limits). The small probability of exceeding the control limits when the process is actually in control has been adopted to avoid issuing numerous "false alarms." If a control chart gave frequent indications that the process was out of control but investigation established that only random variations were present, the chart would quickly cease to be used with any confidence. The probability of making a Type I error is .27, which was considered in Chapter 8 to be a very rigorous test of a hypothesis.

Using a confidence interval on the control chart based on three standard deviations has the disadvantage that it will be slow in warning of a shift in the universe mean, especially when the shift is small. This means that the probability of making a Type II error is high, since there is a large probability of accepting a hypothesis as correct when it is in fact false. The hypothesis that the process is in control is accepted but it is not true; the mean of the universe has shifted without being detected. The use of 3σ control limits is almost universal in this country, which means that a small probability of looking for trouble when none exists is secured at the expense of having a high probability of failing to detect a shift out of control.

Tests for Lack of Control Based on Runs. It is possible to test a process for lack of control without waiting for a sample mean to go outside the control limits. The tests for randomness applied to runs in Chapter 9 may be used with control charts to give warning that an assignable cause may be present. The most practical test to use in detecting shifts in a process average is to use a rule that depends only on extreme runs. It would be valid to suspect that a process average had shifted whenever seven successive points on the control chart were on the same side of the central line. If the process is in control and only random forces cause the sample average to deviate from the center line on the control chart, the probability is .5 that a point will fall above the center line, and .5 that it will fall below the center line. The probability of seven successive points falling on the same side of the center line is the same as the probability of seven successive heads showing when a coin is tossed. This probability is $.5^7 = .078125$. It is considerably larger than the probability of a point falling outside the control line, but the rule may be used to alert management to the fact that something may be wrong even though it does not call for stopping the operation at this time to make a thorough search for an assignable cause.

Computing Control Limits from $\bar{s}$ Instead of $\bar{s}^2$. When the universe is normally distributed, it is possible to compute the control limits of the $\bar{x}$ chart by using the standard deviation (s) to estimate σ instead of using the variance (s^2) as explained previously. Since the universes used in quality control can usually be considered to be normal, the preference has been to use the standard deviation instead of the variance. The mean of the distribution of standard deviations from all possible samples is the expected value of the sample standard deviation, $E(s)$. $E(s)$ is equal to c_2 times the universe standard deviation only when the universe from which the samples are taken is normal. This may be written

$$E(s) = c_2\sigma.$$

The value of c_2 is computed from the formula

$$c_2 = \sqrt{\frac{2}{n}}\,\frac{\left(\frac{n-2}{2}\right)!}{\left(\frac{n-3}{2}\right)!} \tag{10–1}$$

When $n = 4$, the value of c_2 is computed[3]

$$c_2 = \sqrt{\frac{2}{4}}\,\frac{\left(\frac{4-2}{2}\right)!}{\left(\frac{4-3}{2}\right)!} = \sqrt{.5}\,\frac{1!}{.5!}$$

$$c_2 = \sqrt{.5}\,\frac{1}{.5\sqrt{\pi}} = .707107\,\frac{1}{.8862} = .7979.$$

The value of c_2 for each value of n from 2 to 15 is given in Table 10–2; tables for larger values of n are available in books on statistical quality control. It will be noted that the value of c_2 is less than unity but increases as the size of n is increased. The standard deviation of the universe can be estimated from one sample standard deviation, or the mean of several sample standard deviations, designated $\bar{s}$. If as many as 20 standard deviations from small samples are used to compute $\bar{s}$, the estimate would be accurate enough to use. The formula for estimating the universe standard deviation from the average of several sample standard deviations is

$$\hat{\sigma} = \frac{\bar{s}}{c_2}\,. \tag{10–2}$$

The mean of the 20 standard deviations in Table 10–1 is 2.46 ($\bar{s} = \frac{49.2}{20} =$ 2.46). The estimate of the universe standard deviation ($\hat{\sigma}$) is 3.08, found by

[3] The factorial of .5 (written .5!) is $.5 \times \sqrt{\pi}$. The factorial of 3.5 (written 3.5!) is $3.5 \times 2.5 \times 1.5 \times .5 \times \sqrt{\pi}$.

TABLE 10–2

VALUES FOR COMPUTING CONTROL LIMITS FOR $\bar{x}$ CHARTS FROM $\bar{s}$

n	c_2 *	A_1 †
2	.5642	3.760
3	.7236	2.394
4	.7979	1.880
5	.8407	1.596
6	.8686	1.410
7	.8882	1.277
8	.9027	1.175
9	.9139	1.094
10	.9227	1.028
11	.9300	.973
12	.9359	.925
13	.9410	.884
14	.9453	.848
15	.9490	.816

* c_2 computed from Formula 10–1.
† A_1 computed from Formula 10–3.

dividing $\bar{s}$ by c_2 ($\hat{\sigma} = \frac{2.46}{.7979} = 3.08$). This estimate of the universe standard deviation calculated from the mean value of the distribution of sample standard deviations is approximately the same as the estimate secured on page 252, and gives very nearly the same control limits.

$$\hat{\sigma}_{\bar{x}} = \frac{3.08}{\sqrt{4}} = 1.54$$

$$\text{UCL}_{\bar{x}} = 19.0 + 3(1.54) = 23.62$$

$$\text{LCL}_{\bar{x}} = 19.0 - 3(1.54) = 14.38.$$

The calculation of the control limits from $\bar{s}$ may be simplified by using the values of A_1 given in Table 10–2. As stated on page 252, the equation for the upper control limit is $\text{UCL}_{\bar{x}} = \bar{\bar{x}} + 3\,\hat{\sigma}_{\bar{x}}$ and the lower control limit is $\text{LCL}_{\bar{x}} = \bar{\bar{x}} - 3\,\hat{\sigma}_{\bar{x}}$.

Since $$\hat{\sigma}_{\bar{x}} = \frac{\hat{\sigma}}{\sqrt{n}} \text{ and } \hat{\sigma} = \frac{\bar{s}}{c_2}$$

$$\text{UCL}_{\bar{x}} = \bar{\bar{x}} + 3\,\frac{\frac{\bar{s}}{c_2}}{\sqrt{n}} = \bar{\bar{x}} + \frac{3}{c_2\sqrt{n}}\,\bar{s}$$

Letting $$A_1 = \frac{3}{c_2\sqrt{n}} \qquad \textbf{(10–3)}$$

$$UCL_{\bar{x}} = \bar{\bar{x}} + A_1\bar{s} \text{ and } LCL_{\bar{x}} = \bar{\bar{x}} - A_1\bar{s}.$$

The use of the value A_1 from Table 10–2 gives the same control limits as computed previously.

$$UCL_{\bar{x}} = 19.0 + (1.880)(2.46) = 23.62$$
$$LCL_{\bar{x}} = 19.0 - (1.880)(2.46) = 14.38.$$

Computing Control Limits from $\bar{R}$. Since the number of control charts needed in any quality control program is large, any substantial saving in the time required to compute the necessary values is an important consideration. The mean of the ranges ($\bar{R}$), instead of the mean of the standard deviations of the subgroups, can be used in computing the control limits. This is valuable because the range of a sample can be computed in considerably less time than the standard deviation. The relationship between the standard deviation of the universe and the range (R) of a sample is shown by the values d_2 in Table 10–3 for samples with values of n from 2 to 15. The value of d_2 for a given value of n shows the ratio of the average value of the range of all samples of size n to the standard deviation of the universe from which the sample was taken. When $n = 4$, $d_2 = 2.059$, which indicates that the average range of samples of 4 is slightly more than two times the standard deviation of the universe from which the samples were taken. If the value of $\bar{R}$ computed from the data in Table 10–1 is taken as the average value of the distribution of

TABLE 10–3

VALUES FOR COMPUTING CONTROL LIMITS FOR $\bar{x}$ CHARTS FROM $\bar{R}$

n	d_2	A_2
2	1.128	1.880
3	1.693	1.023
4	2.059	.729
5	2.326	.577
6	2.534	.483
7	2.704	.419
8	2.847	.373
9	2.970	.337
10	3.078	.308
11	3.173	.285
12	3.258	.266
13	3.336	.249
14	3.407	.235
15	3.472	.223

Source: American Society for Testing and Materials, *A.S.T.M. Manual on Quality Control of Materials.* (Used with permission.)

ranges from samples of 4, the standard deviation of the universe from which the samples were taken can be estimated by the equation

$$\hat{\sigma} = \frac{\bar{R}}{d_2}.$$

The value of $\bar{R}$ is $\frac{126}{20} = 6.30$, and the estimate of the standard deviation of the universe is

$$\hat{\sigma} = \frac{6.30}{2.059} = 3.06.$$

This estimate of the standard deviation of the universe may be used instead of the estimate based on $\bar{s}$ to compute the control limits for the $\bar{x}$ chart. Because of the savings in time required to calculate R instead of s for each sample, constructing the control limits from $\bar{R}$ is preferred in quality control work. The mean of the standard deviations gives somewhat more accurate values for the control chart; but when the subgroups are small, the ranges give satisfactory values. Just as the calculation of the control limits from $\bar{s}$ is simplified by the use of the values of A_1, the calculation of the control limits from $\bar{R}$ is simplified by the use of the values of A_2 given in Table 10–3.

Since $$\hat{\sigma} = \frac{\bar{R}}{d_2}$$

$$UCL_{\bar{x}} = \bar{\bar{x}} + 3\,\frac{\frac{\bar{R}}{d_2}}{\sqrt{n}} = \bar{\bar{x}} + \frac{3}{d_2\sqrt{n}}\,\bar{R}.$$

Letting $$A_2 = \frac{3}{d_2\sqrt{n}} \qquad \text{(10–4)}$$

$$UCL_{\bar{x}} = \bar{\bar{x}} + A_2\bar{R} \text{ and } LCL_{\bar{x}} = \bar{\bar{x}} - A_2\bar{R}.$$

Using the values of A_2 and $\bar{R}$, the control limits are found to be only slightly different from those computed from A_1 and $\bar{s}$ on page 257.

$$UCL_{\bar{x}} = 19.0 + (.729)(6.30) = 23.59$$
$$LCL_{\bar{x}} = 19.0 - (.729)(6.30) = 14.41$$

R Chart

The *R chart* can be used to show the fluctuations of the ranges of the subgroups about the average range $\bar{R}$. The construction of an R chart follows the same general principle as the $\bar{x}$ chart, using the mean of the ranges of the samples as the centerline and the 3 $\hat{\sigma}_R$ limits. The distribution of the ranges of all possible small samples from a normal universe is not normal, and a larger proportion of the cases may exceed the upper 3 $\hat{\sigma}_R$ limit than would be true for a normal distribution. The lower control limit for small samples in many cases would be negative; but since the range cannot have a negative value, the lower control limit in such situations is set at zero. Although the distribution of the ranges is not normal, the 3 σ control limits indicate when the variability is so great that it may be assumed to be the result of an assignable cause rather than random causes. With no lower control limit, it is possible for the process to go out of control only by exceeding the upper control limit. Such an increase in variability might occur without the mean going out of control, and in such a situation the R chart would give warning of an assignable cause of variation when the $\bar{x}$ chart did not. It is much more common for an assignable cause that increases the variability also to throw the process out of control with respect to the mean, in which case the R chart would not be needed. Sometimes the R chart is not used; but since it can be constructed with very little more work than is needed for the $\bar{x}$ chart, it is generally kept along with the $\bar{x}$ chart.

The control limits for the R chart are

$$\text{UCL}_R = \bar{R} + 3\,\hat{\sigma}_R$$
$$\text{LCL}_R = \bar{R} - 3\,\hat{\sigma}_R.$$

If

$$D_4 = 1 + \frac{3\hat{\sigma}_R}{\bar{R}}$$

and

$$D_3 = 1 - \frac{3\hat{\sigma}_R}{\bar{R}}$$

then

$$\text{UCL}_R = D_4\bar{R}$$

and

$$\text{LCL}_R = D_3\bar{R}.$$

For the data in Table 10–1, the control limits are computed, using the values of D_4 and D_3 from Table 10–4

$$\text{UCL}_R = D_4\bar{R} = (2.282)(6.30) = 14.38$$
$$\text{LCL}_R = D_3\bar{R} = (0)(6.30) = 0.$$

With the value of $\bar{R}$ plotted as the centerline and the control limits just computed, Figure 10–2 is constructed for the 20 subgroups. By extending the chart, the range of each sample may be plotted to determine whether or not the process is in control with respect to variability. This involves computing the range of each subgroup as well as the mean, which is needed for plotting the $\bar{x}$ chart.

TABLE 10-4

VALUES FOR COMPUTING CONTROL LIMITS FOR R CHARTS FROM $\bar{R}$

n	D_3	D_4
2	0	3.267
3	0	2.575
4	0	2.282
5	0	2.115
6	0	2.004
7	.076	1.924
8	.136	1.864
9	.184	1.816
10	.223	1.777
11	.256	1.744
12	.284	1.716
13	.308	1.692
14	.329	1.671
15	.348	1.652

Source: American Society for Testing and Materials, *A.S.T.M. Manual for Quality Control of Materials.* (Used with permission.)

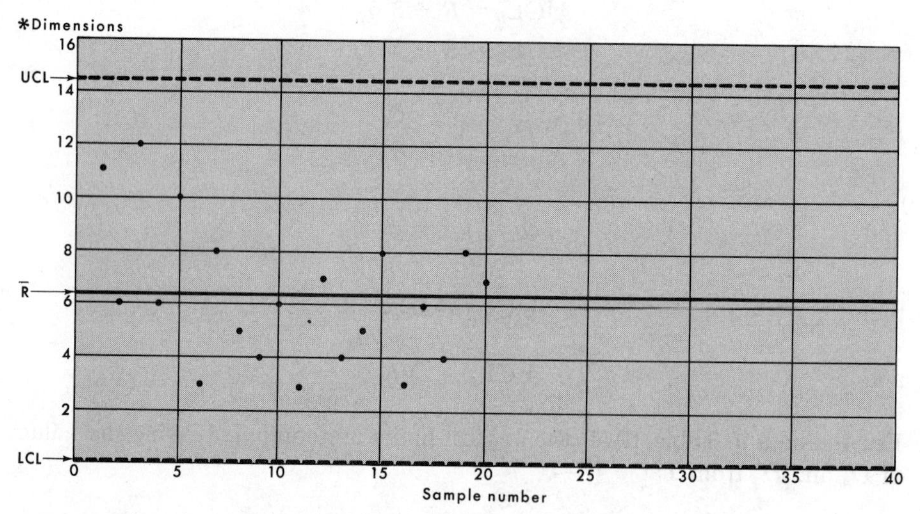

*Unit: .0001 inches in excess of .5000 inches

Source: Table 10-1.

FIGURE 10-2

CONTROL CHART FOR RANGES

CONTROL CHART FOR FRACTION DEFECTIVE

The control chart for variables can be used whenever the quality characteristic can be measured and expressed as a quantity. In many situations, however, the quality characteristics can be expressed only as meeting or not meeting specifications. If an item is either accepted or rejected on the basis of whether or not it passes a given test, a *control chart for fraction defective* (also called a *p chart*) may be used instead of $\bar{x}$ and R charts. Acceptance of an item may be the result of an elaborate analysis, or it may be based on the checking of dimensions by "go" and "not-go" gauges. As long as the inspection of an article results in the classification of the item as accepted or rejected, the fraction defective chart may be used to analyze the data.

Control charts for fraction defective may be based on 100% inspection or on samples, although if sample data are used, it is important that a relatively large sample be used. The small samples that are used with control charts for variables are not satisfactory in an analysis of fraction defective.

The data in Table 10–5 will be used to illustrate the use of the p chart. The output of an electrical component was checked in 100 unit lots and the number of rejected items tabulated. The results of the inspection of the first 25 lots, representing a total of 2,500 items, are given in Table 10–5. The total number of defective items was 270, or .108 of the total produced. If the process were under control, this fraction could be used as an estimate of the process fraction defective (designated as $\bar{p}$). Each of the lots inspected may be considered a sample of 100 from a universe in which the fraction of defective items is .108.

Using Formula 7–10, it is possible to compute the 3 σ confidence interval for a sample of 100 taken from a universe that has a fraction defective of .108. (The correction for a finite universe is omitted since the universe is large in relation to the size of the sample.)

$$3\hat{\sigma}_p = 3\sqrt{\frac{\bar{p}\bar{q}}{n}} = 3\sqrt{\frac{(.108)\,(.892)}{100}} = 3\sqrt{.00096336}$$

$$3\hat{\sigma}_p = (3)(.031) = .093.$$

Using the following values, Figure 10–3 (page 263) was constructed as a control chart and the fraction defective for each of the 25 lots was plotted.

$$\text{Centerline} = \bar{p} = .108$$

$$\text{UCL}_p = \bar{p} + 3\hat{\sigma}_p = .108 + .093 = .201$$

$$\text{LCL}_p = \bar{p} - 3\hat{\sigma}_p = .108 - .093 = .015.$$

Since lot numbers 5 and 20 had a fraction defective higher than the upper control limit, it was assumed that assignable causes were found to have been operating when these lots were processed. In order to compute a better estimate of the universe fraction defective, these two lots were eliminated from the computation. The remaining 23 lots contained 227 defective items out of a total of 2,300 inspected, or a fraction defective of .099. This new value of $\bar{p}$ was used to recompute the control limits.

TABLE 10-5

NUMBER OF DEFECTIVE ELECTRICAL COMPONENTS IN 25 LOTS OF 100

Lot Number	Number Inspected	Number Defective np	Fraction Defective p
1	100	11	.11
2	100	9	.09
3	100	15	.15
4	100	11	.11
5	100	22	.22
6	100	14	.14
7	100	7	.07
8	100	10	.10
9	100	6	.06
10	100	2	.02
11	100	11	.11
12	100	6	.06
13	100	9	.09
14	100	18	.18
15	100	7	.07
16	100	10	.10
17	100	8	.08
18	100	11	.11
19	100	14	.14
20	100	21	.21
21	100	16	.16
22	100	4	.04
23	100	11	.11
24	100	8	.08
25	100	9	.09
Total	2,500	270	

Source: Confidential.

$$3\hat{\sigma}_p = 3\sqrt{\frac{\bar{p}\bar{q}}{n}} = 3\sqrt{\frac{(.099)\,(.901)}{100}} = 3\sqrt{.00089199}$$

$$3\hat{\sigma}_p = (3)(.0299) = .0897$$

$$\text{Centerline} = \bar{p} = .099$$

$$\text{UCL}_p = \bar{p} + 3\hat{\sigma}_p = .099 + .090 = .189$$

$$\text{LCL}_p = \bar{p} - 3\hat{\sigma}_p = .099 - .090 = .009.$$

The new value of $\bar{p}$ and the new control limits were plotted on Figure 10-3 and were extended beyond the first 25 lots to be used to check each succeeding lot for assignable causes of variation in the fraction defective. The

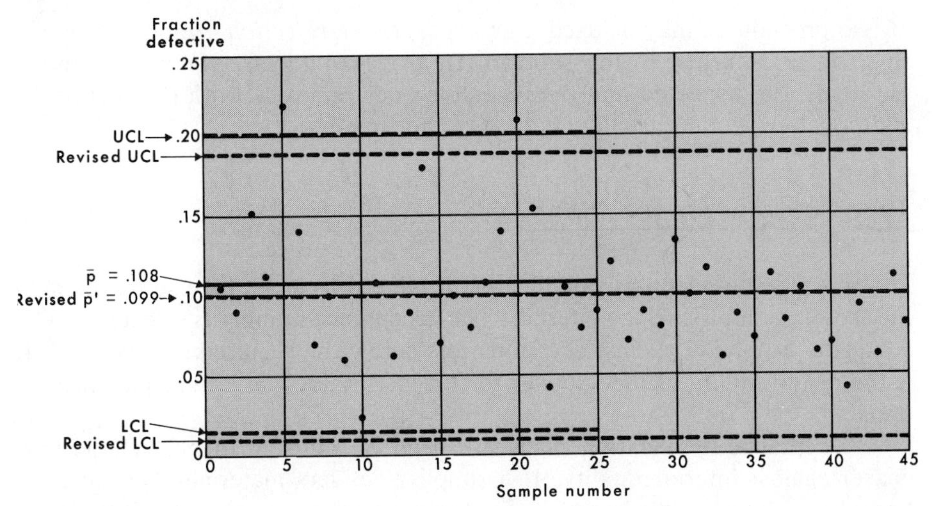

Source: Table 10-5.

FIGURE 10–3

CONTROL CHART FOR FRACTION DEFECTIVE

fraction defective for each of the next 20 lots was plotted and a definite downward trend was evident. On the assumption that a smaller fraction defective could be maintained in the process, a new center line and control limits should be computed and an effort made to keep the process in control at this better quality level. It has been a common occurrence for the fraction defective to drop after a control chart is put into use; and when this happens, the level of $\bar{p}$ is usually revised and an attempt is made to hold the new quality level.

When the subgroup size varies, new control limits should be computed for each subgroup. The control limit will be represented by a broken line, moving farther from the center line when the subgroup is smaller, and closer to the center line for larger subgroups. Except for the additional work of computing the control limits for each subgroup, the interpretation of the p chart based on subgroups of varying sizes offers no new problems. It is possible to compute the control limits for a subgroup of average size and then compute the control limits for individual subgroups only when it is not possible to determine by inspection where the fraction defective for a given lot falls with respect to the control limit.

ACCEPTANCE SAMPLING

The use of sampling inspection by a purchaser to decide whether or not to accept a shipment of product is known as *acceptance sampling*. A sample of the shipment is inspected and if the number of defective items is not more than a stated number, known as the *acceptance number*, the shipment is accepted. If the number of defective items exceeds the acceptance number, one of two procedures may be used. In an *acceptance-rejection* sampling plan, the shipment is returned to the supplier. In an *acceptance-rectification* plan, all the items are inspected and the defective ones removed, with the cost of the additional inspection usually charged to the supplier.

Types of Acceptance Plans

Acceptance-rejection sampling plans were developed for use by the military services in procuring materials. The acceptance number is selected to give adequate protection against accepting lots below the required quality level. It is the responsibility of the supplier to correct any lot that fails to pass inspection.

Acceptance-rectification plans are used in industry to protect the purchaser against inferior quality. If a supplier delivers materials that have too large a percentage of defective items, the purchaser protects himself by 100% inspection. As long as the sampling inspection shows no more defectives than the acceptance number, the buyer takes the lot on the basis of the sample and the supplier is saved the cost of having to make a complete inspection.

Sampling acceptance plans make possible a prompt decision to accept or reject a lot, with knowledge of the probabilities of making a mistake. Deciding on which plan to use requires a decision as to the risk one is willing to take in making two types of errors. The null hypothesis is that the lot is acceptable, but (1) accepting a lot as satisfactory when in fact it is below the quality level is a Type II error, and (2) rejecting a lot as unsatisfactory when it is of acceptable quality is a Type I error.

Sampling acceptance plans may be classified as single, double, multiple, or sequential, depending upon whether one, two, or more samples are taken before reaching a decision. In *single sampling,* the decision is made on the basis of the evidence furnished by one sample. *Double sampling* provides a method for taking a second sample if the results of the first sample are not conclusive. A lot may be rejected on the basis of the information supplied by the first sample if it is bad enough, or it may be accepted after the first sample has been taken if it is good enough. If the first sample does not give a clear enough indication of the quality of the product to make a decision, a second sample is taken. *Multiple sampling* resembles double sampling except that more than two samples may be taken before a final decision is made. *Sequential sampling,* which provides for taking as many samples as needed to reach a decision, results in small samples being used when the quality of the product is either very good or very bad. A larger sample is taken only when the quality of the product is between the two extremes. Sequential sampling plans generally result in less total inspection for a given degree of precision. This type of sampling, developed first in the inspection of manufactured products, is being adapted to all types of situations where the decision between two possible actions is made on the basis of sample data. The following discussion develops fully the concept and procedures of a single-sample acceptance plan. Double, multiple, and sequential plans are not developed in detail for they are not within the scope of an introductory text.

Single-Sample Acceptance Plan

The following example presents a sampling inspection plan to be used by the buyer or a manufactured product. The specifications gave the characteristics of the product and the manufacturer agreed to supply articles that would meet these specifications. He claimed that his process produced no more than 1% defective items and the buyer was satisfied with this quality. In order to assure himself that the product actually was of this quality, the buyer decided to select a random sample of 100 items from each lot shipped to him and to test these items. If not more than 2 items were found to be defective, he would accept the product; but if more than 2 items were defective, he would return the lot to the supplier. Under such an agreement the manufacturer certainly would want to know the probability of the buyer rejecting a lot that was within the standard of not more than 2 defectives. The manufacturer might insist that every lot that met the specifications should be accepted, but as long as he is dealing with samples there is always a probability that a good lot will be rejected. He certainly is entitled to know the probability of this happening. The probability of a good lot being rejected can be computed since the distribution of defective items in samples of a given size drawn from a universe with a known fraction defective is the binomial. The following discussion explains how this probability is computed.

The general term of the binomial expansion, given in Formula 6–4 on page 127, is used to compute each of the probabilities needed. The value of the universe fraction defective is .01, so $\pi = .01$. The sample size is 100, so $n = 100$. Since the lot will be accepted if the number of defectives is 0, 1, or 2, these numbers are the values of r and it will be necessary to compute the probability of securing 0, 1, or 2 defectives in a sample of 100.

$$P(r|n, \pi) = {}_nC_r\, \pi^r\, (1 - \pi)^{n-r}.$$

When $r = 0$ $\quad P(0|100, .01) = {}_{100}C_0\, .01^0\, (1 - .01)^{100-0}$

$$= (1)\,(1)\,(.99)^{100} = (1)\,(1)\,(.3660)$$

$$= .3660.$$

When $r = 1$ $\quad P(1|100, .01) = {}_{100}C_1\, .01^1\, (1 - .01)^{100-1}$

$$= (100)\,(.01)^1\,(.99)^{99} = (100)\,(.01)\,(.3697)$$

$$= .3697.$$

When $r = 2$ $\quad P(2|100, .01) = {}_{100}C_2\, .01^2\, (1 - .01)^{100-2}$

$$= (4{,}950)\,(.01)^2\,(.99)^{98} = (4{,}950)(.0001)(.373464)$$

$$= .1849.$$

The probability of 0, 1, or 2 defective items using the binomial is

$$P(0 \cup 1 \cup 2) = .3660 + .3697 + .1849 = .9206.$$

The probabilities cannot be read directly from the table of the binomial distribution in Appendix F. The probabilities are given for only a few values of n and π because the binomial table requires a great deal of space. Extensive tables of the binomial are not always readily available and it is necessary to compute them as was shown earlier. This process can be rather lengthy so it is worth considering the use of an approximation. The Poisson as an approximation of the binomial is discussed on pages 144 to 145 and using this approximation is recommended whenever the sample size is large and the value of π is small. The calculations below demonstrate that the Poisson in this case is a very good approximation of the binomial and the time saved in making the calculations is considerable.

Formula 6–11, page 139, is used to compute the values of the Poisson. The only term needed to compute the Poisson is the mean, which in this example is 1 defective item. The mean of the Poisson distribution is simply the number of items in the sample times the fraction defective in the universe. If all possible samples of 100 are drawn from a universe, the number of defective items in a sample could vary from 0 to 100. It is possible, though not likely, that all 100 items would be defective. It is also possible that none of them would be defective. The average number of defectives would be 100 times .01 or 1 defective. Therefore, the value of λ in Formula 6–11 is 1. The computation of the probability of a sample containing 0, 1, or 2 defectives is calculated for the Poisson from the formula

$$P(r|\lambda) = \frac{\lambda^r}{r!} e^{-\lambda}.$$

When $r = 0$ $\quad P(0|1) = \frac{1^0}{0!} 2.71828^{-1} = \frac{1}{1} .367880 = .3679.$

When $r = 1$ $\quad P(1|1) = \frac{1^1}{1!} 2.71828^{-1} = \frac{1}{1} .367880 = .3679.$

When $r = 2$ $\quad P(2|1) = \frac{1^2}{2!} 2.71828^{-1} = \frac{1}{2} .367880 = .1839.$

The probability of 0, 1, or 2 defective items using the Poisson is

$$P(0 \cup 1 \cup 2) = .3679 + .3679 + .1839 = .9197.$$

The binomial and the Poisson give the same probability when rounded to 2 decimal places. When carried to 4 decimal places there is a slight difference, but it appears that the Poisson is a satisfactory approximation of the binomial.

Not only are the calculations easier for the Poisson in comparison with the binomial, but the table of Poisson values is much more compact than for the binomial. Appendix J gives values for the Poisson which may be used to check the calculations shown here. Usually it is unnecessary to compute the values of the Poisson since most of the values needed will be found in the table.

If the acceptance plan just described is adopted, it means that 92 out of 100 lots with not more than 1% defective will be accepted when checked with a sample of 100 with a rejection if more than 2 defective items appear. It also means that it can be expected that 8 out of the 100 lots inspected will be rejected even though they actually meet the specification of not more than 1% defective. As long as samples are used it is inevitable that some good lots will be rejected. This is known in quality control as *producer's risk.* In Chapter 8 the term level of significance or α was used to indicate the probability of rejecting a hypothesis that was in fact true. In this chapter the term producer's risk is used to indicate that the producer will have a good lot returned as defective.

While the producer is concerned with the possibility that a good lot will be rejected, the consumer is concerned with the possibility that he will accept a lot that is in fact defective. This is always a possibility with the use of sample inspection and it is important that the probability of this happening be measured. This probability is known as *consumer's risk* and varies with the actual fraction of defective items in a lot. The larger the number of defective items, the smaller the probability that it will be accepted; however, when the fraction defective is not much greater than the fraction agreed upon, there is a high probability of a defective lot being accepted. Consumer's risk is an example of a Type II error described on page 190.

It is possible to compute the probability of accepting a lot with any stated fraction defective using either the binomial or the Poisson distribution. Assume that the lot is submitted with a fraction of 5% defective and a sample of 100 items is checked. If there are 2 or less defectives in the sample, the lot will be accepted and a Type II error has been committed. The probability of making this error is the consumer's risk. In this case the binomial and Poisson give very nearly the same size error. The binomial is the correct value, but the Poisson can be used as a good approximation of it. The probabilities from both distributions are given below to demonstrate that the Poisson will be an acceptable approximation.

The terms of the binomial for $\pi = .05$ and $n = 100$ are computed to be

$$\begin{aligned} P(0|100, .05) &= .0059 \\ P(1|100, .05) &= .0312 \\ P(2|100, .05) &= \underline{.0812} \\ \text{Total} \quad &= .1183 \end{aligned}$$

The terms of the Poisson for $\lambda = (100)(.05) = 5$ are found in Appendix J to be

$$P(0|5) = .0067$$
$$P(1|5) = .0337$$
$$P(2|5) = .0842$$
$$\text{Total} = .1246$$

Both the binomial and the Poisson distributions indicate that there is a probability of approximately .12 that a bad lot will be accepted. The buyer may not want to accept such a lot, but with the use of this sampling plan this is the risk he runs. It is interesting to compute the probability of accepting lots with varying percentages defective to see how the consumer's risk varies. The probability of accepting a lot becomes less as the quality of the lot declines. Table 10–6 shows the probability of acceptance of a lot with various fractions defective ranging from .01 to .07. The first column in the table gives the fraction defective, π. The second column is λ, computed for each value of π ($\lambda = n\pi$). The third column gives P_a, the probability of accepting a lot computed by the Poisson. The probability of accepting a lot is .9197 when λ equals .01, but declines to .0296 when λ equals .07.

TABLE 10–6

PROBABILITY OF ACCEPTING LOTS WITH CERTAIN FRACTIONS DEFECTIVE

Computed from the Poisson $n = 100$
Acceptance Number = 2 or Less Defectives

π	$\lambda = n\pi$	P_a
.010	1.0	.9197
.015	1.5	.8088
.020	2.0	.6767
.030	3.0	.4232
.040	4.0	.2381
.050	5.0	.1246
.060	6.0	.0620
.070	7.0	.0296

The data in Table 10–6 are shown graphically in Figure 10–4. This curve is called the *operating characteristic* or *OC curve* of a sampling plan. The vertical distance from any point on the X axis to the plotted line represents the probability of accepting a lot with the fraction defective represented by that point on the X axis. Assume that the supplier has agreed that his product will not be more than .01 defective; that is, $\pi = .01$. We can set up the null hypothesis that the lot we are testing does not exceed .01 defective. If the lot in fact is not more than .01 defective, we will accept .9197 of such lots offered for inspection. In this case it will be noted that failing to reject the null hypothesis is the equivalent to accepting it. This is contrary to the statement in Chapter 8 that the null hypothesis is either rejected or not rejected, but in acceptance sampling the lot is accepted if it is not rejected by the sample plan.

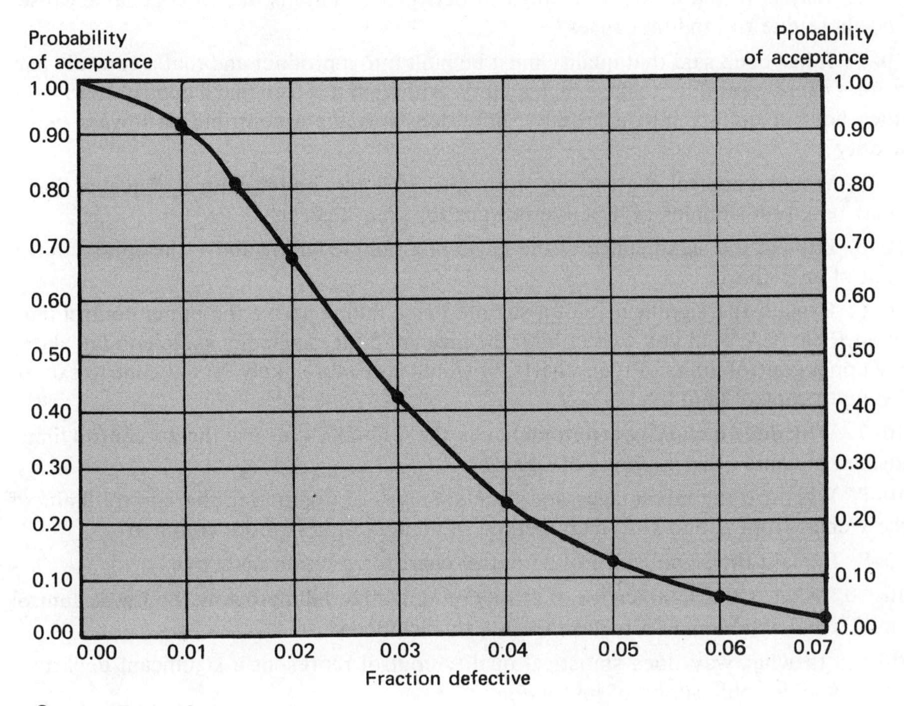

Source: Table 10-6.

FIGURE 10–4

OPERATING CHARACTERISTIC CURVE FOR SAMPLING INSPECTION PLAN

Note that we are not dealing with certainties in acceptance but only with probabilities. The supplier may find that his product was actually only .01 defective, but it has nevertheless been rejected. Although the agreed level was .01 defective, there is a probability of $1 - .9197$ that a good lot will be rejected. This is the value of α, the probability that a true hypothesis will be rejected. If the supplier thinks it is unfair that the probability of rejecting a good lot is this high, a different sample size and acceptance number should be agreed upon.

In conclusion it should be pointed out that acceptance sampling is merely an application of the testing of hypotheses. This discussion could have been included in Chapter 9, but it is presented as a part of statistical quality control, since this specific application of tests of significance is so widely employed in business.

STUDY QUESTIONS

10–1. What is the advantage in using the term "statistical quality control" rather than the shorter term "quality control"?

10–2. Why is it important to distinguish between variations due to assignable causes and those due to random causes?

10–3. It has been said that quality must be built into a product and that a control chart cannot cause a product to have high quality. Although it is true that a control chart itself does not put quality into a product, how does it make a contribution toward better quality?

10–4. Would a control chart based on samples of 8 give better control of quality than a chart based on samples of 4? Give reasons for your answer.

10–5. Explain the significance of the mean of a sample falling above the upper control limit of an $\bar{x}$ chart.

10–6. Explain the significance of a sample range falling above the upper control limit of an R chart. Would you expect both the mean and the range of a sample to fall above the upper control limits of their charts, or would it be more likely for only one to exceed its upper control limit?

10–7. Why does quality control practice in the United States use the 3σ control limits instead of some other multiple of sigma?

10–8. What are the advantages and disadvantages of computing the control limits of the $\bar{x}$ chart from $\bar{R}$ and $\bar{s}$? Which method of computation is more common?

10–9. Explain the significance of a control chart for fraction defective.

10–10. What is the significance of a fraction defective falling below the lower control limit? What action should be taken when this happens?

10–11. In what way does statistical quality control represent a significant departure from the older philosophy of inspection?

10–12. What is meant by the statement that control charts are primarily a diagnostic device?

10–13. Assume that a process can be kept in control with a dispersion about the mean dimension represented by a standard deviation of .0032 inches. The standard deviation was computed from a large sample of items manufactured while the process was in control with respect to the mean dimension. How could this information be used in determining the tolerances to set for this dimension?

10–14. Explain what is meant by the term "consumer's risk."

10–15. Explain what is meant by the term "producer's risk."

10–16. Explain what an operating characteristic curve shows about a sampling plan.

10–17. Describe how an operating characteristic curve is computed.

10–18. What would be the advantage of a double sampling plan over a single sampling plan?

PROBLEMS

10–1. The average length of life of incandescent lamps manufactured by a certain plant was estimated from a large sample to be 1,525 hours, and the standard deviation was estimated at 200 hours. These values may be considered to be the universe mean and the standard deviation. Compute the $3\ \sigma$ upper and lower control limits of an $\bar{x}$ chart based on this information, using samples of 4.

10–2. A total of 10 samples of 8 each was drawn from a universe and the average range $(\bar{R})$ of these 10 samples was found to be .0820 inches. Estimate the standard deviation of the universe from which these samples were drawn.

10–3. The average standard deviation ($\bar{s}$) for the 10 samples described in Problem 10–2 was .0266 inches. Estimate the standard deviation of the universe from which these samples were drawn. How does this estimate compare with that in Problem 10–2?

10–4. Thirty samples of 5 items each were taken from the output of a machine, and a critical dimension measured. The mean of the 30 samples ($\bar{\bar{x}}$) was .6550 inches, and the average range ($\bar{R}$) of the 30 samples was .0040 inches. Compute the upper and lower control limits for the $\bar{x}$ and R charts.

10–5. Measurements made on 15 samples of 5 each for the width of a slot in a forging had an average ($\bar{\bar{x}}$) of .08758 inches. The average range ($\bar{R}$) was .0045 inches. Compute the upper and lower control limits for the $\bar{x}$ and R charts.

10–6. Measurements made on 36 samples of 4 each for the thickness of pads on a half-ring engine mount had an average value ($\bar{\bar{x}}$) of 1.5190 inches. The average range ($\bar{R}$) was .00415 inches. Compute the upper and lower control limits for the $\bar{x}$ and R charts.

10–7. Compute the value of c_2 for the values of n from 3 to 10.

10–8. Compute the value of A_1 for the values of n from 3 to 10.

10–9. Compute the value of A_2 for the values of n from 3 to 10.

10–10. A large sample of a product gave an average fraction defective ($\bar{p}$) of .080. Compute the upper and the lower control limits for lots of the following sizes:

(a) 90	(d) 140	(g) 178	(j) 300
(b) 110	(e) 152	(h) 190	(k) 350
(c) 125	(f) 164	(i) 200	(l) 500

10–11. A total of 20 samples of 6 each is drawn from a universe. The average standard deviation ($\bar{s}$) is found to be 0.25 ozs. Use this value to estimate the standard deviation of the universe.

10–12. Use the data in Problem 10–11 to estimate the standard error of the mean and compute the values of UCL and LCL for a control chart for $\bar{x}$ if $\bar{\bar{x}}$ is found to be 9.5 ozs.

10–13. Assume in Problems 10–11 and 10–12 that $\bar{R}$ is .70 ozs. Use $\bar{R}$ to estimate the universe standard deviation. How does this estimate differ from that in Problem 10–11?

10–14. Use the data in Problem 10–13 to compute UCL and LCL for a control chart for $\bar{x}$.

10–15. Use the data in Problem 10–13 to compute UCL_R and LCL_R for an R control chart.

10–16. A total of 10 samples with 7 observations in each sample produces the following results:

Subgroup	Mean	Range
1	22.9	15
2	38.2	14
3	28.5	22
4	32.7	18
5	25.9	16
6	31.0	17
7	28.8	18
8	30.4	25
9	24.6	20
10	27.3	21
Total	290.3	186

(a) Use $\bar{R}$ to compute $\hat{\sigma}$.

(b) Draw $\bar{x}$ and R charts to determine whether or not the process was under control at the time each of the 10 samples was taken.

10–17. Ten lots of 120 items each were drawn from the output of a process. The number of defective units in each sample are shown below.

Lot Number	Number Inspected	Number of Defects
1	120	5
2	120	4
3	120	3
4	120	5
5	120	4
6	120	2
7	120	2
8	120	6
9	120	1
10	120	4

Draw a control chart for fraction defective and interpret the results.

10–18. The following table gives the results of taking 20 samples of 4 cans of tomatoes from the output of an automatic filling machine. The data represent the weight of each can inspected, carried to hundredths of an ounce less 14.00. For example, the first

Sample Number	Weight of Individual Items x	Σx	Σx^2	$\bar{x}$	R
1	96 81 93 84	354	31,482	88.50	15
2	70 88 82 78	318	25,452	79.50	18
3	89 67 79 60	295	22,251	73.75	29
4	78 84 88 68	318	25,508	79.50	20
5	75 73 67 86	301	22,839	75.25	19
6	92 83 97 63	335	28,731	83.75	34
7	77 88 72 70	307	23,757	76.75	18
8	80 76 75 81	312	24,362	78.00	6
9	82 78 96 77	333	27,953	83.25	19
10	71 89 78 83	321	25,935	80.25	18
11	98 84 86 96	364	33,272	91.00	14
12	84 70 63 65	282	20,150	70.50	21
13	83 79 92 91	345	29,875	86.25	13
14	72 82 89 76	319	25,605	79.75	17
15	72 73 95 70	310	24,438	77.50	25
16	74 76 97 80	327	27,061	81.75	23
17	85 99 79 79	342	29,508	85.50	20
18	85 61 83 74	303	23,311	75.75	24
19	85 89 67 79	320	25,876	80.00	22
20	78 73 81 93	325	26,623	81.25	20
Total		6,431	523,989	1,607.75	395

Source: Company records.

item in sample number 1 is 96, which represents a weight of 14.96 ounces. The decimal has been omitted to simplify calculations.

(a) From the data in the table, compute the trial limits for $\bar{x}$ and R charts. Plot these charts.

(b) Do the measurements show statistical control? If the charts show the process out of control, assume that assignable causes are found and compute revised control limits. If no points are out of control, extend the control limits already computed.

10–19. The following table gives the results of inspecting 10 additional samples of 4 each from the filling machine described in Problem 10–18.

(a) Plot the proper values on the $\bar{x}$ chart and the R chart.

(b) Explain how the charts may be used by management.

Sample Number	Weight of Individual Items			
21	97	82	94	90
22	85	73	60	57
23	87	82	86	86
24	81	66	67	61
25	71	67	59	82
26	99	89	70	94
27	70	96	83	98
28	76	55	93	74
29	86	90	99	85
30	97	82	98	96

Source: Company records.

10–20. Make two estimates of the standard deviation of the universe from which the samples in Problem 10–18 were taken. Base one estimate on $\bar{s}$ and the other on $\bar{R}$. Which of these two estimates do you consider more accurate?

10–21. A manufacturing plant contracts with a supplier of parts for a year's supply of a certain item which is to be supplied in lots of 2,000 at regular intervals. The purchaser proposes to select 40 items at random from each lot supplied. These 40 items will be inspected and if no defects are found, the lot will be accepted. If one or more defects are found, the lot will be returned to the supplier and he will be expected to screen it and replace all defective items with good ones. Assume that the supplier asks you to show him the probability of a lot being accepted. The probability of the lot being accepted will vary with the quality of the lot presented for inspection, so compute this probability for each of the possible fractions defective using the Poisson distribution: .005, .01, .015, .02, .025, .03, .035, .04, .045, .05, .055, and .06. After computing the probability of acceptance for a lot with each of these fractions defective, plot an operating characteristic curve.

10–22. Follow the instructions in Problem 10–21, except that the size of the sample is to be 80 and the lot will be accepted if 0 or 1 defective is found. Plot the operating characteristic curve for the sample of 80 on the chart constructed in Problem 10–21 and state which of the two sampling plans you consider to be the better. Give reasons for your choice.

10-23. The following table gives the number of defective units of a product in 20 lots of 200 units each. Compute the control limits and plot a p chart. Does the chart show statistical control? Assume that assignable causes were found for any points out of control. Set up a control chart which can be used in further production.

Lot Number	Number of Units Inspected	Number of Defective Units
1	200	17
2	200	15
3	200	14
4	200	26
5	200	9
6	200	4
7	200	19
8	200	12
9	200	9
10	200	14
11	200	6
12	200	17
13	200	10
14	200	13
15	200	12
16	200	30
17	200	19
18	200	13
19	200	11
20	200	21

Source: Company records.

10–24. The following table gives the results of inspecting the next 10 lots produced after setting up the control limits for the chart in Problem 10–23. What does this chart indicate as to whether or not the process is in control?

Lot Number	Number of Units Inspected	Number of Defective Units
21	200	8
22	200	15
23	200	17
24	200	13
25	200	20
26	200	16
27	200	24
28	200	14
29	200	9
30	200	11

Source: Company records.

10–25. A sampling plan calls for testing a random sample of 20 items from a large lot. If no more than 1 defective item is found, the lot is accepted; but more than 1 defective will result in the lot being rejected. The probability of a lot being accepted will vary with the value of π, the lot fraction defective. Compute the probability of accepting a lot with the following fractions defective, using the binomial distribution.

.05 .10 .15 .20 .25

10–26. Assume that you are the producer and have agreed to supply lots that contain no more than 2% defective items. If a lot is no more than 2% defective, what is the probability of the buyer mistakenly rejecting it as defective under the sampling plan described in Problem 10–25? What is this probability called? If you were the producer, would you agree to a sampling plan that would have this probability of being rejected even though it met the specifications agreed upon?

10–27. Compute the probabilities for Problem 10–25 using the Poisson distribution as an approximation of the binomial. How good do you consider this approximation?

10–28. Plot the operating characteristic curve for the sampling plan in Problem 10–25 on an arithmetic chart.

10–29. The probability of the buyer accepting a lot that is 25% defective was found in Problem 10–25. If you were the buyer, would you agree to a sampling plan with this probability of accepting a lot that had 25% of the items defective?

11 Simple Linear Regression and Correlation

In preceding chapters the statistical measures discussed were ones used to describe the characteristics of only one series. In this section of the text special attention will be given to a study of the problems involved when the variations in one series are related to the variations in another series or several other series.

When there is a well-established relationship between two or more series, it is possible to use this relationship in making estimates of one variable from known values of the others. Many situations exist where one or more variables can be measured sooner or easier than another, and if there is a close relationship between the variables, an equation of average relationship can be used to estimate a value of the unknown variable from the known values of the others.

In Chapter 12 more than two variables will be considered; in this chapter only two variables will be considered at a time.

For example, the volume of advanced orders for certain durable goods is known by a manufacturer several months ahead of the date the goods are produced, and it can frequently be established that the volume of production of these goods in a given period is related to the volume of orders on hand on an earlier date. In the same way, the amount of activity in the construction industry in a certain area is related to the volume of building permits issued in earlier periods.

In the forecasts just described, one variable is known earlier than the second; but it may be that one variable is known and there is no prospect of obtaining information on the values of the second variable. If the relationship between the two variables can be determined, the unknown variable can be estimated from the known variable. For example, the size of a furniture order for chairs might be used to estimate the cost per chair. If a study is made from sample data to determine the relationship between size of order and cost per unit of production, this relationship can be used to make quick estimates of cost when the size of the order can be anticipated.

Using the relationship between a known variable and an unknown variable to estimate the unknown one is called *regression analysis*. Measurement of the degree of relationship between two or more variables is called *correlation analysis*. The closer the relationship between the two variables, the greater the confidence that may be placed in the estimates. Correlation analysis measures the degree of relationship between the variables, while regression analysis shows how the variables are related.

THE SCATTER DIAGRAM

A *scatter diagram* is a graph representing two series, with the known variable plotted on the X axis and the variable to be estimated plotted on the Y axis, as shown in Figure 11–1. The scale on the X axis represents number of people and the scale on the Y axis represents total income in the county in millions of dollars. It should be noted that one scale is in dollars and the other in number of people. The scales could be in the same units, but they need not be.

Examination of Figure 11–1 shows that there is a fairly close relationship between population and income in the sample of 10 counties under study. The larger the population, the larger the total income in a county, although the relationship is not rigidly fixed. If the plotted points fell in a perfectly straight line, it would be possible to compute income perfectly from a given population. When the relationship is fairly close, as in Figure 11–1, it is possible to make a good estimate of the Y variable if the value of the X variable is known.

The two variables plotted on the scatter diagram must be two characteristics of the same unit. In the example above, both population and income relate to the same time period. A scatter diagram might be made for a number of individuals by plotting the height and weight of each individual as a point on the diagram. In the same manner the annual dividend rate and the price of each of a number of common stocks might be plotted on a scatter diagram to show how prices are related to dividend rates. The score of each employee on an aptitude test and the performance of the employees on a given machine in a factory might be plotted and the scatter diagram used to predict how well a prospective employee might perform.

By convention the known variable, designated as the *independent variable,* is always plotted on the X axis. The variable to be estimated is the *dependent variable* and is plotted on the Y axis. A cause-and-effect relationship may exist between the two; changes in the estimated variable may be the result of changes in the other variable. Establishing the exact nature of a cause-and-effect relationship is a difficult problem, although it is reasonable to assume that if two variables show a high degree of relationship, there is likely to be some causal relationship between them. However, either of the variables might be the cause and the other, the effect; or they might both be the result of some common cause. If the relationship is closer than would be expected to appear by chance,

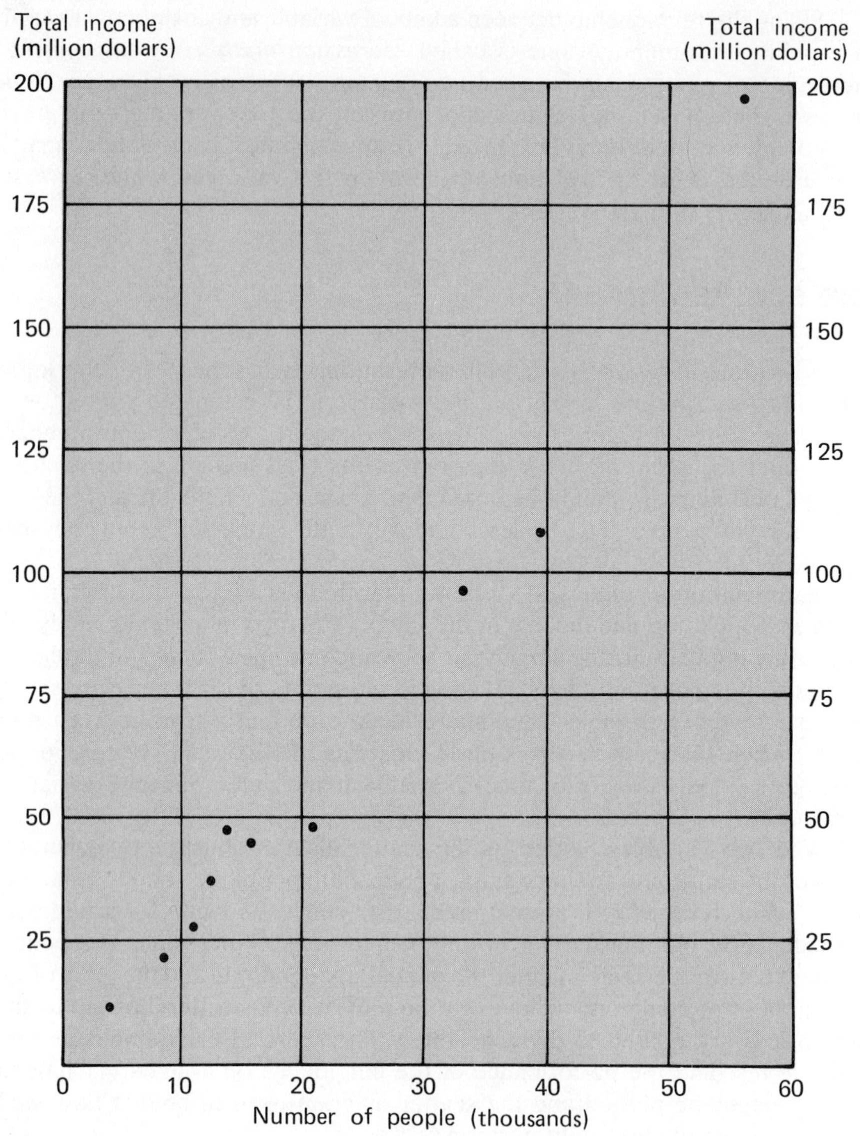

Source: Adapted from *The Dallas Morning News, Texas Almanac, 1974-1975.*

FIGURE 11-1

SCATTER DIAGRAM OF ESTIMATED POPULATION FOR 1972 AND TOTAL INCOME FOR 10 TEXAS COUNTIES

however, it may be assumed that some causal factors are operating. A sample from a universe with no relationship between the two variables could easily show some correlation due to the variation in different samples from the same universe. A test of significance for these samples will be given later.

A scatter diagram may be constructed to show the degree of relationship between two variables, even though it is not intended for use in making estimates. In this situation the designation of the independent variable is rather arbitrary, since neither may be thought of as being dependent upon the other. As pointed out at the beginning of the chapter, the measurement of relationship is referred to as correlation analysis to distinguish it from regression analysis, which is making estimates based on an independent variable.

In regression analysis it is assumed that the X variable is measured without error, and the values are chosen by the experimenter. The value of the Y variable that is associated with each value of the X variable is then determined. The values of the dependent variable are random, but the values of the independent variable are fixed quantities. When making a correlation analysis, however, the values of both variables are random. In this situation it is possible to designate one variable as the independent variable and plot it on the X axis of the scatter diagram. Estimates may be made from values of the X variable, although this method is designed to be used in measuring the degree of relationship between two random variables rather than to make estimates. It should be pointed out that the equation used to estimate the Y variable from values of X should not be used to estimate the X variable from values of Y. Another regression equation that reverses the two values should be used if it is desired to estimate X from values of Y.

The methods of estimating one variable from the known value of another will be discussed first and will be followed by a discussion of the measurement of the degree of relationship between the two variables. Making estimates from a regression equation is a useful device in furnishing information for business decisions. Measures of correlation are used in the study of cause-and-effect relationships and are particularly useful in business and the social sciences when experimentation is impracticable. Methods of correlation analysis make possible the use of data as they are observed in the business world for situations in which experiments cannot be designed.

LINEAR REGRESSION EQUATION

The device used for estimating the value of one variable from the value of another consists of a line through the points on a scatter diagram, drawn to represent the average relationship between the two variables. Such a line is called the *line of regression.* The simplest method of determining the course of this line is to draw it on the diagram freehand so it passes through the center of the points; but such a method is highly subjective. A more objective method of locating the regression line is needed, and the most popular is the method of *least squares.*

The basic problem in selecting a method of fitting a regression line is choosing a criterion for measuring the goodness of fit. If the points to which the line is being fitted will not all fall on a straight line, it is impossible to find any straight line that fits perfectly. Thus, the problem is to define what is meant by the line of "best fit." Some of the points on the scatter diagram will be above the regression line and others will be below. The total deviation of the points from the regression line is the sum of all the vertical distances between the plotted points and the line. One criterion for selecting the best-fitting line might be: Let the sum (disregarding signs) of the deviations of the plotted points from the regression line be less than any other straight line that could be drawn. Another criterion might be: Let the sum of the *squared* deviations from the regression line be less than from any other line. This test of best fit would have an advantage over the criterion that minimizes the sum of the deviations without regard to signs, since it is not strictly logical to ignore the signs. The most common criterion for selecting a straight line to show the average relationship between two variables is that the squared deviations of the actual plotted points from the point directly above or below it on the regression line be a minimum. The name "least squares" is derived from this criterion.

The straight line is represented by the equation

$$y_c = a + bx \tag{11-1}$$

where

y_c = the values on the regression line

a = the y intercept, or the value of the Y variable when $x = 0$

b = the slope of the line, or the amount of change in the Y variable that is associated with a change of one unit in the X variable.

The graph of a linear regression is plotted in Figure 11–2. For the equation $y_c = 2.4 + .5x$, the following tabulation gives the values of y that are associated with values of x from 0 to 6. These values are shown graphically by the straight line in Figure 11–2.

x	y
0	2.4
1	2.9
2	3.4
3	3.9
4	4.4
5	4.9
6	5.4

The problem of fitting a regression line to a set of data by the method of least squares requires the determination of the values a and b for the equation of the straight line that will satisfy the criterion that the sum of the squared differences between the y_c and the y values be a minimum. Formulas 11–2 and 11–3 are used to compute the values of a and b.

$$a = \frac{\Sigma x^2\, \Sigma y - \Sigma x \cdot \Sigma xy}{n \cdot \Sigma x^2 - (\Sigma x)^2} \tag{11-2}$$

$$b = \frac{n \cdot \Sigma xy - \Sigma x \cdot \Sigma y}{n \cdot \Sigma x^2 - (\Sigma x)^2}. \tag{11-3}$$

The value b, called the *regression coefficient,* measures the change in the Y variable that is associated with variations in the X variable. For example, the relationship between the strength of cotton yarn and the strength of the cotton fiber used to make the yarn as shown in Table 11–1 will be computed to illustrate the use of the regression equation in making estimates. This is done by substituting in Formulas 11–2 and 11–3 for a and b.

The summations needed are computed in Table 11–1 and shown below.

$$\Sigma x = 2{,}993 \qquad \Sigma x^2 = 256{,}765$$

$$\Sigma y = 3{,}947 \qquad \Sigma y^2 = 451{,}079$$

$$\Sigma xy = 338{,}829 \qquad n = 35.$$

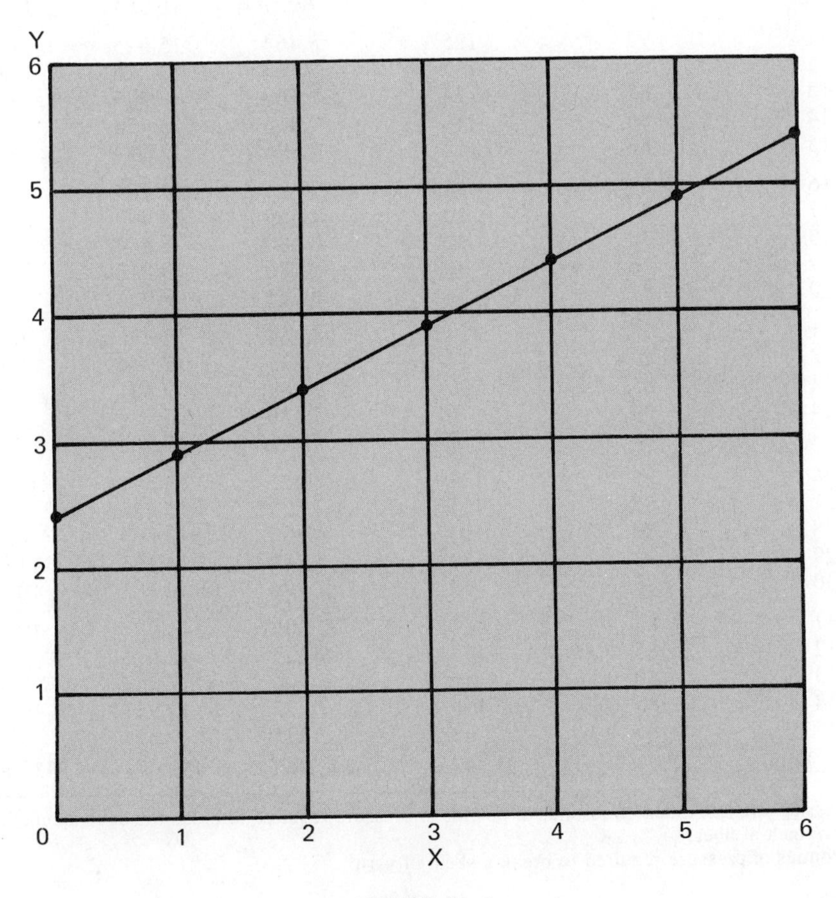

FIGURE 11–2

A STRAIGHT LINE

TABLE 11-1

RELATIONSHIP BETWEEN YARN STRENGTH AND STRENGTH OF COTTON FIBER

Sample Number	Strength of Fiber * x	Yarn Strength ** y	x^2	y^2	xy
1	95	129	9,025	16,641	12,255
2	91	117	8,281	13,689	10,647
3	88	127	7,744	16,129	11,176
4	83	129	6,889	16,641	10,707
5	93	128	8,649	16,384	11,904
6	84	118	7,056	13,924	9,912
7	81	121	6,561	14,641	9,801
8	79	112	6,241	12,544	8,848
9	84	126	7,056	15,876	10,584
10	89	128	7,921	16,384	11,392
11	92	125	8,464	15,625	11,500
12	87	115	7,569	13,225	10,005
13	87	114	7,569	12,996	9,918
14	86	116	7,396	13,456	9,976
15	86	112	7,396	12,544	9,632
16	89	113	7,921	12,769	10,057
17	87	120	7,569	14,440	10,440
18	78	92	6,084	8,464	7,176
19	76	96	5,776	9,216	7,296
20	81	84	6,561	7,056	6,804
21	92	114	8,464	12,996	10,488
22	89	112	7,921	12,544	9,968
23	83	99	6,889	9,801	8,217
24	74	76	5,476	5,776	5,624
25	89	87	7,921	7,569	7,743
26	85	121	7,225	14,641	10,285
27	85	115	7,225	13,225	9,775
28	80	103	6,400	10,609	8,240
29	93	119	8,649	14,161	11,067
30	86	119	7,396	14,161	10,234
31	86	118	7,396	13,924	10,148
32	85	114	7,225	12,996	9,690
33	80	112	6,400	12,544	8,960
34	85	118	7,225	13,924	10,030
35	85	98	7,225	9,604	8,330
Total	2,993	3,947	256,765	451,079	338,829

* Pressley strength index, thousands of pounds of pressure required to break the equivalent of one square inch of fiber.
** Pounds of pressure required to break a skein of yarn.

Source: Cotton Economic Research, The University of Texas at Austin.

When the values on page 281 are substituted in Formulas 11–2 and 11–3, the values of a and b are computed

$$a = \frac{(256{,}765)(3{,}947) - (2{,}993)(338{,}829)}{(35)(256{,}765) - (2{,}993)^2} = \frac{-663{,}742}{28{,}726} = -23.1059667$$

$$b = \frac{(35)(338{,}829) - (3{,}947)(2{,}993)}{(35)(256{,}765) - (2{,}993)^2} = \frac{45{,}644}{28{,}726} = 1.58894381.$$

The resulting regression equation for the data plotted in Figure 11–3 is

$$y_c = -23.106 + 1.589x.$$

Since the strength of the cotton fiber (the X variable) will always be known before the yarn is spun, the strength of the yarn (the Y variable) may be estimated before spinning the yarn. The symbol y_c is used to represent the value of the Y variable computed from the X variable.

ESTIMATES

The regression equation in Formula 11–1 can now be used to estimate the variable from the known value of the X variable. When the average relationship between yarn strength and the strength of the cotton fiber is expressed as an equation, it is possible to substitute in the equation the known strength of the cotton fiber that will be used and to estimate the strength of a lot of yarn before it is spun. Yarn strength is an important quality characteristic and being able to predict what a process will do has numerous uses.

In the regression equation shown above, b equals 1.589. Since this is the slope of the regression line, it means that the estimate of the Y variable, yarn strength, increases 1.589 units with every unit increase in the X variable, cotton fiber strength. Yarn strength is expressed in pounds of pressure required to break a skein of yarn, and cotton fiber strength is measured by the Pressley index. An increase of one point on the Pressley index means that on the average the pressure required to break a skein of yarn will increase by 1.589 pounds. The b value, the *regression coefficient,* measures the change in Y that is associated with a given change in X. The a coefficient locates the level of the regression line and is called the *y intercept.*

Assume that the cotton fiber to be used in spinning a lot of yarn has a Pressley strength index of 91 and it is desired to estimate the strength of the yarn that will result from the use of this cotton. Using the regression equation, it is possible to compute the probable strength of the yarn.

$$y_c = 1.589x - 23.106$$

When $x = 91$

$$y_c = (1.589)\,(91) - 23.106$$

$$y_c = 121.49.$$

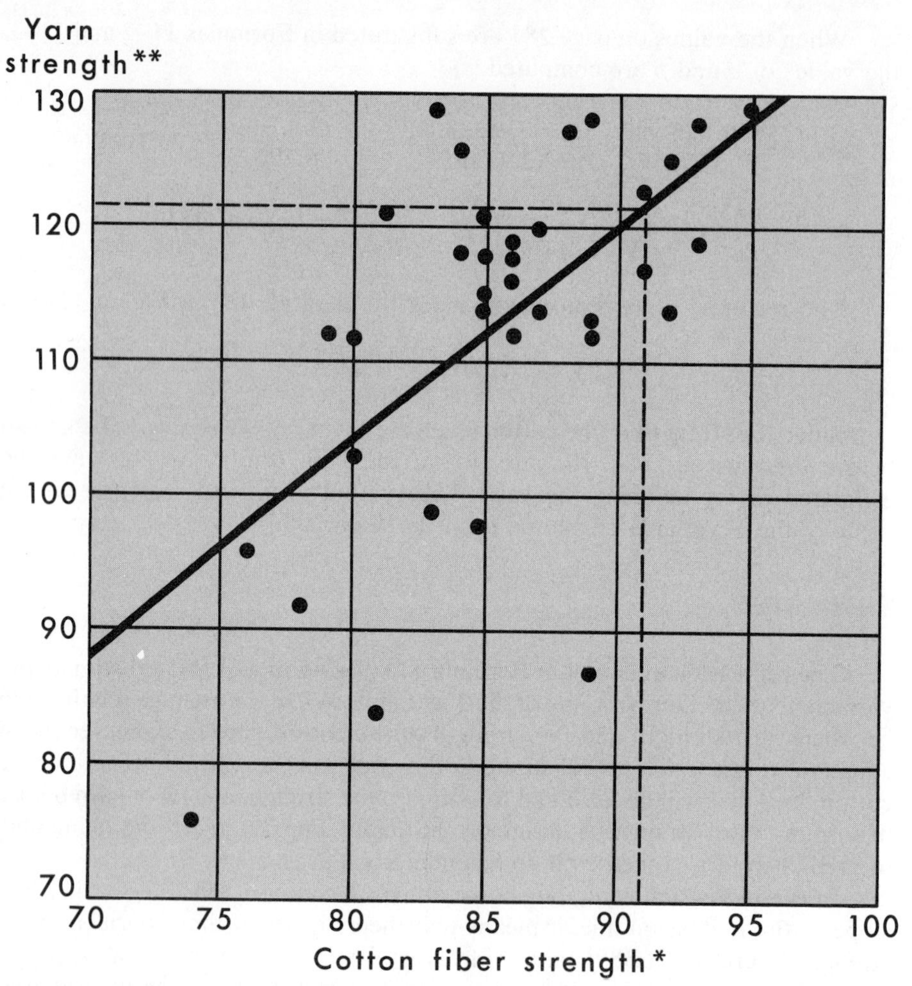

* Pressley strength index, thousands of pounds required to break the equivalent of one square inch of fiber.

** Pounds of pressure required to break a skein of yarn.

Source: Table 11-1.

FIGURE 11-3

SCATTER DIAGRAM OF YARN STRENGTH AND COTTON FIBER STRENGTH

Rounding the estimate of strength to whole pounds as shown in Table 11–1, the estimate of the strength of the yarn is 121 pounds of pressure to break a skein of yarn.

The regression line may be plotted on the scatter diagram as shown in Figure 11–3 and may be used to estimate the strength of the yarn by drawing a perpendicular line at the value $x = 91$, and reading the value on the y scale where the perpendicular intersects the regression line. The dotted lines on

Figure 11–3 show the estimated value of the y variable as slightly greater than 121. The values read from a chart are usually satisfactory estimates.

In making estimates from a regression equation, it is assumed that the relationship has not changed since the regression equation was computed. Also, the relationship shown by the scatter diagram may not be the same if the equation is extended beyond the values used in computing the equation. For example, there may be a close linear relationship between yield of a crop and the amount of fertilizer applied, with the yield increasing as the amount of fertilizer is increased. It would not be logical, however, to extend this equation beyond the limits of the experiment, for it is quite likely that if the amount of fertilizer were increased indefinitely, the yield would level off and eventually decline as too much fertilizer was applied.

RELIABILITY OF ESTIMATES

It was demonstrated in the preceding section that when the relationship between two variables is known, it is possible to compute one when the value of the other is known. The question that immediately arises, however, is: How accurate is the computed value? It can be seen from Figure 11–3 that there is a tendency for the plotted points to fall near the regression line, but not on the line. This means that any estimate based on one of the 35 lots of cotton fiber will be only approximate. Even though there is a fairly close relationship between cotton fiber strength and yarn strength, there are other factors that influence the strength of the yarn. If these factors are not taken into account, it is inevitable that the estimates will not agree with the actual yarn strength in individual lots. The more closely the X and Y variables are related, the more accurate the estimates are likely to be; but as long as there is scatter about the line, an estimate will not be completely accurate.

Computation of Standard Error of Estimate

If a regression equation is used to make estimates, it is desirable to have a measure of the accuracy of these estimates. The more closely the plotted points are grouped around the regression line, the more accurate the estimates based on the regression line are likely to be. Likewise, the more the points scatter, the less reliable an estimate based on the regression line is likely to be. What is needed is a measure which will indicate the degree of scatter of individual items about the regression line. This measure of dispersion can be derived from the sample of 35 lots of cotton yarn and can be used to show the reliability of any estimate made from the computed regression equation.

Table 11–2 shows the y_c values, computed by using the regression equation. If the actual Y values had not been known, these computed values from the regression equation would be the best estimates. Column 5 shows the amount by which each of the estimates deviates from the actual Y value. These deviations from the estimated values can be used to compute a measure of the expected error in the individual estimates.

TABLE 11–2

COMPUTATION OF THE STANDARD ERROR OF ESTIMATE

Cotton Yarn Strength and Cotton Fiber Strength of 35 Lots of Yarn

Sample Number (1)	Cotton Fiber Strength (Pressley Index) x (2)	Yarn Strength (Pounds Required to Break a Skein of Yarn) y (3)	y_c (4)	$y - y_c$ (5)	$(y - y_c)^2$ (6)
1	95	129	127.85	1.15	1.3225
2	91	117	121.49	− 4.49	20.1601
3	88	127	116.73	10.27	105.4729
4	83	129	108.78	20.22	408.8484
5	93	128	124.67	3.33	11.0889
6	84	118	110.37	7.63	58.2169
7	81	121	105.60	15.40	237.1600
8	79	112	102.42	9.58	91.7764
9	84	126	110.37	15.63	244.2969
10	89	128	118.32	9.68	93.7024
11	92	125	123.08	1.92	3.6864
12	87	115	115.14	− .14	.0196
13	87	114	115.14	− 1.14	1.2996
14	86	116	113.55	2.45	6.0025
15	86	112	113.55	− 1.55	2.4025
16	89	113	118.32	− 5.32	28.3024
17	87	120	115.14	4.86	23.6196
18	78	92	100.84	− 8.84	78.1456
19	76	96	97.66	− 1.66	2.7556
20	81	84	105.60	−21.60	466.5600
21	92	114	123.08	− 9.08	82.4464
22	89	112	118.32	− 6.32	39.9424
23	83	99	108.78	− 9.78	95.6484
24	74	76	94.48	−18.48	341.5104
25	89	87	118.32	−31.32	980.9424
26	85	121	111.96	9.04	81.7216
27	85	115	111.96	3.04	9.2416
28	80	103	104.01	− 1.01	1.0201
29	93	119	124.67	− 5.67	32.1489
30	86	119	113.55	5.45	29.7025
31	86	118	113.55	4.45	19.8025
32	85	114	111.96	2.04	4.1616
33	80	112	104.01	7.99	63.8401
34	85	118	111.96	6.04	36.4816
35	85	98	111.96	−13.96	194.8816
Total	2,993	3,947	3,947.19	− .19	3,898.3313

Source: Cotton Economic Research, The University of Texas at Austin.

TABLE 11–2 (concluded)

$$\hat{\sigma}_{yx} = \sqrt{\frac{\Sigma(y - y_c)^2}{n - 2}}$$

$$\hat{\sigma}_{yx} = \sqrt{\frac{3{,}898.3313}{35 - 2}}$$

$$\hat{\sigma}_{yx} = \sqrt{118.1313}$$

$$\hat{\sigma}_{yx} = 10.87.$$

Table 11–2 shows that in sample number 1 the strength of the cotton as measured by the Pressley index was 95. Substituting this value of x in the regression equation gives an estimate of yarn strength of 127.85 pounds required to break a skein of yarn. The actual strength of this lot of yarn was 129 pounds, but if this were not known, the estimate of 128 pounds would be the most probable value. (Since measurements of yarn strength were rounded to the nearest pound, it would be reasonable to round the estimates in the same way, rounding 127.85 to 128, although when used in further calculations it increases the accuracy of the final answer to carry the values to more decimal places than needed in the final results.) The difference between the actual yarn strength of the first sample and the estimate from the regression equation is 1.15 pounds. The amounts in Column 5 represent the deviation of each of the actual yarn strengths of the 35 samples from the estimates made from the regression equation. The amounts are called *residuals,* since the variations of the actual strength from the estimated values are in effect what is left over after taking into account the variations in yarn strength associated with the strength of the cotton fiber used. Presumably these residuals represent variations in yarn strength related to factors other than strength of the cotton fiber.

Finding the expected value of these residuals, which measure the scatter of the actual values about the estimated values, is a problem of averaging the residuals. When the expected value of the residuals has been found, it may be used as an estimate of the error in an individual estimate. Finding the average of the residuals is similar to the problem in Chapter 4 of averaging the deviations from the mean. The sum of the residuals is zero, except for rounding error, just as the deviations of the individual items from their mean total zero. To permit averaging the size of the residuals without regard to the signs, the residuals are squared, averaged, and the square root extracted, just as in the computation of the standard deviation. This measure is called the *standard error of estimate* and is represented by $\hat{\sigma}_{yx}$. It estimates the standard deviation of the amounts that the computed values vary from the actual values, using Formula 11–4.

$$\hat{\sigma}_{yx} = \sqrt{\frac{\Sigma(y - y_c)^2}{n - 2}}. \qquad \textbf{(11–4)}$$

The value of the standard error of estimate is derived from a sample. Referring to pages 160–165, where the estimate of the universe standard deviation from a sample standard deviation was discussed, the sample variance was multiplied by $\frac{n}{n-1}$, since one degree of freedom was lost in estimating the universe mean from the sample before the variance could be computed. The sample standard error of estimate from a regression line is also a biased estimate of the universe value, unless the bias is removed by multiplying by $\frac{n}{n-2}$. The simplest method of calculation is to make this adjustment at the time the standard error of estimate is computed. This can be done by reducing the denominator by the number of degrees of freedom lost, that is, by 2.

The justification for dividing by $n-2$ is that the deviations are taken from a straight line which requires two constants, a and b, to locate. This means that two degrees of freedom are lost in estimating the universe values of these constants by computing them from the sample data. The formula for the standard error of the universe is not given, since the only use for this measure is in making estimates from a sample. If all the values in the universe are known, a scatter diagram is not needed to make estimates.

The Greek lowercase letter sigma (σ) is used because the measure is the root-mean-square; in other words, it is a standard deviation although it is computed from the regression line instead of the mean. The caret ($\hat{}$) shows that it is an estimate of a parameter. The subscript ($_{yx}$) indicates that it is a measure of the dispersion in the estimates of the Y variable made from the known values of the X variable. Since correlation deals with more than one variable, it will be necessary to use subscripts to distinguish between the standard deviations of the distributions. When dealing with two variables, the standard deviation of the X variable estimated from a sample will be represented by $\hat{\sigma}_x$ and, for the Y variable, by $\hat{\sigma}_y$. The sample mean of the X variable will be represented by $\bar{x}$, and the sample mean of the Y variable, by $\bar{y}$. In Chapter 12 the correlation of more than two variables will necessitate modification of the symbols to provide for the identification of each of the variables; these symbols will be introduced as they are needed.

Since $\hat{\sigma}_{yx}$ is a standard deviation of the scatter about the regression line, the error in the individual estimates will sometimes be more than 10.87 and sometimes will be less if all possible estimates were made from the regression equation. If the errors are normally distributed, 68% of the errors in estimates based on the regression equation will be less than or equal to the standard error of estimate. In other words, the standard error of estimate enables one to make a probability statement concerning the reliability of estimates from the regression equation. It may be said that the probability is 68% that an estimate made from the regression equation will not deviate more than 10.87 units from the correct value.

Using the regression equation, the estimate has been made that the strength of yarn spun from cotton with a Pressley index of 91 would be 121 pounds required to break a skein of yarn. How good the estimate is may be answered by the statement that the probability is about 68 out of 100 that the error in the estimate would not be more than 10.87 pounds. This means that the chances are about 68 out of 100 that the strength of this lot of yarn would fall between 111 and 132 pounds required to break a skein. The probabilities are based on the areas of the normal curve, discussed in Chapter 6. If the form of the distribution is not known, Tchebycheff's inequality, described on page 157, may be used to evaluate the probability.

Shorter Method of Computing $\hat{\sigma}_{yx}$

It is possible to use the summations already made to compute the standard error of estimate, without going through the process shown in Table 11–2. Formula 11–5, which uses the values computed in Table 11–1, can be used for this computation.

$$\hat{\sigma}_{yx} = \sqrt{\frac{\Sigma y^2 - a\Sigma y - b\Sigma xy}{n-2}} \tag{11–5}$$

$$\Sigma y = 3{,}947$$
$$\Sigma y^2 = 451{,}079$$
$$\Sigma xy = 338{,}829.$$

The values of the constants in the regression equation were found to be

$$a = -23.1059667$$
$$b = 1.58894381.$$

Substituting the values above in Formula 11–5, $\hat{\sigma}_{yx}$ is found to be 10.87, the same amount secured by Formula 11–4.

$$\hat{\sigma}_{yx} = \sqrt{\frac{451{,}079 - (-23.1059667)(3{,}947) - (1.58894381)(338{,}829)}{35-2}}$$

$$\hat{\sigma}_{yx} = \sqrt{\frac{3{,}898.0084}{33}} = \sqrt{118.1215} = 10.87.$$

The values of a and b are carried out to a large number of decimal places for use in the calculation of $\hat{\sigma}_{yx}$. This is necessary to avoid introducing any rounding error into the calculation. The final figures are relatively small in comparison with the figures used in the calculations. Thus, if the latter figures were rounded as much as is common in most calculations, the resulting error would be unduly large. This calculation gives the same answer to two decimal places for the standard error of estimate as given by Formula 11–4.

EXAMPLES OF THE USE OF REGRESSION ANALYSIS

The Crop Reporting Board in the United States Department of Agriculture issues periodic reports during the growing months, giving estimates of the condition of important crops. The existence of a definite relationship between the estimates of the condition on different dates and the actual yield makes it possible to estimate the yield when the crop report on condition becomes available.

The relationship between the condition of the corn crop on September 1 in Kansas and the final yield per acre is shown by the regression equation

$$y_c = .674 + .3155x.$$

The X variable represents the condition of the crop on September 1, expressed as a percentage, comparing the condition of a growing crop at a given time with a normal condition taken as equal to 100%. The Y variable is the actual yield per acre.

The regression equation expresses mathematically the relationship between the condition of the crop on September 1 and the final yield per acre. An increase of 1% in the condition of the crop is associated with an increase of .3155 bushel. If on September 1 of a given year the condition were reported as 53% of normal, the estimate based on the regression equation would be

$$y_c = .674 + .3155(53)$$

$$y_c = .674 + 16.7215 = 17.4 \text{ bushels per acre.}$$

The standard error of estimate, computed by using Formula 11–5, is 1.31 bushels. This is interpreted as indicating that approximately 68% of the estimates will fall within $\pm$ 1.31 bushels.

The relationship between the sales of nearly all types of consumer goods and disposable personal income is so high that an analysis of the market for nearly all kinds of goods includes a regression analysis of sales and personal income. The analysis can be made between the two variables on a time or geographical basis. Estimates of future sales can be made from projections of future personal income by substituting these estimates in a regression equation describing the relationship between income and sales. Or, estimates of sales in various regions can be made from regression equations reflecting the relationship between sales and income.

Forecasting the prices of farm commodities relies heavily on analysis of the relationship between supply and price as well as between demand and price. Relationships between price and both supply and demand frequently are nonlinear, and more than two variables may need to be studied simultaneously. Methods of analysis to be used in these situations are described in Chapter 12.

In making estimates from the regression equation, the standard error of estimate measures the reliability of these estimates. Therefore, the standard error of estimate also serves as a measure of the degree of relationship between two variables, since the lower the correlation, the larger the scatter of individual values about the regression line. The difficulty with using the standard error of estimate as a measure of the strength of the relationship on correlation is that it is expressed in concrete units appropriate for each situation. It will be remembered that the standard error of estimate is always in the units used for the Y variable. For example, the standard error of estimate for the data plotted in Figure 11–3 is 10.87 pounds required to break a skein of yarn. Whether this is a large or a small degree of scatter depends upon its relative size, which makes it desirable to express the dispersion as a relative measure, divorced from the units in which the variable is expressed. In order to be most useful, a measure of correlation should be independent of the units in which the variables are expressed, so that the measure for one correlation can be compared with any other.

ABSTRACT MEASURES OF RELATIONSHIP

The dispersion of the Y values about the regression line may be compared with the dispersion of the Y values about their mean to give an abstract measure of relationship that is independent of the units of the Y variable. If the values of the Y variable cluster more closely around the regression line than they do around the mean, it indicates that the regression equation explains some of the variation in the Y variable. The variance computed from the arithmetic mean is the average size of the squared deviations from $\bar{y}$, the mean of the Y values. The square of the standard error of estimate, $\hat{\sigma}^2_{yx}$, is also the average of the squared deviations of the same individual values from y_c, the values computed from the regression equation.

If there is no relationship between the X and Y variables, a straight line fitted by the method of least squares will have a slope of zero; that is, $b = 0$. In other words, this line will be parallel to the X axis and large values of Y are as likely to be associated with small values of X as with large values. If an estimate is to be made of a Y value, the arithmetic mean of Y would be the best estimate of this individual value, assuming that nothing is known about the value except the mean of the universe from which the value was selected. A regression equation with a slope of zero does not provide any better estimate of the Y value than the arithmetic mean. In fact, it can be shown that all the estimates from the regression line with a slope of zero will equal the arithmetic mean of the Y variable.

When there is no correlation between the two variables, the variance from $\bar{y}$ is the same as the variance from the values of y_c. If, however, the plotted points on the scatter diagram cluster closely about the regression line and it does not have a slope of zero, the variance computed from the y_c

values will be less than the variance computed from $\bar{y}$. The higher the correlation, the less the scatter about the y_c values and the smaller the value of $\hat{\sigma}^2_{yx}$.

Coefficient of Nondetermination

The preceding concept can be illustrated with the data on strength of yarn and the strength of the cotton fiber from which the yarn was spun, shown as a scatter diagram in Figure 11–3. The regression line has a positive slope, and there is considerable cluster of the plotted points about the regression line. This scatter in the Y values about the regression line was measured by computing the standard error of estimate, found on page 287 to be 10.87, and squaring it to equal 118.1313. The scatter of the Y values from the arithmetic mean can be computed by substituting the summations in Table 11–1 in Formula 4–10 for the variance. When only one variable was analyzed in Chapter 4, the variable was designated as X in Formula 4–10. When variable Y is being analyzed, Formula 4–10 will be written

$$s^2_y = \frac{n\Sigma y^2 - (\Sigma y)^2}{n^2}$$

$$s^2_y = \frac{(35)(451{,}079) - (3{,}947)^2}{35^2}$$

$$s^2_y = \frac{208{,}956}{1{,}225} = 170.5763.$$

Since the variance is computed from a sample, the estimate of the universe variance is made by multiplying the sample variance by $\frac{n}{n-1}$ as described on page 251, giving the following value for the estimated variance of the universe:

$$\hat{\sigma}^2_y = 170.5763 \frac{n}{n-1} = 170.5763 \frac{35}{35-1}$$

$$\hat{\sigma}^2_y = 175.5912.$$

The variance computed from the regression equation on page 287 was 118.1313, which is substantially less than the variance from the mean. The variance from the mean represents the variation in yarn strength due to all factors that might cause variations in strength. The variance from the regression equation represents the variation that is not related to the X variable, the strength of the cotton used. In other words, $\frac{118.1313}{175.5912} = .6728$, which is the fraction of the total variance in yarn strength that is not related to variations in the cotton fiber used to spin the yarn. It is called the *coefficient of nondetermination* and measures the proportion of the total variance that is not related to the X variable. It is generally stated that the coefficient of nondetermination measures the proportion of the total variance in Y that is not explained by variations in X.

These computations may be summarized in Formula 11–6 for the coefficient of nondetermination, which is represented by the square of the lowercase Greek letter kappa (κ^2). The caret (ˆ) indicates that it is an estimate from a sample.

$$\hat{\kappa}^2 = \frac{\hat{\sigma}^2_{yx}}{\hat{\sigma}^2_y} \tag{11-6}$$

$$\hat{\kappa}^2 = \frac{118.1313}{175.5912} = .6728.$$

Coefficient of Determination

Rather than stating the measure of relationship as the proportion of the variance in the Y variable that is *not* related to the variations in X, it is generally preferred to express the measure as the proportion of the variance explained by the variations in X. This measure is the *coefficient of determination*, which is represented by the square of the lowercase Greek letter rho (ρ^2). The computation of ρ^2 is summarized in Formulas 11–7 and 11–8, with the caret (ˆ) indicating that it is an estimate from a sample.

$$\hat{\rho}^2 = 1 - \hat{\kappa}^2 \tag{11-7}$$

or

$$\hat{\rho}^2 = 1 - \frac{\hat{\sigma}^2_{yx}}{\hat{\sigma}^2_y}. \tag{11-8}$$

Substituting in this formula gives the value

$$\hat{\rho}^2 = 1 - \frac{118.1313}{175.5912} = 1 - .6728$$

$$\hat{\rho}^2 = .3272.$$

The interpretation of the coefficient of determination computed here may be summarized by saying that 32.72% of the variance in the strength of cotton yarn may be explained by the variations in the strength of the cotton fiber used in its manufacture. The word "explained" should be used with caution, since it does not mean that the variation in the strength of the yarn was caused by the variations in the strength of the cotton fiber, but that the fluctuations are related to the fluctuations in the strength of the cotton fiber. This means that the remaining percentage of the fluctuations, the coefficient of nondetermination, is related to factors other than the strength of the cotton fiber.

Coefficients of Correlation and Alienation

Two other measures of correlation may be derived by taking the square roots of the coefficient of determination and the coefficient of nondetermination. These measures are known as the *coefficient of correlation* (ρ) and the *coefficient of alienation* (κ). Referring to the analysis of the relationship between strength of yarn and the strength of the cotton fiber used in its manufacture, the following values are found.

$$\hat{\rho}^2 = .3272 \qquad \hat{\rho} = \sqrt{.3272} = .572$$

$$\hat{\kappa}^2 = .6728 \qquad \hat{\kappa} = \sqrt{.6728} = .820.$$

The sum of $\hat{\rho}$ and $\hat{\kappa}$ will not equal 1.00 unless one of the two coefficients is 1.00, in which case the other is 0. In any other circumstances, $\hat{\rho} + \hat{\kappa} > 1.00$.

The nearer $\hat{\rho}$ is to 1.00, the higher the degree of correlation; the smaller $\hat{\rho}$, the less the correlation. When $\hat{\rho}$ equals zero (which means that $\hat{\kappa}$ equals 1.00) there is no correlation. Originally the coefficient of correlation was the only coefficient used, but the coefficient of determination is now considered easier to interpret since it represents the percentage of explained variance. The coefficient of correlation is larger than the coefficient of determination, but it cannot be expressed as a percentage. When $\hat{\rho}^2 = .50$, it can be said that 50% of the variance in the dependent variable is explained by the independent variable. The coefficient of correlation is .707, but this cannot be stated as a percentage of the variance. The larger the coefficient of determination, the closer it comes to the coefficient of correlation, until both coefficients equal 1.00.

Since ρ is the square root of the coefficient of determination, its sign is either plus or minus. When the b value in the regression equation is plus, the correlation is said to be positive. In other words, small X values are associated with small Y values, and large X and large Y values occur together. When the b value is negative, the correlation is said to be negative, and large values of one variable are associated with small values of the other. The coefficient of correlation is given the sign of the b value in the regression equation and in this manner shows whether the correlation is positive or negative.

AN ALTERNATIVE METHOD OF COMPUTATION (PRODUCT-MOMENT)

Formulas other than the ones just given have been derived for computing the various measures of correlation and regression. The method used in the preceding pages gives the values of the coefficients, or abstract measures of relationship, as the last step in the analysis. When only a measure of correlation is desired, or when this measure is wanted first, Formula 11–9 gives the value of r directly from the values computed in Table 11–1.[1]

$$r = \frac{n \cdot \Sigma xy - \Sigma x \cdot \Sigma y}{\sqrt{[n \cdot \Sigma x^2 - (\Sigma x)^2][n \cdot \Sigma y^2 - (\Sigma y)^2]}} . \qquad \textbf{(11–9)}$$

The value of r is the coefficient of correlation and r^2 is the coefficient of determination *computed from sample data,* but they are not unbiased estimates of the parameters. The methods of estimating the universe coefficient of correlation and coefficient of correlation from sample values are discussed below.

If the value of r^2 is needed, it can be found by squaring r or by removing the radical from the denominator and squaring the numerator before making the final division. Since the denominator is a square root, its sign is plus or minus. The positive value is always used, so the sign of the numerator gives the sign of r.

The summations from Table 11–1 are substituted in Formula 11–9, and the value of r computed.

$$n = 35 \qquad \Sigma x^2 = 256{,}765$$

$$\Sigma x = 2{,}993 \qquad \Sigma y^2 = 451{,}079$$

$$\Sigma y = 3{,}947 \qquad \Sigma xy = 338{,}829$$

$$r = \frac{(35)(338{,}829) - (2{,}993)(3{,}947)}{\sqrt{[(35)(256{,}765) - (2{,}993)^2][(35)(451{,}079) - (3{,}947)^2]}}$$

$$r = \frac{45{,}644}{\sqrt{(28{,}726)(208{,}956)}} = \frac{45{,}644}{\sqrt{6{,}002{,}470{,}056}} = \frac{45{,}644}{77{,}475.61} = .58914$$

$$r^2 = .3471.$$

Formula 11–9 gives the sample coefficient of correlation, represented by r, but this value is not an unbiased estimate of the universe coefficient of

[1] This formula is particularly desirable for use on a calculator. Only the values in Formula 11–9 need be written down; all other operations can be performed on the machine. The multiplication of n and Σxy is performed and the product is left in the dials. Σx and Σy are then multiplied with negative multiplication, which will subtract the product of Σx and Σy from the product of n and Σxy that is already in the dials. The remaining number in the dials is the difference between the products and is the numerator of the equation.

The two parts of the denominator should be computed in the same manner. n is multiplied by Σx^2 and the product is left in the dials. Σx is then squared using negative multiplication, which subtracts the square of Σx from the product already in the dials. The result is copied, the machine cleared, and the same calculation made for the second term in the denominator. Care needs to be taken to be certain that the decimals are lined up correctly; but if this precaution is taken, the method is easy to carry out.

Any of the new electronic calculators that have storage units can be used effectively by storing the various products to be recalled as needed further in the calculations.

correlation, nor is r^2 an unbiased estimate of the universe coefficient of determination. The value of $\hat{\rho}$ or $\hat{\rho}^2$ is computed from r using Formula 11–10.

$$\hat{\rho}^2 = 1 - (1 - r^2)\left(\frac{n-1}{n-2}\right). \qquad \textbf{(11–10)}$$

The term $\frac{n-1}{n-2}$ corrects the sample value of the coefficient of determination (r^2) for the bias in using it to estimate the value of ρ^2, the coefficient of determination.[2] This estimate of ρ made from r is designated $\hat{\rho}$.

$$\hat{\rho}^2 = 1 - (1 - .3471)\,\frac{35-1}{35-2}$$

$$\hat{\rho}^2 = 1 - (.6529)\,\frac{34}{33}$$

$$\hat{\rho}^2 = 1 - .673$$

$$\hat{\rho}^2 = .327.$$

This is the same value of the coefficient of determination given by Formula 11–8.

The calculation of the value of r from Formula 11–9 provides all the data needed for the calculation of b by Formula 11–3. The numerator of the equation for b is the same as for r, and the denominator of the equation for b is the first term of the denominator of the formula for r. The value of b can be computed by making one division.

$$b = \frac{n \cdot \Sigma xy - \Sigma x \cdot \Sigma y}{n \cdot \Sigma x^2 - (\Sigma x)^2} = \frac{45{,}644}{28{,}726} = 1.5889.$$

The calculation of the value of a is performed using Formula 11–11.

$$a = \bar{y} - b\bar{x}. \qquad \textbf{(11–11)}$$

[2] Formula 11–10 may be derived from Formula 11–7 as shown.

$$\hat{\rho}^2 = 1 - \frac{s_{yx}^2 \dfrac{n}{n-2}}{s_y^2 \dfrac{n}{n-1}} = 1 - \frac{s_{yx}^2}{s_y^2}\,\frac{\dfrac{n}{n-2}}{\dfrac{n}{n-1}} = 1 - \frac{s_{yx}^2}{s_y^2}\,\frac{n-1}{n-2}$$

$$r^2 = 1 - \frac{s_{yx}^2}{s_y^2}$$

so $$\frac{s_{yx}^2}{s_y^2} = 1 - r^2.$$

Therefore $$\hat{\rho}^2 = 1 - (1 - r^2)\left(\frac{n-1}{n-2}\right).$$

The values of the two arithmetic means can be computed from the summations in Table 11–1 and the resulting values substituted in the equation for a give

$$\bar{x} = \frac{\Sigma x}{n} = \frac{2{,}993}{35} = 85.514$$

$$\bar{y} = \frac{\Sigma y}{n} = \frac{3{,}947}{35} = 112.771$$

$$a = 112.771 - (1.5889)(85.514) = -23.102.$$

The regression equation is

$$y_c = -23.102 + 1.589x.$$

The value of the standard error of estimate ($\hat{\sigma}_{yx}$) can also be calculated from the data already assembled. The equation for the standard error of estimate for the sample is

$$s_{yx} = s_y\sqrt{1 - r^2}. \qquad \textbf{(11–12)}$$

The required value of s_y is 13.06, the square root of the variance of the sample, which was found on page 292 to be 170.5763. Using the value of r, calculated on page 295 to be .58914, the computation of s_{yx} is

$$s_{yx} = s_y\sqrt{1 - r^2} = 13.06\sqrt{1 - .58914^2} = 13.06\sqrt{1 - .3471} = 13.06\sqrt{.6529}$$

$$s_{yx} = (13.06)(.808) = 10.5525.$$

The estimate of the universe standard error of estimate is computed from s^2_{yx}

$$\hat{\sigma}_{yx} = \sqrt{s^2_{yx}\frac{n}{n-2}} = \sqrt{10.5525^2\frac{35}{35-2}}$$

$$\hat{\sigma}_{yx} = \sqrt{111.3553\,\frac{35}{33}} = \sqrt{118.1041}$$

$$\hat{\sigma}_{yx} = 10.87.$$

COEFFICIENT OF RANK CORRELATION

The methods of correlation analysis previously described are based on the assumption that the population being studied is normally distributed. When it is known that the population is not normal or when the shape of the distribution is not known, there is need for a measure of correlation that involves no assumption about the population parameters.

It is possible to avoid making assumptions about the populations being studied by ranking the observations according to size and basing the calculations on the rank values rather than upon the original observations. It does not matter which way the items are ranked; item number 1 may be the largest or it may be the smallest. Using ranks rather than actual observations gives the *coefficient of rank correlation* (r_r), which may be interpreted in the same way as r. The use of ranks has the advantage of making it possible to measure the correlation between characteristics that cannot be expressed quantitatively but that can be ranked.

The ranks may be substituted in Formula 11–9 and the value of the r_r computed in the same manner that r would be. If there are no ties in the rankings, Formula 11–13 gives the same value of the coefficient of rank correlation as Formula 11–9 with considerably less work. Even when ties do occur in the ranking, Formula 11–13 is generally used.

$$r_r = 1 - \frac{6\Sigma d^2}{n^3 - n}. \tag{11–13}$$

Note: The denominator may also be written $n(n^2 - 1)$.

An example of the computation of the rank correlation coefficient is shown in Table 11–3. In this example it is used to measure the correlation between the number of savings and loan associations that are federally chartered in each bank district and the number of associations in each district which are state chartered. Since neither distribution is normal, and since the coefficient of rank correlation does not require a normal distribution, it is logical to use this method for these data.

The first step in the calculations made in Table 11–3 is to rank each bank district, giving the bank district having the largest number of federally chartered associations a rank of 1, and so on. It is not unusual for some values of the variable to be the same, which results in a tie for that rank. In such cases each tied value is assigned the mean of the ranks involved. For example, if the Atlanta and Little Rock districts had the same number of associations, they might be tied for ranks 2 and 3, so each would be given the rank of 2.5. A three-way tie for ranks 4, 5, and 6 would result in all 3 values being given the rank of 5.

$$r_r = 1 - \frac{6\Sigma d^2}{n^3 - n}$$

$$r_r = 1 - \frac{(6)(76)}{12^3 - 12} = 1 - \frac{456}{1{,}716}$$

$$r_r = 1 - .266$$

$$r_r = .734.$$

This procedure is a logical way to handle ties, although when a tie occurs, Formula 11–13 does not give exactly the same answer as computing the coefficient of correlation of the ranks by Formula 11–9. However, if the number of ties is small, the difference is not great enough to invalidate the method.

TABLE 11–3

COMPUTATION OF THE COEFFICIENT OF RANK CORRELATION

Membership of Federal Home Loan Bank Board System at Year-End 1972

Bank District	Federally Chartered		State Chartered		d	d_2
	Number	Rank	Number	Rank	(3) − (5)	$(6)^2$
(1)	(2)	(3)	(4)	(5)	(6)	(7)
Boston	72	12	53	12	0	0
New York	118	8	259	3	5	25
Pittsburgh	146	6	179	6	0	0
Atlanta	460	1	229	4	−3	9
Cincinnati	298	2	202	5	−3	9
Indianapolis	140	7	88	10	−3	9
Chicago	192	4	385	1	3	9
Des Moines	155	5	132	7	−2	4
Little Rock	197	3	318	2	1	1
Topeka	103	9	121	9	0	0
San Francisco	74	11	127	8	3	9
Seattle	89	10	54	11	−1	1
Total						76

Source: U.S. Savings and Loan League, *1973 Savings and Loan Fact Book,* p. 112.

The correlation measured by Formula 11–13 depends upon the aggregate size of d, the difference between the ranks of the paired variables. If the rankings are the same for all pairs, the correlation will be perfect and the values of d will all be zero. Since some of the values of d are negative and some are positive, it is necessary to eliminate the minus signs before summing, just as it is necessary to eliminate the minus signs in averaging the deviations from the mean to compute a measure of dispersion. The same device used in computing the standard deviation is used here; that is, the values of d are squared and then summed. The number of pairs is represented by n as in the coefficient of correlation formulas.

Table 11–3 is presented as a method of giving the values from which r_r is computed, but this form of work sheet would not be used in making the calculations. The easiest method of making the rankings by hand is to enter the values of both variables for each bank district on separate cards. The cards are sorted into an array for the number of federally chartered associations and numbered from 1 to 12, which indicates the rank of each bank district with respect to this variable. Then the cards are sorted into an array for the number of state-chartered associations and numbered again in this order. This gives the rank of each bank district with respect to the second variable.

All the computations with the ranks can then be performed on the cards. The difference between the ranks is computed and entered on each card. The algebraic sum of these differences is zero, which serves as a check on the arithmetic. The square of this difference is entered on each card, the squared values added, and the total of 76 entered in Formula 11–13.

SAMPLING ERROR OF THE COEFFICIENT OF CORRELATION

The formulas for the coefficient of correlation may be used to measure the degree of correlation for two variables in a universe, in which case it is designated ρ. However, it is rather unusual for universe data to be available. It is much more likely that an estimate will be made of the degree of correlation by using a sample. The estimate of the universe correlation made from a sample has been designated $\hat{\rho}$, and since it is an estimate, it is important to have a measure of the sampling error of the estimate. This is the same situation faced in Chapter 7, where the sampling errors of various measures were computed.

For large samples the coefficient of correlation from a normal universe with small or moderate correlation is distributed approximately normally about the universe value, ρ. The standard deviation of the distribution of the coefficient of correlation is given in Formula 11–14.

$$\sigma_r = \frac{1-\rho^2}{\sqrt{n-1}}. \qquad \textbf{(11–14)}$$

When there is a large value of the correlation coefficient, the distribution is approximately normal only for very large samples. For smaller samples it does not take the form of the normal distribution and Formula 11–14 should not be used. The data in Table 11–2 represent a fairly large sample and will be used to illustrate the computation of the confidence interval of the coefficient of correlation. Since the sample is large enough to make the value of $\hat{\rho}$ a reasonably good estimate of ρ, the following computation of the standard error of the coefficient of correlation can be made.

$$\hat{\sigma}_r = \frac{1-r^2}{\sqrt{n-1}} \qquad \textbf{(11–15)}$$

$$n = 35 \qquad r = .5891$$

$$\hat{\sigma}_r = \frac{1-.5891^2}{\sqrt{35-1}} = \frac{1-.3469}{\sqrt{34}} = \frac{.6531}{5.831} = .112.$$

The value of ρ is needed to compute the standard error of the coefficient of correlation, but with a sample of this size there is only a slight inaccuracy in using the value of r as the value of ρ. When the sample is small, however, considerable error is introduced in using r as an estimate of ρ and the resulting computation of the standard error is unsatisfactory.

The interpretation of the confidence interval of the coefficient of correlation follows the treatment in Chapter 7. A confidence interval .300 to .878 may be computed by multiplying the standard error by 2.58.

$$.5891 - (2.58)(.112) = .300$$

$$.5891 + (2.58)(.112) = .878.$$

If the statement is made that the coefficient of correlation in the population (ρ) falls between .300 and .878, it will be correct 99% of the time. In other words, the probability of the statement being correct is .99. Other probabilities may be computed by using the proper multiple of the standard error, such as 1.96 times the standard error. In this case, the probability of the statement being correct is .95. It should be emphasized that Formula 11–15 should be used only for large samples, and even with large samples the distribution departs substantially from normal when the correlation approaches ± 1.

SIGNIFICANCE OF AN OBSERVED CORRELATION

When a correlation is based on a small sample, it is possible to test accurately whether or not the coefficient of correlation differs significantly from zero, even though Formula 11–15 may not be used to estimate the confidence interval. This is valuable information, for if there is a significant degree of correlation, it may be measured more accurately by increasing the size of the sample. If there is no significant correlation, this line of investigation may be dropped.

A test of significance for a value of r requires the calculation of the probability that such a value of r would occur in random sampling from a universe in which there was no correlation. Usually the first question to be asked when a coefficient of correlation has been computed is whether or not the value of r is significant. In other words, is it consistent with the hypothesis that there is no correlation between the two variables in the universe from which the sample was taken? This is another example of the null hypothesis, and the proper procedure is to determine whether or not the facts disprove it. This may be done by the t distribution or the F distribution.

Use of the *t* Distribution

The following value is distributed as t with $n - 2$ degrees of freedom.[3]

$$t = r\sqrt{\frac{n-2}{1-r^2}}. \tag{11–16}$$

In the correlation between the strength of cotton yarn and the strength of the cotton, $r = .5891$, with $n = 35$. The value of t is computed

$$t = .5891\sqrt{\frac{35-2}{1-.3471}} = .5891\sqrt{\frac{33}{.6529}} = .5891\sqrt{50.5437}$$

$$t = (.5891)(7.109) = 4.188.$$

[3] R. A. Fisher, *Statistical Methods for Research Workers* (Edinburgh: Oliver and Boyd, 1970), pp. 195–198.

For 33 degrees of freedom, this value of t indicates that the correlation is significant. According to Appendix K, a value of t equal to 2.750 for 30 degrees of freedom would occur only one time in 100 random samples when drawn from a universe with a value of ρ equal to zero. The probability of a value of t equal to 4.188 for 33 degrees of freedom is extremely small if the value of ρ is zero. The value of t for 33 degrees of freedom is not given in Appendix K, but the value for 30 degrees of freedom is 2.750 and for 40 is 2.704. Interpolation would not give an accurate value of t for 33 degrees of freedom, but it can be seen that the value 4.188 is much larger than any possible value for t, since its value must fall between 2.750 and 2.704. The conclusion, therefore, is that ρ is not equal to zero in the universe from which the sample was taken.

Use of the *F* Distribution

Instead of making use of the t distribution to test a coefficient of correlation for significance, the analysis of variance and the F ratio may be used. The coefficient of determination has been defined as the percentage of the total variance that is explained by the regression equation. The difference between the explained variance and the unexplained variance, expressed as the ratio between two variances, may be tested for a significant difference by the use of analysis of variance.

The computations in Table 11–2 on page 286 give the sum of the squared deviations of the y values from the computed y_c values as 3,898 (rounded to the decimal point). The sum of the total squared deviations from the arithmetic mean of the y distribution can be computed from the summations from Table 11–1 by the use of Formula 4–8 on page 86, substituting n for N, y for X, and $\bar{y}$ for μ.

$$\Sigma(y - \bar{y})^2 = \Sigma y^2 - \frac{(\Sigma y)^2}{n} = 451{,}079 - \frac{3{,}947^2}{35}$$

$$\Sigma(y - \bar{y})^2 = 451{,}079 - 445{,}109 = 5{,}970.$$

The deviations of the computed y values from the mean of the y values can be computed from

$$\Sigma(y_c - \bar{y})^2 = \Sigma(y - \bar{y})^2 - \Sigma(y - y_c)^2$$

$$\Sigma(y_c - \bar{y})^2 = 5{,}970 - 3{,}898 = 2{,}072.$$

This sum can be verified by computing the deviation of each y_c value from $\bar{y}$, but these calculations are not shown.

The three sums of squares make it possible to set up an analysis of variance similar to Tables 8–5 and 8–6. The degrees of freedom associated with the sum of the squared deviations from the mean are 1 less than the number of individual items, or 34. These total degrees of freedom are separated into 33 for the unexplained variance and 1 for the variance explained by the straight line. Two degrees of freedom were lost fitting the straight line to the actual values, since there are 2 constants in the straight line equation. This means that the unexplained variation uses 33 degrees of freedom, which leaves 1 for the explained variation.

TABLE 11–4

TESTING SIGNIFICANCE OF CORRELATION BY *F* RATIO

Type of Variance	Sum of Squared Deviations	Degrees of Freedom	Variance (Estimated)
Explained by regression	2,072	1	2,072.00
Unexplained	3,898	33	118.12
Total	5,970	34	175.59

The F ratio is the ratio between the variance explained by the regression and the variance that is not explained by this equation. It is computed

$$F = \frac{2{,}072}{118.12} = 17.54.$$

Since the value of $F_{.01} = 7.50$, the null hypothesis that the correlation is zero in the universe is rejected. The F value of 17.54 is found to be significant, since it far exceeds the value of $F_{.01}$ with 1 and 32 degrees of freedom. Since the value of F for $d.f._2 = 33$ is not given, the value for $d.f._2 = 32$ is used. If the value of F secured had fallen between the values for 32 and 34, there might have been some question of getting a more accurate value, but since the computed value is so much greater than the value at $d.f._2 = 32$, it is definitely significant.

Since the computed value of F is for 1 and 33 degrees of freedom, it is equal to the value of t^2 computed in testing the correlation for significance by using the t distribution. It was pointed out previously that $F = t^2$ when one of the degrees of freedom equals 1 and the other is the same as was used in the t test.

The coefficient of correlation can be computed from the values in Table 11–4.

$$\hat{\rho} = \sqrt{1 - \frac{118.12}{175.59}} = \sqrt{1 - .673} = \sqrt{.327} = .572.$$

This is the same value of $\hat{\rho}$ found on page 294.

STUDY QUESTIONS

11–1. How does the scatter diagram show the degree of relationship between the two variables plotted?

11–2. Does "measuring the relationship between two variables" mean that one of the variables *causes* the variations in the other? Explain exactly what is meant by this phrase.

11–3. How does the regression equation show the relationship between two variables?

11–4. How can the regression line be used to estimate the value of one variable from the other?

11–5. How can the regression equation be used to estimate the value of one variable from the other?

11–6. A scatter diagram shows a close relationship between consumer income by counties and the sales of a given product. How can this information be used in planning the sales promotion of the product?

11–7. In showing the relationship between two variables on a scatter diagram, how is the decision made as to which variable should be plotted on the X axis, assuming that you do not know how they are related causally?

11–8. Assume that you know the value of one variable and intend to use this variable to forecast the second variable. Which of the variables would you plot on the X axis? Why?

11–9. Could a measure of the reliability of estimates from a regression line be computed by ignoring the signs of the deviations from the line and computing the arithmetic mean of these deviations? Refer to the deviations $(y - y_c)$ in Table 11–2.

11–10. If you used a measure of reliability of estimates by averaging the deviations from the regression by ignoring the signs, what measure of dispersion described in Chapter 4 would the measure resemble?

11–11. The standard error of estimate $(\hat{\sigma}_{yx})$ is preferred as a measure of dispersion over the measure described in Question 11–10. Why is this the case?

11–12. What assurance do you have that an estimate made from a regression equation is reasonably accurate?

11–13. Explain why $\Sigma(y - y_c)^2$, the sum of the squared deviations from the values estimated by the regression line, is divided by $n - 2$ instead of by n in computing $\hat{\sigma}_{yx}$.

11–14. In what sense is the independent variable independent? What distinguishes it from the dependent variable?

11–15. Assume that information is available promptly at the end of each month on the number of credit references checked in the retail credit bureau of a city and the credit sales volume for that month. How could you use this information to estimate the volume of credit sales for the month? Discuss the problems that would be encountered and explain how you would determine the accuracy of the estimates.

11–16. What advantage does an abstract measure of correlation, such as the coefficient of correlation, have over the standard error of estimate in comparing the degree of correlation present in two variables?

11–17. Why is it difficult to compare the degree of correlation in two problems by comparing the values of the standard error of estimate?

11–18. Explain what is meant by saying that the coefficient of determination represents the fraction of the total variance that is explained by the X variable. Does this mean that it represents the fraction of the variance that is *caused* by the X variable?

11–19. Why is it impossible for $\hat{\rho}^2$ to have a value of more than 1.00?

11–20. Explain the significance of a coefficient of correlation of .60.

11–21. Does a coefficient of correlation of .60 mean that 60% of the variance in the Y variable is related to the X variable? Discuss.

11–22. Assume that the value of $\hat{\rho}$ is .40 when studying the relationship between output on a machine and score on a certain aptitude test. When a different test was used, the value of $\hat{\rho}$ was .80. Does this mean that using the second test would give estimates that are twice as accurate as would be obtained from using the first test?

11–23. What indicates whether the relationship in a linear correlation is direct or inverse?

11–24. Why is it important to compute the confidence interval of the coefficient of correlation?

11–25. What are some of the problems encountered in computing the confidence interval of $\hat{\rho}$?

11–26. If the value of σ^2_{yx} is as large as the value of σ^2_y, what does this indicate about the correlation between the X and Y variables? Explain.

11–27. In a study using correlation analysis, one value of $\hat{\rho}$ was found to be $-.73$ and another to be $+.73$. Is the degree of correlation shown by the first value of $\hat{\rho}$ less than the second? Explain.

11–28. If the value of $\hat{\rho}$ is $+93$, is this conclusive evidence that the variations in the Y variable are caused by fluctuations in the X variable? Discuss.

11–29. Under what conditions can Formula 11–15 be used to compute the standard error of the coefficient of correlation?

11–30. Under what circumstances is it important to test the coefficient of correlation for significance?

11–31. If the test of significance indicates that the value of ρ is not zero, does it indicate that the variation in the X variable causes the variation in the Y variable?

11–32. What are the advantages of using r_r instead of r?

11–33. Can the coefficient of rank correlation be used to make estimates of one variable from known values of the other variable? Explain.

11–34. What advantage do you see, if any, in using the F distribution to test for correlation instead of using the t distribution?

PROBLEMS

11–1. The relationship between annual dividend per share and the price per share of investment grade common stocks was computed to be:

$$y_c = 8 + 16x$$
x = annual dividend (dollars)
y = price per share (dollars)

Estimate the price per share of a stock paying a dividend of $2 per share. How would you evaluate the accuracy of this estimate?

11–2. The following equation shows the relationship between corn yield in Kansas and average July temperature:

$$y_c = 154 - 1.90x$$
x = average July temperature (degrees)
y = average yield per acre (bushels)

(a) What would you estimate the average yield per acre to be for a year in which the July temperature averaged 78 degrees? How would you determine how much confidence to put in this estimate?

(b) Would you expect the average yield to be higher or lower if the July temperature were less than 78 degrees?

(c) How would you determine whether or not a lower July temperature would be expected to be associated with a higher or a lower yield?

11–3. The data in the following table represent the results of a study of the average output of 10 employees on an eight-hour shift on a particular machine of a production line, and the score on an aptitude test.

(a) Plot these data on a scatter diagram. Which series should you use as the X variable? Why?

(b) Explain what the scatter diagram shows to be the relationship between the score on the test and performance in the shop.

(c) How can this information be used in hiring employees for this machine?

(d) *Retain this solution for use in Problem 11–4.*

Employee Number	Output (Number of Units)	Score
1	17	110
2	19	125
3	18	140
4	23	150
5	22	165
6	20	170
7	25	190
8	24	200
9	27	210
10	29	215
Total	224	1,675
Sum of squares	5,158	292,375
Σxy	38,700	

11–4. (a) Compute the straight-line regression equation for the data given in Problem 11–3, and plot the regression line on the scatter diagram.

(b) Estimate the output to be expected for the individuals who made the scores 130, 160, 175, and 205.

(c) Does the regression equation give an accurate estimate of the output of an employee? Discuss.

(d) *Retain this solution for use in Problems 11–5 and 11–11.*

11–5. (a) Compute the standard error of estimate for the regression equation found in Problem 11–4.

(b) How accurate do you consider each of the estimates made in Problem 11–4?

(c) *Retain this solution for use in Problems 11–11 and 11–12.*

11–6. The data in the following table represent the record of the receipts of strawberries at a central produce market over a period of 10 days, and the average price of the strawberries.

(a) Plot these data on a scatter diagram. Which series should you use as the X variable? Why?

(b) Explain what the scatter diagram shows to be the relationship between the receipts and the average price received on that day.

(c) How can this information be used in forecasting the price of strawberries in the central market on any given day?

(d) *Retain this solution for use in Problem 11–7.*

Day	Receipts (Pounds)	Average Price per Pound (Cents)
Monday	2,500	47
Tuesday	2,800	45
Wednesday	3,700	39
Thursday	3,700	38
Friday	4,800	32
Monday	5,000	30
Tuesday	4,300	35
Wednesday	3,600	39
Thursday	3,200	42
Friday	2,400	48
Total	36,000	395
Sum of squares	136,960,000	15,937
Σxy	1,372,500	

11–7. (a) Compute the straight-line regression equation for the data given in Problem 11–6 and plot the regression line on the scatter diagram.

(b) Estimate the price of strawberries in the central market on days when the receipts were 2,500, 3,000, 4,200, and 4,900 pounds.

(c) Does the regression equation give an accurate estimate of the price of strawberries in the central market? Discuss.

(d) *Retain this solution for use in Problems 11–8, 11–14, and 11–15.*

11–8. (a) Compute the standard error of estimate for the regression equation found in Problem 11–7.

(b) How accurate do you consider each of the estimates made in Problem 11–7?

(c) *Retain this solution for use in Problems 11–14 and 11–15.*

11–9. A company selling sets of encyclopedias ran an advertising campaign in 25 market areas, offering a free booklet to anyone who sent in a coupon. Salesmen followed up these leads, and at the end of the sales campaign the number of inquiries and the number of sets sold in each area were compiled. The number of inquiries was designated the X variable, and the number of sets sold, the Y variable. The following summations have been made without reproducing the original data for each of the 25 areas.

$\Sigma x = 4{,}000$ $\Sigma x^2 = 1{,}540{,}000$ $\Sigma xy = 224{,}500$
$\Sigma y = 620$ $\Sigma y^2 = 33{,}825$

(a) Compute the regression equation for inquiries and sales.

(b) Estimate the number of sets you would expect to sell in areas from which the number of inquiries received were 30, 150, 250, and 600.

(c) *Retain this solution for use in Problems 11–17 and 11–18.*

11–10. (a) Compute the standard error of estimate and explain how this would be used to evaluate the estimates made in Problem 11–9.

(b) *Retain this solution for use in Problems 11–17 and 11–18.*

11–11. Using the data from Problem 11–3, compute the coefficient of correlation and the coefficient of determination by the product-moment method. Explain the significance of these coefficients. How are they related to the regression equation and the standard error of estimate in Problems 11–4 and 11–5? *Retain this solution for use in Problems 11–12 and 11–13.*

11–12. Compute the coefficient of determination for the data from Problem 11–3 using Formula 11–8. How does this value compare with the coefficient of determination computed in Problem 11–11?

11–13. Test whether or not the coefficient of correlation computed in Problem 11–11 is significantly different from zero. Explain what your results mean.

11–14. Using the data from Problem 11–6, compute the coefficient of correlation and the coefficient of determination by the product-moment method. Explain the significance of these coefficients. How are they related to the regression equation and the standard error of estimate computed in Problems 11–7 and 11–8? *Retain this solution for use in Problems 11–15, 11–16, and 12–2.*

11–15. Compute the coefficient of determination for the data from Problem 11–6 using Formula 11–8. How does this value compare with the coefficient of determination computed in Problem 11–14?

11–16. Test whether or not the coefficient of correlation computed in Problem 11–14 is significantly different from zero. Explain what your results mean.

11–17. Using the data from Problem 11–9, compute the coefficient of correlation and the coefficient of determination by the product-moment method. Explain the significance of these coefficients. How are they related to the regression equation and the standard error of estimate computed in Problems 11–9 and 11–10? *Retain this solution for use in Problems 11–18 and 11–19.*

11–18. Compute the coefficient of determination for the data from Problem 11–9 using Formula 11–8. How does this value compare with the coefficient of determination computed in Problem 11–17?

11–19. Test whether or not the coefficient of correlation computed in Problem 11–17 is significantly different from zero. Explain what your results mean.

11–20. A random sample of 10 employees from the accounting division of a large insurance company is shown below.

Employee Number	Weekly Salary (Dollars)	Years of Service with Company
1	230	2
2	200	3
3	170	1
4	315	5
5	185	1
6	330	7
7	250	4
8	300	7
9	225	6
10	325	9

Years of service is designated the independent variable (X) and weekly salary is designated the dependent variable (Y).

(a) Draw a scatter diagram to determine if there seems to be any relationship between the two variables. Is the pattern linear?

(b) Compute a least-squares regression equation to show the average relationship between Y and X.

(c) If an employee has been with the company for $4\frac{1}{2}$ years, what would you estimate her weekly salary to be?

(d) Compute the standard error of estimate and explain what the value represents.

(e) Compute the coefficient of determination ($\hat{\rho}^2$) and the coefficient of nondetermination ($\hat{\kappa}^2$) and explain what each means.

(f) Compute the coefficient of correlation (r) using the product-moment method. Estimate $\hat{\rho}^2$ from r using Formula 11–10 and compare this value to the one computed in part (e).

(g) Test r for significance using the t distribution and level of significance of 95%.

(h) Test r for significance using the F distribution and a level of significance of 95%. Is the answer here the same as that in (g)?

11–21. A consumer testing bureau tests eight brands of small table radios and ranks them in quality from best (1) to poorest (8). The suggested retail price and quality rating of each brand are shown in the table at the top of page 310. Compute the coefficient of rank correlation (r_r) to show the relationship between the two variables, quality and price. *Hint:* Rank the Y variable first.

Radio	Quality Rating x	Price (Dollars) y
A	3	38.25
B	7	41.00
C	1	20.20
D	8	45.00
E	5	36.00
F	2	42.30
G	6	35.00
H	4	36.67

11–22. The following table gives the daily price of watermelons and the number of cars on track at 8:00 A.M. each morning for 15 days of the watermelon season.

Day	Price per Car (Dollars) y	Number of Cars on Track, 8:00 A.M. x
1	680	35
2	625	39
3	505	63
4	435	90
5	370	115
6	480	86
7	370	89
8	345	95
9	350	114
10	317	121
11	385	123
12	375	124
13	217	162
14	275	143
15	433	92

Source: Hypothetical data.

(a) Plot these data on a scatter diagram and state how closely you think the price of watermelons is related to the supply available on a given day.

(b) Compute the regression equation and plot it on a scatter diagram. Use the regression equation to forecast the price of watermelons on a day when 80 cars were on the track at 8:00 A.M.

(c) Compute the standard error of estimate and summarize how accurate you believe your forecast of the price of watermelons to be.

(d) *Retain this solution for use in Problems 11–23, 11–24, and 12–10.*

11–23. (a) Refer to the data in Problem 11–22 and compute the value of $\hat{\rho}$ by the product-moment method.

(b) Test whether or not the coefficient of correlation differs significantly from zero, and explain what this test means.

11–24. Using the computations made in Problem 11–22, compute the value of $\hat{\rho}$ by Formula 11–8. Do you prefer this method or the product-moment method used in Problem 11–23? Give reasons for your preference.

12 Nonlinear, Multiple, and Partial Correlation

Chapter 11 contained a discussion of simple linear regression and correlation. One of the principal assumptions was that the relationship between two variables could best be described by the equation for a straight line.

This chapter discusses the methods of measuring correlation when the relationship between two variables is best described by a nonlinear regression line, and the methods of measuring the relationship between more than two variables. When measuring the relationship between more than two variables, it may be found that nonlinear regressions describe the relationships better than straight lines. However, the measurement of nonlinear multiple correlation is an involved problem of an advanced nature and will not be discussed in this text. Mathematical methods of measuring nonlinear correlation between several variables are particularly burdensome, and the practical solution to this type of problem is frequently found by employing a graphic method.

NONLINEAR CORRELATION

When a straight line is used to describe the relationship between two variables that are not related linearly, the straight line does not give as good an estimate of the Y variable as can be obtained from a curve that more closely fits the data. The coefficient of correlation (ρ) likewise will reflect this weakness when it is used as a measure of correlation. Nonlinear correlation involves deriving a regression line that is a good fit, computing a measure of reliability of estimates from this regression line, and deriving an abstract measure of the correlation represented by this regression line. The only difference in the nature of nonlinear and linear correlation is the regression line, which is a curve rather than a straight line.

Figure 12–1 is a scatter diagram showing the relationship between bushels per acre of wheat and number of acres irrigated as a percentage of total acres harvested for a sample of 21 Texas counties. The data are shown in Table

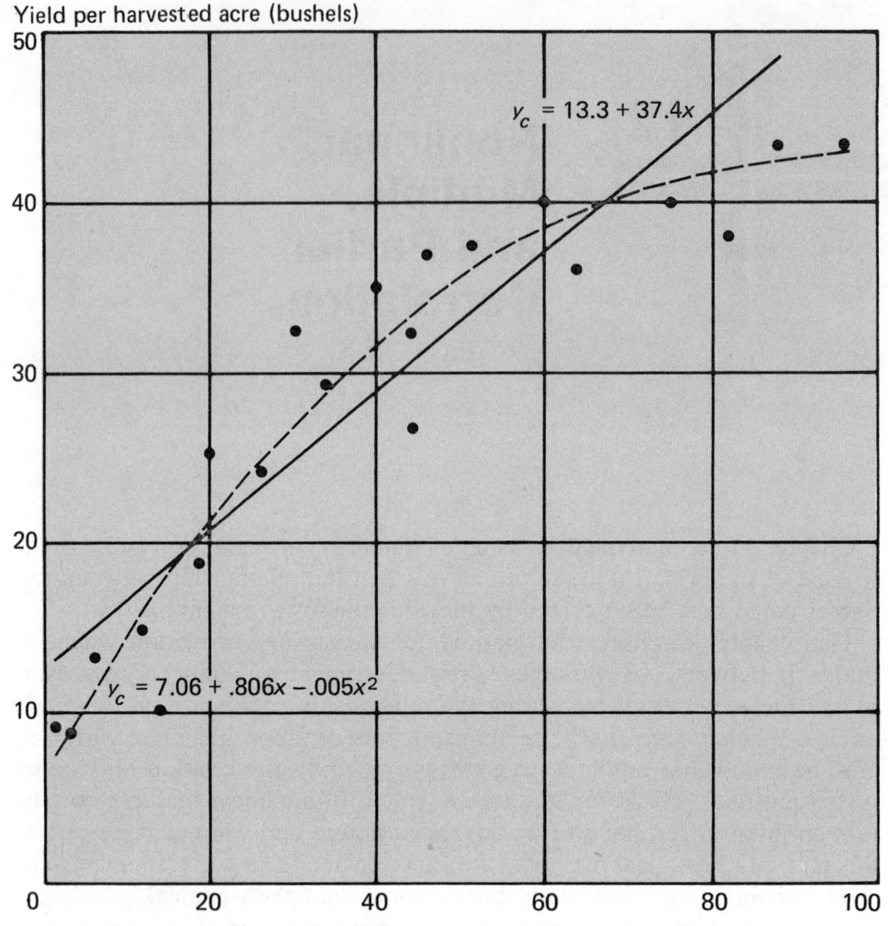

Source: Table 12-1.

FIGURE 12–1

SCATTER DIAGRAM OF BUSHELS PER ACRE OF WHEAT AND NUMBER OF ACRES IRRIGATED AS A PERCENTAGE OF TOTAL ACRES HARVESTED FOR A SAMPLE OF 21 TEXAS COUNTIES

12–1. Table 12–2 on page 314 is a work sheet that is used to show all the sums required to fit both a linear and a nonlinear regression equation to these data.

Regression Line

The straight line plotted in Figure 12–1 was fitted to the data by the method of least squares, using Formulas 11–2 and 11–3 to secure the constants in the equation. The straight-line regression equation is

$$y_c = 13.3025 + .3737x.$$

TABLE 12–1

BUSHELS PER ACRE OF WHEAT AND NUMBER OF ACRES IRRIGATED AS A PERCENTAGE OF TOTAL ACRES HARVESTED FOR A SAMPLE OF 21 TEXAS COUNTIES

County Number	Irrigated Acres as a Percentage of Total Acres Harvested x	Bushel Yield per Harvested Acre y
1	44.3	26.9
2	26.2	24.1
3	63.9	36.2
4	1.0	9.2
5	82.0	38.2
6	51.7	37.6
7	14.0	10.1
8	6.0	13.3
9	88.1	43.5
10	96.0	43.4
11	34.0	29.8
12	11.9	15.0
13	30.1	32.7
14	46.1	36.8
15	20.0	25.3
16	75.0	40.1
17	3.0	8.9
18	44.1	34.3
19	40.1	35.0
20	19.0	19.0
21	60.0	40.0

Source: Adapted from *1972 Texas Small Grains Statistics*, Texas Department of Agriculture.

There are many possible curved regression equations that might be fit to the data. The one used here is a second-degree parabola shown by the equation

$$y_c = a + bx + cx^2. \tag{12–1}$$

Equations in Formula 12–2 may be solved simultaneously to secure the values of the coefficients of the equation shown in Formula 12–1.

$$\begin{aligned} \text{I.} \quad \Sigma y &= na + b\Sigma x + c\Sigma x^2 \\ \text{II.} \quad \Sigma xy &= a\Sigma x + b\Sigma x^2 + c\Sigma x^3 . \\ \text{III.} \quad \Sigma x^2 y &= a\Sigma x^2 + b\Sigma x^3 + c\Sigma x^4 \end{aligned} \tag{12–2}$$

When the values of the summations from Table 12–2 are substituted in the three equations, they appear in the form shown at the top of page 315.

TABLE 12–2

RELATIONSHIP BETWEEN BUSHELS PER ACRE OF WHEAT AND NUMBER OF ACRES IRRIGATED AS A PERCENTAGE OF TOTAL ACRES HARVESTED

x	y	x^2	y^2	xy	x^3	x^2y	x^4
44.3	26.9	1,962.49	723.61	1,191.67	86,938.31	52,790.98	3,851,367.00
26.2	24.1	686.44	580.81	631.42	17,984.73	16,543.20	471,199.87
63.9	36.2	4,083.21	1,310.44	2,313.18	260,917.12	147,812.20	16,672,603.90
1.0	9.2	1.00	84.64	9.20	1.00	9.20	1.00
82.0	38.2	6,724.00	1,459.24	3,132.40	551,368.00	256,856.80	45,212,176.00
51.7	37.6	2,672.89	1,413.76	1,943.92	138,188.41	100,500.66	7,144,340.95
14.0	10.1	196.00	102.01	141.40	2,744.00	1,979.60	38,416.00
6.0	13.3	36.00	176.89	79.80	216.00	478.80	1,296.00
88.1	43.5	7,761.61	1,892.25	3,832.35	683,797.84	337,630.03	60,242,589.79
96.0	43.4	9,216.00	1,883.56	4,166.40	884,736.00	399,974.40	84,934,656.00
34.0	29.8	1,156.00	888.04	1,013.20	39,304.00	34,448.80	1,336,336.00
11.9	15.0	141.61	225.00	178.50	1,685.16	2,124.15	20,053.39
30.1	32.7	906.01	1,069.29	984.27	27,270.90	29,626.53	820,854.12
46.1	36.8	2,125.21	1,354.24	1,696.48	97,972.18	78,207.73	4,516,517.54
20.0	25.3	400.00	640.09	506.00	8,000.00	10,120.00	160,000.00
75.0	40.1	5,625.00	1,608.01	3,007.50	421,875.00	225,562.50	31,640,625.00
3.0	8.9	9.00	79.21	26.70	27.00	80.10	81.00
44.1	34.3	1,944.81	1,176.49	1,512.63	85,766.12	66,706.98	3,782,285.94
40.1	35.0	1,608.01	1,225.00	1,403.50	64,481.20	56,280.35	2,585,696.16
19.0	19.0	361.00	361.00	361.00	6,859.00	6,859.00	130,321.00
60.0	40.0	3,600.00	1,600.00	2,400.00	216,000.00	144,000.00	12,960,000.00
856.5	599.4	51,216.29	19,853.58	30,531.52	3,596,131.97	1,968,592.02	276,521,416.67

Values of y are bushels per acre of wheat, and values of x are number of acres irrigated as a percentage of total acres harvested.

Source: Table 12–1.

$$\text{I.} \quad 599.40 = 21a + 856.50b + 51{,}216.29c$$
$$\text{II.} \quad 30{,}531.52 = 856.50a + 51{,}216.29b + 3{,}596{,}131.97c$$
$$\text{III.} \quad 1{,}968{,}592.02 = 51{,}216.29a + 3{,}596{,}131.97b + 276{,}521{,}416.67c$$

Since there are three unknowns and three equations, the equations may be solved by any of the standard methods. The values of the three coefficients obtained from the solution of the equations are

$$a = 7.06331$$
$$b = .80582$$
$$c = -.00467.$$

The nonlinear regression for predicting bushels per acre is

$$y_c = 7.06331 + .80582x + (-.00467)x^2.$$

The nonlinear regression line is shown as the dashed line in Figure 12–1.

Standard Error of Estimate

The standard error of estimate for the linear regression model is computed using the necessary sums from Table 12–2 and Formula 11–5.

$$\hat{\sigma}_{yx} = \sqrt{\frac{19{,}853.58 - (13.30253)(599.40) - (.3737)(30{,}531.52)}{21 - 2}}$$
$$= 4.98.$$

It is possible to compute the standard error of estimate for the nonlinear model using Formula 12–3.

$$\hat{\sigma}_{yx} = \sqrt{\frac{\Sigma(y - y_c)^2}{n - 3}}. \qquad \textbf{(12–3)}$$

Three degrees of freedom are lost in computing the statistic as there are three constants in the regression equation that is used to compute the y_c values shown in Table 12–3. Using the totals from that table and Formula 12–3, the standard error of estimate for the nonlinear regression model is

$$\hat{\sigma}_{yx} = \sqrt{\frac{207.74}{21 - 3}} = 3.40.$$

Since the standard error of estimate for the curved line is smaller than that for the straight line, the nonlinear regression equation appears to fit the data better than the straight line does.

TABLE 12–3

COMPUTATION OF THE INDEX OF CORRELATION

Bushels per Acre and Irrigated Acres as a Percentage of Total Harvested

y	y_c	$y - y_c$	$(y - y_c)^2$
26.90	33.60	−6.70	44.89
24.10	24.97	− .87	.76
36.20	39.49	−3.29	10.82
9.20	7.86	1.34	1.80
38.20	41.75	−3.55	12.60
37.60	36.24	1.36	1.85
10.10	17.43	−7.33	53.73
13.30	11.73	1.57	2.46
43.50	41.82	1.68	2.82
43.40	41.40	2.00	4.00
29.80	29.06	.74	.55
15.00	15.99	− .99	.98
32.70	27.09	5.61	31.47
36.80	34.29	2.51	6.30
25.30	21.31	3.99	15.92
40.10	41.24	−1.14	1.30
8.90	9.44	− .54	.29
34.30	33.52	.78	.61
35.00	31.87	3.13	9.80
19.00	20.69	−1.69	2.86
40.00	38.61	1.39	1.93
599.40	599.40	0.00	207.74

Values of y are bushels per acre of wheat, and values of y_c are computed using the nonlinear regression equation.

Source: Table 12–1.

The meaning of the standard error of estimate is the same for nonlinear as for linear correlation. It is an average of the deviations of the actual values from the values computed from the regression equation. If these deviations are normally distributed, approximately 68% of the estimates from the regression equation will fall within the range $\pm 1\ \hat{\sigma}_{yx}$. This means that the standard error of estimate serves as a measure of reliability of the estimating equation. Stated in another way, the standard error of estimate measures the goodness of fit of the regression line, whether this line is a curve or a straight line.

Index of Correlation

The same need for an abstract measure of correlation exists for nonlinear correlation as for linear correlation. Since the standard error of estimate is computed in the same basic manner for both types of relationship, the abstract measure of relationship can be computed from $\hat{\sigma}^2_{yx}$ by Formula 12–4, which is essentially the same as Formula 11–8 on page 293. The coefficient of correlation (ρ) is in reality a special case of the index of correlation. In order to emphasize that the regression equation is not a straight line, the letter I is used as the symbol for the measure of correlation, called the *index of correlation*. The term *coefficient of determination* is applied to I_{yx}^2 and is interpreted as the percentage of the variance in the Y variable that is explained by the variations in the X variable. The subscripts are always used with the index of correlation, since the value of I_{yx} is not necessarily the same as I_{xy}. The subscripts are not necessary for linear correlation, since ρ_{xy} is the same as ρ_{yx}.

When using the index of correlation, one must always have information about the regression line, since the value of I has no significance except as related to the measure of average relationship. The sign of the slope of the linear regression line is given to the value of ρ, with a positive slope representing a positive relationship between the variables, and a negative slope, an inverse relationship.

When the regression line is not a straight line, the slope may be positive at some points on the line and negative at other points. For example, the correlation between yield per acre (the dependent variable) and the amount of irrigation water applied might show a positive correlation at some points on the line and a negative correlation at other points. When the amount of water applied is relatively small, the yield per acre is higher for larger amounts of water than for smaller amounts. This means that the correlation is positive. However, farther to the right on the correlation chart, when relatively large amounts of water are applied, yields associated with very large amounts of water are less than for the smaller amounts, giving a negative correlation. In other words, up to a given amount of water, adding more water increases the yield and the correlation is positive. After this optimum amount of water has been applied, a further increase results in a decrease in yield.

Since it is always possible that the relationship in a nonlinear correlation may be positive over part of the range of the variables and negative for other values, it is customary not to attach any sign to the value of a measure of correlation. The simplest way to determine whether the relationship is direct or inverse is to plot the regression line on a scatter diagram.

The formula for the coefficient of determination for the nonlinear model is

$$I_{yx}^2 = 1 - \frac{\hat{\sigma}_{yx}^2}{\hat{\sigma}_y^2}. \tag{12–4}$$

To compute the value of I_{yx} for the example in Table 12–1, it is first necessary to compute the variance of y using the totals given in Table 12–2.

$$\hat{\sigma}_y^2 = \frac{n\Sigma y^2 - (\Sigma y)^2}{n^2}\left(\frac{n}{n-1}\right)$$

$$\hat{\sigma}_y^2 = \frac{(21)(19{,}853.58) - (599.4)^2}{21^2}\left(\frac{21}{20}\right) = 137.25.$$

$$I^2_{yx} = 1 - \frac{3.40^2}{137.25} = 1 - .0842 = .9158$$

$$I_{yx} = \sqrt{.9158} = .957.$$

The coefficient of correlation for the linear model is computed using Formula 11–8 and the sums taken from Table 12–2.

$$\hat{\rho}^2 = 1 - \frac{\hat{\sigma}^2_{yx}}{\hat{\sigma}^2_y}$$

$$\hat{\rho}^2 = 1 - \frac{4.98^2}{137.25} = 1 - .1807$$

$$\hat{\rho}^2 = .8193$$

$$\hat{\rho} = \sqrt{.8193} = .905.$$

The coefficient of correlation is not as high for the straight line as is the index of correlation for the curved line because the nonlinear regression equation is a better fit for these data.

MULTIPLE CORRELATION AND REGRESSION ANALYSIS

The discussions of correlation up to this point have involved measuring the relationship between two variables, although it has been indicated that at times more than two variables should be considered.

In the discussion of the correlation between the strength of the yarn and the strength of the fiber from which it was spun, the scatter diagram in Figure 11–3 (page 284) gave some indication that factors other than the strength of the cotton fiber might have an influence on the strength of the yarn. Table 11–2 (page 286) shows the strength of the lot of yarn number 25 to be 31.32 below the strength estimated from the regression equation. This is the largest deviation of yarn strength from the estimated strength in the 35 samples, and it is natural to ask how this large deviation can be explained. The data in Column x_3 in Table 12–4 (pages 320–321) show the fineness of the 35 samples used to estimate yarn strength from the strength of the cotton fiber. These amounts represent the weight in micrograms of one inch of fiber. The larger the figure reported, the coarser the fiber. Without plotting a scatter diagram, it can be seen that there is a tendency for the stronger yarn to be associated with the finer cotton, and the weaker yarn with the coarser cotton. This suggests that perhaps the rather strong cotton fiber in sample 25 resulted in weaker yarn than should be expected because the fiber was coarser than average. A further check of the points that are substantially below the regression line shows that samples 20 and 24 are also from unusually coarse cotton. The problem of multiple correlation is to relate the variations in the dependent variable to the variations in two or more independent variables, rather than a single variable, as is done in simple correlation.

Since a preliminary examination of the data suggests that the strength of the yarn is influenced by the strength of the cotton fiber and by the fineness of the fiber, it seems desirable to study the way in which the variations in yarn strength are related to variations in both fiber strength and fiber fineness. The data on all three characteristics are given in Table 12–4, where the summations that will be needed are computed.

At this point it is necessary to make a change in the notation that is used in making the calculations. When only two variables are included in a study, the dependent variable is usually designated Y, and the independent variable X. When more than one independent variable is used, it becomes advantageous to distinguish between the variables by means of subscripts and use only the letter X. In this problem of yarn strength, the subscript "1" is used for yarn strength, the dependent variable. The two independent variables are designated X_2 and X_3. This scheme of notation can be expanded to take care of any desired number of independent variables. It is also possible to use this notation in simple correlation by designating the dependent variable x_1 instead of y, and the independent variable x_2 instead of x. When referring to the simple correlation between any of these three variables, it is an advantage to use this notation.

Multiple Regression Analysis

Multiple regression analysis is a method of taking into account simultaneously the relationship between all the variables when two or more independent variables are to be used in making estimates of the dependent variable. When both cotton fiber strength and fineness are taken into account in making estimates of the strength of the yarn, the estimates should be more accurate than when only one of the characteristics of the cotton fiber is used. The use of two or more independent variables is an extension of the basic principles used in two-variable regression analysis. It is necessary to determine the equation for the average relationship between the variables, and then to compute a measure of the accuracy of estimates from this equation.

The Regression Equation. Before discussing the computation of the regression equation for making estimates from two independent variables, it may be worthwhile to review briefly the regression equation for estimating yarn strength from the strength of the cotton fiber. In Chapter 11 the relationship between the strength of cotton yarn and the strength of the cotton fiber used in spinning the yarn was shown by the equation

$$y_c = 1.589x - 23.106.$$

Using x with subscripts for the different values of the variables, the equation can be written

$$x_{1c} = 1.589x_2 - 23.106.$$

TABLE 12-4

RELATIONSHIP BETWEEN YARN STRENGTH AND STRENGTH AND FINENESS OF COTTON FIBER

Sample Number	Yarn Strength* x_1	Strength of Fiber** x_2	Fineness of Fiber*** x_3	x_1^2
1	129	95	3.9	16,641
2	117	91	4.3	13,689
3	127	88	4.3	16,129
4	129	83	4.2	16,641
5	128	93	4.0	16,384
6	118	84	4.4	13,924
7	121	81	4.3	14,641
8	112	79	4.1	12,544
9	126	84	4.3	15,876
10	128	89	3.8	16,384
11	125	92	3.6	15,625
12	115	87	3.0	13,225
13	114	87	3.4	12,996
14	116	86	3.5	13,456
15	112	86	2.9	12,544
16	113	89	3.7	12,769
17	120	87	3.4	14,400
18	92	78	4.3	8,464
19	96	76	4.2	9,216
20	84	81	5.0	7,056
21	114	92	4.6	12,996
22	112	89	4.6	12,544
23	99	83	4.6	9,801
24	76	74	5.1	5,776
25	87	89	6.0	7,569
26	121	85	3.8	14,641
27	115	85	3.5	13,225
28	103	80	4.6	10,609
29	119	93	4.0	14,161
30	119	86	3.5	14,161
31	118	86	4.6	13,924
32	114	85	3.3	12,996
33	112	80	3.5	12,544
34	118	85	4.1	13,924
35	98	85	5.4	9,604
Total	3,947	2,993	143.8	451,079

* Pounds of pressure required to break a skein of yarn.

** Pressley strength index, thousands of pounds of pressure required to break the equivalent of one square inch of fiber.

*** Weight of one inch of fiber in micrograms.

TABLE 12-4 (concluded)

x_2^2	x_3^2	x_1x_2	x_1x_3	x_2x_3
9,025	15.21	12,255	503.1	370.5
8,281	18.49	10,647	503.1	391.3
7,744	18.49	11,176	546.1	378.4
6,889	17.64	10,707	541.8	348.6
8,649	16.00	11,904	512.0	372.0
7,056	19.36	9,912	519.2	369.6
6,561	18.49	9,801	520.3	348.3
6,241	16.81	8,848	459.2	323.9
7,056	18.49	10,584	541.8	361.2
7,921	14.44	11,392	486.4	338.2
8,464	12.96	11,500	450.0	331.2
7,569	9.00	10,005	345.0	261.0
7,569	11.56	9,918	387.6	295.8
7,396	12.25	9,976	406.0	301.0
7,396	8.41	9,632	324.8	249.4
7,921	13.69	10,057	418.1	329.3
7,569	11.56	10,440	408.0	295.8
6,084	18.49	7,176	395.6	335.4
5,776	17.64	7,296	403.2	319.2
6,561	25.00	6,804	420.0	405.0
8,464	21.16	10,488	524.4	423.2
7,921	21.16	9,968	515.2	409.4
6,889	21.16	8,217	455.4	381.8
5,476	26.01	5,624	387.6	377.4
7,921	36.00	7,743	522.0	534.0
7,225	14.44	10,285	459.8	323.0
7,225	12.25	9,775	402.5	297.5
6,400	21.16	8,240	473.8	368.0
8,649	16.00	11,067	476.0	372.0
7,396	12.25	10,234	416.5	301.0
7,396	21.16	10,148	542.8	395.6
7,225	10.89	9,690	376.2	280.5
6,400	12.25	8,960	392.0	280.0
7,225	16.81	10,030	483.8	348.5
7,225	29.16	8,330	529.2	459.0
256,765	605.84	338,829	16,048.5	12,276.0

The estimating equation for a regression analysis with two independent variables is written[1]

$$x_{1c} = a + b_{12.3}x_2 + b_{13.2}x_3. \qquad \textbf{(12–5)}$$

The coefficient $b_{12.3}$ represents the net relationship between variables x_1 and x_2, when the influence of x_3 is taken into account. In making an estimate of x_1, the product of $b_{12.3}$ and x_2 constitutes the contribution of variable x_2 to the estimate. Likewise, the coefficient $b_{13.2}$ represents the net relationship between variables x_1 and x_3, taking into account the influence of x_2. The product of $b_{13.2}$ and x_3 constitutes the contribution of variable x_3 to the estimate.

The b coefficient in simple correlation measures the relationship between the two variables included in the study. But if additional variables have an important influence on the dependent variable, their effect is not isolated but is simply mixed with the influence of the one variable being measured. When the relation between only two variables is measured, there is no way of knowing how much effect other factors may have on the dependent variable. It is possible for the influence of two independent variables to counteract each other to the extent that neither appears to show any correlation with the dependent variable. Yet when the multiple relationship is measured, the correlation may be high.

The three equations in Formula 12–6 may be solved simultaneously to secure the values of the coefficients of the estimating equation in Formula 12–5. It will be seen that the three coefficients are the unknowns in the three equations; the values of the summations may be obtained from Table 12–4, where they were derived from the sample values of the three variables.

$$\begin{aligned}
&\text{I.} \quad \Sigma x_1 = na + b_{12.3}\Sigma x_2 + b_{13.2}\Sigma x_3 \\
&\text{II.} \quad \Sigma x_1x_2 = a\Sigma x_2 + b_{12.3}\Sigma x_2^2 + b_{13.2}\Sigma x_2x_3\,. \\
&\text{III.} \quad \Sigma x_1x_3 = a\Sigma x_3 + b_{12.3}\Sigma x_2x_3 + b_{13.2}\Sigma x_3^2
\end{aligned} \qquad \textbf{(12–6)}$$

Substituting the values of the summations from Table 12–4 gives

[1] The subscripts of the b coefficients are designed to indicate their meaning. The two numbers to the left of the decimal are the *primary subscripts* and show which two variables are related by this coefficient. The subscript to the right of the decimal is the *secondary subscript*. The variable represented by it is included in the study, but its effect is eliminated in determining the relationship of the two variables represented by the primary subscripts. For example, $b_{12.3}$ shows the relationship between variables x_1 and x_2 with the influence of variable x_3 on variable x_1 eliminated. The coefficient $b_{13.2}$ shows the relationship between variables x_1 and x_3 with the influence of x_2 on x_1 eliminated. If there were four independent variables instead of two, the subscript showing the relationship between variables x_1 and x_2 would be written $b_{12.345}$. The influence of variables x_3, x_4, and x_5 on variable x_1 would be eliminated in determining the relationship between variables x_1 and x_2. Since the estimating equation will usually be used with sample values of the variables, it is written using x any y. In referring to the "X variable," capital X will be used as in Chapter 11. It is important to remember that regression analysis is concerned with estimating from a sample much more often than with summarizing universe values.

$$\text{I.} \quad 3{,}947 = 35a + 2{,}993b_{12.3} + 143.8b_{13.2}$$

$$\text{II.} \quad 338{,}829 = 2{,}993a + 256{,}765b_{12.3} + 12{,}276.0b_{13.2}$$

$$\text{III.} \quad 16{,}048.5 = 143.8a + 12{,}276.0b_{12.3} + 605.84b_{13.2}.$$

Since there are three unknowns and three equations, the equations may be solved by any of the standard methods. The values of the three coefficients obtained from the solution of the equations are

$$a = 36.32308812$$

$$b_{12.3} = 1.34066162$$

$$b_{13.2} = -9.29699801.$$

The regression equation for estimating the strength of cotton yarn is

$$x_{1c} = 36.323 + 1.341x_2 - 9.297x_3.$$

Estimates from the Regression Equation. With the regression equation derived from the 35 samples of cotton yarn, it is possible to estimate the strength of a lot of yarn spun from cotton with known fiber strength and fineness. Assume that the cotton being used has a Pressley strength index of 91 and a fineness represented by 3.5 micrograms. An estimate of the strength of the yarn is

$$x_2 = 91 \text{ and } x_3 = 3.5$$

$$x_{1c} = 36.323 + (1.341)(91) - (9.297)(3.5)$$

$$x_{1c} = 36.323 + 122.031 - 32.540 = 125.81.$$

The regression equation computed in Chapter 11 was used to estimate the yarn strength from the strength of the cotton fiber, ignoring the effect of fineness on the yarn. On page 283 the yarn produced from cotton with a Pressley index of 91 was estimated to have a strength of 121.49 pounds required to break a skein. The estimate taking into account both strength of the cotton fiber and its fineness was higher than when fineness was ignored. It can be verified that the average fineness of the 35 samples of cotton was 4.1 micrograms; thus, the sample with a value of 3.5 was finer than average. Since the finer cotton tends to result in stronger yarn, taking this characteristic into account results in a more accurate estimate of the strength of the yarn than if fineness had been ignored. Using only one independent variable resulted in underestimating the strength of the yarn.

Assume that a second lot of yarn was spun from cotton with a Pressley index of 91 and fineness of 5.0 micrograms. Since this cotton is considerably coarser than average, it is to be expected that the yarn spun from it would not be so strong as from the cotton with a Pressley index of 91 and fineness of 3.5 micrograms. An estimate of the strength of the yarn that would be spun from this cotton is

$$x_{1c} = 36.323 + (1.341)(91) - (9.297)(5.0)$$

$$x_{1c} = 36.323 + 122.031 - 46.485 = 111.87.$$

Again, the regression equation based on two independent variables gives a more accurate estimate of yarn strength than if fineness had been ignored. In this case, ignoring fineness in using the simple regression equation results in overestimating the strength of the yarn.

As a third example, assume that the cotton used had a Pressley index of 91 and fineness of 4.1 micrograms, the average of the 35 samples. An estimate of the strength of the yarn that would be spun from this cotton is

$$x_{1c} = 36.323 + (1.341)(91) - (9.297)(4.1)$$

$$x_{1c} = 36.323 + 122.031 - 38.118 = 120.24.$$

When the cotton is average with respect to fineness, the multiple regression equation gives an estimate very close to that given by the simple regression equation, which ignores the effect of fineness. In other words, when the variable added has a value near the average for that characteristic, it has little net effect on the estimate.

The regression equation that uses all the factors which have an influence on the dependent variable can be an extremely useful device for estimating a variable. The chief difficulty with this type of analysis has been the burden of making the calculations. As the number of variables increases, the number of equations to be solved simultaneously and the number of cross products to sum increase to the point that the arithmetic alone makes the analysis extremely burdensome. The computer is ideally adapted to this type of analysis, and with this equipment it is possible to use a large number of variables and a large sample and perform all the calculations in a matter of a few seconds.

Reliability of Estimates. The problem of determining the accuracy of estimates made from the multiple regression equation is basically the same as for estimates from a simple regression equation. Since the correlation is seldom perfect, estimates made from the regression equation will deviate from the correct value of the dependent variable. If an estimate is to be of maximum usefulness, it is necessary to have some indication of its precision. Just as with the simple regression equation, the measure of reliability is the standard deviation of the differences between the actual values of the dependent variable and the estimates made by the regression equation.

The method for determining the standard error of the estimate is to take the square root of the mean of the squared deviations of the actual values of the dependent variable from the computed values. Employing the same rules for subscripts that are used in describing the coefficients of the regression equation, the symbol $\hat{\sigma}_{1.23}$ represents the standard error of estimate. The computed values are based on the multiple relationship between x_1 and the two independent variables, x_2 and x_3. The subscripts of $\hat{\sigma}_{1.23}$ show this fact, since the subscript to the left of the decimal is the dependent variable and the subscripts to the right of the decimal represent the independent variables on which the estimate is based.

Formula 12–7 is used to compute the standard error of estimate for a multiple regression analysis.

$$\hat{\sigma}_{1.23} = \sqrt{\frac{\Sigma(x_1 - x_{1c})^2}{n - m}}. \quad \textbf{(12–7)}$$

The only differences between Formulas 11–4 and 12–7 are that the estimates x_{1c} are based on more than one independent variable, and the denominator is $n - m$ instead of $n - 2$. The letter m represents the number of constants in the regression equation. In this example there are three constants, a, $b_{12.3}$, and $b_{13.2}$. In simple linear regression analysis, there are only two constants, a and b; thus, instead of using m to represent the number of constants, the formula is stated as $n - 2$. The letter m could be used in Formula 11–4, but it is simpler to use m only when the number of constants in the regression equation may vary. In simple linear regression analysis, the denominator is always $n - 2$.

The computation of the standard error of estimate is illustrated in Table 12–5. The values of x_{1c} are calculated and entered in Column 5. The deviations from x_{1c} are computed by subtracting x_{1c} from x_1 and entering the differences in Column 6. These deviations are then squared and entered in Column 7. The sum of Column 7 is $\Sigma(x_1 - x_{1c})^2$ which is used in Formula 12–7 to compute the value of $\hat{\sigma}_{1.23}$.

The difference in accuracy between the estimating equation using one independent variable and the one using two independent variables reflects the influence of the second independent variable. The standard error of estimate for the multiple regression equation is 9.09, compared with the standard error of estimate of 10.87 when the estimates were made from the one variable in Table 11–2. It is presumed that the smaller value of the standard error of estimate obtained when two independent variables were used indicates that taking into account the variations in fineness of the cotton fiber improved the estimates. If the net influence of the additional variable were zero, it would make no difference in the estimate whether it was included or not. But as long as there is any net correlation between an independent variable and the dependent variable, adding the independent variable will improve the estimate. This subject will be developed further in the section on partial correlation.

Formula 11–5 was used to find the standard error of estimate for a simple linear correlation without computing the estimates of the individual values of

TABLE 12–5

THE STANDARD ERROR OF ESTIMATE COMPUTATION OF STRENGTH OF YARN, STRENGTH OF COTTON FIBER, AND FINENESS OF COTTON FIBER

Sample Number (1)	Yarn Strength x_1 (2)	Strength of Cotton Fiber x_2 (3)	Fineness of Cotton Fiber x_3 (4)	x_{1c} (5)	$x_1 - x_{1c}$ (6)	$(x_1 - x_{1c})^2$ (7)
1	129	95	3.9	127.43	1.57	2.4649
2	117	91	4.3	118.35	− 1.35	1.8225
3	127	88	4.3	114.32	12.68	160.7824
4	129	83	4.2	108.55	20.45	418.2025
5	128	93	4.0	123.82	4.18	17.4724
6	118	84	4.4	108.03	9.97	99.4009
7	121	81	4.3	104.94	16.06	257.9236
8	112	79	4.1	104.12	7.88	62.0944
9	126	84	4.3	108.96	17.04	290.3616
10	128	89	3.8	120.31	7.69	59.1361
11	125	92	3.6	126.19	− 1.19	1.4161
12	115	87	3.0	125.07	−10.07	101.4049
13	114	87	3.4	121.35	− 7.35	54.0225
14	116	86	3.5	119.08	− 3.08	9.4864
15	112	86	2.9	124.66	−12.66	160.2756
16	113	89	3.7	121.24	− 8.24	67.8976
17	120	87	3.4	121.35	− 1.35	1.8225
18	92	78	4.3	100.92	− 8.92	79.5664
19	96	76	4.2	99.17	− 3.17	10.0489
20	84	81	5.0	98.43	−14.43	208.2249
21	114	92	4.6	116.90	− 2.90	8.4100
22	112	89	4.6	112.88	− .88	.7744
23	99	83	4.6	104.83	− 5.83	33.9889
24	76	74	5.1	88.12	−12.12	146.8944
25	87	89	6.0	99.86	−12.86	165.3796
26	121	85	3.8	114.95	6.05	36.6025
27	115	85	3.5	117.74	− 2.74	7.5076
28	103	80	4.6	100.81	2.19	4.7961
29	119	93	4.0	123.82	− 4.82	23.2324
30	119	86	3.5	119.08	− .08	.0064
31	118	86	4.6	108.85	9.15	83.7225
32	114	85	3.3	119.60	− 5.60	31.3600
33	112	80	3.5	111.04	.96	.9216
34	118	85	4.1	112.16	5.84	34.1056
35	98	85	5.4	100.08	− 2.08	4.3264
Total	3,947	2,993	143.8	3,947.01	− .01	2,645.8555

$$\hat{\sigma}_{1.23} = \sqrt{\frac{\Sigma(x_1 - x_{1c})^2}{n - m}} = \sqrt{\frac{2{,}645.8555}{35 - 3}} = \sqrt{82.6830} = 9.09$$

the dependent variable. In the same way Formula 12–8 may be used to compute the standard error of estimate for a multiple correlation.

$$\hat{\sigma}_{1.23} = \sqrt{\frac{\Sigma x_1^2 - a\Sigma x_1 - b_{12.3}\Sigma x_1 x_2 - b_{13.2}\Sigma x_1 x_3}{n - m}}. \qquad \textbf{(12–8)}$$

All the values needed for computing the standard error of estimate from this formula are available from previous calculations. It is important to carry out the values of the constants of the regression equation to a large number of decimal places to avoid introducing substantial rounding error into the calculations. Substituting in Formula 12–8, $\hat{\sigma}_{1.23}$ is found to be 9.12. Due to differences in rounding, this value varies slightly from that found in the calculations performed by Formula 12–7.

$$\hat{\sigma}_{1.23} = \sqrt{\frac{451{,}079 - (36.3230)(3{,}947) - (1.3406)(338{,}829) - (-9.2969)(16{,}048.5)}{35 - 3}}$$

$$\hat{\sigma}_{1.23} = \sqrt{\frac{451{,}079 - 143{,}367.2288 - 454{,}255.0360 + 149{,}202.8725}{32}}$$

$$\hat{\sigma}_{1.23} = \sqrt{\frac{2{,}659.6077}{32}} = \sqrt{83.1127} = 9.12.$$

Abstract Measures of Multiple Correlation

The coefficient of determination may be computed for multiple correlation by using Formula 11–8 for simple correlation, with the exception that the value of $\hat{\sigma}^2_{1.23}$ is substituted for the value of $\hat{\sigma}^2_{yx}$.

$$\hat{\rho}^2_{1.23} = 1 - \frac{\hat{\sigma}^2_{1.23}}{\hat{\sigma}^2_1} \qquad \textbf{(12–9)}$$

$$\hat{\rho}^2_{1.23} = 1 - \frac{83.1127}{175.5912} = 1 - .4733$$

$$\hat{\rho}^2_{1.23} = .5267.$$

Therefore, $\hat{\rho}_{1.23} = \sqrt{.5267} = .726.$

The subscripts indicate that the coefficient measures the percentage of the total variance in variable x_1 that is explained by the combined variations in variables x_2 and x_3. The subscripts are written as those of the standard error of estimate are, with the subscript at the left of the decimal representing the dependent variable, and those at the right, the independent variables.

The significance of the coefficient of multiple determination is the same as that of the coefficient of determination for a two-variable correlation. It represents the proportion of the total variance in the dependent variable that is explained by variations in the independent variables. The only difference is that more than one variable is used instead of a single independent variable to explain the variance in the dependent variable.

Significance of a Coefficient of Multiple Correlation

The F ratio may be used to test whether a multiple correlation is significantly different from zero by comparing the variation explained by the regression line with the unexplained variance. This is similar to testing simple linear correlation, except for the fact that the variance explained by the regression equation uses 2 independent variables instead of 1. This means that 2 degrees of freedom are assigned to the variance explained by the regression line, leaving only 32 instead of 33 to be assigned to the unexplained variance. Table 12–6 shows the distribution of the total sum of squared deviations, 5,970.0, to the explained and unexplained variance. Dividing the sum of the squared deviations by the related degrees of freedom gives the variance used to compute the F ratio. The F value of 19.92 is found to be significant, since it far exceeds the value of $F_{.01}$ with 2 and 32 degrees of freedom.

TABLE 12–6

TESTING SIGNIFICANCE OF A MULTIPLE CORRELATION BY THE *F* RATIO

Type of Variance	Sum of Squared Deviations	Degrees of Freedom	Variance (Estimated)
Explained by regression	3,310.4	2	1,655.20
Unexplained	2,659.6	32	83.11
Total	5,970.0	34	175.59

$$F = \frac{1,655.20}{83.11} = 19.92 \qquad F_{.01} = 5.34$$

Comparing Regression Models

Almost all multiple regression and correlation problems must be solved by the use of a digital computer. Excellent simple, nonlinear, and multiple regression programs are available on most computer systems so the analyst need not write his own computer programs.

When computer programs are used, it is possible to experiment with many independent variables and to select only those that can make significant explanations of the variations in the dependent variable not already explained by other independent variables. Under these circumstances it is necessary to know whether or not an independent variable contributes significantly to $\hat{\rho}^2$.

One way to answer this question is to compare the value of $\hat{\rho}_A^2$ for the larger model (Model A), which contains the variable under study, with the value of $\hat{\rho}_B^2$ of the smaller model (Model B), which is identical to the larger model except that it does not include the variable in question. The test can be made using the F distribution and Formula 12–10.

$$F = \frac{\dfrac{\hat{\rho}_A^2 - \hat{\rho}_B^2}{k_A - k_B}}{\dfrac{1 - \hat{\rho}_A^2}{n - k_A}} \qquad (12\text{–}10)$$

where

k_A = the number of variables in Model A

k_B = the number of variables in Model B

n = the number of sets of observations in the study.

For example, suppose that in the previous problem a new variable x_4 was introduced. This new variable might be a measure of humidity present in the spinning room at the time that the yarn is spun. In addition, suppose that $\hat{\rho}^2_{1.234} = .6420$. This value is higher than that for $\hat{\rho}^2_{1.23}$ of .5267 computed for three variables. To determine if the increase in the coefficient of determination is significant, Formula 12–9 might be used.

$$F = \frac{\dfrac{.6420 - .5267}{4 - 3}}{\dfrac{1 - .6420}{35 - 4}} = 9.98.$$

$F_{.01}$ for 1 and 31 degrees of freedom is approximately 7.53, so the addition of the variable x_4 to the regression equation would significantly increase the value of $\hat{\rho}^2$.

PARTIAL CORRELATION

Earlier in the chapter it was stated that the b coefficients of the regression equation measure the net influence of each independent variable on the estimate of the variable x_1. This section will describe a method of measuring the *net correlation* or *partial correlation* between one independent variable and the dependent variable, eliminating the relationship with the other independent variables in the study. This measure of relationship is known as the *coefficient of partial correlation* and is represented by the symbol $\rho_{13.2}$. The primary and secondary subscripts are employed with the same meanings given them previously. The subscripts to the left of the decimal point represent the variables for which the net correlation is being computed; the subscript to the right of the decimal point represents the variable that has been eliminated.

The simple linear correlation between yarn strength and cotton fiber strength gave the regression equation

$$x_{1c} = -23.106 + 1.589x_2.$$

The reliability of estimates from this equation is given by the value of the standard error of estimate. For estimates of yarn strength made from the equation above, $\hat{\sigma}^2_{1\bar{2}} = 118.1313$. The value of $\hat{\sigma}^2_{1\bar{2}}$ represents the variance of variable x_1 that is *not explained* by variable x_2. The standard error of estimate for the multiple regression equation used to estimate yarn strength (x_1) from variables x_2 and x_3 was computed earlier in the chapter and it was found that $\sigma^2_{1.23} =$ 83.1127.

The value of $\sigma^2_{1.23}$ represents the variance in variable x_1 unexplained after variables x_2 and x_3 have both been used in making the estimate of x_1. The reduction in the unexplained variance accomplished by adding variable x_3 measures the net correlation of variables x_1 and x_3 when the influence of x_2 on both variables x_1 and x_3 is taken into account. The unexplained variance was 118.1313 when only independent variable x_2 was considered. When x_3 was added, the unexplained variance dropped to 83.1127. The coefficient of partial correlation is computed from these values.

$$\hat{\rho}^2_{13.2} = 1 - \frac{\hat{\sigma}^2_{1.23}}{\hat{\sigma}^2_{12}} \tag{12-11}$$

$$\hat{\rho}^2_{13.2} = 1 - \frac{83.1127}{118.1313} = .2964$$

$$\hat{\rho}_{13.2} = -.544.$$

The coefficient of net correlation ($\hat{\rho}_{13.2}$) takes the sign of the b coefficient equation, which is minus for variable x_3.

The t distribution may be used to test the significance of the coefficient of partial correlation by reducing the degrees of freedom by the number of variables eliminated. When the number of degrees of freedom have been reduced in this manner, Formula 11–16 may be used to test the significance of the coefficient of partial correlation in the same manner as done for the value of r. With a coefficient of partial correlation of $-.544$, eliminating the effect of one variable, the value of t is found to be -3.667.

$$t = -.544\sqrt{\frac{35-3}{1-.544^2}} = -.544\sqrt{\frac{32}{.704064}} = -.544\sqrt{45.4504}$$

$$t = (-.544)(6.74) = -3.667.$$

For 10 degrees of freedom the value of t is 3.169, so the hypothesis that the correlation is zero must be rejected.

STUDY QUESTIONS

12–1. What is meant by the statement that the relationship between two variables is nonlinear?

12–2. How can you determine whether the relationship between two variables is direct or inverse when it is nonlinear?

12–3. Why is the value of I_{yx}, which was computed on page 317, larger than the value of ρ computed from the same data? Is it to be expected that the value of I_{yx} will generally be larger than ρ?

12–4. Why is it incorrect, for use as a measure of average relationship in a correlation problem, to draw a very irregular line that passes through most of the plotted points?

12–5. Why is simple correlation not adequate for all situations in which correlation analysis techniques may be used?

12–6. In measuring the relationship between yarn strength and the strength of the cotton fiber, it was found that $b = 1.589$ (page 283). In the analysis of multiple correlation, it was found that $b_{12.3} = 1.341$ (page 323). Explain the significance of these two coefficients, pointing out how they differ.

12–7. What is the difference between $\hat{\rho}_{12}$, $\hat{\rho}_{1.23}$, and $\hat{\rho}_{12.3}$?

12–8. What does the difference between $\hat{\sigma}_{12}$ (or $\hat{\sigma}_y$) and $\hat{\sigma}_{1.23}$ measure?

12–9. Give an example (other than given in the text) to illustrate how a measure of the relationship between three or more variables may be used to estimate or forecast one of the variables.

12–10. Explain how it is possible to determine whether an increase in the coefficient of determination which results from the addition of one more independent variable to the regression equation is a significant increase.

12–11. Explain why three degrees of freedom are lost in computing the standard error of estimate for the nonlinear regression equation on page 315.

12–12. Explain what the coefficient of partial correlation is designed to measure.

PROBLEMS

12–1. (a) Using the data in Problem 11–6, compute and draw a nonlinear regression line on the chart prepared in Problem 11–6. How does this curve compare with the straight line fitted by the method of least squares in Problem 11–7?

(b) Read the values of y_c from the regression curve, and from these values compute the standard error of estimate.

(c) Do you consider the straight line or the curve the better measure of average relationship? Estimate the price of strawberries in the central market on days when the receipts were 2,700, 3,200, 4,200, and 4,800 pounds. How do these estimates compare in accuracy with the estimates made in Problem 11–7?

12–2. Compute the index of correlation (I_{yx}) for the data in Problem 11–6. How does this value compare with the value of the coefficient of correlation computed in Problem 11–14?

12–3. The foreman of a production unit is interested in the relationship between the number of days a new employee has been on the job and the number of defective units of output. On a particular day he selects a random sample of 10 new employees and records the number of defective units produced by each. The results are shown in the table at the top of page 332. Treating the number of days on the job as the independent variable, perform the following:

(a) Make a scatter diagram of the observations and compute a nonlinear regression line and plot it on the scatter diagram. Does your curved line appear to fit the data better than would a straight line?

(b) Read the values of y_c from the regression line and use these values to compute the standard error of estimate.

(c) Use a regression line to estimate the average number of defective units in a day which might be expected from an employee who had been on the job only 5 days.

(d) Compute the index of correlation using Formula 12–4. Would you expect this to be higher or lower than r if you had computed r using the product-moment method? Why?

Employee	Number of Defective Units	Days on the Job
A	50	3
B	78	2
C	13	6
D	20	5
E	3	9
F	4	10
G	96	1
H	25	5
I	5	8
J	57	3
Total	351	52

12–4. The following table gives the results of a study made of a sample of 10 employees of a manufacturing plant. The number of units of product turned out in one hour was recorded for each employee in the sample and designated x_1. The score on an aptitude test and the number of years of experience of each employee were also recorded and designated x_2 and x_3 respectively. The summations and cross products that will be needed are given in the table.

Compute the multiple regression equation for estimating output of an employee from a knowledge of his score on the aptitude test and years of experience.

Employee	Output (Units) x_1	Test Score x_2	Experience (Years) x_3
1	32	160	5.5
2	15	80	6.0
3	30	112	9.5
4	34	185	5.0
5	35	152	8.0
6	10	90	3.0
7	39	170	9.0
8	26	140	5.0
9	11	115	.5
10	23	150	1.5
Total	255	1,354	53.0

$\Sigma x_1^2 = 7{,}477$ $\quad \Sigma x_2^2 = 194{,}198$ $\quad \Sigma x_3^2 = 363.00$

$\Sigma x_1 x_2 = 37{,}175$ $\quad \Sigma x_1 x_3 = 1{,}552.0$ $\quad \Sigma x_2 x_3 = 7{,}347.5$

12–5. Using the multiple regression equation computed in Problem 12–4, estimate the hourly output for each of the employees in the table at the top of page 333.

Employee	Test Score	Experience
A	120	3.0
B	175	3.0
C	135	3.0
D	110	8.5
E	125	8.5
F	190	8.5
G	120	5.0
H	140	5.0
I	190	5.0

12–6. For the regression analysis in Problem 12–4, compute the standard error of estimate. What does this show about the accuracy of the estimates made in Problem 12–5?

12–7. Compute the coefficients of partial correlation, $\hat{\rho}_{12.3}$ and $\hat{\rho}_{13.2}$, for the data in Problem 12–4. Explain the significance of these two measures.

12–8. The table below shows for a random sample of 12 statistics students the final course grade in the beginning statistics course, the score on a mathematics aptitude test, and the course load carried by each student for that semester.

(a) Make a scatter diagram of the first two variables, x_1 and x_2. Draw in what you consider to be a reasonable regression line. This should be a straight line and can be done by observation.

(b) Looking at the diagram prepared in (a), does it appear that the third variable, x_3, will offer any explanation of the scatter of the values above and below the regression line?

(c) Compute a regression equation for all three variables using the summations below.

(d) If a student makes 190 on the mathematics aptitude test and is carrying a 17-hour course load, what would you predict his grade in statistics to be?

(e) Compute the standard error of estimate.

(f) Compute the coefficient of determination ($\hat{\rho}^2_{1.23}$) and use F to determine whether or not the value is significant.

Student	Course Grade x_1	Aptitude Test Score x_2	Course Load (Semester Hours) x_3
1	70	170	14
2	48	144	15
3	70	194	18
4	95	216	13
5	65	150	12
6	90	230	17
7	71	212	19
8	38	128	17
9	80	184	13
10	45	120	14
11	55	160	16
12	99	190	12
Total	826	2,098	180

$\Sigma x_1^2 = 61{,}190$ $\Sigma x_2^2 = 380{,}612$ $\Sigma x_3^2 = 2{,}762$

$\Sigma x_1 x_2 = 151{,}008$ $\Sigma x_1 x_3 = 12{,}238$ $\Sigma x_1 x_3 = 31{,}666$

12–9. Assume that 35 observations of three variables (x_1, x_2, and x_3) produce the following two variances:

$$\hat{\sigma}^2_{1.23} = 23.7 \qquad \hat{\sigma}^2_{12} = 65.2.$$

Compute the partial correlation coefficient, $\hat{\rho}_{13.2}$, and explain what this means.

12–10. This problem is based upon the solutions to Problems 11–22 and 11–23.

(a) Compute and then draw a nonlinear regression line on the chart prepared in Problem 11–22. A curve should fit the data better than the straight line computed by the method of least squares in Problem 11–22.

(b) Compute the values of y_c, and from these values compute the standard error of estimate.

(c) Compute the value of the index of correlation (I_{yx}). How does this value compare with the value of $\hat{\rho}$ computed in Problem 11–23.

(d) Do you consider the linear or the nonlinear regression the better device for forecasting the price of watermelons?

(e) Use the regression line to forecast the price of watermelons on a day that has 100 railroad cars on track at 8:00 A.M.

12–11. The following table shows data on four variables for a random sample of 15 observations drawn from a large universe.

Employee Number	Salary (Dollars per Month) x_1	Age (Years) x_2	Education (Years Completed) x_3	Sick Leave (Average per Year) x_4
1	627	27	12	2
2	825	33	13	8
3	540	41	7	0
4	430	19	12	3
5	500	52	6	12
6	2,400	61	16	35
7	450	53	7	0
8	750	29	14	5
9	675	30	12	20
10	900	25	16	15
11	1,500	33	16	0
12	1,200	35	12	4
13	950	22	14	15
14	770	23	15	0
15	650	36	12	6

a. Compute the regression equation and the coefficient of determination using the first three variables. Is the coefficient of determination significant? Explain.

b. Recompute the regression equation and the coefficient of determination using all four variables. Is the coefficient of determination significant?

c. Compare the two coefficients of determination to determine whether the coefficient for the larger model is significantly greater than that for the smaller model.

13 Time Series

Classifying quantitative data on the basis of time periods represents the most efficient method of describing the changes that are constantly taking place in the business situation. The analysis of both internal data and external data, which were discussed in Chapter 2, will be examined in this section.

The basic use of statistical data classified as *time series* is to measure the changes that have taken place in the past. However, the most important reason for keeping this record of the past is the information it yields concerning future changes. Management requires a continuous forecast of both the events outside the business and the course of the business itself. In general, the more precise the record of the past is, the better the basis for a forecast of the future.

A simple example of the use of internal and external data can be seen in the production of raw steel by the United States Steel Corporation and the comparison of these production figures with total raw steel production in the United States. The data classified on a yearly basis would be appropriate for the analysis of change over a considerable period of time. Classification on a monthly or quarterly basis would be needed to analyze short-term fluctuations. Table 13–1 shows the total production of raw steel in the United States from 1956 to 1972. Although there was a steady upward movement in production during this period, it is evident that there were rather pronounced fluctuations from year to year instead of a steady growth. In addition to total steel production in the United States, the table shows the production of the United States Steel Corporation. By comparing the changes in the two time series, a conclusion can be made regarding the ability of the United States Steel Corporation to compete effectively with other steel companies. It appears that the company's trend has been slightly downward compared to an upward trend in the total industry.

The last column in the table shows the percentage of total steel production in the United States that was produced by the company. This simple analysis using percentages plots the relative position of the company in the industry

TABLE 13–1

TOTAL RAW STEEL PRODUCTION IN THE UNITED STATES AND PRODUCTION BY UNITED STATES STEEL CORPORATION, 1956 TO 1972

Year	Total U.S. Production (Million Tons)	United States Steel Corporation	
		Million Tons	Percentage of U.S. Total
1956	115.2	33.4	29.0
1957	112.7	33.7	29.9
1958	85.3	23.8	27.9
1959	93.4	24.4	26.1
1960	99.3	27.3	27.5
1961	98.0	25.2	25.7
1962	98.3	25.4	25.8
1963	109.3	27.6	25.3
1964	127.1	32.4	25.5
1965	131.5	32.6	24.8
1966	134.1	32.8	24.5
1967	127.2	30.9	24.3
1968	131.5	32.4	24.6
1969	141.3	34.7	24.6
1970	131.5	31.4	23.9
1971	120.4	27.2	22.6
1972	133.1	30.7	23.1

Source: American Iron and Steel Institute, *Annual Statistical Report,* and annual reports of United States Steel Corporation.

over a period of years. There has been a rather steady downward trend in its percentage of the total business since 1957.

Figure 13–1 shows the total production of raw steel in the United States and Figure 13–2 shows the production of raw steel by United States Steel Corporation. It can be seen from these charts that production in the United States has had a somewhat sharper upward trend, but that the fluctuations about the trend line have been about the same for the two series. Figure 13–3 shows graphically the ratio of production of United States Steel Corporation to total production in the United States. The downward drift of this ratio clearly shows that United States Steel Corporation gradually produced a reduced proportion of the steel produced in the United States.

TYPES OF FLUCTUATION

The changes that occur in statistical series classified by periods of time are usually grouped into the following four categories, each of which represents a well-defined type of economic change.

1. Secular trend
2. Seasonal variation
3. Cyclical fluctuations
4. Random or erratic fluctuations.

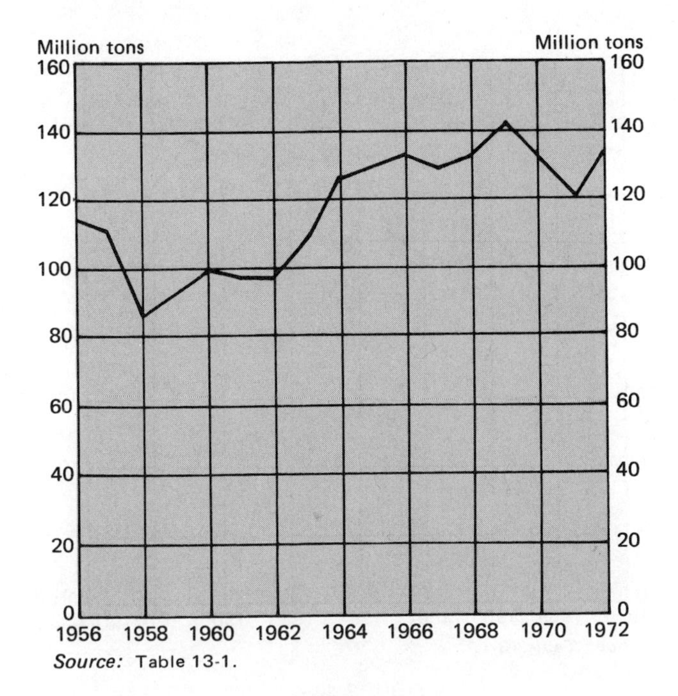

FIGURE 13–1

PRODUCTION OF RAW STEEL IN THE UNITED STATES 1956 TO 1972

Secular trend is long-term movement which reflects the effect of forces that constitute gradual growth or decline. These forces operate over a long period of time and are not subject to sudden reversals in direction. *Seasonal variation* is made up of periodic movements throughout a year. Other types of periodic fluctuations, such as variation throughout a day, represent the same kind of change and are analyzed by the same methods used for annual fluctuations. *Cyclical fluctuations* are recurring changes that do not necessarily occur in a fixed period. This fluctuation is distinguished from seasonal variation by the fact that it does not have a fixed period, although it is a recurring type of change. This phenomenon is the *business cycle.* Intermingled with these three well-defined types of variation are innumerable small variations that are essentially random in nature, resulting from a large number of factors, most of which are relatively unimportant when considered singly. These fluctuations are *random* or *erratic fluctuations.*

The nature of these different forces that interact on data expressed in terms of time units is important to understand, since the subject of time series will be developed around this classification. Methods of measuring each type of fluctuation will be described, and it will be shown how the forecasting of a time series consists of forecasts of the separate types of fluctuation. In the following discussion, examples of data classified on the basis of time are shown

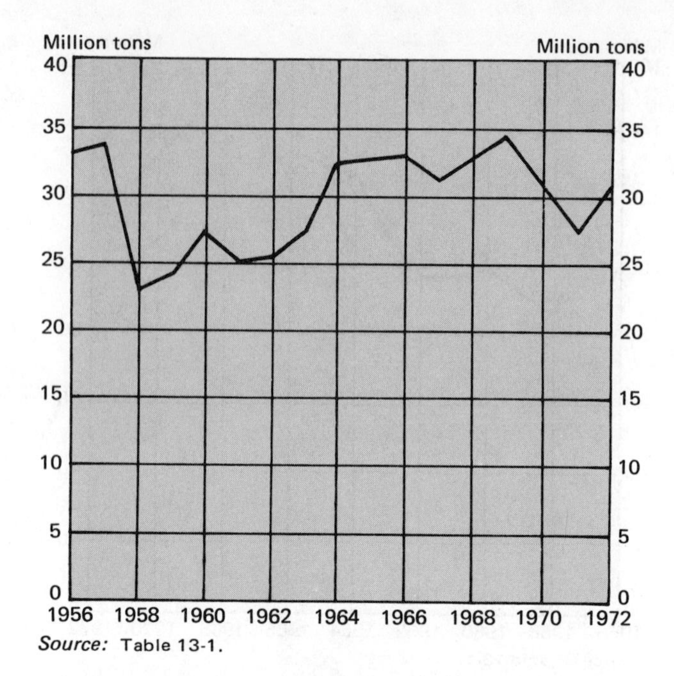

Source: Table 13-1.

FIGURE 13–2

PRODUCTION OF RAW STEEL BY UNITED STATES STEEL CORPORATION, 1956 TO 1972

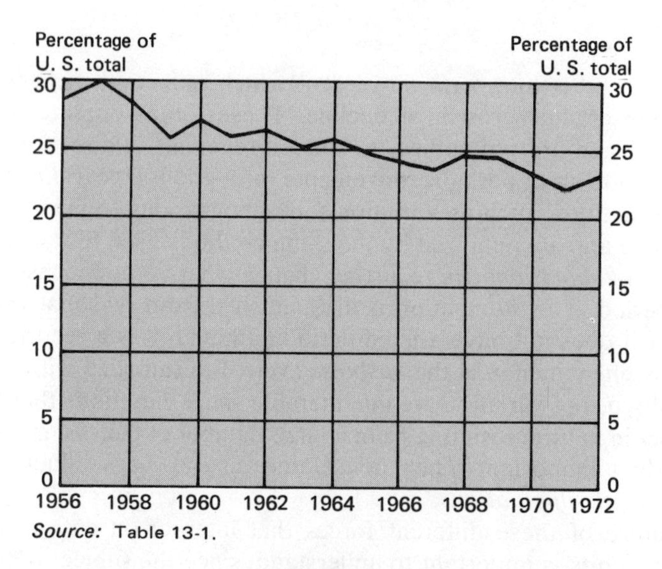

Source: Table 13-1.

FIGURE 13–3

RATIO OF PRODUCTION OF RAW STEEL BY UNITED STATES STEEL CORPORATION TO TOTAL RAW STEEL PRODUCTION IN THE UNITED STATES 1956 TO 1972

in chart form. Since economic change occurs constantly, businessmen use large numbers of charts of time series to keep informed of changes that are taking place in their own firms and in the economic environment outside their businesses.

Secular Trend

Secular trend, or the growth of business and industry, is based primarily on a growing population, and the rate of growth in the total economy reflects directly the growth of the population. Unless a decrease in the standard of living is accepted, a growing population offers a market for an increasing amount of all kinds of commodities. Throughout the history of the United States the growth of population has been accompanied by an increasing per capita production of goods and services, with the result that the growth of total industry has been faster than population growth.

The growth of total industry means that industries and individual business concerns have grown, although the rate of growth has shown wide variations among industries. Synthetic fibers and other materials, household equipment, automobiles, and radio and television are typical examples of industries growing at very rapid rates. Some new industries replace existing industries, while others add their output to the total without actually replacing other products. Natural gas and petroleum have displaced a large amount of coal as a source of energy; motor vehicles have replaced horse-drawn vehicles; and synthetic fibers have replaced a major portion of the silk consumed and are making strong inroads on many uses of cotton.

Figure 13–4 shows factory sales of electronic products from 1953 to 1970, and illustrates the rapid expansion of the sales of a new product.

Figure 13–5 on page 341 shows the monthly sales of passenger cars by factories. Since these data are plotted monthly, the seasonal movements are more pronounced than the secular trend, but it can be seen that a definite upward trend has been present. This chart emphasizes the fact that several different types of fluctuations may occur in the same series.

The changing character of modern industry makes it inevitable that some business units and industries supplying goods and services should grow and others decline. As long as inventions produce new products or new technical methods of producing products, it is to be expected that some industries and businesses will continue to grow. And as long as consuming habits change, it may be expected that certain industries will decline. Any analysis of time series must take this type of change into consideration.

Secular trend is present in price as well as in physical volume series. Prices of commodities tend to decline as cheaper methods of production are developed. On the other hand, if the production of goods depends upon a supply of raw materials that is gradually being exhausted, the price may show a secular trend upward. The long-term movements of the general price level

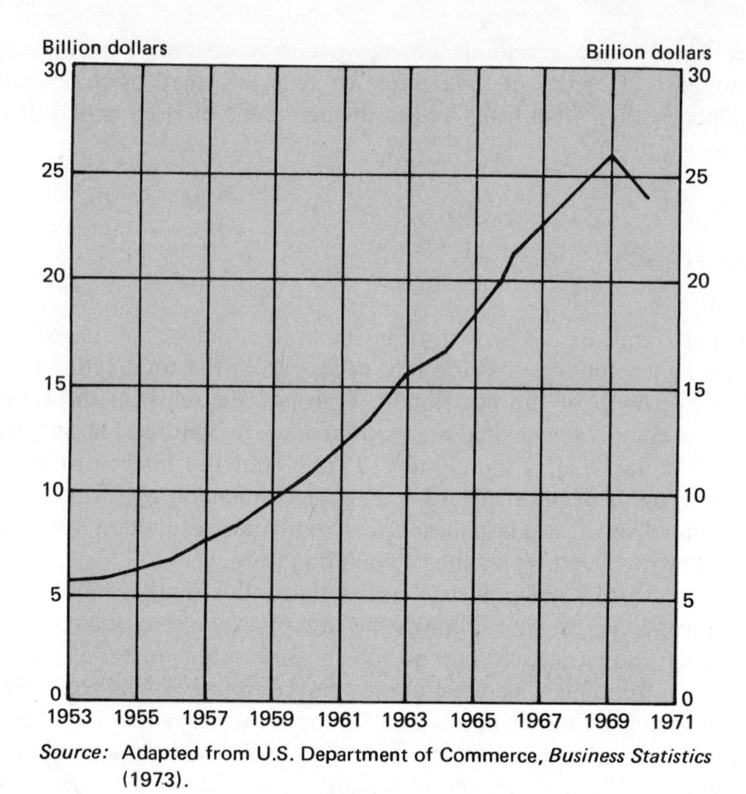

Source: Adapted from U.S. Department of Commerce, *Business Statistics* (1973).

FIGURE 13–4

FACTORY SALES OF ELECTRONIC PRODUCTS, 1953 TO 1970

take on the characteristics of secular trend, although these movements have many of the characteristics of long cyclical fluctuations.

Seasonal Variation

The changes in business and economic activities that result from the changing seasons are true periodicities, and are reflected in changes in industrial production, agricultural production, consumer buying, and prices. Prices and stocks of farm products are influenced by the fact that production is geared to a yearly cycle, while consumption is distributed throughout the year. Consumer needs vary widely from one season to the next. Antifreeze for automobile radiators is needed in the winter, although this pattern is changing as more cars require antifreeze in summer to use as a coolant. The seasonal pattern of the consumption of electricity formerly showed a peak in the winter when the days were short and lights were burned more hours, but the consumption of electricity for air conditioning in the summer has pushed the seasonal peak to the hot months of the year.

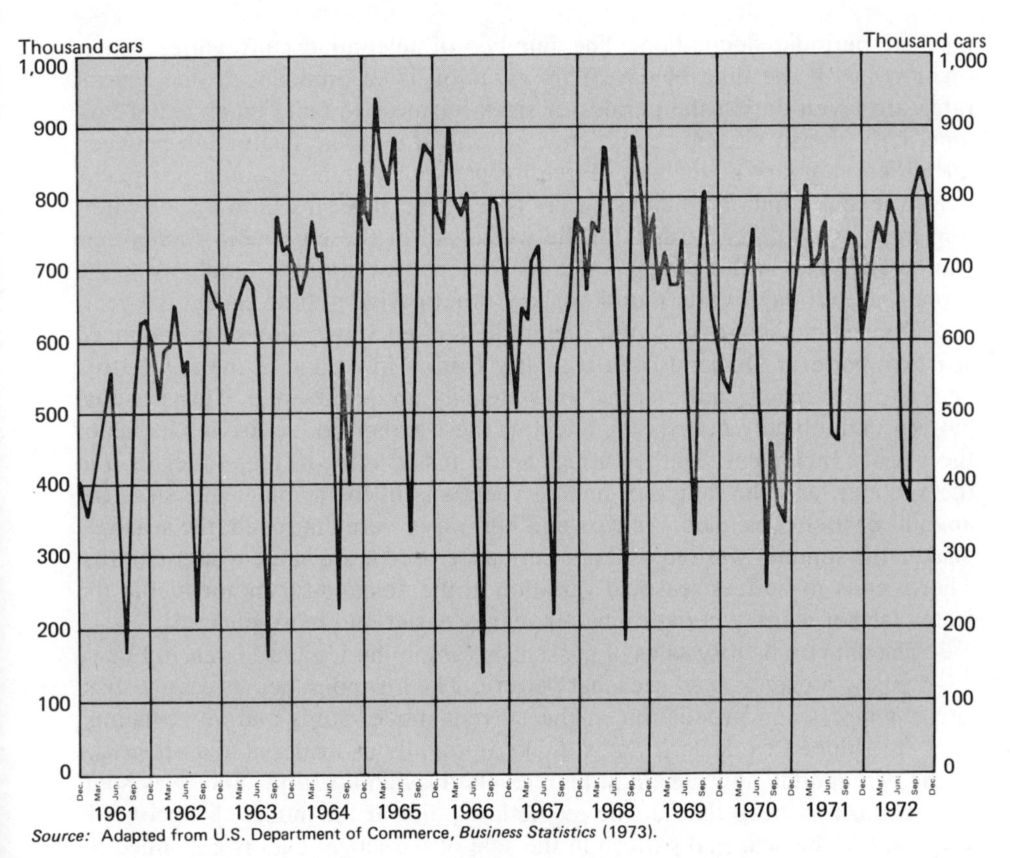

Source: Adapted from U.S. Department of Commerce, *Business Statistics* (1973).

FIGURE 13–5

FACTORY SALES OF PASSENGER CARS IN THE UNITED STATES, 1961 TO 1972

Social customs dictate the time of year many articles are purchased, and distinctive seasonal patterns, independent of changes in the weather, are found. Christmas gifts swell the December retail trade figures; June commencements and weddings increase the business in certain commodities that are suitable for gifts.

Periodic fluctuations in consumer purchasing, of course, cause sales of retailers to be distributed unevenly throughout the year, which in turn affects the regularity of sales of wholesalers and manufacturers. Since many manufacturers produce only when the sale of goods is assured, this bunching of orders causes the rate of production to fluctuate and follow a regular pattern. This seasonal pattern is then reflected in the purchase of materials and supplies and in the employment of labor.

There are other periodicities in business besides those that fluctuate with the seasons; in fact, seasonal variation may be thought of as merely one special type of periodic fluctuation. The number of telephone calls varies rather regularly with the time of day. This variation is so pronounced that special rates are given during the periods of slack business to build up a greater volume. City transit systems experience a decided periodic fluctuation between the different hours of the day. Intercity passenger transportation has peaks over weekends and on holidays. Many types of retail stores show a wide variation in sales on different days of the week. All of these periodic fluctuations can be analyzed in the same manner as seasonal variation by simply using the proper period: day, week, month, or any other period of time except the year.

Domestic demand for motor fuel in the United States shows a pronounced seasonal pattern. Demand rises regularly from a low point in the winter to a peak in midsummer, and then declines steadily until midwinter. This seasonal pattern undoubtedly reflects the effect of the weather on the driving habits of the public. Inclement weather brings about less driving in the winter than in the summer, and the taking of annual vacations in the summer increases the amount of fuel consumed. As cars and highways were improved, the seasonal peak in the summer was reduced and cars were used more in the winter months. There tends to be less seasonal variation in the demand for motor fuel in the states farther south where driving all year is easier and more pleasant.

The data on factory sales of passenger cars in the United States, in Figure 13–5, show a pronounced seasonal pattern. The low point comes each year in late summer when production of the current model stops and all remaining cars are shipped to dealers. Sales then rise rapidly as dealers build up stocks of the new model, and remain at high levels throughout spring and early summer. From this peak the decline to the low point of the model changeover is very rapid. The seasonal pattern in the sale of passenger cars is explained by two factors: (1) the annual introduction of new models tends to produce a regular seasonal fluctuation in sales; (2) there is a pronounced tendency for consumers to buy durable goods just prior to the period of greatest use of these goods, and factory sales of passenger cars reflect this variation in consumer demand. The combined effects of the introduction of new models and the seasonal behavior of consumers produce the seasonal pattern in factory sales.

Cyclical Fluctuations

In many respects the cyclical fluctuations of time series represent the most important type of variation because they do not occur with a fixed period. This makes them much more difficult to anticipate than variations that occur with the seasons, such as seasonal variation.

The cyclical variations in individual series are related to the cycles in the total economy. Business cycles have been described by Burns and Mitchell as successive waves of expansion and contraction that occur at about the same time in many economic activities.[1] Expansion is followed by a recession,

---

[1] Arthur F. Burns and Wesley C. Mitchell, *Measuring Business Cycles* (New York: National Bureau of Economic Research, 1946), p. 3.

which gradually merges into a revival and expansion, with the different phases following each other in regular fashion. These phases are recurrent—but not periodic, as in the case of seasonal variation. The length of a complete cycle varies from a year to approximately 12 years, and the amplitude of the swings varies from one cycle to another.

The cycle in total economic activity may be thought of as the average of the cycles in individual segments of the economy; however, it is very common to consider the cyclical fluctuations in total business activity as a phenomenon distinct from the cycles in the individual series. Regardless of one's theory about the relationship of the individual components of economic activity to the overall level of activity, it is important to measure the cyclical patterns that prevail in the individual series. After this has been done, the pattern in an individual series can be compared with the cycle in total activity and with other individual series. Although there is a considerable degree of relationship between the cyclical fluctuations in individual series, the measurement of the difference in behavior of the individual series is significant. Some series lead at the turning points, such as the peaks and valleys of the business cycle, while others lag. Some series show wide fluctuations through a complete cycle, while others show a very mild variation. The study of specific industries and individual business concerns gives management needed information about the cyclical behavior of their production, sales, prices, and other important areas.

Erratic Fluctuations

In nearly every business organization, sales vary from one day to another because of purely chance factors. In a retail store the weather on any particular day makes a difference in the amount of business done. In a concern where sales are in large units, some days will show no sales and others will show very large sales, for no other reason than that the orders happened to be shipped and billed on one day rather than another.

When data on the amount of sales or production are classified by days, the figures for different days would be expected to show the effect of these irregularities. If the days are added together to make a weekly classification, there would be some tendency for these irregularities to smooth themselves out. The longer the period of time, the greater this tendency would be.

As businessmen have tried to keep closer control over business, they have made use of figures classified on smaller and smaller intervals of time. Yearly data are now considered inadequate. Annual and quarterly data are no longer considered the best possible division, and more data reporting is done on a monthly, weekly, or even daily basis. *The shorter the interval of time on which the data in the form of totals are classified, the more tendency there is for erratic fluctuations to influence the series.*

Some types of data are more affected by these erratic forces than others. In general, the statistical series for an individual business are likely to be more irregular than for an industry as a whole. When a large number of individual business units combine their data, there is a tendency for individual irregularities to cancel out.

In Chapter 10 random variations were assumed to result from a combination of circumstances that cause slight differences in the individual units coming out of a specific production process. When these differences result from a multiplicity of causes, each one of which has only a slight effect on the final result, they are considered to be random. In the analysis of fluctuations in time series described in this chapter the term "erratic" is more commonly used, but the meaning is essentially the same as the term "random variations" in Chapter 10.

Occasionally an irregular fluctuation may be important enough to be given special study, as when retail sales figures in a particular community show the effect of a convention held there or production figures of a company show the effect of a strike. Unless the irregularity is large enough to be especially significant, it is ordinarily pointless to try to find the specific causes, since the causes may not be repeated with a uniformity that aids forecasting. The irregular, "saw-toothed" effect seen in many time series classified on a monthly or weekly basis usually results from the use of a time unit that is so small that the irregular forces affecting the data are not given an opportunity to cancel out. When it is desirable to use a small time unit, such as a week or a month, the irregularities in the data must be ignored, or some simple method of smoothing the data should be used.

TIME SERIES ANALYSIS

The basic purpose of time series analysis is to give the businessman a convenient method of measuring changes in his business over a period of time and relating these changes to those in the economy. The time series that measure changes in one's own business are supplied by the internal records of the company, while information on changes in the whole industry and in business in general will come from various external sources.

Time series generally require fairly extensive analysis before being of maximum usefulness to the business executive, since they may be a composite of any one or all of the four types of fluctuations described previously. The special methods of time series analysis will be given detailed treatment in the following chapters.

Problems of Time Series Analysis

The main problem in time series analysis is the isolation of the different types of fluctuations, particularly trend, seasonal, and cyclical. Data supplied by records of the various aspects of business activity show the composite change in activity; but since several factors are at work, the data show only the net results of these factors. For example, the volume of production of an industry during a certain period might show a net increase that was the result of a sharp cyclical decline combined with a regular seasonal rise that more than offset the cyclical drop.

The seasonal problems of a business are very different from those growing out of the cyclical swings of business. And both of these differ from the problems of long-term trends. The measurement of each of these fluctuations is a basic step in the use of data classified according to time.

When management uses any type of historical data, some problem of forecasting is generally involved. Since business decisions of the planning type must be made well in advance of their execution, it is imperative that decision makers have an indication as to what the future situations will be at the time the decision is actually implemented. The record of the past provides a basis for making estimates of the future.

Just as the measurement of fluctuations in time series requires a separation of the different types of fluctuations, any forecast should be based on a forecast of the different types of variations. Separate forecasts should be made of the trend, seasonal, and cyclical variations, although it is doubtful that a forecast of erratic fluctuations is feasible. Any budgeting in a seasonal business must take into account the seasonal element. A concern with a heavy fixed investment cannot avoid forecasts of the long-term trend, since the business expects to operate for a long time. Very few concerns can afford to ignore the cyclical fluctuations and their inevitable effect on the individual industry and business.

Correlation Analysis of Time Series

Data for two series for the same time period, such as production of steel in the United States and production of United States Steel Corporation, may make use of correlation analysis to compare the variations in the two series. Management can usually profit from comparing the fluctuations in the business with the whole industry, and the coefficient of correlation provides a method of making this comparison.

In the use of correlation analysis of time series, a time series of business data is usually a composite of several distinct movements; so the correlation of two such composite series may turn out to be meaningless. For example, the correlation of two monthly series with strong seasonal variations will result in the correlation being influenced almost entirely by the seasonal fluctuations. On the other hand, if the strongest type of fluctuation in the two series is their rapid growth, the correlation between them will be mostly the result of this factor. In general, it is a sound principle that correlations between time series are most meaningful when they are between the same components of economic change. This will be discussed in the following chapters as each type of fluctuation in time series is considered.

One of the most frequent uses of regression analysis in the study of time series has been in measuring the relationship between series when one series registers cyclical changes earlier than another. The correlation between the two series can be used to forecast one series when the changes in another occur earlier in time. The comparison of two seasonal patterns can be made simply by showing the two indexes of seasonal variation on the same graph as illustrated on page 424. A method of comparing two secular trends is discussed on pages 393 to 396.

ADJUSTMENT OF DATA FOR CALENDAR VARIATION

There are occasions when the varying lengths of the months introduce erratic fluctuations in many time series. For example, daily production in a factory might be almost the same in February and March; but because there were only 20 working days in February and 23 in March, there would be an increase in the March figures.

A method of removing this irregularity is to divide the data for a month by the number of days, or by the number of working days in the month. Whether the correction should be made on the basis of total days in the month or the number of working days depends upon the data. In the example just given it is assumed that production is on a five-day week basis. In industries where production is continuous, the adjustment should be made by dividing production by the number of days in the month. In production series there is a more clearly defined relationship between the number of working days and the amount of output than in such series as retail sales. When evaluating the results of retail business for a month, allowance should be made for the number of days a store is open; but it is somewhat uncertain that the adjustment can be made as accurately as in the case of a manufacturing plant. In retailing certain days of the week are better business days than others; so it is sometimes desirable to make an allowance for the number of Saturdays in a month as well as for the total number of days a store is open.

The Board of Governors of the Federal Reserve System publishes extensive statistical data on production in manufacturing industries and has made detailed studies of the work week used in different industries.[2] The continuous process industries generally use a seven-day week, while in other industries the five-day week predominates. With this information the necessary adjustment for the number of working days in each month can be computed. Table 13-2 shows lumber production for the 12 months of 1972 in millions of board feet, the number of working days in each month, and the average production per working day for each month. The lumber industry uses the five-day week, which means that the smallest number of working days possible in a month is 20 and the largest number is 23. Since 1972 was a leap year, February had 29 days, but there were four Saturdays. Thus, there were 21 working days in February.

The importance of the adjustment for the number of working days in the month is demonstrated in Table 13-2 by the figures for the production of lumber in August and September, 1972. August production was more than in

[2] *Federal Reserve Bulletin,* Vol. 39, No. 12 (December, 1953), pp. 1,280-1,281.

TABLE 13–2

LUMBER PRODUCTION IN THE UNITED STATES, 1972

Total Monthly Production and Production Per Working Day

Month	Production** (Millions of Board Feet)	Number of Working Days*	Production per Working Day (Millions of Board Feet)
January	2,832	21	135
February	3,076	21	146
March	3,383	23	147
April	3,272	20	164
May	3,420	23	149
June	3,301	22	150
July	3,102	20	155
August	3,417	23	149
September	3,303	20	165
October	3,528	22	160
November	3,193	21	152
December	2,664	20	133

* Computed on the assumption that a five-day week is the general practice in the lumber industry and that the following holidays are observed: New Year's Day, Independence Day, Labor Day, Thanksgiving, and Christmas.

Source: ** U.S. Department of Commerce, *Survey of Current Business* (March, 1973).

September, but there were 3 more working days in August than in September. When production is adjusted for the number of working days, September production was 11.1% more than in August, although without this adjustment September production was 3.3% less than in August.

The basic difficulty in adjustment for calendar variation is with the calendar. This difficulty would be eliminated by a change that would make all months the same length. One proposal for calendar reform is to divide the year into 13 months of four weeks each, but this is such a break with tradition that it has very little chance of being accepted. A less drastic revision would be to retain the 12 months but have two 30-day months and one 31-day month in each quarter. This would make all the quarters the same length and would give only a small variation between the lengths of the various months. Since each system would total only 364 days, it is generally proposed that New Year's Day be an extra day between the last day of December and the first day of January and not included in any month. On leap years an extra day could be added between the last day of June and the first day of July. Agitation for calendar reform has been going on for a long time, but the prospects are not promising that it will be accomplished in the near future.

For many years the adjustment for calendar variation took into account the number of holidays in the month, but the present practice generally is not to make this adjustment. The reasoning used is that the holidays occur in the same month every year[3] and their effect can be included in the measurement of seasonal variation, since seasonal variation is defined as the regularly recurring fluctuations within a year.

FIXED-BASE RELATIVES

A time series may be simplified by reducing the data for the various intervals to ratios of the data for one period. This period becomes the base and is usually assigned a value of some multiple of 10. The ratios are called *relatives* and if the base is given a value of 100, they are known as *percentage relatives.* Quantities that are expressed as relatives are also called *index numbers.* The construction of index numbers, however, frequently involves the combining of several series into a composite, and this process may become rather complicated. This subject will be treated in detail in Chapter 17.

An example of percentage relatives is given in Table 13–3. Column 2 shows the production of raw steel in the United States, and Column 3 shows the same information expressed as percentages of the 1956 production. Columns 4 and 5 show similar information for United States Steel Corporation. This series of relatives, which reflects the changes since the base year, is easier to understand than the original data because the comparison with 100 is easier to make. Also, when the comparison is between two years, neither of which is the base year, the relatives are usually easier to compare than the actual data.

Sometimes no year may be considered normal, and instead of using a year as the base, an average of several years may be used. The favorite base year during and immediately after World War I was 1913, the last prewar year. For the decade of the 20's and the first part of the 30's, 1926 was a common base year, although the average of the three years 1923–1925 was also used in many cases. By the end of the 30's, many series of relatives had been shifted to the five-year average, 1935–1939. After World War II government statistical agencies set out to revise their statistical series, and one of the steps in this revision was to shift the index numbers to a postwar base. The average of the three years 1947–1949 was selected as the base period for government index numbers, then the base was shifted to 1957–1959, and now the base is 1967.

When two series are expressed as relatives with respect to the same period, the two may be compared more easily than if they were related to different base periods. For this reason it is of some value to use the same base period as that of well-known published series. Any period, however, may be used if the problems of analyzing the data justify it.

In Table 13–3 a comparison of the two series is easier by using the relatives than by comparing the absolute amounts of steel produced. For the year

[3] The most important exception to this statement is the date of Easter, which may come in March or April. Whenever the date of Easter has an important effect on a series, as it does in department store sales, it becomes important to allow for a different seasonal pattern depending on this factor.

TABLE 13-3

TOTAL RAW STEEL PRODUCTION IN THE UNITED STATES AND PRODUCTION BY UNITED STATES STEEL CORPORATION, 1956 TO 1972

	Production in the United States		Production of United States Steel Corp.	
Year (1)	Millions of Tons (2)	Relatives 1956 = 100 (3)	Millions of Tons (4)	Relatives 1956 = 100 (5)
1956	115.2	100	33.4	100
1957	112.7	98	33.7	101
1958	85.3	74	23.8	71
1959	93.4	81	24.4	73
1960	99.3	86	27.3	82
1961	98.0	85	25.2	75
1962	98.3	85	25.4	76
1963	109.3	95	27.6	83
1964	127.1	110	32.4	97
1965	131.5	114	32.6	98
1966	134.1	116	32.8	98
1967	127.2	110	30.9	93
1968	131.5	114	32.4	97
1969	141.3	123	34.7	104
1970	131.5	114	31.4	94
1971	120.4	105	27.2	81
1972	133.1	116	30.7	92

Source: American Iron and Steel Institute, *Annual Statistical Report,* and annual reports of United States Steel Corporation.

1957 steel production of United States Steel Corporation was relatively higher in relation to 1956 than for the United States as a whole. However, after 1957 the relatives for total United States production were larger than the relatives for the production of United States Steel Corporation.

LINK RELATIVES AND CHAIN RELATIVES

Instead of comparing each period with a fixed period, the comparison wanted may be that of each period with the preceding period. Such relatives are referred to as *link relatives*. Column 3 in Table 13–4 gives the production of steel by United States Steel Corporation for each year expressed as a ratio to the preceding year. For example, production in 1958 was to 1957 as .706 is to 1.000, computed

$$\text{Ratio} = \frac{23.8}{33.7} = .706.$$

TABLE 13-4

COMPUTATION OF LINK RELATIVES AND CHAIN RELATIVES, 1956 = 100

Raw Steel Production by United States Steel Corporation

Year (1)	Production (Million Tons) (2)	Link Relatives (Previous Year = 1.000) (3)	Chain Relatives 1956 = 1.000 (4)	Chain Relatives 1956 = 100.0 (5)
1956	33.4		1.000	100
1957	33.7	1.009	1.009	101
1958	23.8	.706	.712	71
1959	24.4	1.025	.730	73
1960	27.3	1.119	.817	82
1961	25.2	.923	.754	75
1962	25.4	1.008	.760	76
1963	27.6	1.087	.826	83
1964	32.4	1.174	.970	97
1965	32.6	1.006	.976	98
1966	32.8	1.006	.982	98
1967	30.9	.942	.925	93
1968	32.4	1.049	.970	97
1969	34.7	1.071	1.039	104
1970	31.4	.905	.940	94
1971	27.2	.866	.814	81
1972	30.7	1.129	.919	92

Source: Annual reports of United States Steel Corporation.

In the same manner the production for 1957 was found to be related to 1956 as 1.009 to 1.000; in other words, the production for each year is successively the numerator and then the denominator of the ratio.

Since the relatives computed in Table 13–3 and the link relatives computed in Table 13–4 are taken from the same data, there is a fixed relationship between the two. From the link relatives it is possible to compute the fixed-base relatives that were computed directly from the data. Such relatives are referred to as *chain relatives* to distinguish them from the fixed-base relatives computed directly from the original data. The chain relatives shown in Column 4 of Table 13–4 are expressed as ratios to 1.000. In Column 5 of the same table, they are expressed as ratios to 100. A comparison of this column with the fixed-base relatives computed in Table 13–3 demonstrates that they are the same values, although occasionally errors introduced by rounding will make a slight difference. In Table 13–4, the computations were carried out to one more decimal place than the fixed-base relatives in Table 13–3 in order to eliminate some of the error due to rounding of the decimals.

The computations were carried out with ratios to 1.000, or decimal fractions, instead of percentages. In Table 13–4 the chain relative in Column 4 for 1956 was given a value of 1.000, since this year was chosen as the base. The link relative for 1957 is 1.009 in relation to 1956, so this is also the chain relative for 1957. The computation of the chain relative for 1958 is .706, the

1958 link relative, which means that the ratio of 1957 to 1958 is .706 to 1. But 1957 in relation to 1956 is 1.009 to 1, so the ratio of 1958 to 1956 is .706 times 1.009, or .712. This is the chain relative for 1958, with 1956 equal to 1.000. The chain relative for 1959 was computed by multiplying the link relative for 1959 by the chain relative for 1958. Each chain relative was computed by this same method; namely, the link relative for a given year times the chain relative for the preceding year equals the chain relative for the given year.

That the chain relatives and the fixed-base relatives are the same mathematically can be demonstrated by showing all the steps in computing the chain relatives for the first three years.

$$\text{Chain relative 1957} = \frac{1957}{1956} \quad \text{(This is the same as the link relative for 1957.)}$$

$$\text{Chain relative 1958} = \frac{1958}{1957} \times \frac{1957}{1956} = \frac{1958}{1956}$$

$$\text{Chain relative 1959} = \frac{1959}{1958} \times \frac{1958}{1957} \times \frac{1957}{1956} = \frac{1959}{1956}.$$

For a series such as the production of steel by United States Steel Corporation, there is no reason for computing the chain relative when the fixed-base relative can be computed directly from the data. The chain relative does have a very useful application in statistical analysis, however, when only the ratio of one period to the previous period is known. Such a situation may result from the use of sample data to estimate the ratio of one period to the one preceding. If the methods of estimating give unbiased estimates of the ratio from period to period, or estimates that are only slightly biased, the chain relatives will be a satisfactory estimate of the fluctuations in the series in the same way that a fixed-base relative shows the changes in a series. In many situations it is more efficient to use a sample to estimate the ratio between successive periods than to estimate the actual values of the series from samples. It should be recognized, however, that if the ratios are biased, this bias will cumulate rather rapidly as the chain index is extended from one period of time to another.

SEMILOGARITHMIC CHARTS

The presentation of time series in graphic form was discussed in Chapter 2 and examples of bar charts and line charts were given. The principles of construction are essentially the same for all types of bar charts and line charts discussed in Chapter 2. In general, the classes into which the data are grouped are shown on the X axis and the numbers or amounts in the various classes are shown on the vertical scale. All the examples in Chapter 2 use an arithmetic scale for the Y axis, which signifies that a given space on the scale always means a certain magnitude wherever it appears. Thus, a given space that represents $1,000 is the same whether it shows the difference between $2,000 and $3,000, or between $98,000 and $99,000.

When a logarithmic scale is used for the Y axis, it is laid out according to the logarithms of the numbers on the scale. This is a specialized type of chart used for time series, and its purpose is to show the rate of change from one period to another. The ruling on the chart is such that the figures are automatically reduced to a percentage basis when plotted, and *the same vertical distance anywhere on the chart shows the same percentage of change*. A decrease of a given percentage represents the same distance on the scale, whether the decrease be from \$1,000,000 to \$500,000, or from \$10,000 to \$5,000. On the arithmetic scale the first decrease, \$500,000, would take up 100 times as much space as the second decrease, \$5,000, and yet the percentage decrease is the same. Thus, if the interest is in percentage changes in the data, the *semilogarithmic chart* should be used. This type of chart is useful in comparing percentage changes in two series, or in two parts of the same series. If two series are shown on the same chart, the slope of each line shows the *percentage* change in that series. By comparing the slopes of the two lines, it is possible to compare the percentage changes in the two series.

Figure 13–6 shows the production of raw steel in the United States and the production of United States Steel Corporation. The chart is drawn on an arithmetic scale so that the distance of each line from the zero line of the chart shows the amount of production. This chart shows how the production of

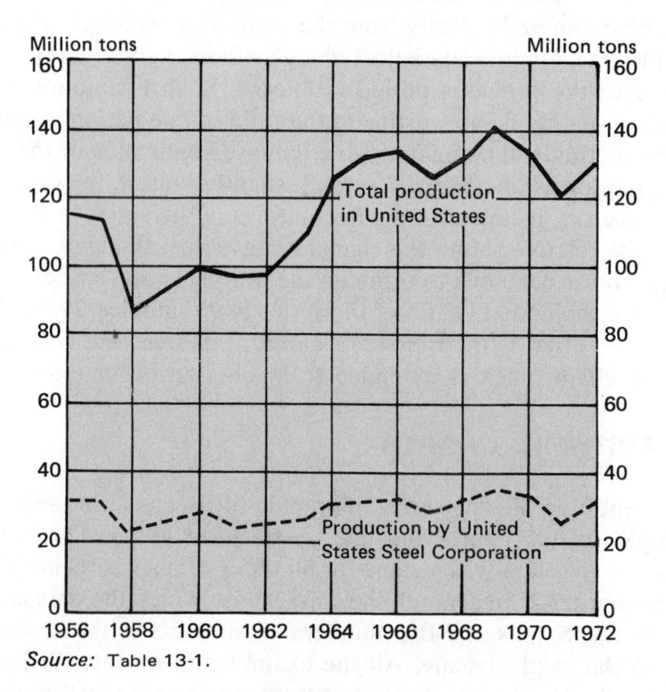

Source: Table 13-1.

FIGURE 13–6

PRODUCTION OF RAW STEEL IN THE UNITED STATES AND PRODUCTION OF RAW STEEL BY UNITED STATES STEEL CORPORATION, 1952 TO 1972

United States Steel Corporation compares with the total volume of production of steel in the United States.

Assume that the chart was drawn to compare the growth of United States Steel Corporation with the steel industry as a whole. The first impression given by the chart is that the industry has grown much faster than United States Steel Corporation. It is difficult to compare the rise in the two lines since the upper line, representing raw steel production in the United States, is based on amounts so much larger than production of United States Steel Corporation. In making this comparison, the important factor is not the *amount* of change in total raw steel production in the United States or production of United States Steel Corporation. A proper comparison must take into account the *relative* change in the two series.

The chart drawn on the arithmetic scale compares the amounts of change, but the data should be presented on a semilogarithmic chart, as in Figure 13–7, so the fluctuations in the two lines can be compared with respect to the relative change in each. Over the period shown in the chart there has been an upward trend in total steel production, while production of United States Steel Corporation has shown a slight downward trend.

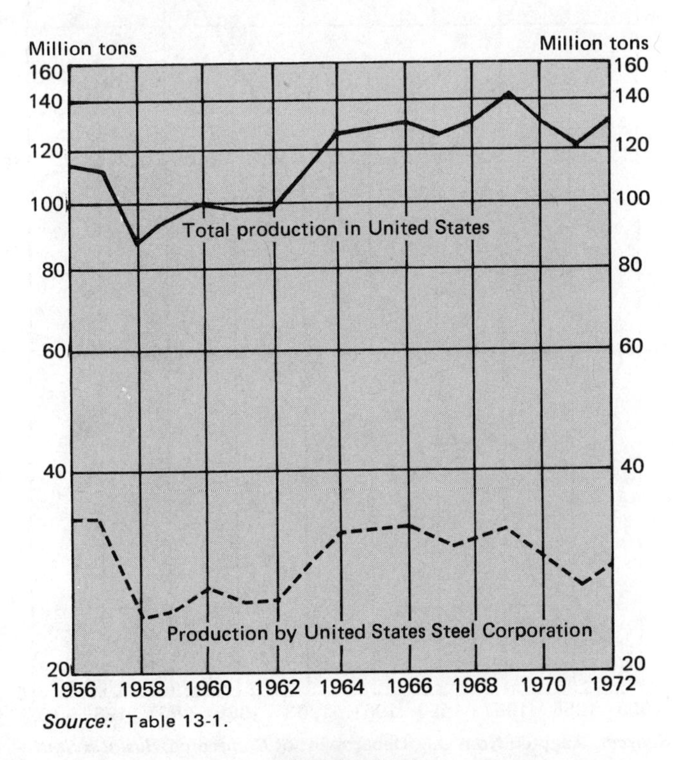

Source: Table 13-1.

FIGURE 13–7

PRODUCTION OF RAW STEEL IN THE UNITED STATES AND PRODUCTION OF RAW STEEL BY UNITED STATES STEEL CORPORATION, 1956 TO 1972

Figure 13–4 shows the growth of a new product, electronic components; but since it was drawn on an arithmetic scale, it indicates the amount of growth in the series. Figure 13–8, drawn on a semilogarithmic chart, shows the rate of growth of this industry. A steady rate of growth was maintained for a few years and then the rate gradually slowed.

Figure 13–9 shows the growth of $1,000 left to accumulate at 6% interest, compounded annually. The money increases at a constant rate—6% per annum. The line is, therefore, a straight line on semilogarithmic paper, and increases to $7,686.09 in 35 years.

The fact that a straight line on semilogarithmic paper shows a constant percentage of change can be used to advantage in computing depreciation

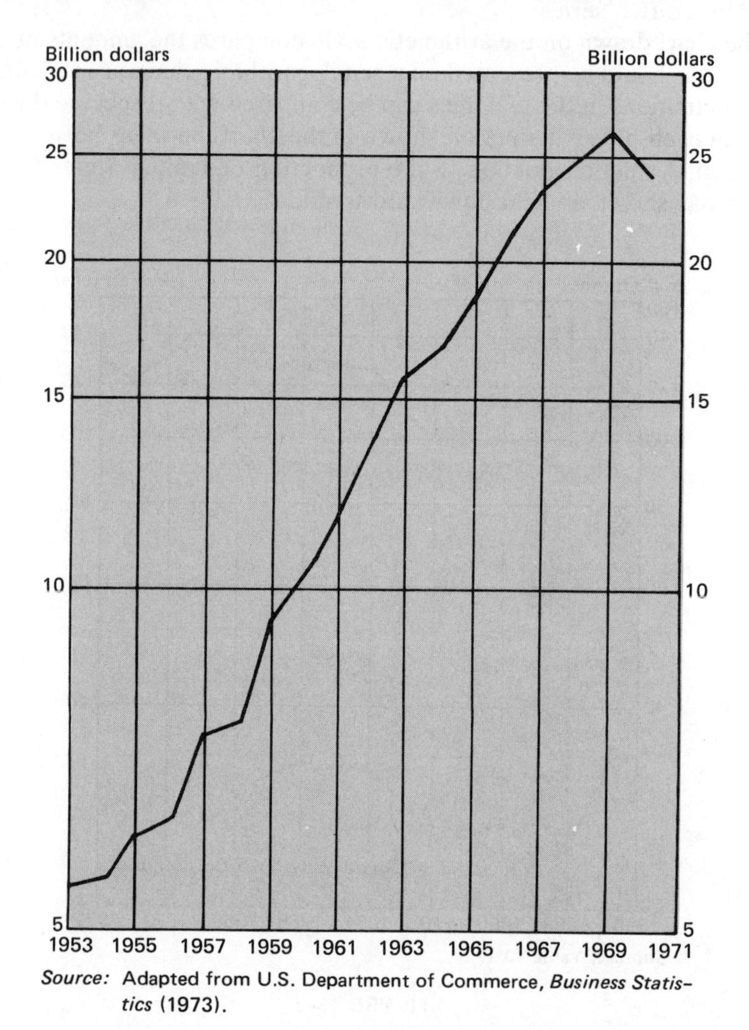

Source: Adapted from U.S. Department of Commerce, *Business Statistics* (1973).

FIGURE 13–8

FACTORY SALES OF ELECTRONIC COMPONENTS, 1953 TO 1970

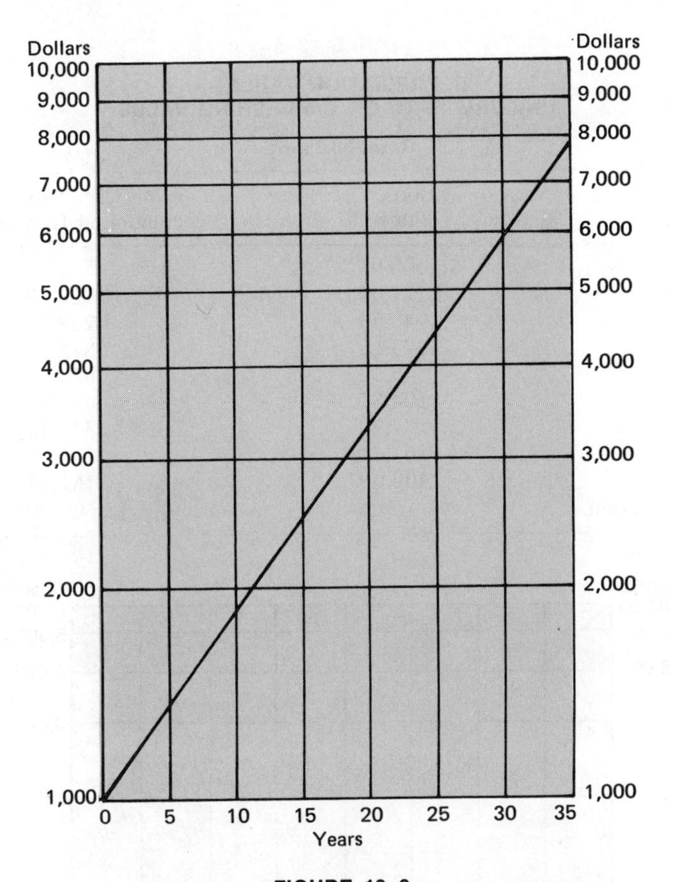

FIGURE 13–9

ACCUMULATION OF $1,000 AT 6% INTEREST COMPOUNDED ANNUALLY OVER 35 YEARS

charges when the method used involves finding a constant rate of depreciation to be applied to the diminishing value of the asset. The value of the asset, therefore, decreases by the same percentage each year. When plotted on a semilogarithmic chart, the net value of the asset is a straight line.

Table 13–5 shows the depreciation on an asset costing $6,000 and having an estimated scrap value of $400 at the end of eight years. The uniform rate on the diminishing value is 28.7166%. Figure 13–10 shows the data on a semilogarithmic chart.

Instead of computing the data in Table 13–5 and then plotting the values of the asset, it is possible to plot the data on the chart first and read the chart to get the book value of the asset at different years. Since the line is a straight line, it is necessary to locate only two points on it, and draw the line connecting these two points. The first and last values are known; in the example above, the cost was $6,000 and the asset was to be depreciated to $400 in eight years. These two figures give two points on the straight line, which make possible

TABLE 13–5

**DEPRECIATION TABLE–
UNIFORM RATE ON DIMINISHING VALUE**

Rate: 28.7166%

Year	Book Value	Credit to Accumulated Depreciation
Original Cost	$6,000.00	
1	4,277.00	$1,723.00
2	3,048.79	1,228.21
3	2,173.28	875.51
4	1,549.19	624.09
5	1,104.32	444.87
6	787.20	317.12
7	561.14	226.06
8	400.00	161.14
Total		$5,600.00

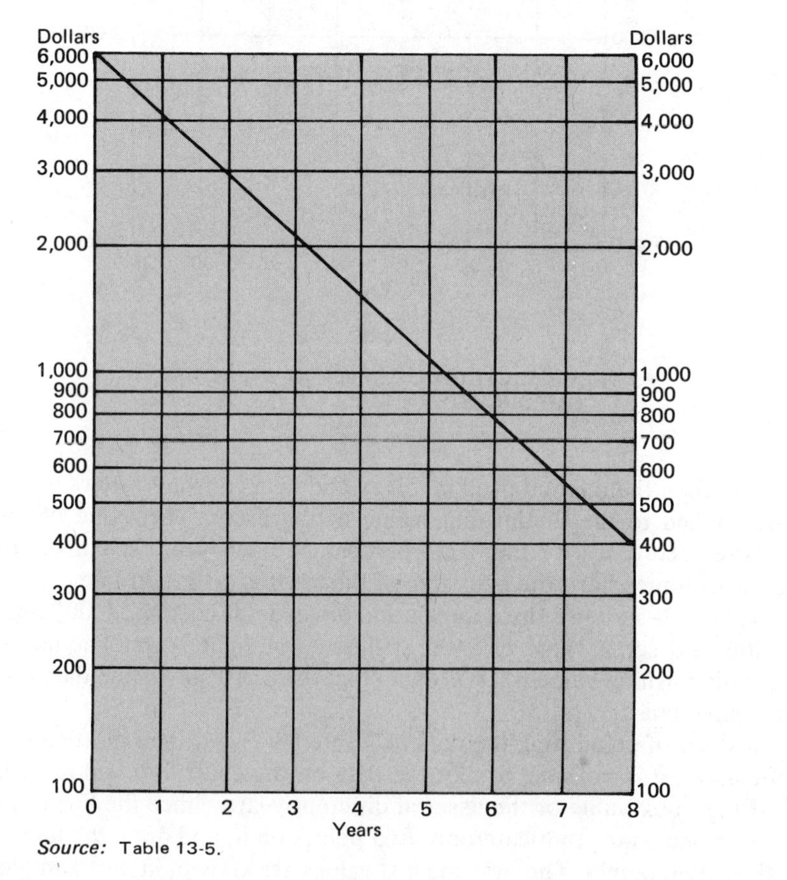

Source: Table 13-5.

FIGURE 13–10

DEPRECIATION AT A UNIFORM RATE ON DIMINISHING VALUE

the drawing of the line. The book value of the asset for any year between the first and the eighth may be read off the chart with sufficient accuracy, if the chart is carefully drawn. The credits to the accumulated depreciation account may then be computed.

One of the objections to the preceding method of determining depreciation charges is the difficulty of computation. Using the semilogarithmic chart provides an easy way of making the computations.

In using the semilogarithmic chart, it must be remembered that its primary purpose is not to compare the sizes of different series. Figure 13–7 gives the impression that United States Steel Corporation production was about one third of the total raw steel production, since the line showing that concern's sales is about one third as far from the bottom of the chart as the line showing the total raw steel production in the United States. Figure 13–6 on the arithmetic scale shows the relative size of the two series, and it can be seen that the total sales are nearly four times as large as the sales of United States Steel Corporation. The purpose of Figure 13–7 is to show *percentage changes in the two series; it is not an accurate picture of the absolute size of the series.* It is, of course, possible to read the figures on the chart and see the absolute size of the data, but the chart does not give an accurate impression of the magnitude of the values at a glance.

Setting Up the Scale

Printed paper on which to construct semilogarithmic charts can ordinarily be obtained wherever graph paper is sold, or it can be constructed for the particular purpose at hand.[4] The size of the paper is designated by the number of cycles. Figure 13–10 shows two-cycle paper. Since the ruling in all cycles is exactly the same, the value of any line in any cycle is always exactly 10 times the value of the corresponding line in the cycle immediately below.

On printed graph paper the major guidelines are numbered, with the first line in a cycle designated 1; the second, 2; and so on to the first line in the next cycle, which is 10. The lines of the second cycle are numbered 10, 20, and so on to 100. Sometimes instead of adding a zero for the numbers of the second cycle, the numbers 1 through 9 are repeated, with the next cycle starting at 1. The value of the first guideline must be assigned, but the other numbered guidelines are computed by multiplying the printed number of the guideline by the value of guideline 1. For example, the first guideline in Figure 13–7 is given the value of 20 million tons, so the second guideline is twice the value of the first line, or 40 million tons. It is necessary, of course, that the first guideline be assigned a value smaller than the smallest amount to be plotted. In Figure 13–10 the major guidelines in the first cycle run from \$100 to \$1,000; \$1,000

[4] A method that is perhaps easier is to plot the logarithms of the actual data on an arithmetic chart.

is the first line of the second cycle. The values for the major guidelines in the second cycle in Figure 13–10 would run from $1,000 to $10,000, but since the largest value is not over $6,000, it is not necessary to use all of the cycle. After the major guidelines of one cycle of a semilogarithmic chart have been designated, additional cycles may be labeled simply by multiplying the corresponding line in the preceding cycle by 10. The range of the data determines the number of cycles that must be used.

On printed semilogarithmic paper the horizontal guidelines are close together to facilitate plotting the data; but when data are plotted on a specially drawn grid, usually only the major guidelines are drawn. This improves the appearance of the chart, although it makes the reading of values rather difficult and may result in slight inaccuracies. It is easier to plot the points if a piece of printed graph paper is used as a plotting scale.

The layout of the semilogarithmic chart is fundamentally no different from that of the other charts discussed previously. Since it does not make any comparison of size, the slope of the line is the only significant part. For this reason the chart starts at whatever point is convenient for showing the data, rather than at zero as in the arithmetic chart. It is impossible to show zero on the semilogarithmic chart. Each line is one tenth of the amount assigned to the corresponding line in the cycle above, so the chart could be extended downward any number of cycles desired and still would not reach zero. This also means that there is never any part of the scale omitted as in the case of the arithmetic chart. Since the first line on the grid is not the base line, it should not be emphasized.

Using Two Scales

When two series are shown on arithmetic charts, it is preferable that they both use the same scale. Since the basic purpose of the arithmetic chart is to show the size of the figures plotted, comparison can be made directly between two lines or parts of the same line. When a different scale is used for each line plotted, this comparison is impossible. Figure 13–11 shows United States Steel Corporation production and the production of raw steel in the United States previously discussed. This chart is similar to Figure 13–6, except that two scales are used on the Y axis of Figure 13–11. It is quite difficult to tell from this chart how much larger the total raw steel production was than the production of steel by United States Steel Corporation. It does make easier, however, the comparison of the fluctuations in the two series. This suggests the principle that two lines can be compared on an arithmetic chart in the same manner as on a semilogarithmic chart if *two scales are used on the arithmetic chart in proportion to the two series to be plotted.* If the two scales are drawn in proportion to the two series, equal percentage fluctuations in the two lines will appear the same on the chart.

When using two scales on an arithmetic chart, it is necessary, first to determine the ratio between the two series, and then to construct two scales that

Source: Table 13-1.

FIGURE 13–11

PRODUCTION OF RAW STEEL IN THE UNITED STATES AND PRODUCTION OF RAW STEEL BY UNITED STATES STEEL CORPORATION, 1956 TO 1972

stand in this ratio to each other. Since there is danger that this may not be done accurately, it is better to use a semilogarithmic chart for this type of comparison.

Two scales can be used also on the semilogarithmic chart, as shown by Figure 13–12. The additional advantage of the semilogarithmic chart is that there is no problem of getting the scales in the same proportion; that is, a curve will have exactly the same shape on a given logarithmic scale regardless of where it is plotted on the chart. This removes the danger of misrepresentation of the facts. In Figure 13–12 the fluctuations in the lines are exactly the same as when they were drawn on a chart with only one scale, as in Figure 13–7. The only effect of using the second scale is to change the position of one line on the chart. In Figure 13–12 the line for the sales of United States Steel Corporation is higher on the chart, but since the distance from the bottom of the semilogarithmic chart is not significant, this makes no difference. And there is the advantage that the two lines are closer together, which makes it easier to compare the fluctuations. Thus, the use of two scales on the semilogarithmic chart involves no disadvantages, and frequently improves the chart.

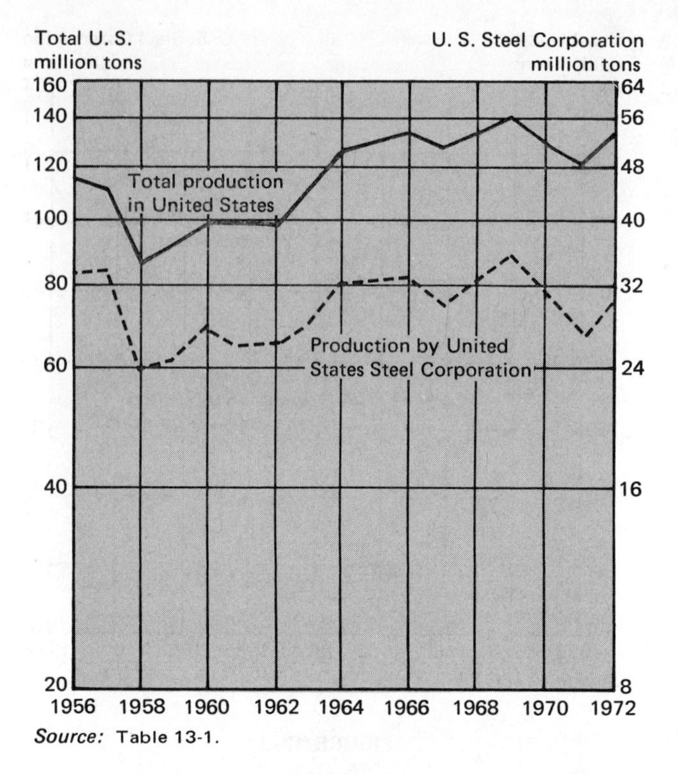

Source: Table 13-1.

FIGURE 13–12

PRODUCTION OF RAW STEEL IN THE UNITED STATES AND PRODUCTION OF RAW STEEL BY UNITED STATES STEEL CORPORATION, 1956 TO 1972

STUDY QUESTIONS

13–1. Why is the classification of statistical data with respect to time important to the business executive?

13–2. Give an example of a situation in which the measurement of the seasonal variation in business would make possible a better decision by a businessman.

13–3. Give three examples of periodic fluctuations in business other than variations related to the seasons. (These examples should be in addition to those given in the text.)

13–4. Some businesses offer price reductions to stimulate business during periods of the day, week, or year when volume is abnormally low. Give three examples of this type of pricing.

13–5. Give an example of an industry in which the measurement of secular trend would be important to the businessman.

13–6. How can the semilogarithmic chart be used to compare the growth of two industries?

13–7. Why does the business executive need information on the cyclical fluctuations of business?

13–8. A number of years ago a businessman was shown the usefulness of a semilogarithmic chart for the analysis of economic data, and he was so much impressed that he decided to use only graphs drawn on this type of grid. Do you consider this decision wise? Why or why not?

13–9. Why is it logical to plot two curves representing series in different units (such as production and prices) on the same semilogarithmic chart? Why is it not so satisfactory to plot two such curves on an arithmetic chart?

13–10. Assume that you want to compare the fluctuations in two time series. Which would be the more effective method of making this comparison: (a) plotting both series on the same semilogarithmic chart, or (b) reducing both series to fixed-base relatives with the same period as the base, and then plotting both relatives on the same arithmetic chart? Explain.

13–11. What conditions might make a chain relative an unsatisfactory measure of the fluctuations in a time series? Explain.

13–12. Many statements have been made that correlation analysis is not valid in comparing the fluctuations in two time series. Under what circumstances do you think it particularly undesirable to measure the degree of correlation between two time series?

PROBLEMS

13–1. The following table gives the sales (in millions of dollars) of Company X, both for the United States and for the territory known as the Southwestern District, for the years 1953 to 1972. Plot these two series on the same scale on an arithmetic chart.

Year	United States	Southwestern District	Year	United States	Southwestern District
1953	295	58	1963	424	84
1954	311	64	1964	394	73
1955	334	67	1965	402	86
1956	339	68	1966	458	94
1957	308	57	1967	493	101
1958	330	65	1968	512	102
1959	371	72	1969	520	100
1960	394	80	1970	498	94
1961	365	79	1971	550	121
1962	410	80	1972	560	129

Source: Hypothetical data.

13–2. The manager of the Southwestern District is using the chart drawn in Problem 13–1 to convince you that she has kept her sales up better than the company did as a whole in 1970. Her district sales declined only \$6 million compared to a decline of \$22 million for the company. What would be your reaction to her statement? Be specific.

13–3. If the sales manager of the company used the chart drawn in Problem 13–1 to convince you that the manager of the Southwestern District did not bring her sales up as much as the company did in 1972, what answer would you give him? Be specific.

13–4. Reduce the two series in Problem 13–1 to relatives using 1953 as the base. Plot these two series on the same arithmetic chart. Does this chart help in answering the questions in Problems 13–2 and 13–3? Would your answers to these questions be the same after looking at this chart of the relatives as it was from looking at the arithmetic chart?

13–5. Compute the percentage that the sales of the Southwestern District are of the total sales of the company for each year from 1953 to 1971. Plot this series of percentages on an arithmetic chart and comment on its significance.

13–6. Plot the two series in Problem 13–1 on the same semilogarithmic chart. What does this chart show about the relative fluctuations in the two series?

13–7. A machine with an expected life of 5 years is to be depreciated at a uniform rate on diminishing value. The original cost of the machine was $10,000 and it is estimated that it will be worth $500 at the end of 5 years.

(a) Compute the credit to the accumulated depreciation account each year.

(b) What difficulty would be encountered if the machine were estimated to have no value at the end of 5 years?

13–8. The following table shows life insurance and disposable personal income per family in the United States from 1963 to 1971.

Year	Life Insurance per Family	Disposable Personal Income per Family
1963	12,200	6,800
1964	13,300	7,300
1965	14,700	7,700
1966	15,900	8,200
1967	17,200	8,700
1968	18,400	9,100
1969	19,500	9,600
1970	20,900	10,200
1971	21,800	10,800

Source: Institute of Life Insurance, *Life Insurance Fact Book* (1972).

(a) Plot the two time series on the same arithmetic chart using the same scale. Describe what these charts tell you about each series.

(b) Plot the two time series on the same semilogarithmic chart. Use two scales so as to bring the two series as close together as possible. Does this chart tell the same story of relative growth as did the first one? Explain.

(c) Show life insurance per family as a link relative. Take a geometric mean of the 8 relatives to show the average relative. What does this figure tell you about the rate of growth of life insurance per family in the United States?

(d) Working with the figures for disposable personal income per family for the years 1963 and 1971 only, what was the average annual rate of increase for that period? (*Hint:* Go back to the section on the geometric mean in Chapter 3.)

13–9. Assume that the population of Central City was 30,000 in 1959 and 84,000 in 1969. If the city grew at a constant rate during that period, what do you estimate the population to have been in 1964? Make this estimate using a semilogarithmic chart and then check your answer arithmetically.

13–10. The following time series shows sales for a six-year period for a hypothetical company.

Year	Sales (Thousands of Dollars)
1967	150
1968	170
1969	200
1970	190
1971	230
1972	260

(a) Express sales as a series of fixed-base relatives with 1967 as the base year.

(b) Express sales as a series of link relatives. Are any of these values the same as those computed in (a)?

(c) Convert the series of link relatives computed in (b) to a series of chain relatives with 1967 as the base year. How does this series compare to the series of fixed-base relatives computed in (a)?

13–11. The following gives the production and manufacturers' stocks of nonfat dry milk (in millions of pounds) in the United States during 1972.

(a) Plot these series on a semilogarithmic chart.

(b) What seasonal fluctuations do you observe in these two series?

(c) Give an explanation of why these seasonal fluctuations appear in the production and cold storage of butter.

Month	Production	Stocks at End of Month	Month	Production	Stocks at End of Month
January	98.5	76.3	July	127.4	107.4
February	100.0	63.8	August	99.4	86.3
March	118.0	62.1	September	77.0	64.7
April	128.9	78.4	October	69.6	47.9
May	153.0	97.1	November	61.6	34.9
June	160.0	106.7	December	75.8	37.9

Source: U.S. Department of Commerce, *Survey of Current Business* (March, 1973).

13–12. The following table gives the production of tissue paper (in thousands of tons) in the United States for each month of the year 1972, during which time the industry operated on a five-day week.

(a) Adjust production for calendar variation and plot the monthly production and production per working day on the same semilogarithmic chart. Do you see any evidence of seasonal fluctuations in the production per working day?

(b) Explain the differences between the two series.

Month	Production	Month	Production
January	327	July	292
February	315	August	330
March	341	September	314
April	330	October	345
May	337	November	330
June	321	December	308

13–13. The following table gives the production of bituminous coal, crude oil, and natural gas in the United States for the years 1954 to 1972.

(a) Reduce these three series to fixed-base relatives with the year 1967 as the base.

(b) Plot all three series of relatives on the same arithmetic chart.

(c) What have been the trends of the three types of fuel?

Year	Bituminous Coal (Million Tons)	Crude Oil (Million Barrels)	Marketed Production of Natural Gas (Billion Cubic Feet)
1954	391.7	2,315	8,742
1955	464.6	2,484	9,405
1956	500.9	2,617	10,064
1957	492.7	2,617	10,680
1958	410.4	2,449	11,030
1959	412.0	2,575	12,046
1960	415.5	2,575	12,771
1961	403.0	2,621	13,254
1962	422.1	2,676	13,387
1963	458.9	2,753	14,747
1964	487.0	2,787	15,462
1965	510.0	2,849	16,042
1966	533.9	3,028	17,207
1967	551.0	3,217	18,171
1968	545.2	3,329	19,332
1969	560.5	3,372	20,698
1970	602.9	3,516	21,921
1971	552.2	3,454	22,493
1972	590.0	3,459	22,720

Source: Bureau of Mines, *Minerals Yearbook.*

13–14. Refer to the data in Problem 13–13 on the production of bituminous coal, crude oil, and natural gas for the years 1954 to 1972.

(a) Plot these series on the same semilogarithmic chart.

(b) Summarize what this chart shows about the trends of the three types of fuel.

(c) Compare this chart with the one drawn in Problem 13–13, pointing out the similarities and the differences.

13–15. The following table gives 1972 data (in millions of barrels) by months for demand, production, and stocks of distillate fuel oil, which is used primarily for heating. Plot these three series on the same arithmetic chart. Explain the interrelationship among the three series.

Month	Demand	Production	Stocks*
January	115.4	78.8	160.1
February	120.8	77.0	122.2
March	107.8	79.6	101.8
April	83.3	74.4	98.3
May	69.8	80.3	112.9
June	65.8	79.8	128.8
July	54.8	78.5	155.6
August	64.0	80.2	174.7
September	66.2	78.8	190.3
October	85.5	84.5	195.6
November	101.5	81.7	182.6
December	131.2	91.2	154.3

* At the end of month.

Source: U.S. Department of Commerce, *Survey of Current Business* (March, 1973).

13–16. The following table gives the annual disposable personal income and personal expenditures for the years 1955 to 1973 (in billions of dollars).

(a) Plot these two series on the same arithmetic chart.

(b) Summarize briefly what this chart shows concerning the cyclical fluctuations in income and expenditures over this period.

(c) What is the significance of the area representing the difference between the two lines plotted on this chart? Shade this area to emphasize it.

Year	Disposable Personal Income	Personal Expenditures
1955	275.3	259.5
1956	293.2	272.6
1957	308.5	287.8
1958	318.8	296.6
1959	337.3	318.2
1960	350.0	333.0
1961	364.4	343.3
1962	385.3	363.7
1963	404.6	384.7
1964	438.1	411.9
1965	473.2	444.8
1966	511.9	479.3
1967	546.3	506.0
1968	591.0	551.2
1969	634.2	596.2
1970	691.7	635.5
1971	746.0	685.8
1972	797.0	747.2
1973	882.5	827.8

Source: U.S. Department of Commerce, *Business Statistics* (1973) and *idem., Survey of Current Business* (March, 1974).

13–17. The following table gives the total consumption of rubber (in thousand of long tons) in the United States, classified as natural and synthetic, for the years 1956 to 1972. Plot the series for the consumption of natural and synthetic rubber on an arithmetic chart.

Year	Natural Rubber	Synthetic Rubber	Total
1956	562	874	1,436
1957	539	926	1,465
1958	485	880	1,365
1959	555	1,073	1,628
1960	479	1,079	1,558
1961	427	1,102	1,529
1962	463	1,256	1,719
1963	457	1,307	1,764
1964	483	1,442	1,925
1965	515	1.541	2.056
1966	550	1,672	2,222
1967	489	1,628	2,117
1968	575	1,893	2,468
1969	598	2,250	2,848
1970	559	2,197	2,756
1971	578	2,241	2,810
1972	640	2,425	3,065

Source: U.S. Department of Commerce.

13–18. Using the data in Problem 13–17 compute the percentage of consumption of natural rubber to total rubber consumption. The following table gives the average price of natural rubber in New York City for each year from 1956 to 1972. Plot the price of natural rubber and the percentage of natural rubber consumption to total rubber consumption on the same arithmetic chart. Do these two series appear to be related? Would you expect any relationship between the fluctuations in these two series?

Year	Dollars per Pound	Year	Dollars per Pound
1956	.343	1966	.236
1957	.311	1967	.199
1958	.282	1968	.198
1959	.365	1969	.262
1960	.385	1970	.218
1961	.296	1971	.180
1962	.285	1972	.181
1963	.263		
1964	.252		
1965	.257		

Source: U.S. Department of Commerce, *Business Statistics* (1973).

13–19. The following table gives the production of automobiles in the United States expressed as percentage relatives with 1967 equal to 100. Express the data for total rubber consumption given in Problem 13–17 as percentage relatives, with the average for 1967 equal to 100. Plot the relatives for automobile production and the relatives for total rubber consumption on the same arithmetic chart. Does there appear to be any relationship between the changes in the two series?

Year	Relatives	Year	Relatives
1956	70.4	1966	116.4
1957	73.5	1967	100.0
1958	50.3	1968	120.0
1959	67.7	1969	111.4
1960	83.6	1970	86.6
1961	71.0	1971	108.3
1962	91.1	1972	112.7
1963	101.4		
1964	103.3		
1965	125.4		

Source: U.S. Department of Commerce, *Business Statistics* (1973).

13–20. Construct a scatter diagram showing the relationship between the percentage relatives for automobile production from Problem 13–19 on the X axis and total rubber consumption from Problem 13–17 on the Y axis. Does this scatter diagram show the relationship between the two series better than the chart drawn in Problem 13–19? Give reasons for your opinion.

14 Secular Trend

The four basic types of time series fluctuations were described briefly in Chapter 13. This chapter and Chapters 15 and 16 discuss the methods that can be used to measure each of these types of fluctuations.

NATURE OF TREND

The secular trend of a time series is best represented graphically by a straight line or a smooth curve, since growth and decline are assumed to proceed gradually without abrupt changes in direction. A straight line on an arithmetic chart indicates that the amount of change was the same each year. Raw steel production shown in Figure 14–1 is an example of a series that has grown approximately the same amount each year. The straight line on the semilogarithmic chart indicates that the *rate* of change was the same each year. Figure 14–2 on page 370 shows that factory sales of television sets have grown at an average annual rate of 8.4%.

Industries producing goods that are being replaced by competitive products may decline, as shown in Figure 14–3 (page 371) by the trend in per capita consumption of cotton in the United States. The shift in consumer demand that has brought about an increase in synthetic fiber consumption has also resulted in a decline in the demand for cotton.

If neither the arithmetic nor the semilogarithmic straight line describes the underlying movement of the series, a curve on either an arithmetic or a semilogarithmic chart might be used. It is difficult for a constant rate of growth to be maintained over a long period of time; the more common pattern is for the rate of growth to decrease gradually. This means that the trend line on a semilogarithmic chart rises less and less sharply until it approaches a horizontal line. Figure 14–4 on page 372 shows the demand for distillate fuel oil. The secular trend has been rising at a decreasing rate; and although the trend is still rising, it would not be unreasonable to conclude that it is approaching the

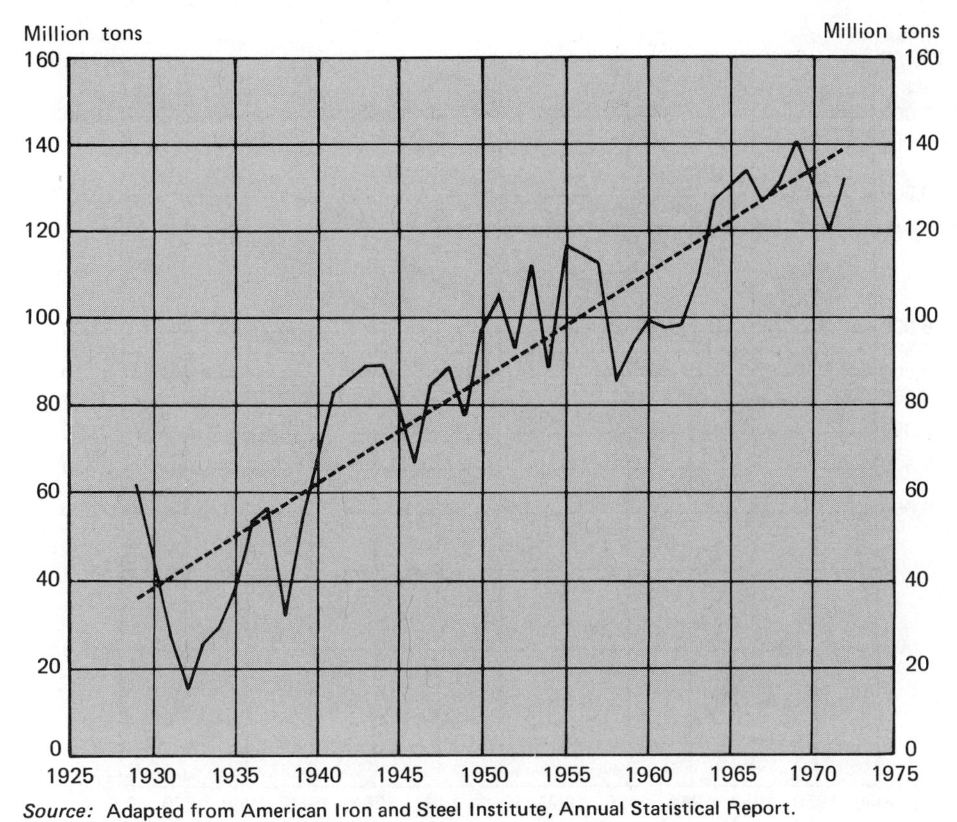

Source: Adapted from American Iron and Steel Institute, Annual Statistical Report.

FIGURE 14–1

RAW STEEL PRODUCTION IN THE UNITED STATES, 1929 TO 1972

point where there will be no further dramatic growth in demand. Distillate fuel oil is an important fuel for heating. In the past it has made deep inroads in the market for coal, but in more recent years has been meeting severe competition from natural gas.

Since the rate of growth usually changes slowly, several straight lines can sometimes be used, each one covering a portion of the whole period. For example, it can be seen in Figure 14–4 that a straight line describes the trend in demand for distillate fuel oil reasonably well for the period 1947 to 1956; however, from 1956 to 1972 the rate of growth was considerably less, and another straight line with a different slope was used. For the first period, the growth was at the annual rate of 7.8%; and for the second period, 2.6%. Instead of using two straight lines with different slopes, it would be possible to use a curve that rises at a gradually decreasing rate. Probably it is more logical to describe the growth in a series by a trend line that increases at a decreasing rate, but for many purposes it is desirable to have a measure of the average annual rate of growth for a shorter period. The use of several straight lines makes possible

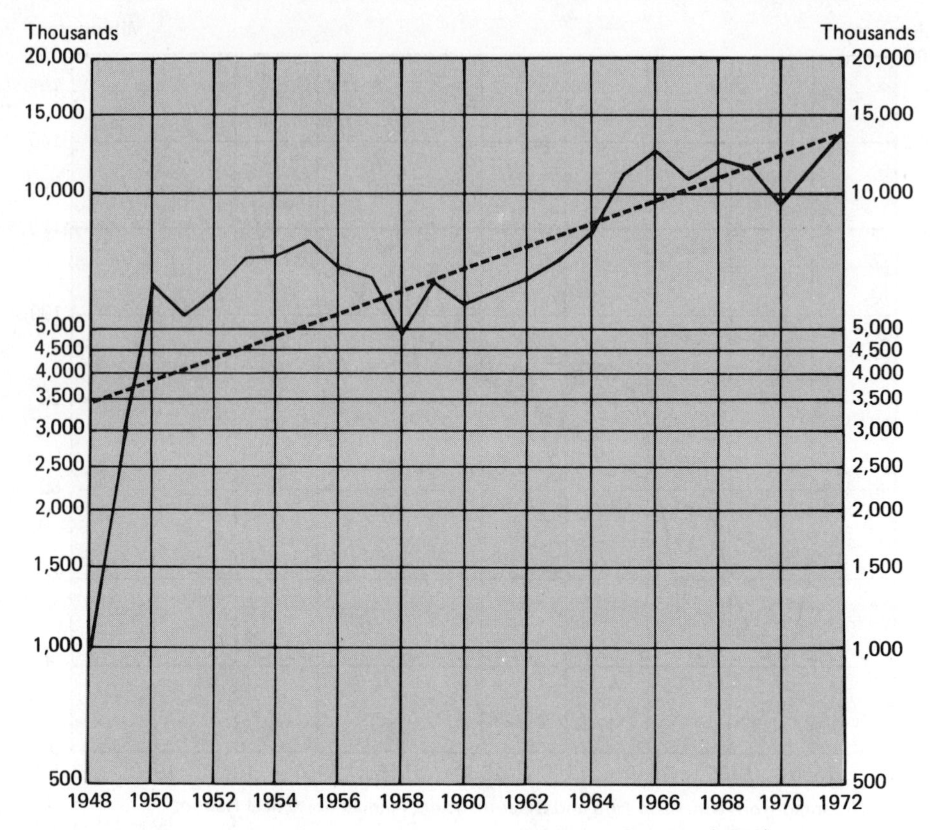

Source: Adapted from U.S. Department of Commerce, *Business Statistics* (1973).

FIGURE 14–2

FACTORY SALES OF TELEVISION SETS, 1948 TO 1972

a comparison of the rates of growth in the different periods. It will be seen also in the following discussion that the work of measuring the trend is much easier when a straight line is used instead of a curve.

In each of the charts shown in this chapter, the trend line cuts between the high and the low points of the line plotted from the actual data, with the result that the average of the trend values for a given series is approximately equal to the average of the actual values. In other words, the trend line represents the course the series might have taken if disturbing influences had not pushed it higher in some periods and lower in others. In this sense the trend represents an average value for the series, balancing high and low cyclical and seasonal values.

A different concept sometimes advanced is that the secular trend should represent full capacity, and deviations from trend would occur when operations fall below full capacity. Instead of the line representing actual data fluctuating above and below the trend line, the actual data line would appear to hang from

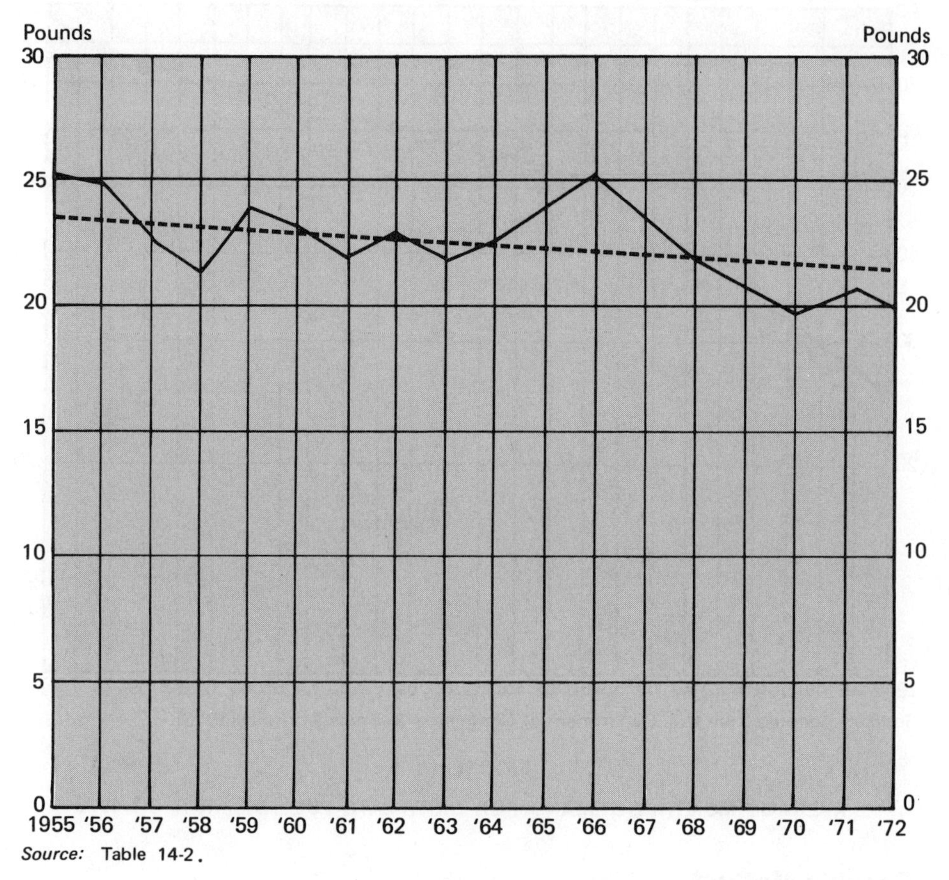

Source: Table 14-2.

FIGURE 14–3

PER CAPITA COTTON CONSUMPTION IN THE UNITED STATES, 1955 TO 1972

the trend line in loops. Periods of operation at full capacity would push the data line on a chart up to the trend line, and any operations at less than full capacity would pull the line below the trend. This concept of trend as full capacity is much less common than the treatment of trend as an average that cuts through the high and the low points of the series. In the analysis that follows it will be assumed that the line describing the trend graphically will cut between the high and the low points of the data.

MEASURING SECULAR TREND

This section describes three methods of measuring secular trend: (1) the graphic method, (2) the method of semiaverages, and (3) the method of least squares.

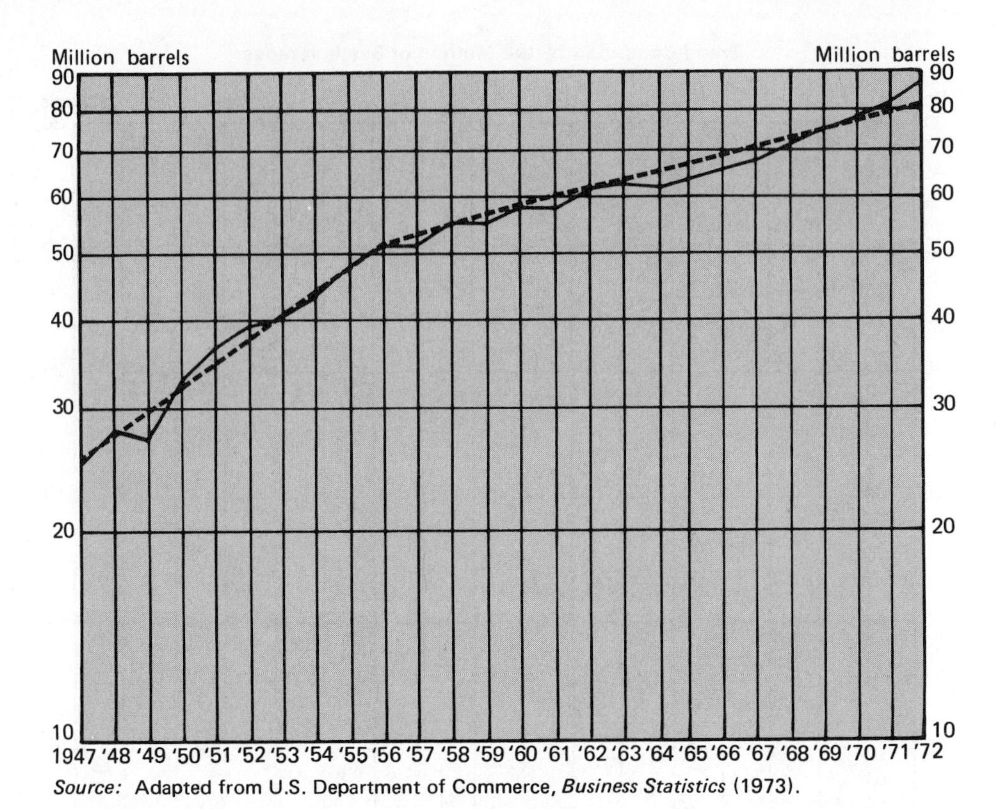

Source: Adapted from U.S. Department of Commerce, *Business Statistics* (1973).

FIGURE 14–4

AVERAGE MONTHLY DEMAND FOR DISTILLATE FUEL OIL, 1947 TO 1972

Graphic Method

The simplest method of determining the trend values of a time series is to draw through the data a straight line that describes the underlying, long-term movement in the series and ignores the movements of a cyclical nature that reverse after a short period. If the trend is being determined from annual data, there will be no seasonal fluctuations to obscure the underlying trend movement. If monthly data were used, it would be necessary to avoid being influenced by the regularly recurring seasonal fluctuations; however, the measurement of secular trend seldom makes use of data classified on the basis of a time period shorter than a year.

Figure 14–5 illustrates a linear trend computed graphically. The annual refinery runs of crude oil by Texaco, Inc., were plotted on an arithmetic chart and the trend line drawn so that it followed the gradual increase in runs, below the peaks but above the low years. It is essential that the trend line follow the underlying course of the actual data without being influenced by the forces that cause the alternating periods of expansion and contraction. Since refinery runs of crude oil appear to have increased by the same amount per year throughout

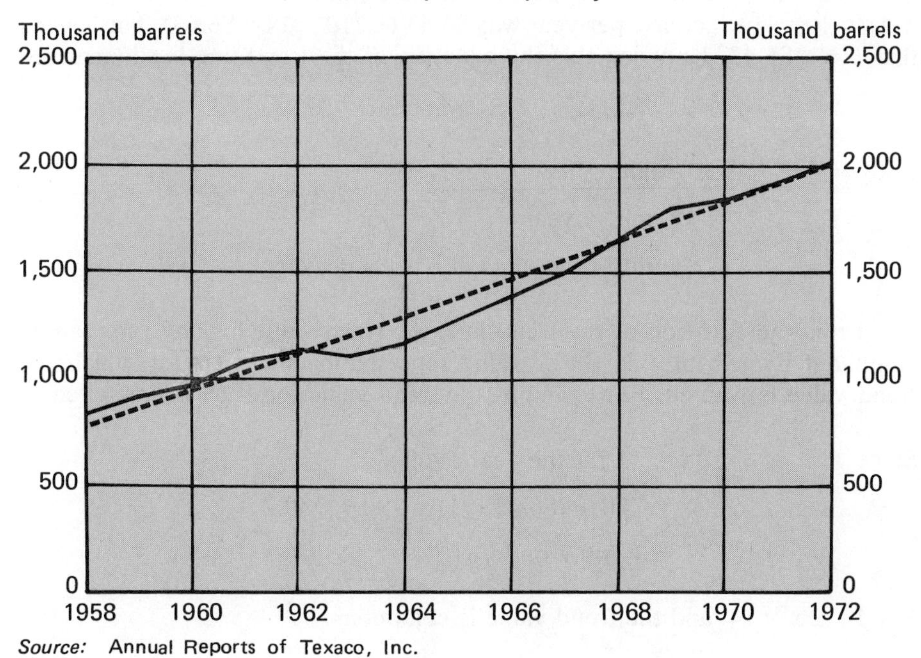

Source: Annual Reports of Texaco, Inc.

FIGURE 14–5

REFINERY RUNS OF CRUDE OIL BY TEXACO, INC., 1958 TO 1972

the period since 1958, a straight line is a good description of the underlying movement in the series.

The following equation for a straight line is given on page 280, Formula 11–1, where it is used to describe the relationship between the variables in a correlation analysis.

$$y = a + bx. \tag{14–1}$$

The measurement of trend by a straight line is in effect the measurement of the correlation between time and the variable for which trend is being measured, in this case, the refinery runs of crude oil by Texaco, Inc. Since time is always plotted on the X axis of a chart, it is designated as the X variable, and the series for which the trend is measured is the Y variable. Since it is inconvenient to use an origin too far from the present, it is normal practice to choose a recent year as the origin and assign it the value of zero. For the series on refinery runs of crude oil, the first year, 1958, can be used as the origin and the x value will be zero. The year 1959 will be $x = 1$, and so on throughout the data.

The trend value of 790 for 1958 can be read from the chart. The value of the trend in the origin year is the a value in the formula for a straight line; by letting 1958 be the origin, $a = 790$. Again reading from the chart, the trend line was 2,000 in 1972, an increase of 1,210 units (thousands of barrels per day) in 14 years. The increase per year was 86.43 ($1{,}210 \div 14 = 86.43$). This means that $b = +86.43$. Knowing this the equation of the trend line is summarized

$$y_c = 790 + 86.43x$$

origin: 1958

x unit: 1 year

y unit: thousands of barrels per day.

From the equation of the trend line, the trend value for any year can be computed by substituting the x value representing the year for which the trend value is wanted. For example, the trend value for 1969 is computed

Since $\quad x = 11$ for the year 1969

$$y_c = 790 + (86.43)(11) = 790 + 950.7$$
$$= 1{,}740.7 \text{ or } 1{,}741.$$

For 1972 $x = 14$, and the trend value is computed

$$y_c = 790 + (86.43)(14) = 790 + 1{,}210.0$$
$$= 2{,}000.0.$$

The trend value for each year from 1958 to 1972 may be read from the chart or computed from the trend equation. These trend values are given in Table 14–1.

The graphic determination of a trend line is particularly useful when the deviations from the underlying trend are small, with the result that the path of the trend is clearly defined and very little doubt exists as to its proper location.

Instead of locating the line entirely by inspection, the average of the original data may be used as the trend value at the middle of the time period. The average of the actual data should equal the average of the trend values. The average of the straight-line values will be the value for the middle year. Establishing one point on the line in this manner eliminates one subjective element in the graphic solution. The slope of the trend line must still be decided, however.

If a straight line does not fit the data, any type of curve may be used. The criterion for goodness of fit remains the same as for the straight line; that is, the trend line must follow the general course of the data, cutting between the high and low points so that approximately the same amount of area between the plotted line and the trend line is above the trend as below it. The secular trend is assumed to measure the gradual growth or decline of the series. This assumption requires that the trend line follow the course of the data even though it must gradually change its direction.

TABLE 14–1

TOTAL REFINERY RUNS OF CRUDE OIL BY TEXACO, INC, 1958 TO 1972

Trend Computed by the Graphic Method

Year	x	Runs (Thousands of Barrels per Day)	Trend Values (Thousands of Barrels per Day)
1958	0	839	790
1959	1	917	876
1960	2	964	963
1961	3	1,009	1,049
1962	4	1,045	1,136
1963	5	1,115	1,222
1964	6	1,179	1,309
1965	7	1,291	1,395
1966	8	1,399	1,481
1967	9	1,504	1,568
1968	10	1,671	1,654
1969	11	1,801	1,741
1970	12	1,850	1,827
1971	13	1,937	1,914
1972	14	2,007	2,000

Source: Annual Reports of Texaco, Inc.

Method of Semiaverages

Although the graphic method locates the average level of the trend line at the average of the actual data, the slope is determined by the person drawing the line. Since one of the major contributions of statistical methods is the substitution of objective measurements for subjective judgments, the graphic method is not highly regarded. The use of the average of actual data to locate one point on the trend line substitutes measurement for one subjective decision, but an element of judgment still remains in the determination of the slope of the line. On the other hand, the method of semiaverages uses a simple objective method to compute the slope of the line as well as to locate its level.

The data for which the trend values are to be computed are divided into two equal periods, and the average is computed for each period. The per capita consumption of cotton for 1955 to 1972 shown in Table 14–2 is divided into two periods of 9 years each, 1955 to 1963 and 1964 to 1972. The total consumption for the first half is designated S_1; the total consumption for the second period is designated S_2. The average annual consumption for each period is computed by dividing each of the subperiod totals by 9, the number of years in each of the subperiods. The average annual consumption for the first period of 9 years was 23.1 pounds per capita; and for the second period, 22.0 pounds. The middle year of the first period is 1959, and the trend value is the average for the period, 23.1 pounds. The average for the second period, 22.0 pounds, is the trend value for the middle year of the second period, 1968.

TABLE 14-2

PER CAPITA COTTON CONSUMPTION IN THE UNITED STATES, 1955 TO 1972

Trend Values Computed by Method of Semiaverages

Year (1)	x (2)	Consumption Per Capita (Pounds) (3)	Subperiod Totals (4)	Trend Values (Pounds) (5)
1955	−4	25.3		23.6
1956	−3	25.0		23.5
1957	−2	22.6		23.4
1958	−1	21.3		23.3
1959	0	24.0		23.1
1960	1	23.3		23.0
1961	2	22.0		22.9
1962	3	22.9		22.8
1963	4	21.8	208.2	22.6
1964	5	22.5		22.5
1965	6	24.0		22.4
1966	7	25.1		22.3
1967	8	23.5		22.1
1968	9	21.9		22.0
1969	10	20.9		21.9
1970	11	19.7		21.8
1971	12	20.5		21.6
1972	13	20.0	198.1	21.5
Total			406.3	

Source: Textile Organon (March, 1972).

$$a = \frac{208.2}{9} = 23.133$$

$$b = \frac{4(198.1 - 208.2)}{18^2} = \frac{-40.4}{324} = -.1247$$

$$y_c = 23.133 - .1247x$$

origin: 1959

x unit: 1 year

y unit: pounds of cotton

With two points on the straight line located, the slope of the line is computed using simple arithmetic. The trend fell from 23.1 in 1959 to 22.0 in 1968, a decline of 1.1 pounds. Since the decline occurred over a period of 9 years, the average annual decline was −.1247 pounds. This is the b value of the trend equation, and is computed

$$b = \frac{23.133 - 22.011}{9} = \frac{1.122}{9} = -.1247.$$

Either 1959 or 1968 may be used as the origin. If 1959 is designated the origin, $a = 23.1$, since the trend value of 1959 is 23.1. With 1959 as the origin, the equation of the trend is

$$y_c = 23.133 - .1247x$$

origin: 1959

x unit: 1 year

y unit: pounds of cotton per capita.

Since 1959 is the origin on the time scale (X axis), the years prior to 1959 are minus; for example, 1958 is −1, 1957 is −2, and so on. To compute the trend values, the proper value of x is substituted in the trend equation and the value y_c is determined. The trend value for each year is given in Table 14–2.

When the period is an even number of years, Formula 14–2 may be used to compute the slope of the trend line.

$$b = \frac{4(S_2 - S_1)}{N^2} \qquad \textbf{(14–2)}$$

where

S_1 = sum of the y values for the first half of the period

S_2 = sum of the y values for the second half of the period

N = number of years covered by the series.

This formula is used in Table 14–2 to compute the slope of the trend line. The trend values computed by the method of semiaverages are plotted in Figure 14–3, page 371.

Method of Least Squares

The method of least squares is a widely used method of fitting a curve to data and is the most popular method of computing the secular trend of time series. The method locates the trend value for the middle of the time period at the average for the data to which the trend line is being fitted. This means that the least squares method gives the same value at midrange as the two methods previously described. The slope of the line, however, is computed differently, so all the other trend values will be different from those given by the other two methods.

The basic problem of selecting a method of fitting a trend line is to decide on a criterion for measuring goodness of fit. If the points to which the line is being fitted will not all fall on a straight line, then no straight line will fit perfectly. Thus, it becomes a problem of deciding which line fits best. The actual values of the data will deviate from the trend values in all or most of the years. One criterion that might be used to select the best-fitting line is to decide whether the sum (disregarding signs) of the deviations of the actual values from the trend values is a minimum. If it is, this line might be regarded as the line of best fit since, in total, the trend values come closer to the actual values than any other line.

Another criterion might be to decide if the sum of the squared deviations is a minimum. This test of best fit would have an advantage over the criterion that minimizes the sum of the deviations without regard to signs, since it is not strictly logical to ignore the signs. The most common criterion for selecting a straight line to show the trend is to determine if this is the line from which the squared deviations of the actual values from trend values are a minimum. The name *least squares* is derived from the use of this criterion.

The method of least squares gives the most satisfactory measurement of the trend of a series when the distribution of the deviations is approximately normal, but judgment must be used in deciding whether such conditions are present. If there are some extremely large deviations from the trend, the least squares method will tend to overweight these deviations, resulting in a trend that does not follow the general course of the series. Since some time series are subject to extreme fluctuations, the method of least squares should not be used in every situation. The only real criterion for the selection of a method for measuring trend is judging how well the resulting trend line follows the general movement of the series. However, when the least squares method is appropriate, it is probably the most satisfactory method of making the calculations. The least squares method also has the virtue of being impersonal.

When using the method of least squares to compute a trend line, it is convenient to use the middle of the time series as the origin. If the series consists of an odd number of years, the origin is at the middle of the middle year. If an even number of years is used, the origin falls between the two middle years. Although the method works equally well for both situations, the details of calculation for an even number of years differ slightly from those used for an odd number of years. Following the computation of a trend line for an odd number of years, an illustration will be given for the method applied to an even number of years.

Odd Number of Years. The problem of fitting a trend by the method of least squares requires the determination of the values a and b for the equation $y_c = a + bx$ that satisfy the criterion that the sum of the squared differences between the y_c and the y values be a minimum.

Formulas 11–2 and 11–3 give the values of a and b in the least squares regression, so using these formulas with time as the x variable will give the least squares trend line. In this situation it is desirable to assign values arbitrarily to the x variable so that the sum of the x values will be zero. Since time is an arithmetic progression, any arithmetic progression may be assigned to the x series. In Table 14–1 (page 375) the first year in the series was designated as zero and the following years numbered in sequence. But since any year may be used as the origin, with the value of zero, it saves time to assign the middle year of an odd number of years as zero. The sum of the years with a minus sign will exactly equal the sum of those with a plus sign, which means that the sum of the x values will always be zero. This is done in Column 2 of Table 14–3.

TABLE 14–3

AVERAGE MONTHLY FACTORY SALES OF DOMESTIC PASSENGER CARS, 1962 TO 1972

Trend Computed by the Method of Least Squares (Odd Number of Years)

Year (1)	x (2)	Thousands y (3)	xy (4)	Trend Values (Thousands) yc (5)
1962	−5	562.8	−2,814.0	629.5
1963	−4	620.3	−2,481.2	632.6
1964	−3	629.5	−1,888.5	635.7
1965	−2	758.4	−1,516.8	638.8
1966	−1	694.8	− 694.8	641.9
1967	0	589.2	0	645.0
1968	1	700.6	700.6	648.0
1969	2	650.5	1,301.0	651.1
1970	3	515.6	1,546.8	654.2
1971	4	676.8	2,707.2	657.3
1972	5	696.0	3,480.0	660.4
Total		7,094.5	340.3	

Source: U.S. Department of Commerce, *Business Statistics* (1973).

$$a = \frac{\Sigma y}{N} = \frac{7{,}094.5}{11} = 644.9545$$

$\Sigma x^2 = 110$ **(from Appendix C)**

$$b = \frac{\Sigma xy}{\Sigma x^2} = \frac{340.3}{110} = 3.0936$$

$y_c = 644.9545 + 3.0936x$

origin: July 1, 1967

x unit: 1 year

y unit: average monthly factory sales of domestic passenger cars

When $\Sigma x = 0$ in Formulas 11–2 and 11–3, they will reduce to Formulas 14–3 and 14–4, which are normally used for computing the constants of a straight line equation.

$$a = \frac{\Sigma x^2 \Sigma y - 0 \cdot \Sigma xy}{N \cdot \Sigma x^2 - (0)^2} = \frac{\Sigma x^2 \cdot \Sigma y}{N \cdot \Sigma x^2}$$

$$a = \frac{\Sigma y}{N}. \quad \textbf{(14–3)}$$

$$b = \frac{N \cdot \Sigma xy - 0 \cdot \Sigma y}{N \cdot \Sigma x^2 - (0)^2} = \frac{N \cdot \Sigma xy}{N \cdot \Sigma x^2}$$

$$b = \frac{\Sigma xy}{\Sigma x^2}. \quad \textbf{(14–4)}$$

If any years are missing or if only selected years are used so that the x variable is not an arithmetic progression, this method cannot be used. When using a computer to fit a trend line, it may be easier to use Formulas 11–2 and 11–3 with the first year as the origin, rather than Formulas 14–3 and 14–4, since the saving in time required for the calculations is so small that it may be simpler to use the standard regression formulas for two variables.

The value of y is taken from Column 3 in Table 14–3 and the value of a is computed

$$a = \frac{\Sigma y}{N} = \frac{7{,}094.5}{11} = 644.9545.$$

The trend value for the middle year when the origin is at midrange is the arithmetic mean of all the values of the y variable. This is the same value for the middle year as given by the two methods previously discussed.

Σxy is the sum of the products of the individual x and y values in Table 14–3, computed in Column 4; Σx^2 is given in Appendix C as 110 when $N = 11$.[1] These amounts are substituted in Formula 14–4 and the value of b is computed

$$b = \frac{\Sigma xy}{\Sigma x^2} = \frac{340.3}{110} = 3.0936.$$

With the a and b values computed, the trend equation is written

[1] If the table in Appendix C is not available, the value of Σx^2 may be computed from Formula 14–5 for any odd-numbered value of N.

$$\Sigma x^2 = \frac{N(N^2 - 1)}{12}. \quad \textbf{(14–5)}$$

When $N = 11$, the value of Σx^2 is computed

$$\Sigma x^2 = \frac{11(11^2 - 1)}{12} = \frac{11(121 - 1)}{12} = \frac{1{,}320}{12} = 110.$$

$y_c = 644.9545 + 3.0936x$

origin: July 1, 1967

x unit: 1 year

y unit: average monthly factory sales of domestic passenger cars.

Since the trend value for July 1, 1967, is 644.9545, the trend value a year later will be 3.0936 more, or 648.0. For each successive trend value later than 1967, the origin is found by adding 3.0936; and for each trend value earlier than 1967, by subtracting 3.0936. The trend values computed and entered in Column 5 of Table 14–3 are plotted in Figure 14–6, on which the original data are also plotted. Instead of computing the trend values by successively adding or subtracting the b value, it is possible to compute the trend value for any given year by substituting the proper x value in the trend equation. For example, the trend value for 1962 is computed as follows.

Since $x = -5$

$$y_c = 644.9545 + (3.0936)(-5) = 629.5.$$

Even Number of Years. When the number of years to which the trend line is to be fitted is even, only a slight modification of the approach followed in Table 14–3 is necessary. The midrange now falls between two years instead of on the middle of the middle year. The arbitrary values assigned to the years are odd numbers, which makes the common difference 2 instead of 1. Column 2 of Table 14–4 (page 383) shows these values: 1963 is −9, 1964 is −7, and so on. The origin, 0, is between 1967 and 1968, or January 1, 1968.

The arithmetic mean of the y values is the trend value at the origin; in this case, January 1, 1968. Therefore,

$$a = \frac{6{,}351.7}{10} = 653.17.$$

The value of b is found by using Formula 14–4. The common difference of the arithmetic progression assigned to the x values is 2, which means that one x equals one-half year. $\Sigma xy = -223.1$, and $\Sigma x^2 = 330$ (from Appendix C).[2] The value of b is computed

[2] If the table in Appendix C is not available, the value of Σx^2 may be computed from Formula 14–6 for any even-numbered value of N.

$$\Sigma x^2 = \frac{N(N^2 - 1)}{3}. \quad \textbf{(14–6)}$$

When $N = 10$, the value of Σx^2 is computed

$$\Sigma x^2 = \frac{10(10^2 - 1)}{3} = \frac{10(100 - 1)}{3} = \frac{990}{3} = 330.$$

$$b = \frac{-223.1}{330} = -.6761.$$

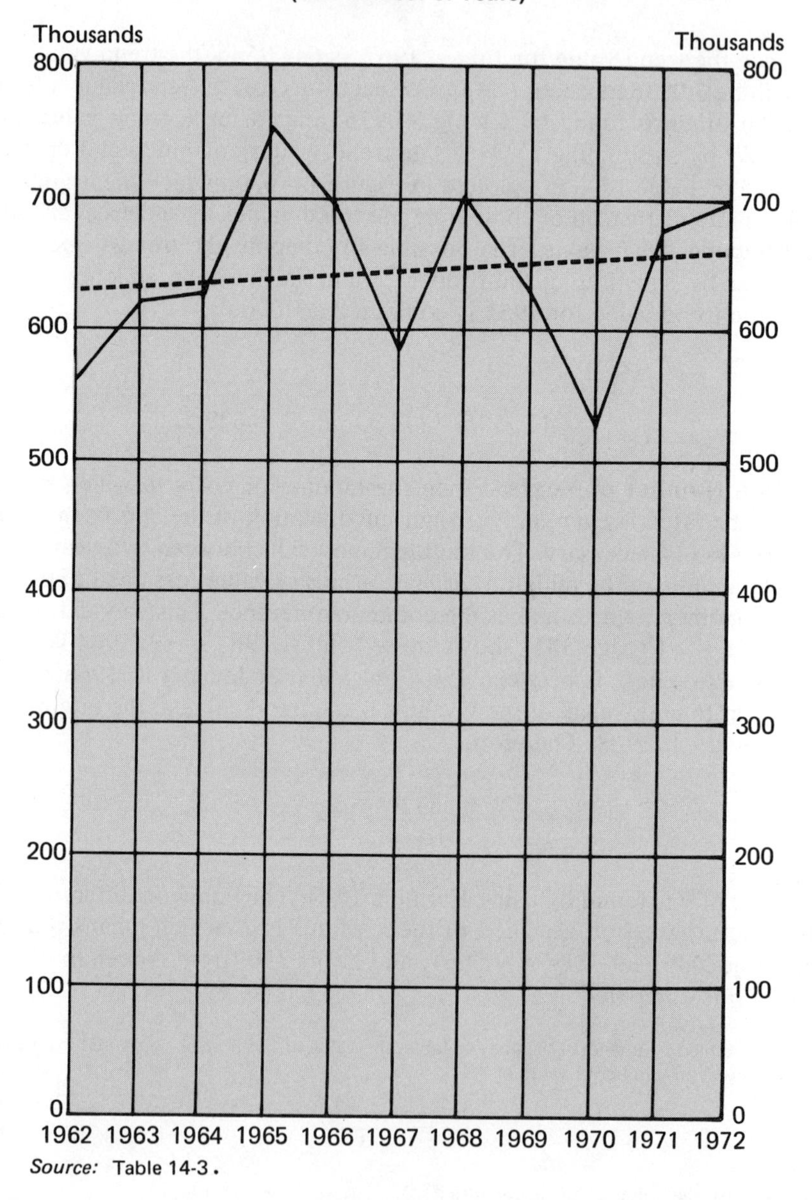

Source: Table 14-3.

FIGURE 14–6

AVERAGE MONTHLY FACTORY SALES OF DOMESTIC PASSENGER CARS 1962 TO 1972

TABLE 14-4

AVERAGE MONTHLY FACTORY SALES OF DOMESTIC PASSENGER CARS 1963 TO 1972

Trend Computed by the Method of Least Squares (Even Number of Years)

Year (1)	x (2)	Thousands y (3)	xy (4)	Trend Values (Thousands) y_c (5)
1963	−9	620.3	−5,582.7	659.3
1964	−7	629.5	−4,406.5	657.9
1965	−5	758.4	−3,792.0	656.6
1966	−3	694.8	−2,084.4	655.2
1967	−1	589.2	− 589.2	653.8
1968	1	700.6	700.6	652.5
1969	3	650.5	1,951.5	651.1
1970	5	515.6	2,578.0	649.8
1971	7	676.8	4,737.6	648.4
1972	9	696.0	6,264.0	647.1
Total		6,531.7	− 223.1	

Source: U.S. Department of Commerce, *Business Statistics* (1973).

$$a = \frac{\Sigma y}{N} \frac{6,531.7}{10} = 653.17$$

$\Sigma x^2 = 330$ **(from Appendix C)**

$$b = \frac{\Sigma xy}{\Sigma x^2} \frac{-221.1}{330} = .6761$$

$y_c = 653.17 - .6761x$

origin: January 1, 1968

x unit: 1/2 year

y unit: average monthly factory sales of domestic passenger cars

The trend equation is

$y_c = 653.17 - .6761x$

origin: January 1, 1968

x unit: ½ year

y unit: average monthly sales of domestic passenger cars.

The calculation of the trend for an even number of years is the same as for any other straight line. The individual x values are substituted in the trend equation and the y_c value is computed. For 1968 the x value is 1 and the trend value is computed

$$y_c = 653.17 + (-.6761)(1) = 652.5.$$

The x value for 1969 is 3 and the calculation of the trend value is made by substituting the value 3 in the trend equation.

$$y_c = 653.17 + (-.6761)(3) = 651.1.$$

The trend value for each year can be computed in this manner. After one value has been determined, however, the easiest method of computing the remaining values is to add the annual increment for the later years and subtract it for earlier years. The yearly increase in the trend line is two times the b coefficient, since b is the amount of change in y when one x equals one-half year. Successive additions of -1.3522 to the trend value of the preceding year give the trend values for the years 1969 to 1972. Successive subtractions of -1.3522 from the 1968 trend value give the values for earlier years. (Since the sign of b is minus, subtracting a minus value gives increasing trend values in the earlier years.)

The trend line fitted to average monthly sales of domestic passenger cars for the period 1963 to 1972 differs somewhat from the trend fitted to the period 1962 to 1972, as shown in Figure 14–7. Omitting the year 1962 changes the slope of the line from plus to minus. This example points up the fact that substantial variations in the values near the end of a series will cause considerable difference in the values a and b computed by the method of least squares. The fact that 1962 was lower than the following years has the effect of changing the trend line for the period from a rising trend to a declining trend. Since the period used in computing the trend can have so much effect on the trend line, the decision as to the period to be used is extremely important.

MEASURING NONLINEAR TREND

The preceding discussion was devoted chiefly to the computation of straight-line trends, but many times a straight line will not describe accurately the long-term movement of a time series. Figure 14–2 (page 370) depicted the trend of factory sales of television sets as a straight line on semilogarithmic paper, since the growth seemed to be at a constant rate. If the same series and the same trend value were shown on an arithmetic grid, the trend would be a curve rather than a straight line. Figure 14–4 (page 372) shows the demand for distillate fuel oil on a semilogarithmic chart. Two straight lines on the semilogarithmic chart are needed to describe the trend in this series: one for the period 1947 to 1956, and the other for the period 1956 to 1972, when the rate of growth was slower than for the earlier years.

Straight Line Fitted to Logarithms by Method of Least Squares

This section describes the methods of computing the logarithmic straight line—a nonlinear trend on an arithmetic chart, but a straight line on a semilogarithmic chart. Since the most common method of computing the line is to

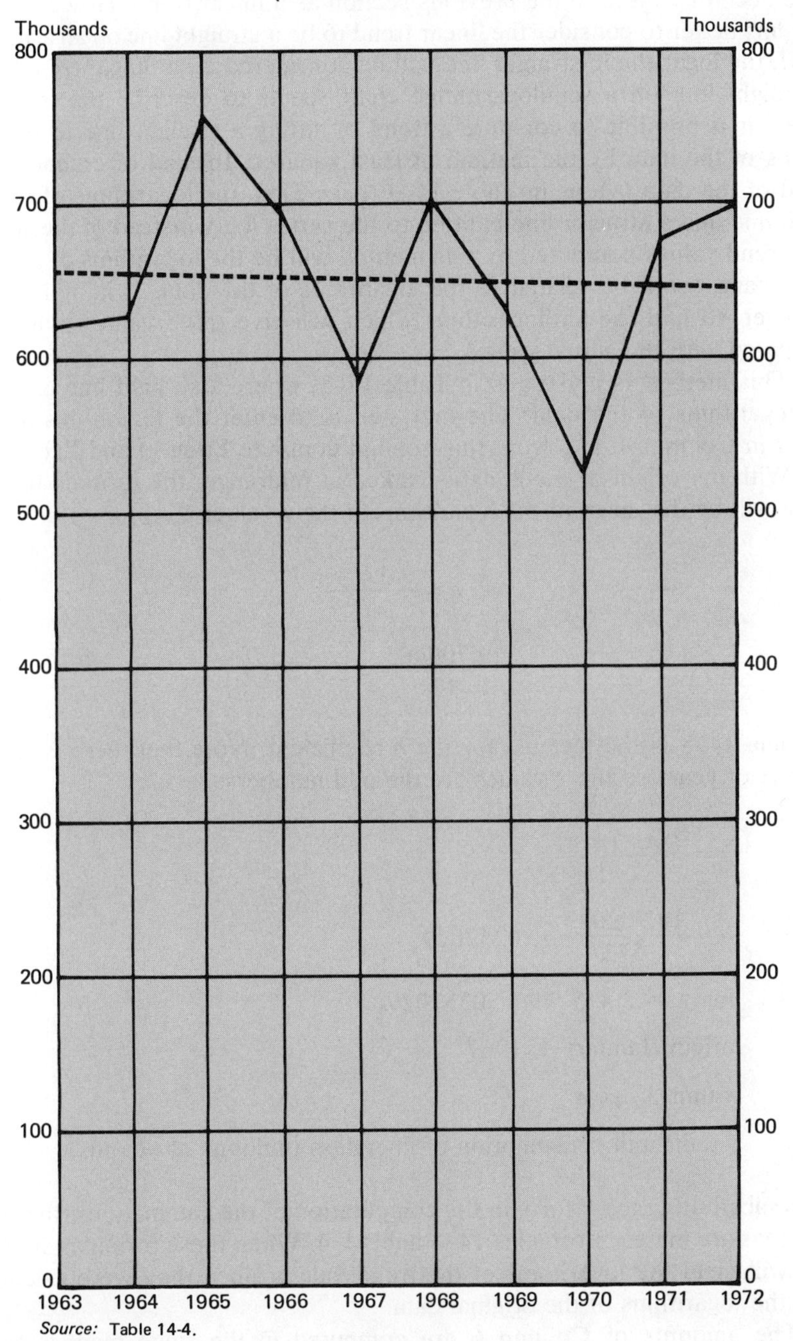

Source: Table 14-4.

FIGURE 14–7

AVERAGE MONTHLY FACTORY SALES OF DOMESTIC PASSENGER CARS 1963 TO 1972

fit a straight line to the logarithms of the data, the subject might logically have been discussed in the previous section as a linear trend. However, since it seems better to consider the linear trend to be a straight line on an arithmetic chart, the logarithmic straight line will be considered a nonlinear trend. When a straight line on a semilogarithmic chart seems to describe the trend of a series, it is possible to compute a trend by fitting a straight line to the logarithms of the data by the method of least squares. Instead of computing the trend of the data (y), as in Table 14–3 (page 379), the logarithms of the data are found and a straight line is fitted to the series $\log y$ instead of the series y. The trend values computed by this method will be the logarithms of the trend values and could be related to the logarithms of the data. It is more useful, however, to find the antilogarithms which will give trend values that can be compared with the actual values.

This method is illustrated in Table 14–5, where a straight line is fitted to the logarithms of the data. The first step is to enter the logarithms of the y series in Column 4, and from this column compute $\Sigma(\log y)$ and $\Sigma(x \cdot \log y)$.

With the origin of the equation taken at midrange, the formula for the a value will be the mean of the logarithms of the y values.

$$a = \frac{\Sigma(\log y)}{N} \qquad \textbf{(14–7)}$$

$$a = \frac{29.709465}{12} = 2.475789.$$

Formula 14–8 is the formula for the b coefficient. Note that there is an even number of years so the x values are the odd numbers.

$$b = \frac{\Sigma(x \cdot \log y)}{\Sigma x^2} \qquad \textbf{(14–8)}$$

$$b = \frac{14.702077}{572} = .0257029$$

$\log y_c = 2.475789 + .0257029x$

origin: January 1, 1967

x unit: $\frac{1}{2}$ year

y unit: mill consumption of fiberglass (millions of pounds).

After substituting $\log y$ for y in the computation of the summations, these two equations are merely Formulas 14–3 and 14–4. When these formulas are used, they will yield the logarithms of the trend values since they were calculated from the logarithms of the original data.

The amounts of Column 6 are computed in the same manner as the amounts in the last column of Table 14–4 (page 383). The value of $\log y_c$ for 1967 is found by adding the value of b (.0257029) to the value of a (2.475789).

TABLE 14–5

MILL CONSUMPTION OF FIBERGLASS IN THE UNITED STATES, 1961 TO 1972

Computation of Semilogarithmic Straight Line

Year (1)	x (2)	Consumption (Millions of pounds) y (3)	Log Y (4)	$x \cdot$ Log y (5)	Log y_c (6)	y_c (7)
1961	−11	147.2	2.167905	−23.846955	2.193057	155.9
1962	− 9	177.5	2.249198	−20.242782	2.244463	175.5
1963	− 7	182.0	2.260071	−15.820497	2.295869	197.5
1964	− 5	220.6	2.343606	−11.718030	2.347275	222.4
1965	− 3	268.2	2.428459	− 7.285377	2.398680	250.4
1966	− 1	315.3	2.498724	− 2.498724	2.450086	282.2
1967	1	303.7	2.482445	2.482445	2.501492	317.3
1968	3	383.8	2.584105	7.752315	2.552898	357.3
1969	5	460.5	2.663230	13.316150	2.604304	402.2
1970	7	404.9	2.607348	18.251436	2.655709	453.9
1971	9	466.7	2.669038	24.021342	2.707115	509.8
1972	11	569.3	2.755341	30.308751	2.758521	573.9
Total			29,709470	14.702074		

Source: Textile Organon (March, 1973).

$$a = \frac{\Sigma(\log y)}{N} = \frac{29.709470}{12} = 2.475789$$

$\Sigma x^2 = 572$ from Appendix C

$$b = \frac{\Sigma(x \cdot \log y)}{\Sigma x^2} = \frac{14.702074}{572} = .0257029$$

$\log y_c = 2.475789 + .0257028x$

origin: January 1, 1967

x unit: ½ year

y unit: mill consumption of fiberglass in the United States (millions of pounds)

Log y_c for each succeeding year is found by adding the value of $2 \times b$; for each earlier year, by subtracting the value of $2 \times b$. The trend values in Column 7 are derived by finding the antilogs for the logarithms in the preceding column.

The trend values are plotted along with the original data in Figure 14–8, a semilogarithmic chart. The fact that the trend is a straight line on semilogarithmic paper means that it increases at a constant rate. This trend characteristic is of considerable significance. That each of the trend values computed in Table 14–5 is 12.6% more than the preceding year can be verified by computing the percentage change between two trend values. These computations are unnecessary, however, since the $2b$ value is the logarithm of the ratio

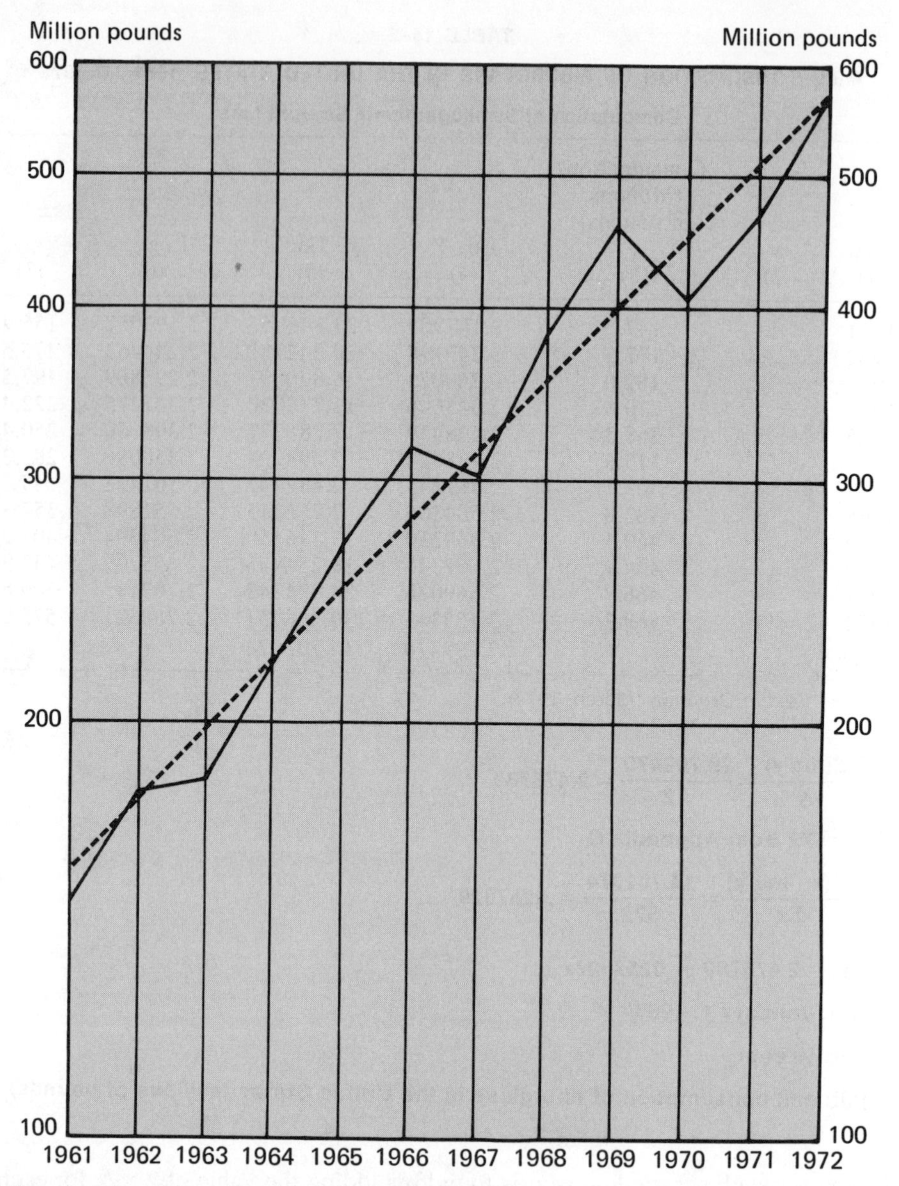

Source: Adapted from *Textile Organon* (March, 1973).

FIGURE 14-8

MILL CONSUMPTION OF FIBERGLASS IN THE UNITED STATES, 1961–1972

of each year to the preceding year. In Table 14–5 the value of $2b$ was found to be .0514058, which is the logarithm of 1.126. Adding .0514058 to the logarithm of any trend value multiplies that trend value by 1.126, which is increasing it by 12.6% (1.126 − 1.000 = .126 or 12.6%). The rate of increase for the trend can be determined from the b value of any logarithmic trend. If the x value is 1 year, the annual rate of increase is found by subtracting 1 from the antilog of b.

Second-Degree Parabola

The examples of trend lines given previously illustrate that the trend may not continue in a given direction at either a constant amount or a constant rate. A series may have an upward trend, but the amount of the increase may diminish. It is not uncommon for the trend of such a series to turn down finally.

In the preceding discussion the method of least squares was applied to the fitting of a straight line to the data, and a straight line to the logarithms of the data. Other types of curves can be fitted by the method of least squares, thus making possible the use of this method when a more flexible trend than a straight line is required. The simplest of these nonlinear trend lines is the *second-degree parabola* shown in Figure 14–9, page 391, which is computed

$$y_c = a + bx + cx^2. \quad \textbf{(14–9)}$$

When numerical values for a, b, and c have been derived, the trend value for any year may be computed by substituting in the equation the value of x for that year. Formulas 14–10 and 14–11 are used for computing these constants, on the condition that the years are represented by an arithmetic progression with the origin at midrange, as already demonstrated in Table 14–3 (page 379).

$$c = \frac{N \cdot \Sigma x^2 y - \Sigma x^2 \Sigma y}{N \cdot \Sigma x^4 - (\Sigma x^2)^2}. \quad \textbf{(14–10)}$$

$$a = \frac{\Sigma y - c\Sigma x^2}{N}. \quad \textbf{(14–11)}$$

Formula 14–4 (page 380) is used to calculate the value of b.

$$b = \frac{\Sigma xy}{\Sigma x^2}.$$

The values to be substituted in these formulas are derived in Table 14–6 in the same manner as for a straight line.[3] Since the computations in Table 14–6 are for an odd number of years, the years are assigned values of a progression with a common difference of 1. No example is given for an even number of years since no new principle is involved.

[3] If the table in Appendix C is not available, the value of Σx^4 may be computed from the formula

$$\Sigma x^4 = \frac{3N^5 - 10N^3 + 7N}{240}. \quad \textbf{(14–12)}$$

TABLE 14–6

AVERAGE MONTHLY FACTORY SALES OF TRUCKS AND BUSES IN THE UNITED STATES, 1950 TO 1972

Computation of a Second-Degree Parabola

Year (1)	x (2)	x^2 (3)	Average Monthly Sales (Thousands) y (4)	xy (5)	x^2y (6)	y_c (7)
1950	−11	121	98.9	−1,087.9	11,966.9	96.3
1951	−10	100	100.3	−1,003.0	10,030.0	91.4
1952	− 9	81	88.0	− 792.0	7,128.0	87.3
1953	− 8	64	88.9	− 711.2	5,689.6	84.0
1954	− 7	49	70.6	− 494.2	3,459.4	81.6
1955	− 6	36	88.0	− 528.0	3,168.0	80.1
1956	− 5	25	74.9	− 374.5	1,872.5	79.4
1957	− 4	16	74.6	− 298.4	1,193.6	79.5
1958	− 3	9	58.2	− 174.6	523.8	80.5
1959	− 2	4	78.5	− 157.0	314.0	82.3
1960	− 1	1	81.5	−81.5	815.0	84.9
1961	0	0	77.1	0	0	88.4
1962	1	1	91.9	91.9	91.9	92.7
1963	2	4	109.7	219.4	438.8	97.9
1964	3	9	114.8	344.4	1,033.2	103.9
1965	4	16	134.7	538.8	2,155.2	110.8
1966	5	25	133.9	669.5	3,347.5	118.5
1967	6	36	117.9	707.4	4,244.4	127.0
1968	7	49	147.1	1,029.7	7,207.9	136.4
1969	8	64	148.4	1,187.2	9,497.6	146.6
1970	9	81	130.5	1,174.5	10,570.5	157.7
1971	10	100	159.5	1,595.0	15,950.0	169.6
1972	11	121	191.2	2,103.2	23,135.2	182.3
Total	0	1,012	2,459.1	3,958.7	123,099.5	

Source: U.S. Department of Commerce, *Business Statistics* (1973) and *idem.*, *Survey of Current Business* (March, 1973).

$$c = \frac{N \cdot \Sigma x^2 y - \Sigma x^2 \Sigma y}{N \cdot \Sigma x^4 - (\Sigma x^2)^2}$$

From Appendix C

$\Sigma x^2 = 1{,}012$

$\Sigma x^4 = 79{,}948$

$$c = \frac{(23)(123{,}099.5) - (1{,}012)(2{,}459.1)}{(23)(79{,}948) - (1{,}012)^2} = \frac{342{,}679.3}{814{,}660} = .420641$$

$$a = \frac{\Sigma y - c\Sigma x^2}{N}$$

TABLE 14-6 (concluded)

$$a = \frac{2{,}459.1 - (.420641)(1{,}012)}{23} = \frac{2{,}459.1 - 425.688692}{23} = \frac{2{,}033.411308}{23}$$

$$= 88.4092$$

$$b = \frac{\Sigma xy}{\Sigma x^2} = \frac{3{,}958.7}{1{,}012} = 3.91176$$

$$y_c = 88.4092 + 3.91176(x) + .42064(x^2)$$

origin: July 1, 1961

x unit: 1 year

y unit: average monthly factory sales of trucks and buses in the United States.

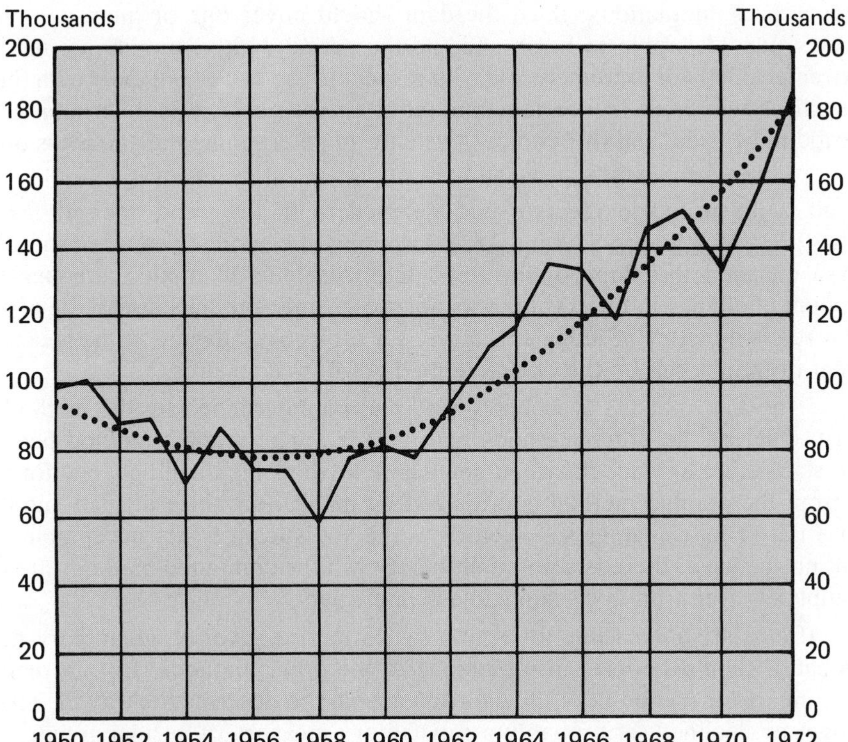

Source: Table 14-6.

FIGURE 14-9

FACTORY SALES OF TRUCKS AND BUSES IN THE UNITED STATES, 1950 TO 1972

CHOICE OF TREND LINE

There is no positive objective test that can be applied to a line to determine whether or not it is a satisfactory measure of the secular trend of a series. Within limits, it is possible to decide by inspection of a chart whether a given line is a reasonable measure of the trend. No matter what other method is used to decide if a trend is satisfactory, the plotted line must give a satisfactory fit to a graph of the series to which the trend is fitted.

It is usually true that a number of lines will fit the data and describe the secular trend reasonably well. In such a case it is necessary to make a choice among them, and the decision may well be made on the basis of the method of computation used. The method of least squares is probably the most favored of all those described in the preceding pages. The line from which the square of the deviation is a minimum usually fits the data in a satisfactory manner, though not always. If none of the lines described will fit satisfactorily, it is possible to use a different period; the choice of the period is somewhat arbitrary. However, the movements, other than trend, most likely to affect the data are cyclical fluctuations; thus, the data should cover one or more complete cycles. It is desirable to begin and end the period with years that are neither extremely high nor extremely low with respect to the cycle. But even with these specifications, there is a certain amount of choice possible in determining the period to be used, and this choice is a factor in determining the goodness of fit.

The importance of variations in the time period to which the trend line is fitted is illustrated in the two periods used to fit the trend lines in Tables 14–3 and 14–4 (pages 379 and 383). Dropping the earliest year used in Table 14–3 changed the slope of the trend line from plus to minus, although the amount of change in the b value was not great. Since the selection of the period covered is a matter of judgment, there is a subjective element in the measurement of trend even by the objective method of least squares.

If no line that fits reasonably well can be determined by the method of least squares, the other methods may be tried. The graphic method permits the statistician to place the trend line where he thinks it should go, and for this reason, the graphic method is criticized by many. But since a trend line that does not fit in a manner satisfactory to the statistician is not acceptable and will not be used, there is a possibility that even the computed methods are not completely free from a certain subjective element.

If a least squares line fitted to a logical period gives a satisfactory fit, it probably should be used in preference to the other methods. But, in reality, any line that is a good fit of the data and appears to describe the way they have grown or declined, is probably a satisfactory measure of secular trend. It is difficult to get away from the goodness of fit as observed on a graph as the basis for deciding on a trend line.

The trend lines fitted by the method of least squares or the method of semiaverages have an inflexibility that often makes them a poor fit of the series.

For example, an unusually high or low year at the end of the series has a pronounced effect on the least squares trend line. This is inevitable, since the equations used minimize the squared deviations from the line. A large increase or decrease in any year, and particularly in several years, will pull the line in the direction of the extreme amounts. In doing so, it will tilt the straight line along its whole length.

USING A MEASURE OF SECULAR TREND

The management of a growing business should have a clear picture of that growth as part of its basic information. The secular trend of a series gives a better picture of the growth than the actual data, since the actual data contain the effect of the business cycle on the series as well as the effect of the secular trend.

If it is desired to compare the growth of two series, the best comparison can be made between the trends rather than between the two series themselves. This is particularly true if any pronounced cyclical fluctuations are present in the data.

Although it is not commonly done, it would emphasize the changes in the trends of two series if the actual data were not plotted on the chart to make easier the comparison of the two trend lines. For example, in tracing the changes in the trend of consumption of different kinds of food, it might be easier to plot the trend lines instead of the actual consumption figures for each year. In this way the attention could be focused on the underlying shifts in demand and not obscured by the short-term variations in demand.

In the following pages some of the ways in which the measures of secular trend can be used in making business decisions are discussed.

Comparison of Two Logarithmic Trend Lines

Since the rates of increase of two logarithmic straight lines are percentages, they can be compared even though the actual values of the trends are entirely different and not comparable. For example, if we want to compare the rate of growth of cigarette consumption with the rate of growth of gasoline consumption, the rates of increase of the trend lines can be used. Figure 14–10 shows cigarette consumption and gasoline consumption for 1952 to 1972. The equations of the two straight lines fitted to the logarithms of the two series are:

Cigarette consumption

$\log y_c = 2.70582 + .00879317x$

origin: July 1, 1962

x unit: 1 year

y unit: consumption (billions of cigarettes)

Demand for gasoline

$\log y_c = 3.20805 + .0141139x$

origin: July 1, 1962

x unit: 1 year

y unit: demand (millions of barrels)

It was shown on page 388 that the rate of increase in the logarithmic straight line is represented by the b value of the trend equation. From the equations just given, the rates of increase are

Cigarette consumption: $b = .000879317$, which is the logarithm of 1.020. This represents a 2.0% increase.

Demand for gasoline: $b = .00141139x$, which is the logarithm of 1.033. This represents a 3.3% increase.

These two trend lines represent the growth in the two industries since 1952. The two percentages compare the rates of growth of the two series and show that the consumption of these two products grew at nearly the same rate. The slopes of straight lines fitted to the logarithms of the data can always be compared in this manner. The two trend lines in Figure 14–10 are very nearly parallel although the trend for the demand for gasoline is slightly steeper.

Instead of comparing the trends of two different series, the trend lines may be used to compare the rate of growth of the same series in two different periods. Figure 14–4 (page 372) shows the demand for distillate fuel oil from 1947 to 1972. The rate of growth was much greater in the first part of the period, from 1947 to 1956, than in the later part, from 1956 to 1972. The trend lines plotted on the chart are straight lines fitted to the logarithms of the two series: demand for distillate fuel oil from 1947 to 1968 and from 1956 to 1972. The equations of the two trend lines are

Trend equation, 1947 to 1956

$\log y_c = 1.562973 + .032887x$

origin: January 1, 1952

x unit: 1 year

y unit: average monthly demand for distillate fuel oil, 1947–1956 (millions of barrels)

Trend equation, 1956 to 1972

$\log y_c = 1.781356 + .011732x$

origin: July 1, 1964

x unit: 1 year

y unit: average monthly demand for distillate fuel oil, 1956–1972 (millions of barrels).

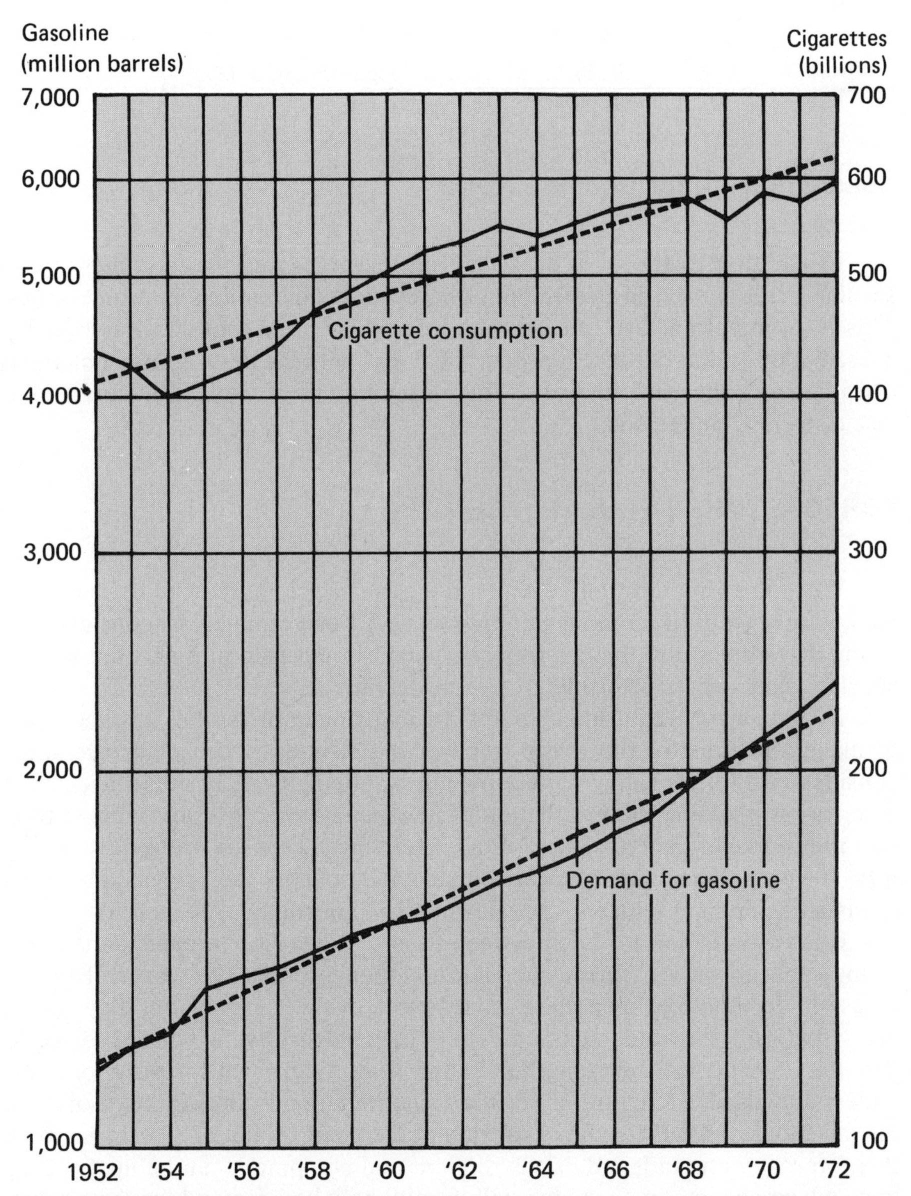

Source: Adapted from U.S. Department of Commerce, *Business Statistics* (1973) and *idem., Survey of Current Business* (March, 1973).

FIGURE 14–10

CIGARETTE CONSUMPTION AND DOMESTIC DEMAND FOR GASOLINE IN THE UNITED STATES, 1952 TO 1972

The b values in the equation represent the rates of increase of the trends, which are

1947 to 1956: $b = .032887$, which is the logarithm of 1.079.
This represents a 7.9% increase.

1956 to 1972: $b = .011732$, which is the logarithm of 1.027.
This represents a 2.7% increase.

Adjusting Data for Secular Trend

In addition to the uses described on the preceding pages, a measure of secular trend is used in determining the cyclical fluctuations in a time series. This is an important use of a measure of secular trend, but it will not be discussed at this point because it properly belongs with the description of methods of isolating cyclical fluctuations. The measurement of cyclical fluctuations is discussed in Chapter 16.

FORECASTING SECULAR TREND

When the long-term movements in a series are important to the businessman, a forecast of these movements is needed. For example, any consideration of the investment possibilities of an industry should take into account its past secular trend and its probable growth or decline.

In an industry requiring a large fixed investment of capital, it is necessary to make a forecast of the future trend of the business when plans are under consideration for replacing old equipment or buying new. If an electric power company is planning the installation of new generators, it is advisable to forecast the course of the consumption of current during the life of the new generators. If a pronounced growth is anticipated, it would probably be good business to install equipment with a larger capacity than needed at present.

It is too common for businessmen to base a decision to expand manufacturing capacity on the current demand for their product. However, if the increase in demand has been due primarily to a cyclical expansion, they should know that from the fundamental nature of the business cycle it can be expected that the demand will not continue at the same high level for very long. An understanding of the nature of secular trend and a conscious forecast of it are the most logical bases for the expansion of the fixed capital of a business. With this forecast as a guide, additions to plant and equipment can be made at the most advantageous times, which will usually be when demand for the product of the concern is below the peak. At such times prices are lowest and the expansion can be carried out under the most favorable conditions. The fact that demand at this time is not sufficient to utilize the full capacity of the plant is not the most important factor to consider; the forecast of the long-term trend of the business should be the guiding factor. As long as cyclical fluctuations are present, it is inevitable that the concern will operate well below peak capacity at certain times. These cyclical fluctuations represent a separate problem that the businessman must take into account. But when the problem is a forecast of the long-term trend, it is highly important not to permit the results to be influenced by the cyclical movements.

The contribution of statistical methods to the forecasting of secular trend consists primarily in the measurement of events in the past. Since the only basis for a forecast of the future is a knowledge of past happenings, the first step in the forecast of secular trend is normally the measurement of the trend in the immediate past.

The use of an equation to describe the trend of a series has suggested that a simple method of forecasting trend would be by *extrapolation,* that is, projection of this fitted trend into the future. This process should be used only when it has been decided that the trend in the future will follow the same pattern as it did in the period to which the trend line was fitted. If the trend grew the same amount each year, a straight-line trend would be projected with the same increase each year in the future that it had shown in the past. If a straight line were fitted to the logarithms of the data, it would mean that the *rate* of increase was the same each year. If this line were extrapolated, it would mean that the trend would continue to increase at the same *rate.*

It is dangerous to assume that the increase or decrease in the trend for a relatively short period (or even for a long period) in the past will continue for any fixed time in the future. Many different factors would normally have an effect on the growth or decline of a series, and it is reasonable to expect that these factors will not stay the same and that their importance with respect to each other will not remain fixed.

Projecting a past trend assumes that the factors causing the change in the secular trend *will continue to have the same influence in the future as in the past.* A more logical approach to the problem of forecasting trend is to determine as accurately as possible what factors caused the growth or decline, to consider the development of any new factors, and then to evaluate the future effect of each of these. Often the separate components may be forecast with more accuracy than the composite trend, since fewer causal factors would be affecting each component.

The first step to take in applying this method is to measure the secular trend, using one or more of the methods described earlier. It would be important, if possible, to measure the trend in the several elements. It might even be possible to separate the trend of one or more of the components into subcomponents, which might be forecast individually. Each situation would be a separate problem, and the use of statistical methods would be chiefly to measure certain aspects of the problem and to break it into as many simple factors as possible. From this point the forecast would not use statistical methods so much as economic analysis of the factors in the specific problem.

STUDY QUESTIONS

14–1. What is meant by the statement that the measurement of secular trend is in many aspects a problem of averages?

14–2. What are the advantages and disadvantages of computing a measure of secular trend graphically? Arithmetically?

14–3. Discuss the problem of deciding upon the proper period of time to which a trend line should be fitted.

14–4. If the trend is being fitted to a time series that has a very pronounced erratic fluctuation, such as that caused by an extended strike, would you consider leaving this year out of the calculation of the trend? If it did seem advisable to leave such an unusual year out of the calculation, how would you fit a straight line by the method of least squares?

14–5. What are the advantages and disadvantages of the method of semiaverages for computing a straight-line trend as compared with the method of least squares?

14–6. Discuss the problem of deciding which of several computed trend lines is the best measure of the long-term growth in a series.

14–7. What are the major forces that cause economic time series to exhibit a long-term movement called secular trend.

14–8. The method of semiaverages, described for a time series consisting of an even number of years, permits dividing the series into two equal parts. Usually it is possible to vary the period used for fitting the trend enough to insure having an even number of years. However, it if were very important that all of a period consisting of an odd number of years be used, describe how you would fit a trend line by the method of semiaverages.

14–9. Give an example of why it might be necessary to use an odd number of years in fitting a trend line by the method of semiaverages. (Any series with an odd number of years can be made into one with an even number of years by dropping one year, presumably the earliest.)

14–10. Why does the straight line fitted to the logarithms of the data permit the comparison of the b values of two equations fitted to series in entirely different units? (For example, in the text the comparison was made between the b values of the equations representing cigarette consumption and gasoline consumption.)

14–11. If you used a computed trend to forecast the trend of a series, you would be making some assumptions. What are some of these basic assumptions?

14–12. How do you explain the fact that cigarette consumption and gasoline consumption increased at nearly the same rate, as shown in Figure 14–10 (page 395)?

14–13. What criterion should be used to decide whether or not to fit a nonlinear trend line to a series? (Remember that it is always possible to break the series into segments small enough so that a straight line could be made to fit.)

PROBLEMS

14–1. The table at the top of page 399 gives the yearly sales of three companies (in millions of dollars) for the years 1961 to 1973.

Year	Company A	Company B	Company C
1961	14	10	12
1962	16	14	15
1963	20	11	11
1964	22	13	8
1965	27	17	10
1966	20	11	6
1967	22	15	11
1968	28	21	14
1969	30	26	14
1970	31	24	11
1971	29	20	10
1972	33	27	12
1973	37	29	13

(a) Fit a straight line by the method of least squares to each of the series.

(b) Plot each of the series and the computed trend values on an arithmetic chart.

14–2. The following table gives the average monthly sales of three companies (in millions of dollars) for the years 1960 to 1973.

Year	Company D	Company E	Company F
1960	6	18	24
1961	9	16	18
1962	8	20	23
1963	10	29	16
1964	12	22	12
1965	13	28	14
1966	10	34	7
1967	12	30	9
1968	14	32	12
1969	16	36	15
1970	17	38	12
1971	19	42	8
1972	18	40	7
1973	16	43	7

(a) Fit a straight line by the method of least squares to each of the series.

(b) Plot each of the series and the computed trend values on an arithmetic chart.

14–3. Compute the trend values by the method of semiaverages for each of the series in Problem 14–2. Plot these trend values on the charts drawn in Problem 14–2. How much do these trend values differ from those computed in Problem 14–2 by the method of least squares? Which method do you consider gives the better measure of trend?

14–4. The table at the top of page 400 gives total crude oil runs to stills in the United States, and the crude oil runs to stills (both in thousands of barrels per day) by the Exxon Company, U.S.A., and the Phillips Petroleum Company from 1963 to 1972.

(a) Plot these series on the same arithmetic chart.

(b) Plot these series on the same semilogarithmic chart.

(c) Fit a straight line to each of the series and plot the trend values on the arithmetic chart drawn in (a).

(d) Fit a straight line to the logarithms of each of the series and plot the trend values on the semilogarithmic chart drawn in (b).

(e) Compute the average rate of growth in the trend of each of the three series.

Year	U.S. Total	Exxon Company U.S.A.	Phillips Petroleum Company
1963	8,687	806	241
1964	8,831	782	248
1965	9,043	805	249
1966	9,444	816	292
1967	9,815	912	335
1968	10,341	943	356
1969	10,629	992	359
1970	10,870	989	364
1971	11,199	976	371
1972	11,730	1,029	386

Source: U.S. Department of Commerce, *Business Statistics* (1973), and the annual reports of Exxon Company, U.S.A., and Phillips Petroleum Company.

14–5. The following table gives the gross national product (in constant dollars) in the United States for the years 1954 to 1972.

(a) Plot the series on an arithmetic chart.

(b) Compute the equation of the straight-line trend and compute the trend values for all the years by the method of least squares.

(c) Plot the trend values on the chart with the actual values plotted in (a).

Year	Billions of 1958 Dollars	Year	Billions of 1958 Dollars
1954	407	1964	578
1955	438	1965	617
1956	446	1966	657
1957	453	1967	673
1958	447	1968	707
1959	476	1969	727
1960	488	1970	722
1961	497	1971	742
1962	530	1972	790
1963	550		

Source: U.S. Department of Commerce, *Business Statistics* (1973).

14–6. The data on gross national product in constant dollars given in Problem 14–5 will be used in this problem.

(a) Plot this series on a semilogarithmic chart.

(b) Compute the equation of the straight line to the logarithms by the method of least squares, and from the equation compute the trend value for all years.

(c) Plot the trend values on the chart drawn in (a).

14–7. The following table gives the consumption of aluminum and magnesium (in thousands of short tons) in the United States from 1957 to 1971.

Year	Aluminum	Magnesium
1957	1,776	44.4
1958	1,811	35.4
1959	2,146	41.6
1960	1,753	37.1
1961	2,048	45.4
1962	2,359	46.0
1963	2,591	55.0
1964	3,216	54.7
1965	3,734	69.9
1966	4,002	82.7
1967	4,009	98.2
1968	4,663	86.4
1969	4,710	95.1
1970	4,519	93.5
1971	5,074	99.1

Source: United States Bureau of Mines.

(a) Plot these series on the same semilogarithmic chart. State briefly how the trends of the two series compare.

(b) Compute the equation of the straight line to the logarithms by the method of least squares, and from the equation compute the trend values for all years, for each series.

(c) Plot the trend values for each series on the chart with the actual values plotted in (a).

(d) Compute the average percentage increase in the trend of each series. Compare these percentages with your conclusions in (a).

14–8. The following table gives the factory sales of semiconductors (discrete devices) from 1958 to 1972.

Year	Millions of Dollars	Year	Millions of Dollars
1958	236	1966	931
1959	408	1967	810
1960	560	1968	800
1961	565	1969	905
1962	583	1970	769
1963	533	1971	621
1964	658	1972	720
1965	761		

Source: U.S. Department of Commerce, *Business Statistics* (1973) and *idem.*, *Survey of Current Business* (March, 1973).

(a) Plot this series on an arithmetic chart.

(b) Compute the equation of the second-degree parabola by the method of least squares, and from the equation compute the trend values for all years.

(c) Plot the trend values on the chart drawn in (a).

15 Seasonal Variation

Any time series that shows a persistent tendency for certain months to be particularly high or low probably contains a definite seasonal movement, which can be measured by statistical methods. The measures of seasonal variation are usually referred to as *indexes of seasonal variation.* Because seasonal fluctuations in today's economy are so pronounced, it is important that the economist and the business executive understand the significance and use of these indexes and the methods of computation discussed in this chapter. One of the most important uses of an index of seasonal factors—the removal of the fluctuations due to seasonal factors from a time series—will be discussed in the next chapter.

INDEXES OF SEASONAL VARIATION

The computer facilitates lengthy computations involved in compiling an index of seasonal variation. In fact, refinements in methodology can be used which would not be feasible if computations were done on a desk calculator.

Probably the simplest method of showing seasonal variation is to express the figure for each month as a percentage of the total for the year. For example, residential consumption of electric power in Texas shows a pronounced seasonal increase in summer, resulting from the use of air conditioning. The winter months are all below the average monthly consumption for the year. Or to state it another way, more than one twelfth of the total electricity is consumed in each of the months from June through October, and less than one twelfth of the total in each of the remaining seven months. Computing these percentages results in a table of percentages such as Table 15–1. If there were no seasonal variation in the data, each of these monthly percentages would be $8\frac{1}{3}\%$, which is just one twelfth of the total for the year. The amount that each of them differs from $8\frac{1}{3}\%$ expresses the effect of the seasonal factor on the data.

TABLE 15–1

PERCENTAGE OF YEARLY TOTAL CONSUMPTION OF RESIDENTIAL ELECTRIC POWER IN TEXAS IN EACH MONTH, 1963 TO 1972

Month	Percentage of Yearly Total	Month	Percentage of Yearly Total
January	7.2	July	12.2
February	6.6	August	13.4
March	6.1	September	12.3
April	5.7	October	8.4
May	6.2	November	6.5
June	8.9	December	6.5
		Total	100.0

Source: Bureau of Business Research, The University of Texas at Austin.

Another method that perhaps makes it easier to see the effect of seasonal variation on the different months is to multiply each of the percentages by 12, which makes their total 1,200. If there were no seasonal variation, each of the monthly percentages would be 100, with a total of 1,200 for the year. The amount by which any one month differs from 100 is the effect of seasonal variation on that month. Thus, the data of Table 15–1 would be written as in Table 15–2. This means that since December is 78.0%, the effect of the seasonal variation on December is −22.0%. The difference between each percentage and 100 is the effect that seasonal variation has on each month.

A method of showing such a seasonal variation even more clearly is to express each month not as a percentage, but by giving the amount, plus or minus, by which a particular month deviates from 100. The Table 15–2 index would then be expressed as in Table 15–3. These index numbers show the amount by which each month is above or below an average month; an average month is one in which no seasonal variation is present.

These three methods of expressing the seasonal variation give exactly the same information, expressing it in each case in a slightly different manner.

TABLE 15–2

INDEX OF SEASONAL VARIATION FOR CONSUMPTION OF RESIDENTIAL ELECTRIC POWER IN TEXAS, 1963 TO 1972

Month	Index Number	Month	Index Number
January	86.4	July	146.4
February	79.2	August	160.8
March	73.2	September	147.6
April	68.4	October	100.8
May	74.4	November	78.0
June	106.8	December	78.0
		Total	1,200.0

Source: Table 15–1.

TABLE 15–3

INDEX OF SEASONAL VARIATION FOR CONSUMPTION OF RESIDENTIAL ELECTRIC POWER IN TEXAS, 1963 TO 1972

Month	Index Number	Month	Index Number
January	−13.6	July	+46.4
February	−20.8	August	+60.8
March	−26.8	September	+47.6
April	−31.6	October	+ .8
May	−25.6	November	−22.0
June	+ 6.8	December	−22.0
		Total	0

Source: Table 15–2.

The second method, as illustrated in Table 15–2, is the one used most extensively by statisticians.

The three indexes were computed from monthly data ranging over a period of years, and represent the average fluctuation in the monthly data for that period. This average variation is taken as typical of the effect that seasonal variation has on the series. Specific instructions on how to compute an index of seasonal variation are given later in this chapter.

Probably the simplest method of arriving at an index of seasonal variation would be for an individual who is familiar with a particular industry or business to estimate the percentage of the total business of the year normally transacted in each month. These percentages would be an index of seasonal variation, if made to total 100 as the illustration given in Table 15–1. The accuracy of the index would depend entirely upon the knowledge of the individual making the estimate, and could not be expected to approach any high degree of precision. The owner of a retail store would know that December is the month in which his store sells the largest amounts of goods, but he probably would not know whether it is typically 40%, 50%, or 60% above the average month's sales.

A method almost as simple would be to choose a year in which the seasonal variation is thought to be typical of all years, and express each month of that year as a percentage of the total for that year. This would constitute an index expressed in the form used in Table 15–1. Its accuracy would depend upon how representative the particular year actually was. Since there is considerable danger that no one year will be highly representative, the method cannot be recommended. The basic reason for substituting statistical measures for opinions and general impressions is the fact that properly computed measures have a better chance of being accurate.

MEASURING SEASONAL VARIATION—A PROBLEM OF AVERAGES

The basic approach to the measurement of the effect of seasonal forces on a time series has been the use of averages. The idea of a seasonal pattern is, in reality, based on the concept of an average. The seasonal forces do not have exactly the same effect in different years. For example, a late spring will cause the seasonal rise in retail trade to come later than in other years. An absolutely rigid seasonal pattern will not hold in any business. However, if the effects of the seasonal factors are reasonably similar from year to year, it is valid to compute an average seasonal pattern. The problem of dispersion in data and the effect of dispersion on the use of an average are also involved, which means that some method of measuring the dispersion must be used to evaluate any measure of seasonal variation. The more variations in the seasonal patterns from year to year, the less typical, and therefore the less reliable, the average.

Another situation would be for the seasonal pattern to show a gradual shift over the years. In this situation it would be correct to say that the seasonal pattern showed a trend, and instead of computing an average of the values for a given month, it would be appropriate to compute the measure of secular trend for the seasonal pattern of each month. Methods will be developed first to measure the stable seasonal pattern, and then a method of measuring the seasonal pattern when it is changing.

Table 15–4 shows the factory sales of domestic passenger cars in the United States by months for the years 1963 to 1972. In Figure 13–5 (page 341), the factory sales of domestic passenger cars are plotted by months for these years. The pronounced seasonal fluctuation can be seen, and the pattern appears regular enough to justify computing an index of seasonal variation. The simplest method of measuring the seasonal variation is to find the average January value over a period of years, the average February value, and so on for each of the 12 months. The average factory sales for each month are shown in Table 15–5.

The arithmetic mean of August factory sales is 267.47 thousand cars, while June factory sales average 761.86 thousand cars. These two months represent the months of lowest and highest sales of cars, and we may assume that the two averages measure the relative volume of sales in the two months. Average factory sales in the other 10 months may in the same manner be taken as measures of the seasonal level of those months. The 12 means may together be used as an index of the seasonal fluctuations. To be of maximum usefulness, these 12 means should be reduced to one of the forms given in Tables 15–1 through 15–3. This has been done in Column 4 of Table 15–5, where the mean for each month is expressed as a percentage of the mean of the 12 monthly means.

SPECIFIC SEASONALS

In the preceding section an index of seasonal variation was computed simply by averaging factory sales of passenger cars for each month over a period of 10 years. It was assumed that the differences between the averages for the various months were due entirely to seasonal forces. Since trend and

TABLE 15–4

FACTORY SALES OF DOMESTIC PASSENGER CARS IN THE UNITED STATES

By Months, 1963 to 1972

(Thousands)

Year	Jan.	Feb.	March	April	May	June	July	Aug.	Sept.	Oct.	Nov.	Dec.
1963	658.0	592.8	637.1	671.8	695.1	672.9	649.4	165.1	463.0	779.2	726.2	733.0
1964	709.0	665.4	700.9	770.2	719.5	726.7	562.2	230.8	563.8	394.7	648.4	862.4
1965	782.8	753.1	937.9	846.9	819.3	880.9	745.6	330.4	438.5	825.4	878.7	861.3
1966	780.4	748.8	902.0	793.9	771.2	802.5	480.0	136.4	592.4	797.7	791.2	740.5
1967	625.0	501.9	647.4	628.3	713.4	732.3	410.6	218.3	570.6	608.8	645.2	768.5
1968	747.2	668.2	764.0	747.8	876.2	781.6	605.4	182.6	620.0	889.5	831.0	693.7
1969	782.0	676.7	721.0	676.9	678.0	740.4	446.9	329.5	706.5	815.6	644.0	588.8
1970	545.0	528.4	594.4	627.2	684.4	758.4	464.3	254.0	454.2	365.4	341.1	570.6
1971	678.1	719.0	815.9	703.6	716.7	761.3	468.9	456.6	712.0	758.6	736.6	593.2
1972	666.0	716.1	765.2	736.9	798.0	761.6	393.6	371.0	808.8	841.7	827.4	666.2
Total	6,973.5	6,570.4	7,485.8	7,203.5	7,471.8	7,618.6	5,226.9	2,674.7	5,929.8	7,076.6	7,069.8	7,078.2

Source: U.S. Department of Commerce, *Business Statistics* (1973).

TABLE 15-5

INDEX OF SEASONAL VARIATION

Based on the Arithmetic Mean of Actual Monthly Data
Factory Sales of Domestic Passenger Cars in the United States, 1963 to 1972

(Thousands)

Month (1)	Total (2)	Arithmetic Mean (3)	Index of Seasonal Variation (4)
January	6,973.5	697.35	106.8
February	6,570.4	857.04	100.6
March	7,485.8	748.58	114.6
April	7,203.5	720.35	110.3
May	7,471.8	747.18	114.4
June	7,618.6	761.86	116.6
July	5,226.9	522.69	80.0
August	2,674.7	267.47	41.0
September	5,929.8	592.98	90.8
October	7,076.6	707.66	108.3
November	7,069.8	706.98	108.2
December	7,078.2	707.82	108.4
Total	78,379.6	7,837.96	1,200.0

Source: Table 15–4.

cyclical fluctuations are also present in this series, the assumption was probably not correct. It is usually more accurate to isolate the effect of the seasonal forces on each month before averaging a number of years to secure a measure of the typical seasonal variation.

The measurement of the effect of the seasonal forces on a given month will result in a figure called a *specific seasonal.* The different methods of computing these specific seasonals give a number of different methods of computing an index of seasonal variation. Since the methods of averaging the specific seasonals are the same no matter how the specific seasonals were derived, the difference between methods is chiefly a difference in the ways of securing the values to be averaged.

Ratio to 12-Month Moving Average Centered at Seventh Month

The most widely used method of measuring the seasonals is a ratio to a 12-month moving average or total. The basic principle of this method is that an average of the 12 months of a year cannot be affected by the seasonal influences, since each month was included in the total, forcing the seasonal effects to average out. Therefore, it is possible to compare an individual month with this average to isolate the effect of the seasonal forces on that individual month. For example, as shown in Table 15–6, the sales for June, 1964, were

TABLE 15–6

FACTORY SALES OF DOMESTIC PASSENGER CARS IN THE UNITED STATES, 1964

Month	Sales (Thousands)
January	709.0
February	665.4
March	700.9
April	770.2
May	719.5
June	726.7
July	562.2
August	230.8
September	563.8
October	394.7
November	648.4
December	862.4
Total	7,554.0
Monthly average (arithmetic mean)	629.5

Source: Table 15–4.

726.7 thousand cars, or 15.4% above 629.5 thousand, the average month of 1964 (computed by dividing total sales for 1964 by 12). What can be assumed to be the cause of June sales being this much above the average month for the year? It was stated on page 336 that the forces affecting a time series may be classified as secular trend, seasonal, cyclical, and random; so the value for the month of June (as well as any other month) will be determined by the combination of the four factors. The average monthly sales of a given year cannot be influenced by seasonal factors, as mentioned before. By averaging the 12 months most of the effect due to random fluctuations would be eliminated, since random movements are movements without any definite pattern. Trend and cyclical fluctuations, however, affect an average for a year, so it is not valid to assume that this average is free of trend and cyclical influences. The difference between the sales for June and the average for the year is due to seasonal or random fluctuations, or both.

There is probably no satisfactory method of separating the random and seasonal variations, but when the combined effects of the random and seasonal movements are averaged for several values of a given month, the random movements tend to average out, leaving only the seasonal fluctuations in the data.

The total of the 12 months of 1964 is 7,554.0 thousand cars and the monthly average is 629.5 thousand cars. July sales were 562.2 thousand cars or 89.3% of the monthly average for 1964. In 1970 the total number of cars was 6,187.4 thousand or an average of 515.6 thousand per month (see Table 15–7). July, 1970, sales were 464.3 thousand or 90.1% of the monthly average for 1970. This ratio to the 12-month average is not greatly different from the

TABLE 15-7

FACTORY SALES OF PASSENGER CARS IN THE UNITED STATES, 1970

Month	Sales (Thousands)
January	545.0
February	528.4
March	594.4
April	627.2
May	684.4
June	758.4
July	464.3
August	254.0
September	454.2
October	365.4
November	341.1
December	570.6
Total	6,187.4
Monthly average (arithmetic mean)	515.6

Source: Table 15-4.

89.3% in July, 1964, even though the number of cars sold in the two months were considerably different. In other words, the ratios for July, 1964, and July, 1970, are not affected by trend or cyclical factors active during those years. There is always the possibility of random factors being present, but averaging a number of July ratios will be relied on to remove the effect of these random variations.

If the monthly average for the year can be used as the base of the ratio for July, can it be used as the base of the ratios for the other months of the year? Since both June and July are near the center of the period averaged, these two months might be compared with the average for the 12 months of the year, but other months are farther from the middle of the period covered by the data and this creates a problem. Since June and July are near the center of the 12 months averaged, the effect of the trend and cycle on the months preceding and following these center months balance each other. The fact that there are six months in one direction and only five in the other direction cannot be avoided, although a method of eliminating some of this difference is discussed further on pages 412 and 416.

The months preceding June and following July will be increasingly influenced by the trend and cycle in the series as months farther from the center of the year are used. To avoid running the risk that these months are influenced by the trend and cycle, it is better to compute a new monthly average of 12 months to be used in computation of the ratio for each month. For example, the sales for July could be compared to the average of the 12 months, January through December, 1964. The sales for August would be compared to the average of the 12 months, February, 1964, through January, 1965. Each following month would move the period for which the average is computed ahead one month so the specific month for which the seasonal was determined would be compared with the 12 months of which it was the seventh month. The comparison might just as well have been made with June and the 12 months of 1964 since it is as near the center of the year as July. Neither point is exactly in the middle of the period but it makes no difference which is used as long as the average is moved ahead one month for each comparison.

Table 15–8 represents a systematic method of carrying out the computations. A 12-month total must be taken to compare with each month for which data are to be used. The total for January through December, 1964, is written

TABLE 15–8

FACTORY SALES OF DOMESTIC PASSENGER CARS IN THE UNITED STATES, 1964 AND 1965

Specific Seasonals Computed by Ratio to the 12-Month Moving Average Centered at the Seventh Month

Year and Month	Factory Sales (Thousands)	12-Month Moving Total Centered at Seventh Month	12-Month Moving Average	Ratio to Moving Average
1964				
Jan.	709.0			
Feb.	665.4			
March	700.9			
April	770.2			
May	719.5			
June	726.7			
July	562.2	7,554.0	629.5	89.3
Aug.	230.8	7,627.8	635.6	36.3
Sept.	563.8	7,715.5	643.0	87.7
Oct.	394.7	7,952.5	662.7	59.6
Nov.	648.4	8,029.2	669.1	96.9
Dec.	862.4	8,129.0	677.2	127.3
1965				
Jan.	782.8	8,283.2	690.3	113.4
Feb.	753.1	8,466.6	705.5	106.7
March	937.9	8,566.2	713.8	131.4
April	846.9	8,440.9	703.4	120.4
May	819.3	8,871.6	739.3	110.8
June	880.9	9,101.9	758.5	116.1
July	745.6	9,100.8	758.4	98.3
Aug.	330.4			
Sept.	438.5			
Oct.	825.4			
Nov.	878.7			
Dec.	861.3			

Source: Table 15–4.

opposite July of that year. This amount is 7,554.0 thousand cars. The total to be used for August, found by adding the 12 months from February, 1964, through January, 1965, is 7,627.8. The moving totals for each of the months of 1964 are entered in Column 3 of Table 15–8. Instead of adding the 12 months each time, it saves computations to subtract the earliest month of the 12-month total and add the next month.

In order to compute the ratios for the individual months, it is necessary to divide each 12-month total by 12 to get the monthly average, which is then divided into the individual months to compute each month's percentage of the average. This is the specific seasonal for this month and measures the effect of seasonal (and sometimes random) variations on that month. Computations for only the two years, 1964 and 1965, are shown, since other years would be computed the same way.

Ratio to 2-Item Average of 12-Month Moving Average

When the 12-month average was related to July, six of the months preceded July and five months were later than July (page 410). It would be more logical if there were the same number of months preceding and following the month with which the average was compared. If 13 months were used, not all the seasonal influence would be removed. For example, if the 13 months from January, 1964, through January, 1965, were used, the seasonal effect of January would be included twice; but the seasonal effects of the other months would be included only once. An average of 13 months might be computed by giving each of the two months that were farthest away from the center a weight of $\frac{1}{2}$. For example, July, 1964, sales would be compared with the average of the sales shown in Table 15–9.

The ratio of July sales to the 12-month average computed in Table 15–9 is 88.9%. (Ratio $= \frac{562.2}{632.57}\,100 = 88.9$.) In the preceding section this ratio was found to be 89.3%, based on the 12 full months from January through December. The specific seasonals computed by using 13 months, with the first and last months weighted $\frac{1}{2}$, are somewhat more precise measures of the seasonal factor, but the increase in the accuracy of the index of seasonal variation is usually so small that it is unimportant. This method, however, is preferred by most statisticians in spite of the increased clerical work. If a computer is used, this increased work is negligible.

Table 15–10 shows the easiest method of computing the ratios described. A 12-month moving total is taken and centered between the sixth and seventh months. The total for 1964 is 7,554.0, and it is entered between June and July. The total for February, 1964, through January, 1965, is 7,627.8 and is entered between July and August. These are the same totals that were found in Table 15–8. The next step is to add the total that was entered between June and July to the total between July and August. This new total entered opposite July consists of 24 months; but January, 1964, and January, 1965, have each been

TABLE 15-9

FACTORY SALES OF DOMESTIC PASSENGER CARS IN THE UNITED STATES

January, 1964, to January, 1965

Month and Year	Sales (Thousands)
1964	
January (½ of 709.0)	354.5
February	665.4
March	700.9
April	770.2
May	719.5
June	726.7
July	562.2
August	230.8
September	563.8
October	394.7
November	648.4
December	862.4
1965	
January (½ of 782.8)	391.4
Total	7,590.9
Monthly average (arithmetic mean)	632.57

Source: Table 15–4.

TABLE 15-10

FACTORY SALES OF DOMESTIC PASSENGER CARS IN THE UNITED STATES, 1964 TO 1972

Specific Seasonals Computed by Ratio to 2-Item Average of 12-Month Moving Average

Year and Month (1)	Factory Sales (Thousands) Y (2)	12-Month Moving Total Centered (3)	2-Item Moving Total (4)	Moving Average (5)	Specific Seasonal (6)
1963					
Jan.	658.0				
Feb.	592.8				
March	637.1				
April	671.8				
May	695.1				
June	672.9	7,443.6			
July	649.4	7,494.6	14,938.2	622.43	104.33
Aug.	165.1	7,567.2	15,061.8	627.58	26.31
Sept.	463.0	7,631.0	15,198.2	633.26	73.11
Oct.	779.2	7,729.4	15,360.4	640.02	121.75
Nov.	726.2	7,753.8	15,483.2	645.13	112.57
Dec.	733.0		15,561.4	648.39	113.05
		7,807.6			

(continued)

TABLE 15–10 (continued)

Year and Month (1)	Factory Sales (Thousands) Y (2)	12-Month Moving Total Centered (3)	2-Item Moving Total (4)	Moving Average (5)	Specific Seasonal (6)
1964					
Jan.	709.0		15,528.0	647.00	109.58
		7,720.4			
Feb.	665.4		15,506.5	646.10	102.99
		7,786.1			
March	700.9		15,673.0	653.04	107.33
		7,886.9			
April	770.2		15,389.3	641.22	120.11
		7,502.4			
May	719.5		14,927.0	621.96	115.68
		7,424.6			
June	726.7		14,978.6	624.11	116.44
		7,554.0			
July	562.2		15,181.8	632.57	88.87
		7,627.8			
Aug.	230.8		15,343.3	639.30	36.10
		7,715.5			
Sept.	563.8		15,668.0	652.83	86.36
		7,952.5			
Oct.	394.7		15,981.7	665.90	59.27
		8,029.2			
Nov.	648.4		16,158.2	673.26	96.31
		8,129.0			
Dec.	862.4		16,412.2	683.84	126.11
		8,283.2			
1965					
Jan.	782.8		16,759.8	697.91	112.16
		8,466.6			
Feb.	753.1		17,032.8	709.70	106.12
		8,566.2			
March	937.9		17,007.2	708.63	132.35
		8,440.9			
April	846.9		17,311.5	721.35	117.40
		8,871.6			
May	819.3		17,972.5	748.90	109.40
		9,101.9			
June	880.9		18,202.7	758.45	116.15
		9,100.8			
July	745.6		18,199.2	758.30	98.33
		9,098.4			
Aug.	330.4		18,192.5	756.02	43.59
		9,094.1			
Sept.	438.5		18,152.3	756.35	57.98
		9,058.2			
Oct.	825.4		18,063.4	752.64	109.67
		9,005.2			
Nov.	878.7		17,962.3	748.43	117.41
		8,957.1			
Dec.	861.3		17,835.8	743.16	115.90
		8,878.7			
1966					
Jan.	780.4		17,491.8	728.82	107.08
		8,613.1			
Feb.	748.8		17,032.2	709.67	105.51
		8,419.1			
March	902.0		16,992.1	708.00	127.40
		8,573.0			
April	793.9		17,118.3	713.26	111.31
		8,545.3			
May	771.2		17,003.1	708.46	108.86
		8,457.8			
June	802.5		16,794.4	699.78	114.68
		8,337.0			
July	480.0		16,518.6	688.27	69.74
		8,181.6			
Aug.	136.4		16,116.3	671.51	20.31
		7,934.7			
Sept.	592.4		15,614.8	650.62	91.05
		7,680.1			
Oct.	797.7		15,194.6	633.11	126.00
		7,514.5			
Nov.	791.2		14,971.2	623.80	126.84
		7,456.7			
Dec.	740.5		14,843.2	618.47	119.73
		7,386.5			
1967					
Jan.	625.0		14,703.6	612.65	102.02
		7,317.1			
Feb.	501.9		14,716.1	613.17	81.85
		7,399.0			
March	647.4		14,776.4	615.67	105.15
		7,377.2			
April	628.3		14,565.5	606.90	103.53
		7,188.3			
May	713.4		14,230.6	592.94	120.32
		7,042.3			
June	732.3		14,112.6	588.02	124.54
		7,070.3			

TABLE 15–10 (continued)

Year and Month (1)	Factory Sales (Thousands) Y (2)	12-Month Moving Total Centered (3)	2-Item Moving Total (4)	Moving Average (5)	Specific Seasonal (6)
July	410.6	7,192.5	14,262.8	594.28	69.09
Aug.	218.3	7,358.8	14,551.3	606.30	36.01
Sept.	570.6	7,475.4	14,834.2	618.09	92.32
Oct.	608.8	7,594.9	15,070.3	627.93	96.95
Nov.	645.2	7,757.7	15,352.6	639.69	100.86
Dec.	768.5	7,807.0	15,564.7	648.53	118.50
1968					
Jan.	747.2	8,001.8	15,808.8	658.70	113.44
Feb.	668.2	7,966.1	15,967.9	665.33	100.43
March	764.0	8,015.5	15,981.6	665.90	114.73
April	747.8	8,296.2	16,631.7	679.65	110.03
May	876.2	8,482.0	16,778.2	699.09	125.33
June	781.6	8,407.2	16,889.2	703.72	111.07
July	605.4	8,442.0	16,849.2	702.05	86.23
Aug.	182.6	8,450.5	16,892.5	703.85	25.94
Sept.	620.0	8,407.5	16,858.0	702.42	88.27
Oct.	889.5	8,336.6	16,744.1	697.76	127.50
Nov.	831.0	8,138.4	16,475.0	686.46	121.06
Dec.	693.7	8,097.2	16,235.6	676.48	102.55
1969					
Jan.	782.0	7,938.7	16,035.9	668.16	117.04
Feb.	676.7	8,085.6	16,024.3	667.68	101.35
March	721.0	8,172.1	16,257.7	677.40	106.44
April	676.9	8,098.2	16,270.3	677.93	99.85
May	678.0	7,911.2	16,009.4	667.06	101.64
June	740.4	7,806.3	15,717.5	654.90	113.06
July	446.9	7,569.3	15,375.6	640.65	69.76
Aug.	329.5	7,421.0	14,990.3	624.60	52.75
Sept.	706.5	7,294.4	14,715.4	613.14	115.23
Oct.	815.6	7,244.7	14,539.1	605.80	134.63
Nov.	644.0	7,251.1	14,449.5	603.99	106.62
Dec.	588.8	7,269.1	14,520.2	605.01	97.32
1970					
Jan.	545.0	7,268.5	14,555.6	606.48	89.86
Feb.	528.4	7,211.0	14,497.5	604.06	87.47
March	594.4	6,958.7	14,179.7	590.40	100.68
April	627.2	6,508.5	13,467.2	561.13	111.77
May	684.4	6,205.6	12,714.1	529.75	129.19
June	758.4	6,187.4	12,393.0	516.37	146.87
July	464.3	6,320.5	12,507.9	521.16	89.09
Aug.	254.0	6,511.1	12,831.6	534.65	47.51
Sept.	454.2	6,732.6	13,243.7	551.82	82.31
Oct.	365.4	6,809.0	13,354.6	564.23	64.76
Nov.	341.1	6,841.3	13,650.3	568.76	59.97
Dec.	570.6	6,844.2	13,685.5	570.23	100.07

TABLE 15–10 (concluded)

Year and Month (1)	Factory Sales (Thousands) Y (2)	12-Month Moving Total Centered (3)	2-Item Moving Total (4)	Moving Average (5)	Specific Seasonal (6)
1971					
Jan.	678.1	6,848.8	13,693.0	570.54	118.85
Feb.	719.0	7,051.4	13,900.2	579.17	124.14
March	815.9	7,309.2	14,360.6	598.36	136.36
April	703.6	7,702.4	15,011.6	625.48	112.49
May	716.7	8,097.9	15,800.3	658.35	108.86
June	761.3	8,120.5	16,218.4	675.77	112.66
July	468.9	8,108.4	16,228.9	676.20	69.34
Aug.	456.6	8,105.5	16,213.9	675.58	67.59
Sept.	712.0	8,054.8	16,160.3	673.35	105.74
Oct.	758.6	8,088.1	16,142.9	672.62	112.78
Nov.	736.6	8,169.4	16,257.5	677.40	108.74
Dec.	593.2		16,339.1	680.80	87.13
1972		8,169.7			
Jan.	666.0	8,094.4	16,264.1	677.67	98.28
Feb.	716.1	8,008.8	16,103.2	670.97	106.73
March	765.2	8,105.6	16,114.4	671.43	113.97
April	736.9	8,188.7	16,629.43	678.93	108.54
May	798.0	8,279.5	16,468.2	686.17	116.30
June	761.6	8,352.5	16,632.0	693.00	109.90
July	393.6				
Aug.	371.0				
Sept.	808.8				
Oct.	841.7				
Nov.	827.4				
Dec.	666.2				

Source: Table 15–4.

included once, and the months between have each been included two times. This gives a total approximately double the total in Table 15–9. To compute the moving average, this total must be divided by 24 instead of 12.

$$\text{Monthly average} = \frac{15{,}181.8}{24} = 632.57 \text{ (the same as found in Table 15–9).}$$

Table 15–10 shows the calculations for each month of the period from 1963 to 1972. The ratios in the last column differ very little from those computed for the same months in Table 15–8.

AVERAGING SPECIFIC SEASONALS

The problem of measuring seasonal variation is a problem of averages (page 405). The previous section has described methods of isolating specific seasonal fluctuations for individual years. This section will discuss the problem of how to average the specific seasonals that have been derived.

Table 15–10 gives the ratios of factory sales of domestic passenger cars in the United States to 2-item average of the 12-month moving average for the complete years 1964 through 1971, and parts of 1963 and 1972. If the percentages showing the specific seasonal for each month are to be averaged to find a typical seasonal, a significant concentration of the individual items for each month is necessary. If there is too much dispersion among the percentages, it is incorrect to use any average as typical of the whole group. (The dispersion of items being averaged was discussed in Chapter 4.) In order to find out whether there is sufficient concentration to warrant averaging the specific seasonals, the ratios for each month were put into an array and a chart drawn of the array. Table 15–11 shows the arrays, and Figure 15–1 presents them graphically. If the graph shows enough concentration for an average to be significant, the next step is to find an average of the ratios for each month.

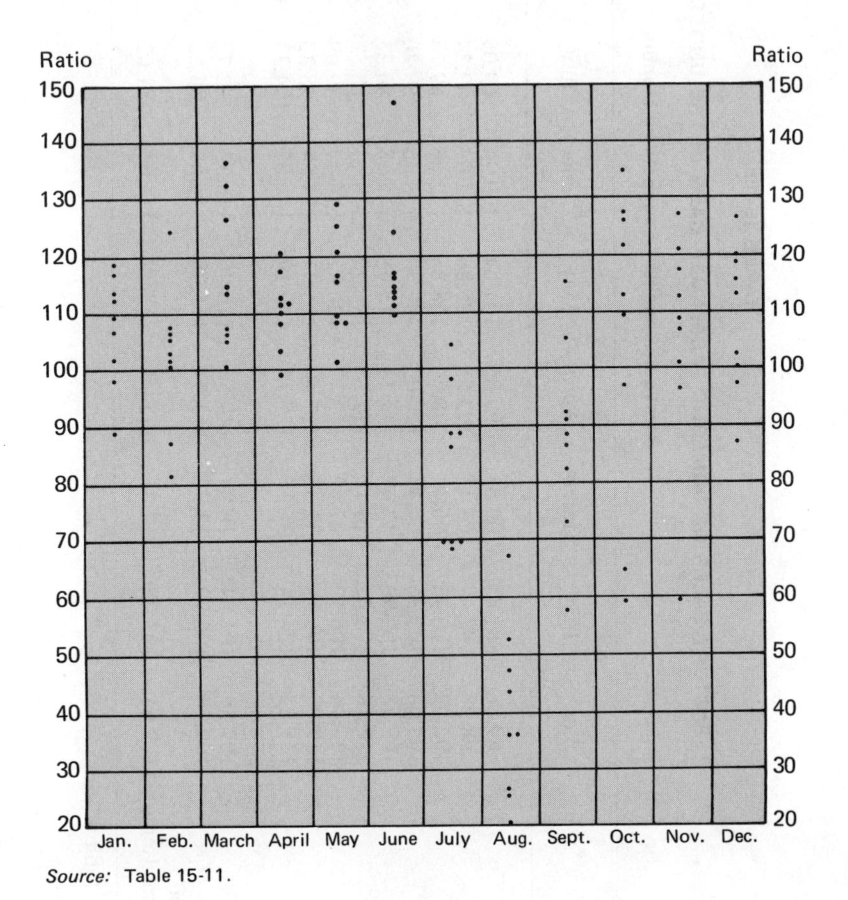

Source: Table 15-11.

FIGURE 15–1

MONTHLY ARRAYS OF RATIOS TO 2-ITEMS OF 12-MONTH MOVING AVERAGE

TABLE 15–11

FACTORY SALES OF DOMESTIC PASSENGER CARS IN THE UNITED STATES, 1963 TO 1972

Arrays of Specific Seasonals by Months

	Jan.	Feb.	March	April	May	June	July	Aug.	Sept.	Oct.	Nov.	Dec.
	89.86	81.85	100.68	99.85	101.64	109.90	69.09	20.31	57.98	59.27	59.97	87.13
	98.28	87.47	105.15	103.53	108.86	111.07	69.34	25.94	73.11	64.76	96.31	97.32
	102.02	100.43	106.44	108.54	108.86	112.66	69.74	26.31	82.31	96.95	100.86	100.07
	107.08	101.35	107.33	110.03	109.40	113.06	69.76	36.01	86.36	109.67	106.62	102.55
	109.58	102.99	113.97	111.31	115.68	114.68	86.23	36.10	88.27	112.78	108.74	113.05
	112.16	105.51	114.73	111.77	116.30	116.15	88.87	43.59	91.05	121.75	112.57	115.90
	113.44	106.12	127.40	112.49	120.32	116.44	89.09	47.51	92.32	126.00	117.41	118.50
	117.04	106.73	132.35	117.40	125.33	124.54	98.33	52.75	105.74	127.50	121.06	119.73
	118.85	124.14	136.36	120.11	129.19	146.87	104.33	67.59	115.23	134.63	126.84	126.11
Mean	107.59	101.84	116.04	110.56	115.06	118.37	82.75	39.57	88.04	105.92	105.60	108.93
Median	109.58	102.99	113.97	111.31	115.68	114.68	86.23	36.10	88.27	112.78	108.74	113.05
Mean of 3 Central Items	109.61	103.28	112.01	111.04	113.79	114.63	81.62	38.56	88.56	114.73	109.31	110.50

Source: Table 15–10.

Arithmetic Mean

Chapter 3 states that the arithmetic mean is the best known and most generally useful average. Therefore, the arithmetic mean of the ratios has been computed first. The arithmetic mean of the ratios for each month has been entered in Column 2 of Table 15–12. These means total 1,200.27 instead of 1,200.0. The monthly indexes of seasonal variation should total 1,200.0 after rounding to one decimal place; and the simplest method to do this is to decrease two values, April and May, after rounding. For a larger difference a somewhat more complex method should be used, which is described in the following section.

TABLE 15–12

COMPUTATION OF INDEX OF SEASONAL VARIATION

Arithmetic Means of Specific Seasonals
Factory Sales of Domestic Passenger Cars in the United States, 1963 to 1972

Month (1)	Mean of Ratios to 2-Item Average of 12-Month Moving Average (2)	Index of Seasonal Variation Average Month = 100 (3)
January	107.59	107.6
February	101.84	101.8
March	116.04	116.0
April	110.56	110.5
May	115.06	115.0
June	118.37	118.4
July	82.75	82.7
August	39.57	39.6
September	88.04	88.0
October	105.92	105.9
November	105.60	105.6
December	108.93	108.9
Total	1,200.27	1,200.0

Source: Table 15–11.

Median

Since there is always danger that an extreme value will have an undue amount of influence on the arithmetic mean, the median is frequently used instead of the mean to average specific seasonals. In Table 15–13 the median for each month has been entered from Table 15–11 and totaled. Since their total is 1,213.38, each of the monthly indexes should be reduced proportionately to make their total 1,200 and the monthly average 100. Since the desired total of 1,200 is .98897 times the actual total of 1,213.38, the level of the index is reduced by multiplying each monthly value by .98897. The adjusted index in Column 3 of Table 15–13 totals 1,200 after the leveling factor has been applied. The median cannot logically be used in further calculations (page 66). However, the fact that the median may be more nearly typical than the mean frequently outweighs this disadvantage.

TABLE 15–13

COMPUTATION OF INDEX OF SEASONAL VARIATION

Median of Specific Seasonals
Factory Sales of Domestic Passenger Cars in the United States, 1963 to 1972

Month (1)	Median of Ratios to 2-Item Average of 12-Month Moving Average (2)	Index of Seasonal Variation Average Month = 100 (3)
January	109.58	108.4
February	102.99	101.9
March	113.97	112.7
April	111.31	110.1
May	115.68	114.4
June	114.68	113.4
July	86.23	85.3
August	36.10	35.7
September	88.27	87.3
October	112.78	111.5
November	108.74	107.5
December	113.05	111.8
Total	1,213.38	1,200.0

Source: Table 15–11.

$$\textbf{Leveling factor} = \frac{1,200}{1,213.38} = .98897.$$

Modified Mean

Because there is always a possibility that a few unusually large or small items may unduly influence the mean, a modification is sometimes used to average specific seasonals. Since the number of specific seasonals being averaged is small, an exceptionally large or small one in only a few months could have considerable effect on the index of seasonal variation computed by the use of the arithmetic mean. The *modified mean* used is an arithmetic mean of only the central items, eliminating an equal number of the largest and the smallest values to be averaged. If 9 specific seasonals were being averaged, the largest 2 and the smallest 2 might be eliminated, and the average based on the central 5. This average has some of the characteristics of the median, since it will be less influenced by extremely large or small items than the mean. Since it is the arithmetic mean of several central items, it is less likely to be erratic than the median.

The choice of the number of central items to be used is arbitrary, but what number is taken makes very little difference in the result. Averaging the central three or four items is ordinarily satisfactory. In Table 15–14 the means of the three central ratios have been computed. These modified means were then leveled to total 1,200 in the same way as in Table 15–13.

TABLE 15–14

COMPUTATION OF INDEX OF SEASONAL VARIATION

Modified Mean of Specific Seasonals
Factory Sales of Domestic Passenger Cars in the United States, 1963 to 1972

Month (1)	Modified Mean of Ratios to 2-Item Average of 12-Month Moving Average (2)	Index of Seasonal Variation Average Month = 100 (3)
January	109.61	108.9
February	103.28	102.6
March	112.01	111.3
April	111.04	110.3
May	113.79	113.1
June	114.63	114.0
July	81.62	81.1
August	38.56	38.3
September	88.56	88.0
October	114.73	114.0
November	109.31	108.6
December	110.50	109.8
Total	1,207.64	1,200.0

Source: Table 15–11.

$$\textbf{Leveling factor} = \frac{1{,}200}{1{,}207.64} = .993674.$$

EVALUATION OF THE DIFFERENT METHODS OF MEASURING SEASONAL VARIATION

Comparing the three indexes of seasonal variation from Tables 15–12, 15–13, and 15–14 indicates that the method of averaging the specific seasonals was not an important factor in this case. There is so little difference in the results that no one method could be chosen as outstandingly better. The method of isolating the specific seasonals, however, seems to have had more effect on the index. In Table 15–15 the index based on the means of the actual monthly items is compared with the index based on the means of the ratios to the 12-month moving average; and there is considerable variation between the two.

Because of the greater accuracy of the specific seasonals when the ratios to the 12-month moving average are used, this method is much preferred to averaging the actual monthly data. It is possible to remove the effect of the trend from the averages instead of from the specific seasonals before they are averaged, and methods of doing this have been worked out by statisticians. The methods that use some form of the ratio to the 12-month moving average are so generally satisfactory that they have become standard for measuring seasonal variation.

TABLE 15–15

COMPARISON OF TWO INDEXES OF SEASONAL VARIATION

Factory Sales of Domestic Passenger Cars in the United States, 1963 to 1973

Month (1)	Index Based on Means of Actual Monthly Sales (2)	Index Based on Means of Ratios to 2-Item Average of 12-Month Moving Average (3)	Difference Between Indexes Column 2 Minus Column 3 (4)
January	106.8	107.6	− .8
February	100.6	101.8	−1.2
March	114.6	116.0	−1.4
April	110.3	110.5	− .2
May	114.4	115.0	− .6
June	116.6	118.4	−1.8
July	80.0	82.7	−2.7
August	41.0	39.6	1.4
September	90.8	88.0	2.8
October	108.3	105.9	2.4
November	108.2	105.6	2.6
December	108.4	108.9	− .5

Source: Tables 15–5 and 15–12.

CHANGING SEASONAL PATTERN

The discussion of measuring seasonal variation up to this point has dealt with the problem of computing the average seasonal pattern. This method is appropriate when the seasonal pattern is relatively stable and an average of several years will be typical of the underlying seasonal movements during these years. Seasonal variation that is basically related to the changing seasons can be measured by this method with considerable success, but seasonal fluctuations that result from customs, habits, or business practices may be changed abruptly when something happens to affect those practices. The change may be abrupt, for example, in the production and the sales of industries that regularly introduce new models on an annual basis. The automobile industry is the outstanding example, and changes in the seasonal pattern of production and sales can be traced to changes in the dates for introducing new models.

Other changes in seasonal pattern may come about gradually as new uses for a product, or conditions under which it is used, change. The increase in summer air conditioning has changed the seasonal pattern of electric power consumption. For many years the peak of electric power use by residential customers came in the winter when the days were shortest, since the major domestic use of power was for lighting. Air conditioning has increased the use of power in the summer until the peak of electric power consumption in Texas now comes in the summer instead of in the winter.

Better cars and highways have over the years reduced the peak of gasoline consumption in the summer months by shifting a larger proportion of total consumption into the winter months. Table 15–16 and Figure 15–2 give two indexes of seasonal variation in gasoline consumption in the United States that show the extent to which demand has tended to be spread more evenly throughout the year in the postwar period in comparison with the years before World War II.

TABLE 15–16

COMPARISON OF INDEXES OF SEASONAL VARIATION*
1930 TO 1944 AND 1962 TO 1972

Domestic Demand for Motor Fuel in the United States

	Index of Seasonal Variation	
Month	1930 to 1944	1962 to 1972
January	83.5	92.0
February	76.7	86.0
March	92.6	98.0
April	99.8	99.4
May	107.7	104.1
June	111.6	106.1
July	112.0	109.3
August	115.2	108.5
September	107.5	98.4
October	105.8	102.7
November	97.0	96.3
December	90.6	99.2

* Indexes computed by method of ratios to 2-item average of 12-month moving average, using the modified mean.

Source: U.S. Department of Commerce, *Survey of Current Business* (March, 1973).

Because business is constantly changing, it is normal for the seasonal pattern to shift over a period of years. Many industries make an effort to level out the seasonal fluctuations by trying to stimulate business in the periods that are seasonally slow, or by adding lines with different seasonal peaks in order to spread their business more evenly throughout the year. In developing methods for measuring the seasonal fluctuations in business, it has been necessary to provide methods for checking whether or not an appreciable shift has occurred in the pattern. It is important to detect both abrupt and gradual changes

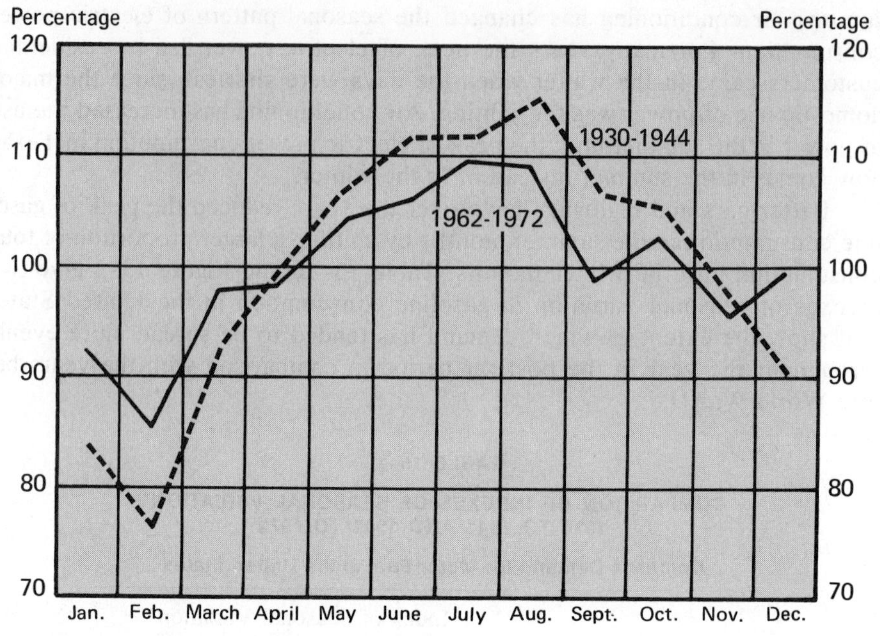

Source: Table 15-16.

FIGURE 15–2

INDEXES OF SEASONAL VARIATION IN DEMAND FOR MOTOR FUEL IN THE UNITED STATES 1930 TO 1944 AND 1962 TO 1972

in the seasonal pattern, and to measure as accurately as possible what these changes are.

Figure 15–1 is designed to show the amount of dispersion in the specific seasonals for the different months before they are summarized by an average. It is equally important to check the specific seasonals for each month to determine whether there appears to be any significant shift in the measures of seasonal variation. The simplest device that can be used for this purpose is to plot the specific seasonals for each month chronologically, as in Figure 15–3. These 12 graphs give a picture of the changes that have taken place in the seasonal pattern of factory sales of domestic passenger cars for each of the months of the year. Whenever such a series of charts gives evidence that the seasonal pattern is changing, it is desirable to summarize each month by a trend line instead of by an average.

The shift in the specific seasonals for a given month is in the nature of a trend; the methods of measuring trend described in Chapter 14 may be used to measure this shift. The straight line trend equation was computed for each of the 12 months and the trend values plotted for each month on Figure 15–3 were computed from the resulting trend equations. The equation for each of the 12 months is given below, based on nine specific seasonals. In Figure 15–3 the months January through June have no specific seasonal for the first year,

Factory Sales of Domestic Passenger Cars in the United States, 1963 to 1972

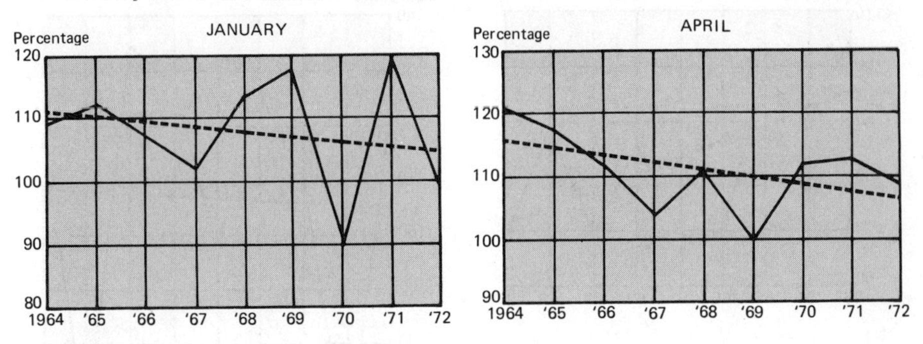

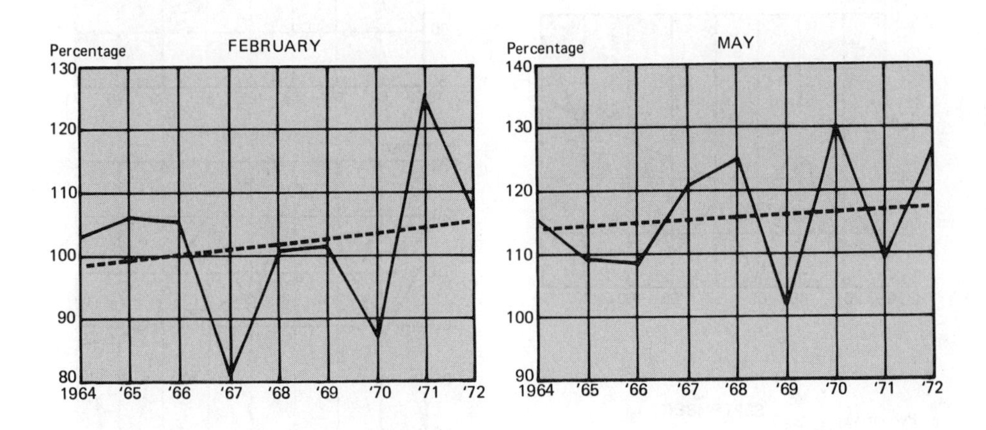

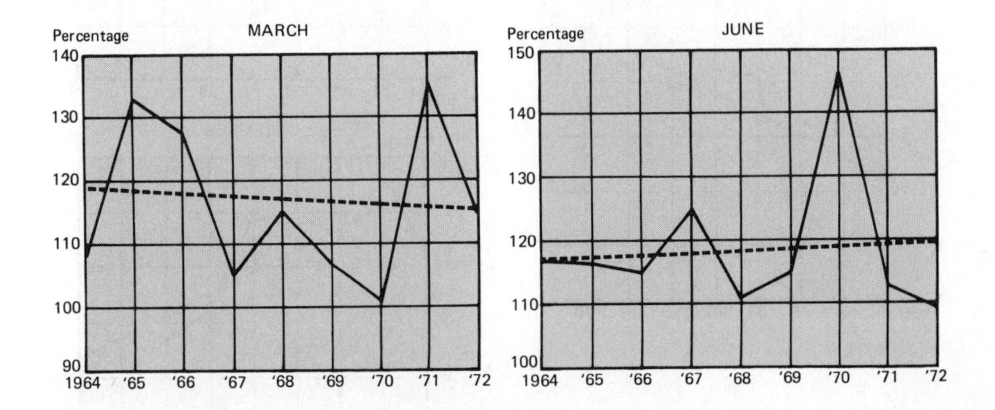

FIGURE 15–3

CHANGING SEASONAL INDEX

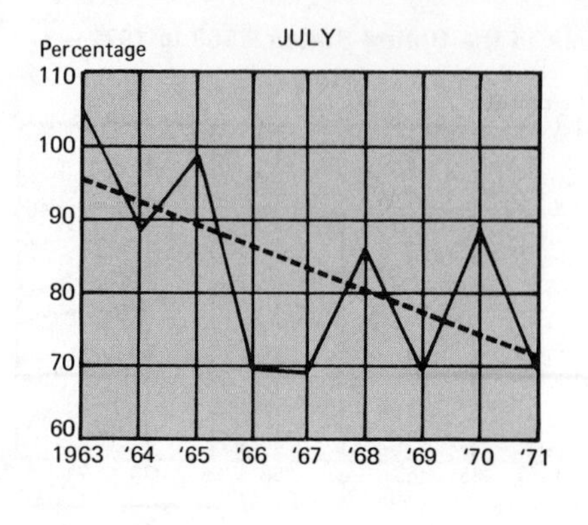

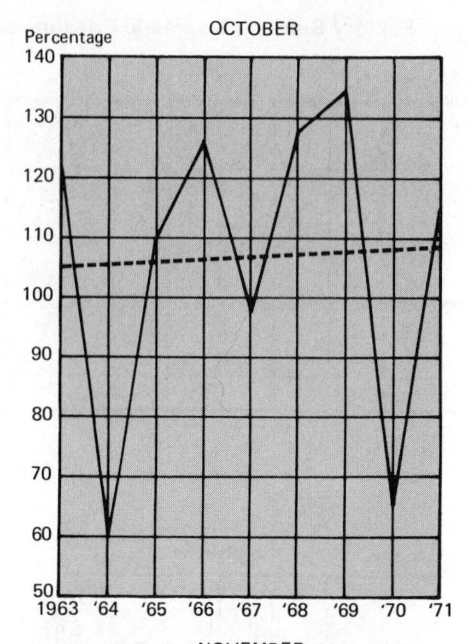

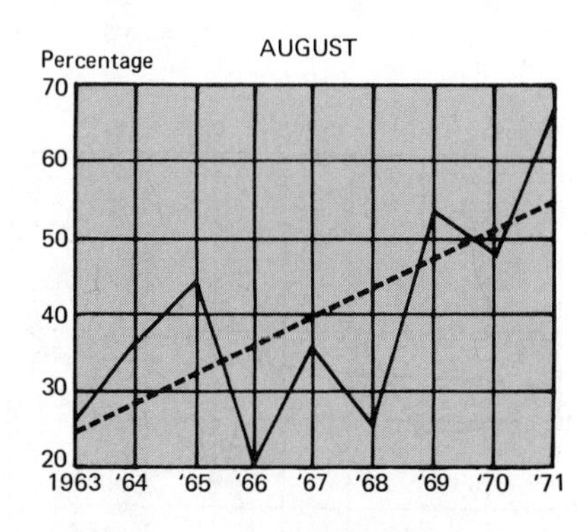

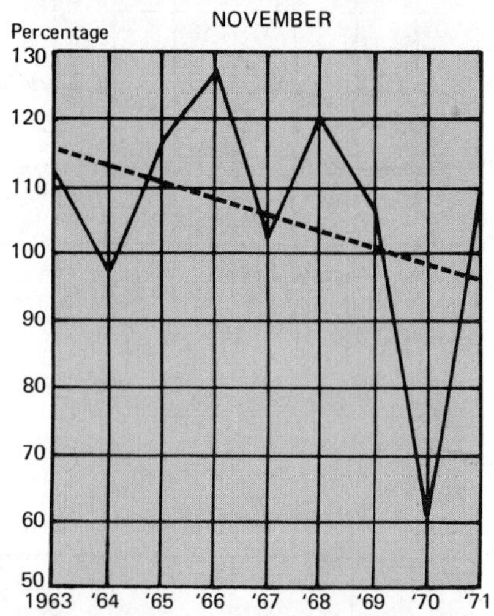

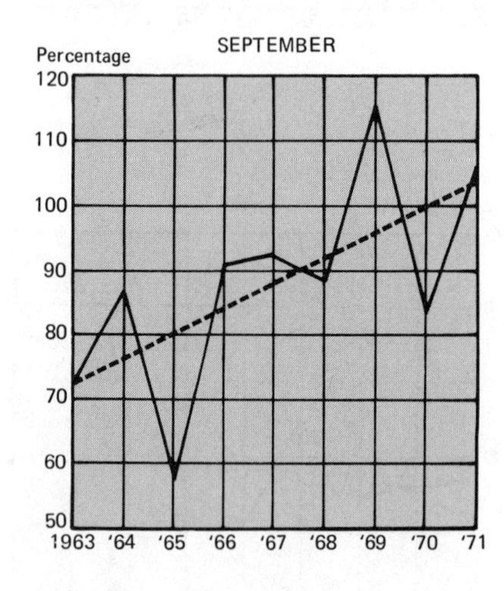

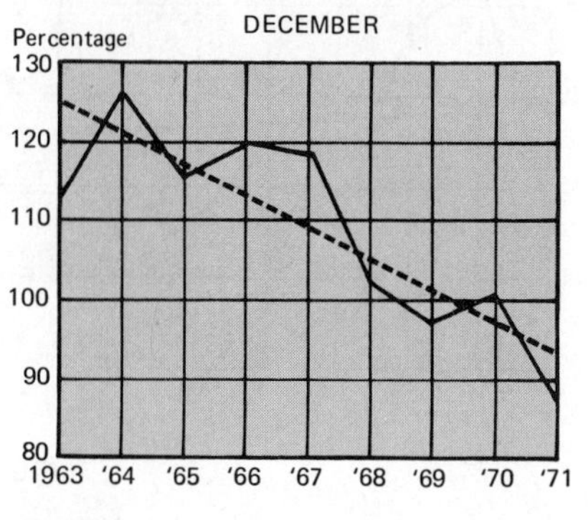

Source: Table 16-8.

FIGURE 15–3 (concluded)

1963. The months July through December have no specific seasonal for the last year, 1972. This results from the fact that the 2-item average of the 12-month moving average does not provide values for the first six months or the last six months of the period for which the moving average is computed. The origin of each of the equations for the first six months is the middle of 1968 and for the last six months is the middle of 1969. The x unit one year and the y unit is the specific seasonal computed by finding the ratio of a given month to the 2-item average of the 12-month moving average.

January:	seasonal index = 107.59 − .743 x
February:	seasonal index = 101.84 + .874 x
March:	seasonal index = 116.04 − .226 x
April:	seasonal index = 110.56 − 1.063 x
May:	seasonal index = 115.06 + .381 x
June	seasonal index = 118.37 + .271 x
July:	seasonal index = 82.75 − 3.00 x
August:	seasonal index = 39.57 + 3.722 x
September:	seasonal index = 88.04 + 3.845 x
October:	seasonal index = 105.92 + .534 x
November:	seasonal index = 105.60 − 2.528 x
December:	seasonal index = 108.93 − 3.936 x

Figure 15–3 shows that the seasonal pattern of factory sales of passenger cars in the United States has not been constant throughout the period 1963 to 1972. For the first six months of the year, there appears to be only a small amount of trend in the specific seasonals, and although trend lines have been drawn for these six months, it appears that averaging the specific seasonals would be just about as satisfactory as a trend line. In the third quarter of the year, however, the seasonal fluctuations have undergone great changes in the 10 years.

Since consumers tend to buy expensive durable goods just before the period of maximum use of those goods, it appears that the seasonal behavior of the spring and the early summer months is relatively stable. All of these months enjoy factory sales above the average for the year, and the shifts in this pattern are not large. The third quarter of the year, however, reflects the influence of the new models on factory sales. July sales have shown a sharp decline relative to the other months of the year, but August sales have been increasing. Most of the shift to the new models is made before the end of August. With the changeover to the new models being accomplished more completely in August now than in earlier years, factory sales in September have been increasing. Apparently dealers are well stocked by the end of October, since November and December sales decline somewhat.

The trends for the various months shown in Figure 15–3 are entered in Table 15–17 for each year from 1963 through 1972. Since the indexes for a year should total 1,200, the values may be leveled to make them total this amount. This adjustment may be made by shifting the curves on the chart

TABLE 15–17

INDEXES OF SEASONAL VARIATION*

Factory Sales of Domestic Passenger Cars in the United States, 1963 to 1972

Year	Jan.	Feb.	Mar.	April	May	June	July	Aug.	Sept.	Oct.	Nov.	Dec.
1963	(111.3)**	(97.5)	(117.5)	(115.9)	(113.2)	(117.0)	94.8	24.7	72.7	104.8	115.7	124.7
1964	110.6	98.3	116.9	114.8	113.5	117.3	81.8	28.4	76.5	104.3	113.2	120.7
1965	109.8	99.2	116.7	113.7	113.9	117.6	88.8	32.1	80.4	104.9	110.7	116.8
1966	109.1	100.1	116.5	112.7	114.3	117.8	85.8	35.8	84.2	105.4	108.1	112.9
1967	108.3	101.0	116.3	111.6	114.7	118.1	82.8	39.6	88.0	105.9	105.6	108.9
1968	107.6	101.8	116.0	110.6	115.1	118.4	79.8	43.3	91.9	106.5	103.1	105.0
1969	106.8	102.7	115.8	109.5	115.4	118.6	76.8	47.0	95.7	107.0	100.5	101.1
1970	106.1	103.6	115.6	108.4	115.8	118.9	73.8	50.7	99.5	107.5	98.0	97.1
1971	105.4	104.5	115.4	107.4	116.2	119.2	70.8	54.5	103.4	108.1	95.5	93.2
1972	104.6	105.3	115.1	106.3	116.8	119.5	(67.8)	(58.2)	(107.2)	(108.6)	(93.0)	(89.3)

* Computed from the trend equations for the 12 months given on pages 425–426.
** Values in parentheses have been extrapolated.

slightly, or the difference between 1,200 and the actual total of the indexes may be spread proportionately as done in Table 15–13. Since the total for each year is so near 1,200, no adjustment has been made in this example.

MAKING USE OF A MEASURE OF SEASONAL VARIATION

One use of an index of seasonal variation is in making the monthly plans of a business. (An index of seasonal variation on a quarterly or a weekly basis is sometimes used, but it is the most common practice to make such plans on a monthly basis.) In a business that suffers no seasonal variation, planning the operations for the coming year is much less complicated than it is in a highly seasonal business when working capital needs fluctuate widely with different months and plans must be made to have adequate funds on hand when they are needed. Part of the basic information an executive should have about his business is the seasonal pattern of the whole industry and the degree of resemblance or difference that his own business exhibits. Quotas, sales plans, advertising campaigns, financial budgets, and production schedules are all based on this information.

The seasonal pattern in advertising is usually made to coincide with the seasonal fluctuations in sales, perhaps with advertising running a little in advance. This is done on the theory that the time to put forth the greatest sales effort is during that part of the year when people want to buy the product. Sometimes an advertising campaign is undertaken to stimulate sales in the dull periods. Reducing prices in seasons characterized by declining sales may be used as a method of increasing business.

How a measure of seasonal variation should be computed depends on the regularity with which the seasonal movement occurs and on the importance of this factor in the operations of the business. A well-informed executive no doubt knows with considerable accuracy the seasonal pattern in the fluctuations of his business. Sometimes, however, he only thinks he knows; it is notoriously true that general impressions may be far wrong.

The consumption of many products shows a seasonal pattern that influences retail distribution. Gasoline is one of these commodities. Figure 15–4 shows the seasonal variation in the retail distribution of gasoline in all states and in Texas. There is a more pronounced variation in the United States total than in Texas, resulting from the fact that automobiles can be used more generally the year round in Texas than in all of the United States.

Figure 15–5 shows the seasonal variation in gasoline consumption in Florida and Maine, one a winter resort state and the other a summer resort state. The fluctuations in Maine are particular striking.

Figure 15–6 shows the seasonal variation in farm income in the West North Central States and in the South Central states. The fact that the West North Central states produce a large percentage of the livestock and livestock products in the country tends to make for stability of farm income in the different months of the year. The farmers in the South Central states receive a

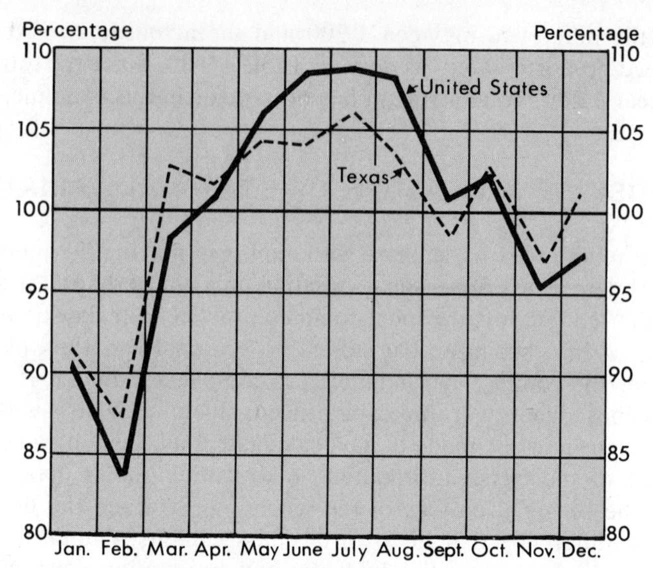

Source: Adapted from U.S. Department of Commerce, *Survey of Current Business* (March, 1973), and Bureau of Business Research, The University of Texas, Austin.

FIGURE 15-4

INDEXES OF SEASONAL VARIATION IN DEMAND FOR MOTOR FUEL IN THE UNITED STATES AND TEXAS

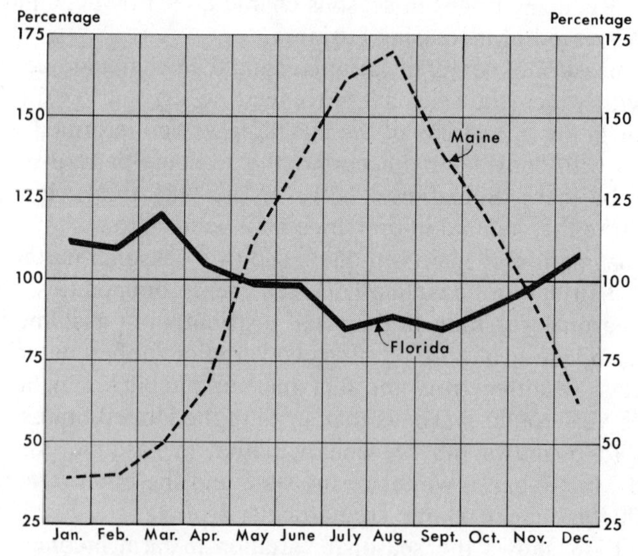

Source: Adapted from Simon Kuznets, *Seasonal Variations in Industry and Trade* (1933).

FIGURE 15-5

INDEXES OF SEASONAL VARIATION IN GASOLINE CONSUMPTION IN FLORIDA AND MAINE

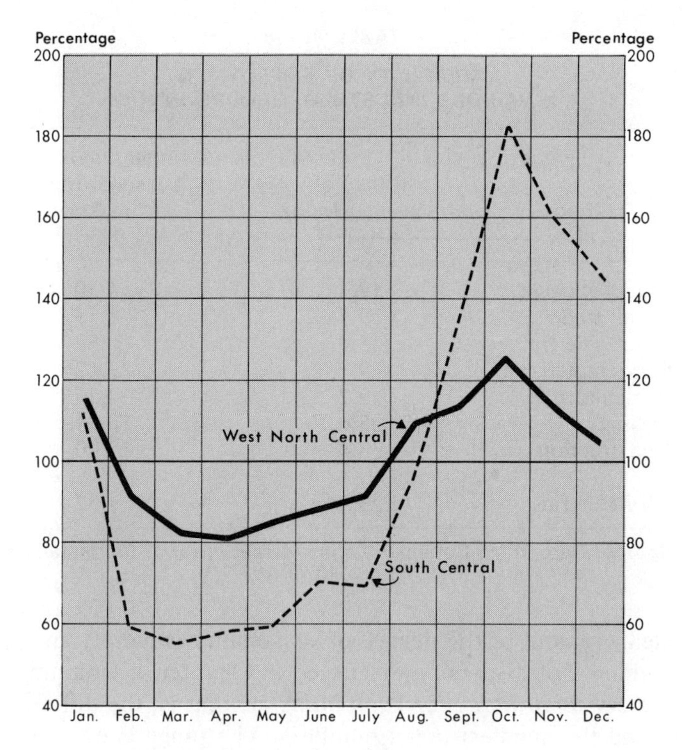

Source: Indexes computed from U.S. Department of Agriculture, *Farm Income Situation* (1973).

FIGURE 15–6

INDEXES OF SEASONAL VARIATION IN FARM CASH INCOME IN WEST NORTH CENTRAL AND SOUTH CENTRAL REGIONS

predominant proportion of their income from crops, such as cotton, which are sold as soon as they are picked. This causes the pronounced peak of income in the fall. Advertising and sales plans of companies selling in the farm market should be adjusted to the seasonal pattern of the farmers' income in different sections. In some sections farm income is more evenly distributed throughout the year than in others. In this connection it may be enough for a sales executive to have fairly accurate general knowledge, but it is better to have indexes showing this information in a specific quantitative form.

Measuring Differences in Seasonality

At times it is important to have a method of concisely summarizing the extent of the seasonal variation in different series. Variation in employment is a serious problem for many industries. A study of this matter would ordinarily require some method of measuring the degrees of fluctuation in various industries. Table 15–18 shows the differences in Iowa, as revealed by a study of the labor market in that state.

TABLE 15–18

VARIABILITY OF EMPLOYMENT IN VARIOUS INDUSTRIAL GROUPS IN IOWA

Group (1)	Range of Seasonal Index (2)	Average Deviation of Seasonal Index from Mean (3)
Manufacturing	17.4	4.30
Retail trade	9.7	1.78
Wholesale trade	15.0	3.92
Service industries	9.7	2.62
Agriculture	43.2	14.30
Mining	28.0	10.35
Construction	53.4	17.00
All industries	23.3	7.57

Source: Herbert W. Bohlman, *Labor Market in Iowa: Characteristics and Trends* (Des Moines, Iowa, 1937), p. 31.

The measurement of the degree of seasonality is merely an application of the measurements of dispersion discussed in Chapter 4. Column 2 shows the *range of the seasonal index for each industry,* that is, the difference between the largest and the smallest index numbers. The range is one measure of dispersion, but it is based on only two items and so it tells nothing about the scatter in the remaining items. Column 3 shows the *average deviation of the index numbers, computed from the mean of the 12 indexes for the year.* Since the mean of the 12 indexes is 100, it is easy to find the deviation of the individual months from the mean, and then to compute the mean of these deviations.

Table 15–19 illustrates the computation of the average deviation of the index from the mean, using the index of seasonal variation computed for factory sales of passenger cars from Table 15–14. The range of this index is 75.7, the difference between the August index and the October and June index. The average deviation of 15.5 computed in Table 15–19 gives the average amount that the indexes deviate from 100 and represents the degree of seasonality in the series.

Using a Measure of Seasonal Variation in Forecasting

It was stated in Chapter 13 that one important reason for measuring the changes in time series is the need for forecasting. This is true of seasonal fluctuations in data, since all budgeting and planning done on a monthly basis must make provision for the variations from month to month due to the seasonal changes.

If the seasonal pattern shows no indication of undergoing a sudden change, it is usually safe to assume at a given data that the following year will be the same as the preceding year. If there is any indication that an abrupt change

TABLE 15-19

COMPUTATION OF THE AVERAGE DEVIATION OF THE INDEX OF SEASONAL VARIATION

Factory Sales of Passenger Cars in the United States

Month (1)	Index of Seasonal Variation (2)	Deviations from the Mean (100)* (3)
January	108.9	8.9
February	102.6	2.6
March	111.3	11.6
April	110.3	10.3
May	113.1	13.1
June	114.0	14.0
July	81.1	18.9
August	38.3	61.7
September	88.0	12.0
October	114.0	14.0
November	108.6	8.6
December	109.8	9.8
Total	1,200.0	185.5
Arithmetic mean	100.0	15.5

* Signs ignored.

Source: Table 15-14.

has occurred, this of course will not be true, and other methods must be employed to predict what the new seasonal pattern will be. A basic assumption in the computation of an index of seasonal variation was that the seasonal pattern is stable enough to be averaged in order to secure the typical seasonal. When this is true and there is no evidence that any pronounced change is occurring, this typical seasonal is a reasonable forecast of the variation for the coming year.

If the pattern shows gradual change, it will be somewhat more accurate to use a changing index, which means a different index each year. When the change from one year to the next is slight, however, it is usually sufficient to use the most recent index as a forecast of the following year. For example, the index for 1972 in Table 15-17 might be used as a forecast of the index for 1973.

When a forecast of the index has been made, this index may be used to put the seasonal movement into the forecast for periods of less than a year. For example, if a forecast of passenger car sales of 9.6 million was made for 1974, the index of seasonal variation could be used to break this down into a forecast of monthly sales. On the basis of our knowledge of the seasonal pattern of factory sales of passenger cars, we would not forecast that each month would be one twelfth of the annual total; January typically accounts for more than one twelfth and July for less than one twelfth.

Using the index for 1972 from Table 15-17 as a forecast of the seasonal pattern in 1974, the calculations in Table 15-20 are made as a forecast of the

TABLE 15–20

FORECAST OF MONTHLY FACTORY SALES OF DOMESTIC PASSENGER CARS IN THE UNITED STATES FOR 1969, Based on Estimated Annual Total of Thousand Passenger Cars

Month (1)	Index of Seasonal Variation (Forecast for 1974) (2)	Estimated Monthly Sales (Index of Seasonal Variation Times 800,000 Cars) (3)
January	105.3	842,400
February	106.0	848,000
March	116.0	928,000
April	107.0	856,000
May	117.6	940,800
June	120.4	963,200
July	68.3	546,400
August	58.6	468,800
September	107.9	863,200
October	109.4	875,200
November	93.6	748,800
December	89.9	719,200
Total	1,200.0	9,600,000

Source: Table 15–17.

monthly sales if the total for the year 1974 was 9.6 million cars. If there were no seasonal variation, a reasonable forecast for each month would be 800,000 cars, or one twelfth of the annual total. January is expected to be 104.6% of the average month. Thus, the forecast for January is 104.6% of 800,000 cars, or 836,800. The forecast for each month is made in the same way, with the result that the total of 9.6 million cars is divided among the months on the basis of the forecast of the seasonal pattern for 1974.

A somewhat different use of the index of seasonal variation in forecasting or estimating is made in the following case. Assume that a statewide highway planning survey is under way and data are being collected on the volume of traffic. After a year, the seasonal fluctuations on different types of roads are measured and indexes are computed. The traffic count is made with an automatic counting device, but the cost of leaving such a device on every road for 12 months is high. Instead of counting for a full year on every road, it would be satisfactory in many cases to count for only a month, and from this count estimate the total annual volume. However, when traffic varies with the seasons, the seasonal pattern should be taken into account in making the estimate.

Assume that the traffic on a particular road showed 3,685 cars for July. What would be reasonable estimate for a year? Certainly it would not be satisfactory to multiply 3,685 by 12, since traffic for July is undoubtedly higher than for the average month. The effect of the seasonal variation should be taken into account, and if the index for this particular type of road shows July traffic was typically 25% above the average, it is possible to estimate the average monthly traffic from the July count. From this average the annual volume can then be computed.

The statement that the July traffic is 25% above the average means that the index of seasonal variation for July is 125. The count of 3,685 cars represents 125% of the average month, and the problem is to compute the average month.

$$125\% = 3{,}685$$
$$1\% = 3{,}685 \div 125 = 29.48$$

therefore

$$100\% = 29.48 \times 100 = 2{,}948.$$

If the traffic in the average month is 2,948, the total traffic for the year is 2,948 times 12, or 35,376.

An estimate of the volume of traffic for each month also could be made from the index of seasonal variation, in the same way as the estimates of the factory sales of passenger cars for the 12 months were made in Table 15–20.

STUDY QUESTIONS

15–1. Why is the measurement of seasonal variation a problem of averages?

15–2. What factors should be considered in choosing the length of the period to use in computing an index of seasonal variation?

15–3. Define specific seasonals.

15–4. What are the advantages of determining specific seasonals by expressing each month as a percentage of the 12-month moving average instead of expressing all the months in a year as percentages of the average for that year?

15–5. What is the advantage of using as specific seasonals the ratios to the 12-month moving average, centered by taking a 2-item moving average, over the 12-month moving average centered at the seventh month? How significant do you consider this advantage?

15–6. What are the relative advantages of the following methods of averaging specific seasonals to find the index of seasonal variation?

(a) Arithmetic mean.

(b) Median.

(c) Modified arithmetic mean.

15–7. How can one determine whether to use a changing index of seasonal variation instead of computing an average index for a number of years?

15–8. Assume that a manufacturing concern is trying to reduce the amplitude of the seasonal swings in its production rate by adding new products with different seasonal patterns. After a substantial volume of business has been built up for the new products, how would you determine whether or not the diversification of products has succeeded in reducing the seasonal swings of the business?

15–9. How accurately do you think a businessman can estimate the seasonal pattern prevailing in the demand for his product? For example, do you believe that the businessman has an accurate enough knowledge of the seasonal pattern (without making any calculations) to set up a monthly budget that takes into account the seasonal fluctuations?

PROBLEMS

15–1. A retail store has the following index of seasonal variation in sales.

January	65	July	83
February	70	August	73
March	95	September	100
April	110	October	115
May	105	November	124
June	103	December	157

The merchandise manager of the store forecasts that sales for the coming year will total $60 million. On the basis of this forecast of total sales, make a forecast of sales for each month.

15–2. (a) January sales of the store described in Problem 15–1 were $3,315,000. If this level of sales prevails for the remaining 11 months of the year, what will be the total annual sales for this year?

(b) Assume that February sales for the year described in Problem 15–1 were $3,710,000. If the level of sales for the first 2 months prevails for the remaining 10 months of the year, what will be the total annual sales for this year?

15–3. The following table gives the average cold-storage holdings of frozen fruits and vegetables on the first of the month (in millions of pounds) for a six-year period.

(a) From these averages compute the indexes of seasonal variation for cold-storage holdings of frozen fruits and vegetables.

(b) Plot both indexes on one chart and describe briefly the two seasonal patterns.

Date	Frozen Fruits	Frozen Vegetables
Jan. 1	324	531
Feb. 1	299	477
Mar. 1	266	425
Apr. 1	228	384
May 1	189	351
June 1	187	333
July 1	228	352
Aug. 1	322	414
Sept. 1	466	509
Oct. 1	383	589
Nov. 1	389	621
Dec. 1	375	603

Source: United States Department of Agriculture.

15–4. Compare the amplitude of the seasonal fluctuations in cold-storage holdings of frozen fruits and vegetables by computing the average deviation from the mean for the two indexes of seasonal variation found in Problem 15–3. Summarize briefly what the measures of dispersion show.

15–5. The following index of seasonal variation reflects the changing volume of business of a mountain resort hotel that caters to the family tourist in the summer and the skiing enthusiast during the winter months.

January	120	July	149
February	138	August	155
March	92	September	91
April	41	October	69
May	49	November	82
June	126	December	88

(a) If the hotel has 600 guest days in January and if the normal seasonal pattern holds, how many guest days could be expected in February?

(b) What would you estimate the total number of guest days to be for the entire year?

15–6. Given the following figures on the seasonal index and department store sales, would you say that business is better or worse in July as compared to June, after adjustment for seasonal variation?

Month	Index of Seasonal Variation	Sales
June	117	\$2,319,452
July	87	1,750,333

15–7. The following table shows sales for a company (in millions of dollars) by quarters.

	1969	1970	1971	1972	1973
Quarter 1	15	17	18	20	21
2	22	23	27	27	30
3	24	26	29	30	32
4	35	38	40	45	46
Total	96	104	114	122	129

(a) Compute an index of seasonal variation by quarters by averaging quarterly sales. What is the weakness inherent in this method?

(b) Compute an index of seasonal variation by quarters by using a ratio to 2-item average of a 4-quarter moving average. This can be done by following the same procedure explained in this chapter for monthly data.

15–8. Using the index of seasonal variation given in Problem 15–5 compute the average deviation about the median. What does this measure tell you about the seasonal pattern of business for the hotel?

15–9. Using the data in Problem 15–7 and the index of seasonal variation computed in (b), estimate sales for the third quarter of 1975 if the projected annual sales for that year total \$150 million.

15–10. Stocks of nonfat dry milk held by manufacturers show a pronounced seasonal variation, as measured by the following index of seasonal variation based on eleven years, 1962 to 1972.

The production of milk fluctuates widely with the seasons, but the consumption of fresh milk is distributed throughout the year much more evenly. The surplus of milk during the months of large production is absorbed in the manufacture of products that

Month	Index	Month	Index
January	80.3	July	134.5
February	76.5	August	122.0
March	73.4	September	104.0
April	91.4	October	90.8
May	120.6	November	81.4
June	139.0	December	86.1

can be stored, thus shifting the seasonal surplus into stocks. This tends to put wide seasonal swings into the production of the products made from surplus milk. Nonfat dry milk is one of the products that can be stored, and production tends to take on a seasonal pattern that is determined by the availability of surplus milk. The seasonal pattern of the production of nonfat dry milk is revealed in the data given in the following table (in millions of pounds) for each month from 1962 through 1972.

	1962	1963	1964	1965	1966	1967	1968	1969	1970	1971	1972
January	186	176	177	189	131	135	125	118	104	116	99
February	180	168	181	181	123	130	126	114	104	112	100
March	206	195	207	202	144	146	143	129	128	131	118
April	217	214	218	214	165	175	166	147	138	149	129
May	259	249	250	239	186	193	191	176	171	175	153
June	243	240	236	223	191	204	189	176	169	178	160
July	191	182	182	172	131	160	152	140	141	137	127
August	159	145	148	131	111	123	119	113	118	118	99
September	132	119	122	100	87	98	89	83	88	92	77
October	142	121	126	102	93	99	88	80	90	94	70
November	144	128	133	106	93	97	89	72	81	77	62
December	173	158	171	130	124	119	114	103	109	95	76

Source: U.S. Department of Commerce, *Business Statistics* (1973).

(a) Construct an index of seasonal variation by the method of ratio to the 2-item average of the 12-month moving average.

(b) On the same chart plot the index of seasonal variation in production and the index of seasonal variation in stocks.

(c) Summarize the relationship between the two seasonal patterns.

(d) Compare the amplitude of the seasonal fluctuations in stocks and production of nonfat dry milk by computing the average deviation from the mean for the two indexes of seasonal variation. Summarize briefly what the measures of dispersion show.

(e) Assume that a forecast has been made that total production of nonfat dry milk for the coming year will be 2,500 million pounds. Forecast the monthly production of evaporated milk that you would consider reasonable on the basis of the seasonal pattern determined above.

(f) *Retain this solution for use in Problem 16–16.*

15–11. The following table gives the index of seasonal variation for the stocks of distillate fuel oil at the end of the month for the years 1962 and 1972. The level of fuel oil inventories is at its highest point in October, which is the beginning of the heating season, since this grade of fuel oil is used mainly for space heating. The level of stocks declines from October to a low point in March, after which date stocks rise.

Month	Index	Month	Index
January	91.9	July	105.2
February	76.0	August	119.6
March	65.2	September	130.1
April	67.3	October	136.4
May	75.0	November	131.6
June	88.1	December	113.6

The data below give the monthly domestic demand for distillate fuel oil (in millions of barrels) from 1962 to 1972. It can be seen that demand for fuel oil is higher in the winter months than in the other months of the year, but this problem will be to measure the variation in demand precisely.

(a) Construct an index of seasonal variation by the 2-item average of the 12-month moving average for the demand for distillate fuel oil.

(b) Plot on the same chart the index of seasonal variation in demand for fuel oil and the index of seasonal variation in stocks given above.

(c) Summarize the relationship between the two seasonal patterns and explain why it exists.

	1962	1963	1964	1965	1966	1967	1968	1969	1970	1971	1972
January	101	103	96	93	96	93	119	119	127	125	115
February	82	88	82	87	88	90	103	96	97	108	121
March	76	72	73	84	77	91	87	91	96	100	108
April	54	49	60	61	63	58	61	67	74	79	83
May	45	48	47	46	53	60	57	59	60	66	70
June	40	40	44	42	49	49	49	52	53	60	66
July	41	40	41	44	43	48	46	50	50	55	55
August	37	43	41	48	51	46	50	51	53	57	64
September	44	51	48	50	50	47	54	58	59	61	66
October	51	50	57	57	59	60	62	62	70	67	86
November	72	61	66	72	75	81	78	83	79	91	102
December	90	102	95	93	93	93	108	112	110	113	131

Source: U.S. Department of Commerce, *Business Statistics* (1973).

15-12. The following table gives the index of seasonal variation, 1962 to 1972, for stocks of gasoline, which are at their highest point in March and then decline until late in the fall.

Month	Index	Month	Index
January	106.5	July	94.2
February	110.6	August	91.3
March	111.1	September	93.1
April	106.9	October	92.5
May	102.3	November	94.6
June	97.8	December	99.1

Plot the index of seasonal variation in gasoline stocks on an arithmetic chart with the index of seasonal variation in distillate fuel oil stocks. Why do the seasonal patterns of the stocks of these two petroleum products show such different seasonal patterns?

15–13. The following table gives the monthly domestic demand for gasoline (in millions of barrels) from 1962 to 1972.

(a) Construct an index of seasonal variation by the 2-item average of the 12-month moving average.

(b) Plot the index of seasonal variation in demand for gasoline on an arithmetic chart with the index of seasonal variation in stocks of gasoline given in Problem 15–12.

(c) Summarize the relationship between the two seasonal patterns and explain why it exists.

(d) *Retain this solution for use in Problem 16–15.*

	1962	1963	1964	1965	1966	1967	1968	1969	1970	1971	1972
January	122	125	131	125	133	137	148	158	164	165	173
February	109	115	122	120	126	129	145	145	151	155	167
March	130	128	136	140	146	152	156	160	173	183	200
April	129	138	141	141	147	146	163	169	172	188	190
May	141	143	145	150	154	161	169	178	184	185	201
June	141	141	154	155	165	166	166	173	187	195	206
July	143	150	157	157	160	163	181	188	195	201	208
August	147	151	150	154	165	171	179	185	190	197	217
September	127	134	146	142	150	153	160	171	180	184	195
October	137	142	148	147	151	161	170	177	185	189	199
November	133	133	132	140	148	154	158	164	168	185	196
December	126	134	146	149	150	151	162	175	182	189	199

Source: U.S. Department of Commerce, *Business Statistics* (1973).

16 Cyclical Fluctuations in Time Series

Since practically every type of business activity is affected by the cyclical swings of total business, information on these changes is extremely important to business management, particularly in their planning efforts. An individual business concern has little influence on the cyclical swings of the economy; however, if its management is well informed on the status of these cycles, it will be better able to make intelligent decisions regarding present and future operations of the organization.

IMPORTANCE OF THE BUSINESS CYCLE

When economic conditions are approaching a peak of prosperity, inventories should be held down, expansion of plant equipment should be little or none at all, extension of credit should be made very carefully, and the individual firm should be put into a position to lose as little money as possible when the inevitable decline in prices and volume occurs. When recovery is beginning to get under way, it is time to increase inventories, make needed expansion and modernization of equipment, increase advertising budgets and sales efforts, and get the firm in a position to take advantage of improved conditions.

When a business firm is caught with a large, high-priced inventory in a period of declining prices, there is usually little to do but take a loss. If plant and equipment were enlarged and improved at high prices just before a decline in volume, the concern may not be able to survive. Usually the new equipment can be justified on the basis of a continuation of business at a level approaching the high point of prosperity. If volume falls off, the money spent for expansion may be a complete loss. If the expansion was financed with borrowed funds, the interest charges may be the factor that puts the concern into bankruptcy.

At the bottom of a recession, no future commitments are usually made until it is certain that recovery is under way. If recovery comes suddenly after a long recession, a concern may find itself unable to step up production fast enough to take full advantage of the increased demand.

The cyclical fluctuation in business is an ever-present, though not fully understood, force that must be constantly monitored. Sometime in the future the business cycle may be brought under control, but at present businessmen are limited to managing their firms with as complete knowledge as possible of what is happening, trying to minimize possible losses and to remain sensitive to opportunities for additional profit. For example, an executive cannot prevent a period of prosperity from ending in a sudden drop in prices and business activity; but if he can foresee the event, he can get his firm into the best possible condition to go through this period. At the bottom of a recession he can get his business concern into a condition to earn as large a profit as possible when conditions do improve. Proper management enabled a few concerns to show a profit even during the depression of the 1930's. Others came through in sound condition, though they may not have made money every year.

Even though it seems that the cyclical swings in business are at present inevitable, a great deal of study has been made in recent years of the possibility of controlling these fluctuations. Enough progress has been made in the use of some devices, notably credit control and government fiscal policy, to make it desirable that the business statistician consider the effects that these stabilization efforts have on the economy when he is forecasting the future.

This chapter describes the statistical methods that may be employed to measure the purely cyclical fluctuations in a time series. This is the first step in supplying information on such fluctuations. The method may be applied to the internal statistical data of an individual company, to the data for an industry, or to series showing changes in any of the elements of economic conditions that interest the businessman.

If an executive wants to see how the cyclical fluctuations in the sales of his concern compare with the cyclical fluctuations in the industry, it is necessary to secure a measure that expresses the cyclical fluctuations in the two series so that they can be compared. The same principle applies to the comparison of an individual business firm or an industry with general economic conditions.

MEASURING CYCLICAL FLUCTUATIONS

No completely satisfactory method of measuring directly the cyclical swings in a time series has been developed. The irregular nature of the fluctuations makes impossible any attempt to find an average cycle that could be used to represent the effect of the cycle on the series. For this reason the methods used to measure seasonal variation cannot be used for cyclical fluctuations. The length of the seasonal movement is fixed at 12 months. The cyclical fluctuations, on the other hand, show a wide variation in the length of the cycle as well as in the amplitude of the variations. Business volume expands and contracts in an oscillating movement that varies greatly in the time required to make a complete cycle. Professor W. C. Mitchell computed the average length of the business cycle to be four years, but this average was computed from cycles ranging in length from one year to nine years.

The best approach to the problem of measuring the cyclical fluctuations in a time series has been the indirect method of removing the variation in the series that results from seasonal forces and secular trend. The remaining fluctuations are then considered to be cyclical and erratic movements. Frequently no attempt is made to separate these two types of fluctuation, since no very satisfactory technique has been developed.

In the following sections the isolation of cyclical fluctuations is described for several types of data. Since seasonal variation and secular trend can be measured with reasonable accuracy, it is possible to calculate what the fluctuations in a series would have been if the trend and seasonal factors had not been present. This technique is based on the assumption that the trend, seasonal, and cyclical variations are independent of each other and therefore the cyclical fluctuations would have been the same if there had been no trend and seasonal variation in the series. It is believed that this assumption is generally valid, but the extent to which it is invalid is reflected in the resulting measure of cyclical fluctuations.

Adjusting Annual Data for Secular Trend

A time series consisting of period data on a yearly basis may show the effect of trend, cyclical, and erratic fluctuations but cannot show seasonal variations. Since each yearly amount includes all 12 months, any seasonal characteristics the data might possess are lost in the yearly total, and it is also usually true that a year is long enough for the erratic fluctuations to cancel out. Occasionally there will be nonrecurring events of sufficient magnitude for their effects to appear in annual data, but the minor factors that cause data classified on the basis of short periods to fluctuate erratically will ordinarily not influence annual data.

In an annual series that is affected only by trend and cyclical movements, the measurement of secular trend can be used to remove the effect of trend and leave only cyclical fluctuations. In Figure 14–1 on page 369 it can be seen that forces other than growth have influenced the production of raw steel in the different years. The fluctuations of the series about the line of trend have a wavelike swing that has given this movement the name *cyclical*. However, the trend has risen so rapidly that it tends to obscure the cyclical fluctuations. These movements can be seen more clearly in Figure 16–1, in which the trend factor has been taken out, leaving only the cyclical fluctuations.

The most commonly used method of separating the effect of the cycle from the trend is to express the data for each year as a percentage of the trend value for that year. These calculations appear in Table 16–1 for each year from 1929 to 1972, and the results are shown graphically in Figure 16–1. The logic underlying this method of separating the effects of trend and cycle is to

TABLE 16–1

RAW STEEL PRODUCTION IN THE UNITED STATES, 1926 TO 1968

Adjustment for Secular Trend

Year (1)	Production (Millions of Tons) (2)	Trend (Millions of Tons) (3)	Percentage of Trend (4)
1929	61.7	36.1	170.9
1930	44.6	38.5	115.8
1931	28.6	40.9	69.9
1932	15.1	43.2	35.0
1933	25.7	45.6	56.4
1934	29.2	48.0	60.8
1935	38.2	50.4	75.8
1936	53.5	52.8	101.3
1937	56.6	55.2	102.5
1938	31.5	57.6	54.7
1939	52.8	60.0	88.0
1940	67.0	62.4	107.4
1941	82.8	64.8	127.8
1942	86.0	67.2	128.0
1943	88.8	69.6	127.6
1944	89.6	72.0	124.4
1945	79.7	74.4	107.1
1946	66.6	76.7	86.8
1947	84.9	79.1	107.3
1948	88.6	81.5	108.7
1949	78.0	83.9	93.0
1950	96.8	86.3	112.2
1951	105.2	88.7	118.6
1952	93.2	91.1	102.3
1953	111.6	93.5	119.4
1954	88.3	95.9	92.1
1955	117.0	98.3	119.0
1956	115.2	100.7	114.4
1957	112.7	103.1	109.3
1958	85.3	105.5	80.9
1959	93.4	107.9	86.6
1960	99.3	110.3	90.0
1961	98.0	112.6	87.0
1962	98.3	115.0	85.5
1963	109.3	117.4	93.1
1964	127.1	119.8	106.1
1965	131.5	122.2	107.6
1966	134.1	124.6	107.6
1967	127.2	127.0	100.2
1968	131.5	129.4	101.62
1969	141.3	131.8	107.2
1970	131.5	134.2	98.0
1971	120.4	136.6	88.1
1972	133.1	139.0	95.8

Source: American Iron and Steel Institute, *Annual Statistical Report* (1973).

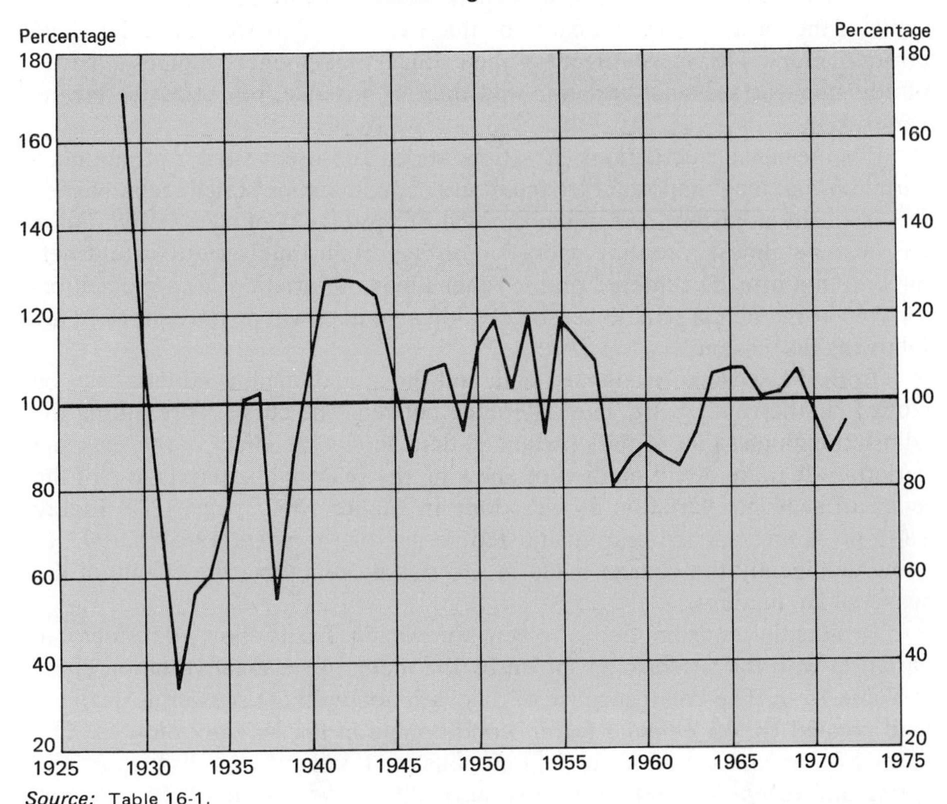

Source: Table 16-1.

FIGURE 16–1

RAW STEEL PRODUCTION IN THE UNITED STATES, 1929 TO 1972

assume that each movement is essentially independent of the other and that the forces causing the cyclical fluctuations have approximately the same relative effect on the series regardless of the level of the trend. The most severe depression of the century pushed raw steel production to a level of 35.0% of the trend in 1932, and a much milder recession in 1938 dropped production to 54.7% of trend. However, the amount of decline from the trend in 1938 was almost as much as the decline from the trend in 1932, since the trend had risen during the six years. The relative fluctuation appears generally to be more significant than the amount of the fluctuation when there is also secular trend in the series.

Adjusting Monthly Data for Seasonal Variation

In a series with neither upward nor downward trend, the fluctuations will be the result of cyclical and erratic variations, when classified on a yearly basis.

When classified into periods of less than a year, usually months or quarters, some of the variation will probably be due to seasonal as well as cyclical and erratic factors. The measurement of the cyclical movements becomes a matter of adjusting for seasonal variation and then, if possible, removing the erratic variations.

The seasonal fluctuations in a time series represent such a pronounced variation that most important business and economic time series are published in seasonally adjusted form. The removal of the effects of seasonal variation has become almost a routine operation on important time series, since it permits attention to be centered on the other kinds of variation. The procedures for converting actual data to seasonally adjusted data will be introduced in the following discussion.

Both the seasonally adjusted data and the actual unadjusted data may be plotted on the same chart. The difference between the curve representing the adjusted data and that of the unadjusted data shows the effect of the seasonal variation. A more direct method of showing the seasonal pattern is to plot the index of seasonal variation as was done in Figure 15–2 (page 424). Figure 16–2 gives an adjusted and unadjusted series for the years 1968 to 1972 to show graphically the change made in a series by removing the effects of the seasonal fluctuations.

Residential consumption of electric power in Texas shows considerable variation with the seasons, as shown by the index of seasonal variation given in Table 16–2. The greatest cause of this variation with the seasons is the peak load created by the demand for air conditioning in the summer months. The index of seasonal variation shows that consumption increases throughout the spring and reaches a peak in August, normally the hottest month in the year. Since natural gas is the favorite fuel for heating in Texas, the demand for electricity for heating does not fill the gap left by the decline in the use for air conditioning at the end of the summer. The shorter days cause some increase in demand for lighting in the winter months, but this is not nearly large enough to offset the reduction in demand caused by the decline in consumption for air conditioning.

Comparing the consumption of electricity in two months requires that an allowance be made for the seasonal pattern. For example, consumption in February, 1972, declined 4.6% from January, but this does not necessarily mean that February was a poorer month for the power companies. The index of seasonal variation indicates that consumption in January is 86.4% of the average month, while February is only 79.2% of the average month. This is a decline in expected consumption of 8.3%. The actual decline in 1972 was only 4.6%, so consumption in February was actually higher than in January if one takes into consideration the regular behavior of consumption in these two months. February, 1972, consumption declined less than the normal decline between January and February. It is possible to adjust each of these monthly figures for seasonal variation and compare the change in these seasonally adjusted figures.

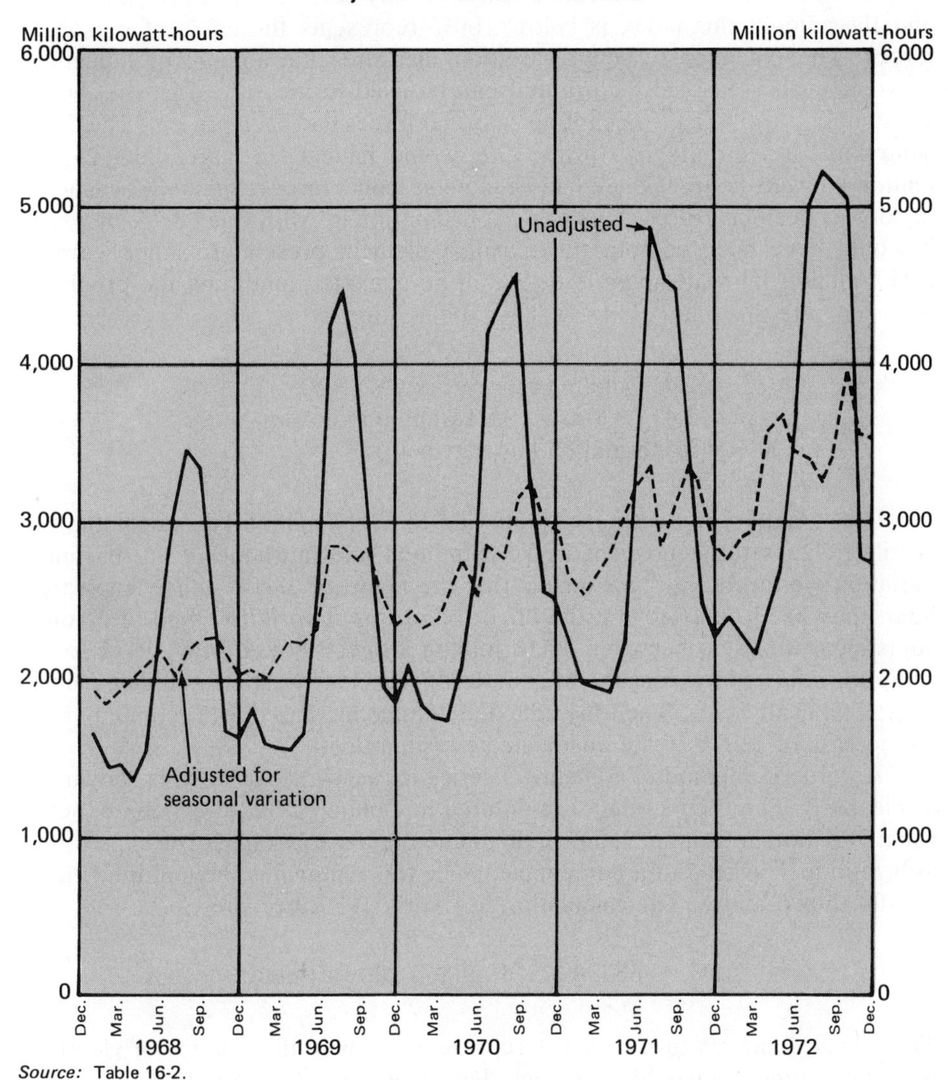

Source: Table 16-2.

FIGURE 16–2

RESIDENTIAL ELECTRIC POWER CONSUMPTION IN TEXAS, 1968 TO 1972

When it is desired to compare the level of consumption among many months, the procedure shown above becomes too cumbersome, and it is desirable to have a systematic method of making allowance for the effect of the fluctuations due to seasonal variation. The residential consumption of electricity in Texas is at the seasonal low in April, partly because April has only 30 days and also because there is little demand for air conditioning in that month. Due to these factors, consumption is only 68.7% of the average month, and the amount this index is below 100% represents the effect of seasonal forces. The index of seasonal variation measures the amount of influence exerted by the seasonal factors and can be used to measure what the April consumption would have been if no seasonal forces had been present. The consumption of electricity in April, 1972, would have been larger than 2,417 million kilowatt-hours if there had been no seasonal forces. Since the seasonal factor has been measured to be 68.7, consumption in April was 68.7% of what it would have been without the seasonal element present. In other words, 2,417 million kilowatt-hours is 68.7% of an average month and the problem is to compute the value of 100%. The computation is

$$\begin{aligned} 68.7\% &= 2{,}417 \text{ million kilowatt-hours} \\ 1\% &= 2{,}417 \div 68.7 = 35.18 \text{ million kilowatt-hours} \\ 100\% &= 3{,}518 \text{ million kilowatt-hours.} \end{aligned}$$

Consumption of electricity amounting to 2,417 million kilowatt-hours in April, 1972, is the equivalent of 3,518 million kilowatt-hours if no seasonal variation were present. This means that the figure of 2,417 million kilowatt-hours has been adjusted for the effect of *seasonal variation,* and after this adjustment its value becomes 3,518 million kilowatt-hours. May, 1972, consumption adjusted by using the May index of seasonal variation would be 3,696 million kilowatt-hours. The May adjusted volume increased 5.1% compared to an increase of 13.6% in the unadjusted consumption.

An orderly method of adjusting a series for seasonal variation is shown in Table 16–2. The original data are entered in Column 2 and the index of seasonal variation in Column 3. The index of seasonal variation may be expressed as a ratio to 1 instead of a percentage, since this eliminates the multiplication by 100 shown above. The calculation for April, 1972, becomes

$$2{,}417 \div .687 = 3{,}518 \text{ million kilowatt-hours.}$$

The adjusted figures in Column 4 represent the way the volume of electric power consumption would have been distributed over the months if all forces affecting the data remained the same except that there were no seasonal influences.

In Table 16–2 the same index of seasonal variation is used to adjust the data for each year from January, 1968, to December, 1972. This is done on the assumption that the seasonal pattern has not shown any appreciable change over this period. The eleven years 1961 to 1971 were used to compute the index of seasonal variation, but the index has been used to adjust the monthly data for 1972 also. The seasonal index can be used to adjust data beyond the period on which it is based on the assumption that it still measures the seasonal change even though the most recent years were not used in its computation. If this assumption cannot be made, it is not possible to adjust current data for seasonal variation.

TABLE 16–2

RESIDENTIAL ELECTRIC POWER CONSUMPTION IN TEXAS, 1968 TO 1972

Adjusted for Seasonal Variation

(Millions of Kilowatt-Hours)

Year and Month (1)	Millions of Kilowatt-Hours (2)	Index of Seasonal Variation (3)	KWH Adjusted for Seasonal Variation (4)
1968			
January	1,666	86.4	1,928
February	1,477	79.2	1,865
March	1,450	73.0	1,986
April	1,375	68.7	2,001
May	1,568	74.4	2,110
June	2,351	106.8	2,205
July	3,010	146.4	2,053
August	3,476	160.8	2,170
September	3,310	147.6	2,252
October	2,288	101.4	2,256
November	1,687	78.3	2,155
December	1,606	78.1	2,056
1969			
January	1,817	86.4	2,103
February	1,601	79.2	2,021
March	1,585	73.0	2,171
April	1,555	68.7	2,263
May	1,658	74.3	2,231
June	2,572	106.6	2,413
July	4,200	146.6	2,865
August	4,522	160.2	2,823
September	4,035	147.0	2,745
October	2,881	101.4	2,841
November	1,953	78.3	2,494
December	1,832	78.1	2,346
1970			
January	2,115	86.4	2,448
February	1,848	79.2	2,333
March	1,749	73.0	2,396
April	1,735	68.7	2,525
May	2,069	74.3	2,785
June	2,684	106.6	2,518
July	4,169	146.6	2,844
August	4,437	160.2	2,770
September	4,558	147.0	3,101
October	3,278	101.4	3,233
November	1,998	78.3	2,552
December	1,954	78.1	2,502
1971			
January	2,261	86.4	2,616
February	1,988	79.2	2,510
March	1,953	73.0	2,675

(continued)

TABLE 16–2 (concluded)

Year and Month (1)	Millions of Kilowatt-Hours (2)	Index of Seasonal Variation (3)	KWH Adjusted for Seasonal Variation (4)
April	1,922	68.7	2,798
May	2,321	74.3	3,124
June	3,495	106.6	3,279
July	4,867	146.6	3,320
August	4,546	160.2	2,838
September	4,495	147.0	3,058
October	3,436	101.4	3,389
November	2,460	78.3	3,142
December	2,258	78.1	2,891
1972			
January	2,400	86.4	2,778
February	2,289	79.2	2,890
March	2,149	73.0	2,944
April	2,417	68.7	3,518
May	2,746	74.3	3,696
June	3,704	106.6	3,475
July	5,002	146.6	3,412
August	5,233	160.2	3,267
September	5,130	147.0	3,490
October	4,026	101.4	3,970
November	2,776	78.3	3,545
December	2,672	78.1	3,536

Source: Bureau of Business Research, The University of Texas at Austin.

If there is so much change in the seasonal pattern that the index cannot be projected for several years, it is generally preferable to use a changing index described on pages 422 to 429. If a fixed index of seasonal variation is used beyond the period for which is was computed, it should be checked after the end of the year to see if any substantial change in the index is required. Statistical organizations normally maintain a regular system for keeping the indexes of seasonal variation up-to-date, and if a computer is available they may be revised regularly.

ADJUSTING MONTHLY DATA FOR BOTH SEASONAL VARIATION AND TREND

Monthly data can be adjusted for secular trend in the same manner as annual data were adjusted in Table 16–1, except that a trend value must be secured for each month and compared with the monthly data. If the series has no seasonal variation, the data adjusted for trend would then contain cyclical and erratic movements; but if there is a seasonal movement in the series, adjustment must be made for seasonal variation as well as trend in order to isolate the cyclical and erratic fluctuations.

Most monthly data do show a seasonal movement which calls for adjustment. In Table 16–3 the factory sales of passenger cars are adjusted for seasonal variation, and then these figures are further adjusted for trend. These computations combine the procedures used in Tables 16–1 and 16–2. The index of seasonal variation was computed by the method of ratios to the 12-month moving average. The secular trend was computed for the years 1963 to 1972 in Table 14–4 (page 383).

TABLE 16–3

FACTORY SALES OF DOMESTIC PASSENGER CARS IN THE UNITED STATES, 1963 TO 1972

Adjusted for Seasonal Variation and Secular Trend

Year and Month (1)	Factory Sales (Thousands) (2)	Index of Seasonal Variation (3)	Factory Sales Adjusted for Seasonal Variation (Thousands) (4)	Trend (Thousands) (5)	Sales Adjusted for Trend and Seasonal Variation (6)
1963					
January	658.0	107.56	611.7	659.9	92.71
February	592.8	101.82	582.2	659.8	88.25
March	637.1	116.02	549.1	659.6	83.25
April	671.8	110.53	607.9	659.5	92.15
May	695.1	115.04	604.2	659.4	91.63
June	672.9	118.34	568.6	659.3	86.24
July	649.4	82.73	784.9	659.2	119.07
August	165.1	39.56	417.4	659.1	63.33
September	463.0	88.02	526.0	659.0	79.83
October	779.2	105.90	735.8	658.9	111.68
November	726.2	105.57	687.9	658.7	104.42
December	733.0	108.90	673.1	658.6	102.19
1964					
January	709.0	107.56	659.1	658.5	100.10
February	665.4	101.82	653.5	658.4	99.26
March	700.9	116.02	604.1	658.3	91.77
April	770.2	110.53	696.8	658.2	105.87
May	719.5	115.04	625.5	658.1	95.04
June	726.7	118.34	614.1	658.0	93.33
July	562.2	82.73	679.5	657.8	103.30
August	230.8	39.56	583.5	657.7	88.71
September	563.8	88.02	640.5	657.6	97.40
October	394.7	105.90	372.7	656.5	56.69
November	648.4	105.57	614.2	657.4	93.43
December	862.4	108.90	791.9	657.3	120.48
1965					
January	782.8	107.56	727.8	657.2	110.74
February	753.1	101.82	739.6	657.1	112.57
March	937.9	116.02	808.4	656.9	123.06

(continued)

TABLE 16-3 (continued)

Year and Month (1)	Factory Sales (Thousands) (2)	Index of Seasonal Variation (3)	Factory Sales Adjusted for Seasonal Variation (Thousands) (4)	Trend (Thousands) (5)	Sales Adjusted for Trend and Seasonal Variation (6)
1965					
April	846.9	110.53	766.2	656.8	116.65
May	819.3	115.04	712.2	656.7	108.45
June	880.9	118.34	744.4	656.6	113.37
July	745.6	82.73	901.2	656.5	137.28
August	330.4	39.56	835.2	656.4	127.25
September	438.5	88.02	498.2	656.3	75.91
October	825.4	105.90	779.4	656.2	118.79
November	878.7	105.57	832.3	656.0	126.87
December	861.3	108.90	790.9	655.9	120.58
1966					
January	780.4	107.56	725.5	655.8	110.63
February	744.8	101.82	735.4	655.7	112.16
March	902.0	116.02	777.5	655.6	118.59
April	793.9	110.53	718.3	655.5	109.58
May	771.2	115.04	670.4	655.4	102.29
June	802.5	118.34	678.1	655.3	103.49
July	480.0	82.73	580.2	655.1	88.56
August	136.4	39.56	344.8	655.0	52.64
September	592.4	88.02	673.0	654.9	102.77
October	797.7	105.90	753.3	654.8	115.04
November	791.2	105.57	749.4	654.7	114.47
December	740.5	108.90	680.0	654.6	103.88
1967					
January	625.0	107.56	581.1	654.5	88.78
February	501.9	101.82	492.9	654.4	75.33
March	647.4	116.02	558.0	654.2	85.29
April	628.3	110.53	568.4	654.1	86.90
May	713.4	115.04	620.2	654.0	94.82
June	732.3	118.34	618.8	653.9	94.63
July	416.6	82.73	496.3	653.8	75.91
August	218.3	39.56	551.9	653.7	84.42
September	570.6	88.02	648.3	653.6	99.19
October	608.8	105.90	574.9	653.4	87.98
November	645.2	105.57	611.2	653.3	93.54
December	768.5	108.90	705.7	653.2	108.03
1968					
January	747.2	107.56	694.7	653.1	106.36
February	668.2	101.82	656.3	653.0	100.50
March	764.0	116.02	658.5	652.9	100.86
April	747.8	110.53	676.5	652.8	103.64
May	876.2	115.04	761.7	652.7	116.70
June	781.6	118.34	660.5	652.5	101.21
July	605.4	82.73	731.7	652.4	112.16
August	182.6	39.56	461.6	652.3	70.76

TABLE 16–3 (continued)

Year and Month (1)	Factory Sales (Thousands) (2)	Index of Seasonal Variation (3)	Factory Sales Adjusted for Seasonal Variation (Thousands) (4)	Trend (Thousands) (5)	Sales Adjusted for Trend and Seasonal Variation (6)
1968					
September	620.0	88.02	704.4	652.2	108.00
October	889.5	105.90	840.0	652.1	128.81
November	831.0	105.57	787.1	652.0	120.73
December	693.7	108.90	637.0	651.9	97.72
1969					
January	782.0	107.56	727.0	651.8	111.55
February	676.7	101.82	664.6	651.6	101.99
March	721.0	116.82	621.5	651.5	95.38
April	676.9	110.53	612.4	651.4	94.01
May	678.0	115.04	589.4	651.3	90.49
June	740.4	118.34	625.6	651.2	96.08
July	446.9	82.73	540.2	651.1	82.96
August	329.5	39.56	833.0	651.0	127.96
September	706.5	88.02	802.7	650.9	123.32
October	815.6	105.90	770.2	650.7	118.35
November	644.0	105.57	610.0	650.6	93.76
December	588.8	108.90	540.7	650.5	83.11
1970					
January	545.0	107.56	506.7	650.4	77.90
February	528.4	101.82	519.0	650.3	79.80
March	594.4	116.82	512.3	650.2	78.80
April	627.2	110.53	567.4	650.1	87.29
May	684.4	115.04	595.0	650.0	91.54
June	758.4	118.34	640.8	649.8	98.62
July	464.3	82.73	561.2	649.7	86.37
August	254.0	39.56	642.1	649.6	98.84
September	454.2	88.02	516.0	649.5	79.45
October	365.4	105.90	345.1	649.4	53.13
November	341.1	105.57	323.1	649.3	49.76
December	570.6	108.90	524.0	649.2	80.71
1971					
January	678.1	107.56	630.4	649.1	97.13
February	719.0	101.82	706.2	648.9	108.82
March	815.9	116.02	703.3	648.8	108.39
April	703.6	110.53	636.6	648.7	98.13
May	716.7	115.04	623.0	648.6	96.06
June	761.3	118.34	643.3	648.5	99.20
July	468.9	82.73	566.8	648.4	87.41
August	456.6	39.56	115.4	648.3	178.05
September	712.0	88.02	808.9	648.2	124.80
October	758.6	105.90	716.4	648.0	110.54
November	736.6	105.57	697.7	647.9	107.69
December	593.2	108.90	544.7	647.8	84.08

TABLE 16–3 (concluded)

Year and Month (1)	Factory Sales (Thousands) (2)	Index of Seasonal Variation (3)	Factory Sales Adjusted for Seasonal Variation (Thousands) (4)	Trend (Thousands) (5)	Sales Adjusted for Trend and Seasonal Variation (6)
1972					
January	666.0	107.56	619.2	647.7	95.59
February	716.1	101.82	703.3	647.6	108.60
March	765.2	116.02	659.6	647.5	101.87
April	736.9	110.53	666.7	647.4	102.98
May	798.0	115.04	693.7	647.3	107.17
June	761.6	118.34	643.6	647.1	99.45
July	393.6	82.73	475.7	647.0	73.53
August	371.0	39.56	937.9	646.9	144.98
September	808.8	88.02	918.9	646.8	142.07
October	841.7	105.90	794.8	646.7	122.91
November	827.4	105.57	783.7	646.6	121.21
December	666.2	108.90	611.7	646.5	94.63

Source: Tables 15–4 and 15–17.

Figure 16–3 shows the unadjusted demand and the trend line for the years 1963 to 1972. Figure 16–4 (page 456) shows the sales adjusted for trend and seasonal variation, expressed as a percentage of the secular trend. The fluctuations remaining after the series has been adjusted for trend and seasonal variation can be considered the cyclical and erratic movements of the series.

The computation of the monthly trend values in Column 5 of Table 16–3 was not explained in Chapter 14, since normally the only occasion for deriving a trend value for each month is the desire to adjust the series for trend to isolate the cyclical fluctuations. The computation of monthly trend values is described in this section for the most important types of trends.

When the trend line has been located by the graphic method, the monthly trend values may be secured in the same manner as the annual trend values. The monthly figures to which the trend is to be fitted should first be plotted, and the trend located graphically. The monthly values may then be read from the chart in the same manner as the annual values. For other methods of computation, annual trend values are ordinarily computed first and monthly values derived from the annual values. For a straight line fitted to the original data or to the logarithms of the data, the following steps are needed to get monthly trend values from annual values:

1. Determine that the trend was fitted to average monthly data or, if fitted to total annual data, divide the annual trend values by 12 to secure the average monthly trend values.
2. Determine the increase or decrease from one year to the next. One twelfth of this amount will be the increase or decrease in the trend *from one month to the next.*

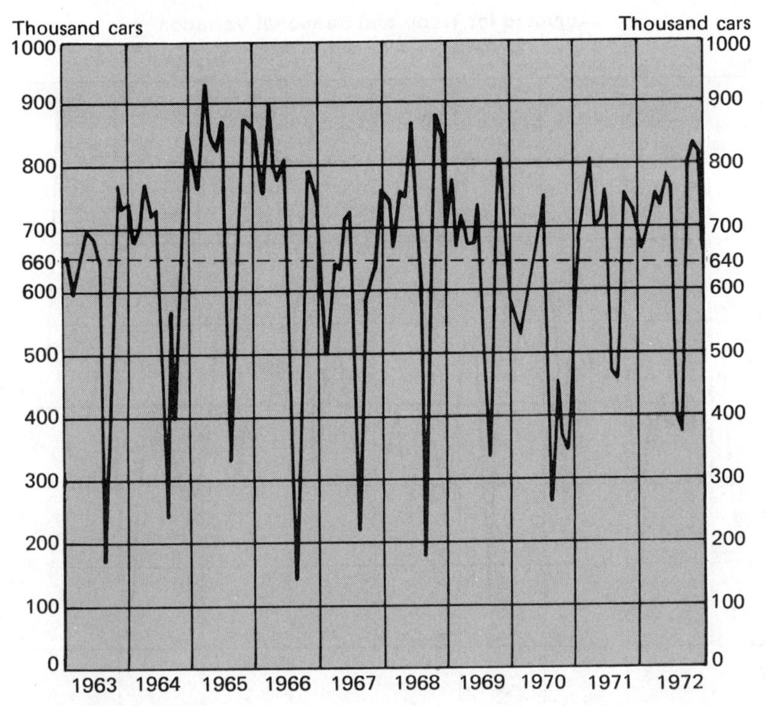

Source: Table 16-3.

FIGURE 16–3

FACTORY SALES OF DOMESTIC PASSENGER CARS IN THE UNITED STATES, 1957 TO 1968

3. The average monthly trend for a year is the level of the trend at July 1 of that year. The trend for the month of July, however, will be the level of the trend at July 15.[1] Since the rise or fall of the trend in a month was found in step 2 above, the change occurring in one half a month can be computed by merely dividing by two. This amount added to the average monthly trend for the year will give the trend value at July 15.
4. After the trend for one month has been found, the trend for all other months can be computed by adding or subtracting the monthly change, determined in step 2. This statement applies only to the straight line, fitted either to the data or to the logarithms. For any trend other than a straight line, the monthly increase or decrease will *not* be the same for all months. This is discussed further below.
5. Check the accuracy of the calculations by computing the average of the twelve monthly trend values for each year. These averages should agree exactly with the average monthly trend values previously computed.

In computing the monthly trend values for factory sales of passenger cars, for instance, step 1 consisted merely of verifying that the trend values in Table

[1] This is an application of the principle stated on page 34 that period data are always considered as falling on the center of the period. If the trend of point data is being used, the same general principles of computing the monthly trend would be used; but each case would be a special problem depending on the date within the month for which trend values were wanted.

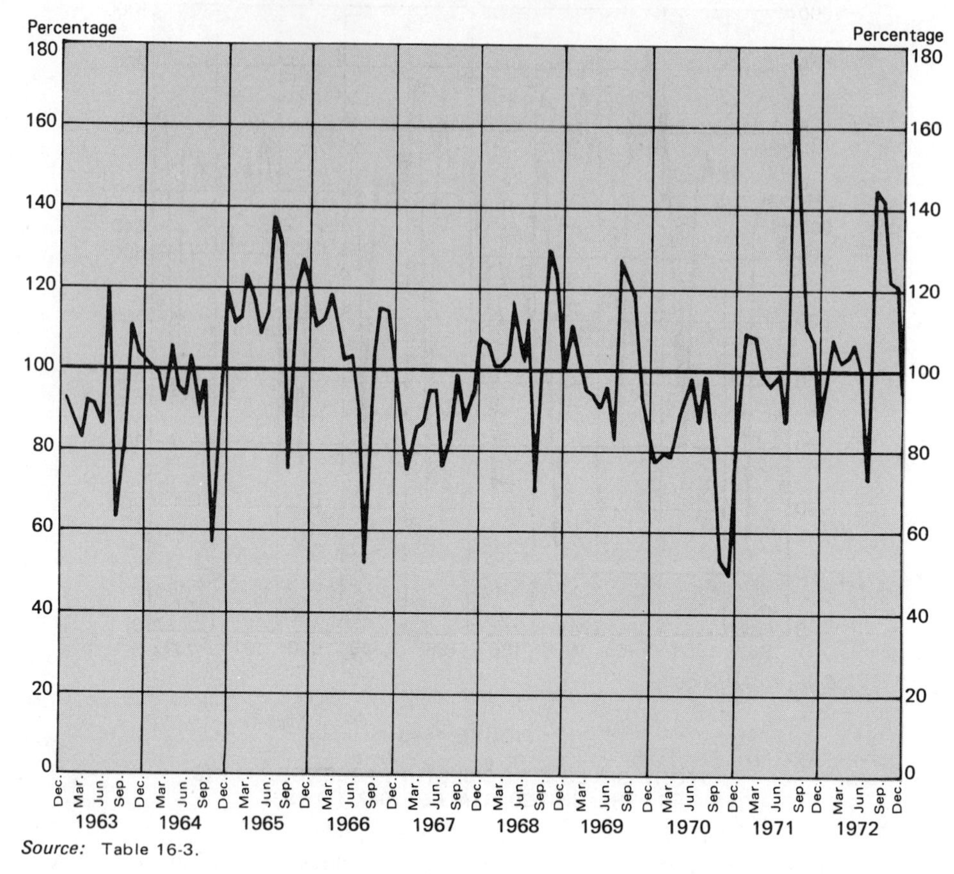

Source: Table 16-3.

FIGURE 16–4

FACTORY SALES OF DOMESTIC PASSENGER CARS IN THE UNITED STATES 1963 TO 1972

14–4 (page 383) were computed from a series of average monthly data and therefore represented average monthly trend values.

Step 2 was to find the amount by which the trend increased in one year. Since the straight line was fitted to an even number of years by the method of least squares, the b constant of the trend equation represents the decrease in one-half year or 6 months. On page 382 b was found to equal −.6761, and the monthly decrease in the trend was therefore one sixth of this amount, or −.11268.

In step 3 the trend for the year 1963 was found to be 659.3, and since the value for a year is the value at midyear, this amount is the trend on July 1, 1963. In step 2 it was determined that the trend declined .1127 per month, which means that between July 1 and July 15 it would decline one half this amount, or .05634. The trend for July 15, 1963, is therefore 659.2, a figure reached by subtracting .05634 from the trend for July 1.

Step 4 consists of deducting the monthly decrease of .11268 from each monthly trend value to secure the trend value for the next month, which was then rounded to one decimal place.

In adjusting current data for secular trend, it is necessary to project the trend into the current year without using that year for its computation. After complete data are available for a year, it is wise to check to see if these actual data make it necessary to revise the trend. In many instances it is safe to project the monthly trend values for several years before recomputing the trend.

When computing the monthly trend values for the straight line fitted to the logarithms of the actual data, the steps performed above are carried out with the a and b values to compute the logarithms of the monthly trend values. Then, the additional step of finding the antilogs must be carried out.

When computing the monthly trend values for the second-degree parabola, it is satisfactory to interpolate on a straight line between the annual trend values of the parabola.

FORECASTING CYCLICAL FLUCTUATIONS IN TIME SERIES

Forecasting the cyclical fluctuations in industry and trade is a major problem of management, for many decisions made in individual business concerns depend upon the course of activity in a particular industry or in business as a whole. This chapter has described in some detail the methods of measuring these cyclical fluctuations. Probably the chief reason for wanting to measure them is for the purpose of forecasting them. However, the contributions of statistical methods to the subject of forecasting have been confined chiefly to furnishing a more precise measurement of these changes. Purely statistical methods of forecasting cyclical fluctuations have so far not been discovered.

Nevertheless, business plans must be made for the future, and forecasting is inevitable even though it is inaccurate. The only basis for the future is the history of what has happened in the past. Thus, the more accurately we can measure the past, the better equipped we are to forecast. The causes of fluctuations are so complex that probably no simple method of forecasting will ever be found. Actually only the beginnings have been made at the present time. But the information made available on the fluctuations in individual time series represents part of the factual basis for a forecast of changes in general business activity.

STUDY QUESTIONS

16-1. The fact that seasonal fluctuations occur over a fixed period of a year makes their measurement somewhat less difficult than the measurement of cyclical fluctuations in business. Do you agree with this statement? Give reasons for your answer.

16–2. A very large percentage of the published measures of important elements of economic activity that are classified on the basis of periods less than a year are adjusted for seasonal variation. What types of fluctuations may still be present in these adjusted series?

16–3. A very small proportion of published time series are adjusted for secular trend. Why is the adjustment for seasonal variation made so much more frequently than for secular trend?

16–4. Determining the cyclical fluctuations of a time series by removing the effects of the types of fluctuations that show a regular pattern is not considered ideal by those who use it. Since the method is not held in high regard, how do you explain its extensive use? What are the chief objections to this method of measuring the cyclical fluctuations in business?

16–5. Are there any situations likely to arise where it would be better to adjust for seasonal variation by subtracting the amount of the seasonal influence rather than dividing by the seasonal adjustment factor?

16–6. Why would you expect that the adjustment of a series by means of a changing index of seasonal variation would give a more precise measure of the cyclical fluctuations than by means of a fixed index of seasonal variation?

16–7. Business activity during the summer months is below the average monthly volume in enough lines of business to cause the index of seasonal variation for total business activity to decline during the summer and to rise sharply in the fall. When business is in a period of cyclical decline, it is very common for reports on economic activity during the summer to predict that business will improve "as soon as the fall upturn" is felt.

(a) Does this "fall upturn" refer to a cyclical or a seasonal movement?

(b) Is a cyclical upturn in business more likely to occur at the same time as a seasonal rise, or is the cyclical pattern independent of the seasonal pattern?

(c) Is it possible that business analysts are inclined to confuse the seasonal and the cyclical fluctuations in business? Discuss.

16–8. Assume that during a period of 12 months the statistical measures of economic activity declined 25% but that the measures were still above the peak reached during the most prosperous period 15 years earlier. Does this mean that business is not in reality depressed, even though a decline of 25% has been registered?

PROBLEMS

16–1. Assume that the values of the index of seasonal variation for sales of a certain company are 85 for August and 110 for September. Sales in August were $245,650 and in September, $316,800.

(a) How did the cyclical position of the business in September compare with that of August?

(b) If the manager of the business estimated correctly in August that the sales in September would be approximately $316,800, would he have been justified in predicting that an improvement in September business was expected? (Should he have explained further what he meant by "improvement"?)

16–2. A certain retailer makes, on the average, 12.5% of his annual sales in December and only 6% of his annual sales in January. If sales in January were $649,440 compared with $1,230,000 in the preceding December, does he have reason to worry about the underlying trend of business? Or can he consider his business in January to be reasonably good in comparison with the volume in December?

16–3. The following table gives the value of shipments of motor vehicles and parts in the United States for each month in 1972 and the index of seasonal variation for each month.

(a) Plot the shipments for 1972 on an arithmetic chart.

(b) Adjust the shipments for seasonal variation and plot the adjusted data on the chart drawn in (a).

Month	Shipments (Millions of Dollars)	Index of Seasonal Variation
January	4,733	99.2
February	5,338	109.1
March	5,390	109.8
April	5,426	104.5
May	5,489	108.2
June	5,530	116.3
July	3,410	64.0
August	3,596	67.4
September	5,855	110.9
October	6,203	109.6
November	6,248	107.2
December	5,167	93.8

Source: U.S. Department of Commerce, *Survey of Current Business* (March, 1973).

16–4. The production of a manufacturing plant was 390,000 units in March, 1973. The index of seasonal variation for March was 130, and the trend value for March, 1973, was 250,000. Adjust the March, 1973, production for seasonal variation and for secular trend. Explain what each of these adjusted figures means.

16–5. The production of the manufacturing plant described in Problem 16–4 was 341,000 units in April, 1973. The index of seasonal variation for April was 110, and the trend value for April, 1973, was 251,000. Adjust the April, 1973, production for seasonal variation and for secular trend. Compare the results with the comparable figure secured in Problem 16–4.

16–6. This problem will use the trend values computed in Problem 14–1. If the solution to Problem 14–1 has not been retained, it will be necessary to compute the trend values for the three time series before working this problem.

(a) Adjust each of the three series in Problem 14–1 for secular trend and plot the adjusted series on one arithmetic chart.

(b) What does this chart show about nontrend fluctuations in the three series?

16–7. This problem will use the trend values computed in Problem 14–2. If the solution to Problem 14–2 has not been retained, it will be necessary to compute the trend values for the three series before working this problem.

(a) Adjust each of the three series in Problem 14–2 for secular trend and plot the adjusted data for the three series on one arithmetic chart.

(b) What does this chart show about nontrend fluctuations in the three series?

16–8. Plot the following sales figures on an arithmetic chart and comment on the kinds of fluctuations you can see in the data.

Month	Sales (Dollars)	Month	Sales (Dollars)
January	279,900	July	304,515
February	257,920	August	336,600
March	264,600	September	368,115
April	237,440	October	394,800
May	273,000	November	424,580
June	305,910	December	448,560

16–9. Using the index of seasonal variation shown below, adjust the data in Problem 16–8 to remove the seasonal influences. Plot the adjusted series on the chart drawn in the previous exercise.

Month	Index	Month	Index
January	90	July	90
February	80	August	100
March	80	September	110
April	70	October	120
May	80	November	130
June	90	December	140

16–10. Use the data in Problems 16–8 and 16–9 also in this problem. Assume that a least-squares trend equation has been computed for these data and other years not shown here. The trend value for January is $305,000 and the average monthly increase is $5,000. Compute the other 11 trend values and adjust the 12-month series computed in Problem 16–9 for trend. Plot the resulting series of index numbers on an arithmetic chart. What does this chart tell you about the effects of the business cycle on sales for the year under study?

16–11. Compare sales for the following two time periods in terms of the business cycle after making adjustments for seasonal variation and trend.

	May	June
Sales	34,177	35,677
Trend	31,055	31,065
Index of seasonal variation	89	105

16–12. This problem will use the trend values computed in Problem 14–4. If the solution to Problem 14–4 has not been retained, it will be necessary to compute the trend values for refinery runs of crude oil in the United States before working this problem.

Adjust the refinery runs of crude oil in the United States for secular trend and plot the adjusted data on an arithmetic chart.

16–13. This problem will use the trend values for refinery runs of crude oil by Exxon Company and by Phillips Petroleum Company computed in Problem 14–4. If the solution to Problem 14–4 has not been retained, it will be necessary to compute the trend values for these two series before working this problem.

(a) Adjust each of these series for secular trend and plot the two adjusted series on one arithmetic chart.

(b) What does this chart show with respect to the relationship between the fluctuations of the two series?

16-14. The following table shows average monthly demand for gasoline in the United States from 1962 to 1972.

(a) Plot this series on an arithmetic chart.

(b) Compute the trend values by the method of least squares and plot on the chart with the original data.

(c) Adjust the series for secular trend and plot the adjusted series on an arithmetic chart.

Year	Millions of Barrels
1962	132
1963	136
1964	142
1965	143
1966	150
1967	154
1968	163
1969	170
1970	178
1971	185
1972	195

Source: U.S. Department of Commerce, *Business Statistics* (1973) and *idem., Survey of Current Business* (March, 1973).

16-15. (a) From the trend line fitted to the demand for gasoline in Problem 16-14, derive the monthly trend values for each month from January, 1962, to December, 1972.

(b) Use the monthly trend values with the index of seasonal variation given in Problem 15-13 to adjust the data for trend and seasonal variation from January, 1962, to December, 1972.

(c) Plot the adjusted series on an arithmetic chart.

(d) How well does the chart measure the cyclical fluctuations in demand for gasoline?

16-16. Problem 15-10 gives the production of nonfat dry milk by months from January, 1962, to December, 1972.

(a) Using the index of seasonal variation for production of nonfat dry milk computed in Problem 15-10, adjust the production for seasonal variation and plot the unadjusted and the adjusted series on the same arithmetic chart.

(b) After making allowance for seasonal variation, summarize briefly the changes that have taken place in the production of nonfat dry milk.

Index Numbers

The methods of time series analysis described in preceding chapters have been applied to individual series, with the objective of isolating specific types of fluctuations, either secular trend, seasonal variation, or cyclical movements. The methods of index-number construction discussed in this chapter deal with the techniques of combining time series that otherwise cannot be added because they are not in comparable units.

NATURE OF AN INDEX NUMBER

Items of the same kind may usually be added if all the measurements are expressed in the same units. The annual production of wheat can be measured simply by totaling the output of the individual producers or the total amount moving through the markets. Even though there may be different grades of the product, as is generally true for commodities, all bushels of wheat are nearly enough alike to make the total production a significant amount.

When a measurement is to be taken of the composite changes in the production of a number of commodities that are not expressed in the same units and cannot therefore be added or averaged, the methods of index-number construction must be used. It is preferable to reduce different items to comparable units and then add them. For example, in measuring the total production of fuel, it would not be satisfactory to add coal, petroleum, and natural gas output; but it would be possible to convert the production of each fuel into units of heat, which could be added. In this case the average price for each Btu (British thermal unit) could also be computed and over a period of time this would measure the change in fuel prices. Even this computation, however, can be criticized on the ground that the efficiency with which the different fuels are used varies. As a result, it would be necessary to reduce the various fuels to Btu content after taking into consideration the effective energy that could be obtained from each fuel, using the equipment that is currently available. Although on a few occasions it might be possible to reduce the different items to comparable units and then add them, such an approach has only limited applicability. When this cannot be done, the methods of index-number construction should be utilized.

The basic device used in all methods of index-number construction is to average the *relative change* in either quantities or prices, since relatives are comparable and can be added even though the data from which they were derived cannot themselves be added. Pounds of cotton and bushels of wheat cannot properly be added; but if wheat production was 110% of the previous year's production and cotton production was 106%, it is valid to average these two percentages and to say that the volume of these commodities produced was 108% of the previous year. This assumes they are of equal importance, since each is given the same weight; but if cotton production is six times as important as wheat production, the percentages should be weighted 6 to 1. The average relative secured by this process is an *index number.*

When data can be added and the single series reduced to a fixed-base series of relatives, the relatives are called index numbers by some statisticians; others reserve the term "index number" exclusively for an average of relatives derived from series that cannot properly be added. Since many relatives based on a single series are widely used as measures of business conditions, it seems simpler to use "index number" to describe both simple relatives and averages of relatives. The term will be used in this chapter to refer to relatives derived in both ways.

If it is desired that the index number measure the cyclical fluctuations, it is necessary to adjust for trend and seasonal variation by the methods described in Chapter 16 when either of these fluctuations is present in the data. When a number of series are averaged to secure a composite index, the individual series may be adjusted before they are averaged or the composite index may be adjusted. The latter method requires less work and is often satisfactory.

SIMPLE RELATIVES

Index numbers computed as simple relatives may be fixed-base relatives (pages 348 to 349) or chain relatives (pages 349 to 351). If these relatives are on a monthly basis, they may be adjusted for seasonal variation when necessary. It makes no difference whether the basic data were adjusted for seasonal variation before being reduced to relatives or whether the relatives were used in the computation of the index of seasonal variation.

When a series has been adjusted for both seasonal variation and secular trend, such as factory sales of passenger cars in Table 16–3, the series becomes an index. The base in this case is not one year or a period of years, but is the computed normal or trend value for each period. The base in such a series of relatives may be described as normal = 100%.

AVERAGES OF RELATIVES

Some of the well-known methods of computing composite index numbers will be illustrated by using the data on prices and production of four building materials given in Tables 17–1 and 17–2. In each case the index number will be an average of relatives rather than an average of the original data. For this reason the series representing the quantity of building materials produced and the prices for which they were sold have been reduced to relatives, with 1967 equal to 100 in each series. These relatives are given in Tables 17–3 and 17–4.

The computation of indexes performed by averaging these relatives in different ways will be illustrated in the remainder of this chapter. The following symbols are used:

Q_i = a quantity of index for period i, which may be for any time unit—day, week, month, or year.

P_i = a price index for period i, which may be any time unit—day, week, month, or year.

p_o = the prices in the base period.

p_i = the prices in period i, which may be any time unit—day, week, month, or year.

q_o = the quantities in the base period. The quantities may be production, consumption, marketings, or any other information in physical units.

q_i = the quantities in period i, which may be any time unit—day, week, month, or year.

n = the number of series in the index.

The period for which an index is computed may be identified by the subscript. For example, the price and the quantity indexes for 1972 are written P_{72} and Q_{72}.

TABLE 17–1

PRODUCTION OF FOUR BUILDING MATERIALS IN THE UNITED STATES, 1967 TO 1972

Year	Western Pine (Million Board Ft.)	Portland Cement (Million Barrels)	Brick, Unglazed (Millions)	Linseed Oil (Million Pounds)
1967	10,180	374.0	7,117	370.6
1968	10,851	397.4	7,534	306.6
1969	9,999	409.8	7,290	291.8
1970	9,378	389.8	6,494	314.5
1971	10,019	420.2	7,570	412.2
1972	10,436	440.1	8,397	439.7

Source: U.S. Department of Commerce, *Business Statistics* (1973).

TABLE 17-2

PRICES OF FOUR BUILDING MATERIALS IN THE UNITED STATES, 1967 TO 1972

Year	Western Pine (Dollars per Thou. Board Ft.)	Portland Cement (Dollars per Barrel)	Brick, Unglazed (Dollars per Thousand)	Linseed Oil (Dollars per Lb.)
1967	71.95	3.17	38.30	.129
1968	87.72	3.25	37.29	.127
1969	107.18	3.35	41.29	.120
1970	83.79	3.67	42.97	.109
1971	96.44	3.95	44.96	.089
1972	130.91	4.18	46.76	.092

Source: U.S. Department of Commerce, *Business Statistics* (1973).

TABLE 17-3

PRODUCTION OF FOUR BUILDING MATERIALS IN THE UNITED STATES, 1967 TO 1972

Relatives, 1967 = 100

Year	Western Pine	Portland Cement	Brick, Unglazed	Linseed Oil
1967	100.0	100.0	100.0	100.0
1968	106.6	106.3	105.9	82.7
1969	98.2	109.6	102.4	78.7
1970	92.1	104.2	91.3	84.9
1971	98.4	112.4	106.4	111.2
1972	102.5	117.7	118.0	118.7

Source: Table 17-1.

TABLE 17-4

PRICES OF FOUR BUILDING MATERIALS IN THE UNITED STATES, 1967 TO 1972

Relatives, 1967 = 100

Year	Western Pine	Portland Cement	Brick, Unglazed	Linseed Oil
1967	100.0	100.0	100.0	100.0
1968	121.9	102.5	97.4	98.5
1969	149.0	105.7	107.8	93.0
1970	116.5	115.8	112.2	84.5
1971	134.0	124.6	117.4	69.0
1972	181.9	131.9	122.1	71.3

Source: Table 17-2.

Unweighted Arithmetic Mean of Relatives

Table 17–5 gives the ratio of production in 1972 to production in 1967 for each building material. There is considerable variation in these ratios for the different materials. The problem is to average the relatives to secure a typical ratio of 1972 to 1967, which will be the index number for 1972. The index number for any year may be computed in the same manner.

TABLE 17–5

COMPUTATION OF INDEX OF BUILDING MATERIAL PRODUCTION, 1972

1967 = 100

Arithmetic Mean of Relatives

Building Materials	Relatives 1967 = 100
Western pine	102.5
Portland cement	117.7
Brick, unglazed	118.0
Linseed oil	118.7
Total	456.9
Arithmetic mean	114.2

Source: Table 17–3.

If the products were assumed to be equally important, an unweighted arithmetic mean of the relatives would be acceptable. This mean is computed by adding the relatives and dividing the sum by the number of relatives, which gives an average relative of 114.2%. This means that the volume of building materials produced in 1972 was 114.2% of 1967, the base year.

To find the index of prices of these building materials, an arithmetic mean of the price relatives is computed in the same manner in Table 17–6. The sum of the relatives is 507.2, which is divided by 4 to give the average relative of 126.8%. This index of prices means that the level of building material prices in 1972 was 126.8% of the level of prices in 1967.

The computations just performed can be summarized by formulas based on the symbols described on page 464. The quantity relatives for the year 1972 are represented by $\frac{q_i}{q_o}100$, since q_i represents the quantities produced in 1972 and q_o, the quantities produced in 1967, the base year. The index of quantities produced was computed using Formula 17–1.

$$Q_i = \frac{\sum\left(\frac{q_i}{q_o}100\right)}{n} \qquad \textbf{(17–1)}$$

$$Q_{72} = \frac{456.9}{4} = 114.2.$$

TABLE 17-6

COMPUTATION OF INDEX OF BUILDING MATERIAL PRICES, 1972

1967 = 100

Arithmetic Mean of Relatives

Building Materials	Relatives 1967 = 100
Western pine	181.9
Portland cement	131.9
Brick, unglazed	122.1
Linseed oil	71.3
Total	507.2
Arithmetic mean	126.8

Source: Table 17-4.

The price relatives for 1972 are represented by $\frac{p_i}{p_o}100$, and the index of building material prices is computed using Formula 17-2.

$$P_i = \frac{\sum\left(\frac{p_i}{p_o}100\right)}{n} \tag{17-2}$$

$$P_{72} = \frac{507.2}{4} = 126.8.$$

Unweighted Geometric Mean of Relatives

The index of volume of building material production may be computed by using a geometric mean of the relatives instead of the arithmetic mean. Using the data in Table 17-7 it is computed

$$\text{Log } Q_i = \frac{\sum \log\left(\frac{q_i}{q_o}100\right)}{n} \tag{17-3}$$

$$\text{Log } Q_{72} = \frac{8.227834}{4} = 2.056958$$

$$Q_{72} = 114.0.$$

TABLE 17–7

COMPUTATION OF INDEX OF
BUILDING MATERIAL PRODUCTION, 1972

1967 = 100

Geometric Mean of Relatives

Building Materials	Relatives 1967 = 100	Logarithms of Relatives
Western pine	102.5	2.010724
Portland cement	117.7	2.070777
Brick, unglazed	118.0	2.071882
Linseed oil	118.7	2.074451
Total		8.227834
Arithmetic mean of logarithms		2.056958
Geometric mean		114.0

Source: Table 17–3.

The geometric mean of the price relatives is computed by Formula 17–4 and is found to be 120.2 for 1972.

$$\text{Log } P_i = \frac{\sum \log\left(\frac{p_i}{p_o} 100\right)}{n} \tag{17–4}$$

$$\text{Log } P_{72} = \frac{8.319884}{4} = 2.079971$$

$$P_{72} = 120.2.$$

This formula is the same as Formula 17–3, except the price relatives are substituted for the quantity relatives. The computations are shown in Table 17–8.

The Problem of Averages

A great deal of attention has been given to the selection of the proper average to use in the computation of an index number. The arithmetic mean, the geometric mean, and the harmonic mean have all been suggested by statisticians, but the geometric mean has certain points of decided superiority over the other two. The geometric mean is always less than the arithmetic mean of the same data (page 56); and in the construction of index numbers, it is generally agreed that the arithmetic mean has an upward bias to the extent that it is larger than the geometric mean. At one time this led to the conclusion that only the geometric mean should be used in index-number construction, since it alone is free from bias. However, the average to be used is related to the weighting system, which will be discussed before any conclusion is stated as to which average should be used.

TABLE 17–8

COMPUTATION OF INDEX OF BUILDING MATERIAL PRICES, 1972
1967 = 100

Geometric Mean of Relatives

Building Materials	Relatives 1967 = 100	Logarithms of Relatives
Western pine	181.9	2.259833
Portland cement	131.9	2.120245
Brick, unglazed	122.1	2.086716
Linseed oil	71.3	1,853090
Total		8.319884
Arithmetic mean of logarithms		2.079971
Geometric mean		120.2

Source: Table 17–4.

The Problem of Weights

The unweighted arithmetic mean of the relatives was computed on page 466, assuming that each of the building materials was of the same importance. However, the building materials are not of equal importance, and it is reasonable to assume that they will seldom be equally important. For this reason a weighted average of the relatives will normally be computed, and a number of weighting schemes have been developed. Any system of weights must meet two requirements:

1. The amounts used as weights must measure the comparative importance of the different relatives to be averaged.
2. They must be amounts that can properly be added, since the formula for a weighted average involves dividing by the sum of the weights.

The weights should be computed from actual data that measure the relative importance of the different series, although if no data were available, weights might be estimated. We might assign western pine a weight of 30; Portland cement a weight of 50; and so on. But since the basic reason for using statistical methods is to replace opinions with objectively derived facts, computed weights should always be used if any kind of information can be obtained to measure the relative importance of the items included in the index.

Many different kinds of data may represent the basis for the computation of weights, but the most generally used is some measure of the *value* of the different items. For the index of building material prices, the relative importance of each material might be determined from the value of each material produced in a typical year or in several years. Values of different products expressed in dollars are comparable and so they can be added—a condition that must be met by any system of weights. Table 17–9 gives the value of building materials produced for each year from 1967 to 1972.

TABLE 17–9

VALUE OF BUILDING MATERIALS PRODUCED, 1967 TO 1972

Millions of Dollars

Year	Western Pine	Portland Cement	Brick, Unglazed	Linseed Oil	Total
1967	732.5	1,185.6	272.6	47.8	2,238.5
1968	951.8	1,291.6	280.9	38.9	2,563.2
1969	1,071.7	1,372.8	301.0	35.0	2,780.5
1970	785.8	1,430.6	279.0	34.3	2,529.7
1971	966.2	1,659.8	340.3	36.7	3,003.0
1972	1,366.2	1,839.6	392.6	40.5	3,638.9

Source: Tables 17–1 and 17–2.

The next question is that of the period for which the values should be computed, since the values of the building materials produced in any one year or combination of years might be used as the weights. It has been confirmed by a considerable amount of research that values for certain years are biased when used as weights either for price or for quantity indexes. Generally the use of the values of the base period *gives too much weight to the items with the smallest rise in price, or the largest drop if prices are falling*. This results in the index being too low; or, in other words, this scheme of weights has a *downward bias*. This is true for both a price index and a quantity index.

Total values for the whole period covered by the index will not suffer from this type of bias, but such a weighting scheme requires a great deal of computation. The scheme can be used for constructing an index over a past period of time, but the chief interest in most indexes is in current changes. This requires the use of a weighting scheme based on past data and one that does not need to be changed frequently. The ease with which they can be used is an important recommendation for the values for the base period.

The arithmetic mean of the relatives gives the index an *upward bias* (page 468). Base-period values used as weights result in an index that is too low; that is, they give the index a *downward bias*. While these two biases are generally not equal, they are in opposite directions. Sometimes the upward bias of the arithmetic mean is larger and sometimes the downward bias of the weight scheme is larger. The result of this approximate offsetting of the biases is that the arithmetic mean of relatives weighted with base-year values is not completely correct, but it is not biased since sometimes it is too large and sometimes it is too small. The amount of error is small enough that the formulas in the following discussion are satisfactory to use. This method of weighting is the simplest that can be used, and the arithmetic mean is the simplest average.

Arithmetic Mean of Relatives Weighted with Base-Year Values

In Table 17–10 the weighted arithmetic mean is computed for the production of building materials, using 1967 as the base year and the values of the base-year production as weights. All the necessary data, both the relatives and the data needed as weights, are given in Tables 17–3 and 17–9. The computations simply consist of finding a weighted average of the relatives in Column 2, using the values in Column 3 as weights. The computation of a weighted average was explained in Chapter 3 and demonstrated in Table 3–5. The only difference in Table 17–10 is that a different set of symbols is used to identify the quantities entering into the computation. The relatives to be averaged in Column 2 are multiplied by the weights in Column 3. The products are entered in Column 4 and totaled. This total is the numerator of the equation, and is divided by the sum of the weights in Column 3. The formula for a weighted arithmetic mean of quantity relatives using base-year values as weights is

$$Q_i = \frac{\Sigma\left(p_o q_o \frac{q_i}{q_o} 100\right)}{\Sigma(p_o q_o)} \tag{17–5}$$

$$Q_{72} = \frac{252{,}467.03}{2{,}238.5} = 112.8.$$

TABLE 17–10

COMPUTATION OF INDEX OF BUILDING MATERIAL PRODUCTION, 1972

1967 = 100

Weighted Arithmetic Mean of Relatives, Base-Year Values as Weights

Building Materials (1)	Relatives 1967 = 100 $\frac{q_i}{q_o}100$ (2)	Weights (Base-Year Values)* $p_o q_o$ (3)	$p_o q_o \frac{q_i}{q_o} 100$ (4)
Western pine	102.5	732.5	75,081.25
Portland cement	117.7	1,185.6	139,545.12
Brick, unglazed	118.0	272.6	32,166.80
Linseed oil	118.7	47.8	5,673.86
Total		2,238.5	252,467.03
Index			112.8

* Millions of dollars.

Source: Tables 17–3 and 17–9.

The computation of an index of building material prices would be made in the same manner as the computation of the index of building material production, except that the relatives averaged in this case would be computed from the *prices* of the individual materials. The weights used should be the same as for the index of building material production. The details of computation of the price index are given in Table 17–11, which differs from Table 17–10 only in that price relatives instead of quantity relatives are entered in Column 2.

In the equation used for computing the index, the price relative $\frac{p_i}{p_o}$ is substituted for the quantity relative $\frac{q_i}{q_o}$.

$$P_i = \frac{\Sigma\left(p_o q_o \frac{p_i}{p_o} 100\right)}{\Sigma(p_o q_o)} \tag{17-6}$$

$$P_{72} = \frac{326{,}314.99}{2{,}238.5} = 145.8.$$

TABLE 17-11

COMPUTATION OF INDEX OF BUILDING MATERIAL PRICES, 1972

1967 = 100

Weighted Arithmetic Mean of Relatives, Base-Year Values as Weights

Building Materials (1)	Relatives 1967 = 100 $\frac{p_i}{p_o}100$ (2)	Weights (Base-Year Values)* $p_o q_o$ (3)	$p_o q_o \frac{p_i}{p_o} 100$ (4)
Western pine	181.9	732.5	133,241.75
Portland cement	131.9	1,185.6	156,380.64
Brick, unglazed	122.1	272.6	33,284.46
Linseed oil	71.3	47.8	3,408.14
Total		2,238.5	326,314.99
Index			145.8

* Millions of dollars.

Source: Tables 17-4 and 17-9.

Weighted Geometric Mean

The computation of a weighted geometric mean raises the question of the bias in the weights computed from base-year values. Since the geometric mean is not biased, there is no upward bias in the average to offset the downward bias in the weights, as in the case of the arithmetic mean weighted with base-year values. When the geometric mean is used to compute an index number, it is necessary that an unbiased system of weights be used.

One possible system of unbiased weights that can be used with the index for any given year is derived from the average values for the given year and the base year. The values of the current year will be biased upward to the same degree that the values for the base year are biased downward, with the result that the average of the two values will be unbiased. Table 17-12 gives the percentage weights derived from the 1967 and 1972 values, and the average values for the two years, 1967 and 1972. There was a substantial change in the relative importance of the different building materials between 1967 and 1972 and neither distribution is completely satisfactory as a measure of the relative importance of the individual building materials in the two years. Since the index number compares one year with the base year, 1967, the weight scheme should be a measure of the importance of each building material in both years It appears that an average of the two distributions is the most logical solution to the problem of finding an unbiased weight scheme for the 1972 index.

TABLE 17–12

VALUE OF BUILDING MATERIAL PRODUCED 1967, 1972, AND AVERAGE OF 1967 AND 1972

Used as Weight Schemes for Index Numbers

	1967		1972		Average of 1967 and 1972	
Building Materials	Millions of Dollars	Percentage of Total	Millions of Dollars	Percentage of Total	Millions of Dollars	Percentage of Total
Western pine	732.5	32.7	1,366.2	37.5	1,049.35	35.7
Portland cement	1,185.6	53.0	1,839.6	50.6	1,512.60	51.5
Brick, unglazed	272.6	12.2	392.6	10.8	332.60	11.3
Linseed oil	47.8	2.1	40.5	1.1	44.15	1.5
Total	2,238.5	100.0	3,638.9	100.0	2,938.70	100.0

Source: Table 17–9.

Indexes of quantities and prices are computed in Tables 17–13 and 17–14, using the geometric mean with the average values of 1967 and 1972 as weights. The amounts have been calculated by Formulas 17–7 and 17–8.

$$\text{Log } Q_i = \frac{\sum w\left(\log \frac{q_i}{q_o} 100\right)}{\sum w} \quad \textbf{(17–7)}$$

$$\text{Log } Q_{72} = \frac{204.951807}{100} = 2.049518$$

$$Q_{72} = 112.1.$$

$$\text{Log } P_i = \frac{\sum w\left(\log \frac{p_i}{p_o} 100\right)}{\sum w} \quad \textbf{(17–8)}$$

$$\text{Log } P_{72} = \frac{216.227181}{100} = 2.162282$$

$$P_{72} = 145.3.$$

TABLE 17-13

COMPUTATION OF INDEX OF BUILDING MATERIAL PRODUCTION, 1972

1967 = 100

Weighted Geometric Mean of Relatives

Building Materials (1)	Relatives 1967 = 100 (2)	Logarithms of Relatives (3)	Weights (Percentage of Average Value 1967 and 1972) (4)	$w\left(\log \frac{q_i}{q_o} 100\right)$ (5)
Western pine	102.5	2.010724	35.7	71.782847
Portland cement	117.7	2.070777	51.5	106.645016
Brick, unglazed	118.0	2.071882	11.3	23.412267
Linseed oil	118.7	2.074451	1.5	3.111677
Total			100.0	204.951807
Weighted geometric mean of logarithms				2.049518
Index				112.1

Source: Tables 17-7 and 17-12.

TABLE 17-14

COMPUTATION OF INDEX OF BUILDING MATERIAL PRICES, 1972

1967 = 100

Weighted Geometric Mean of Relatives

Building Materials (1)	Relatives 1967 = 100 (2)	Logarithms of Relatives (3)	Weights (4)	$w\left(\log \frac{p_i}{p_o} 100\right)$ (5)
Western pine	181.9	2.259833	35.7	80.676038
Portland cement	131.9	2.120245	51.5	109.192618
Brick, unglazed	122.1	2.086716	11.3	23.579891
Linseed oil	71.3	1.853089	1.5	2.779634
Total			100.0	216.228181
Weighted geometric mean of logarithms				2.162282
Index				145.3

Source: Tables 17-8 and 17-12.

FACTORS TEST

This section will present the *factors test,* or *factor reversal test,* that can be applied to a given situation to determine how well a formula measures the changes in prices and in quantities. The basis of the factors test is the fact that the value is the product of price and quantity. The value of $1,366.2 million for western pine production in 1972 may be factored into the quantity produced, 10,436 million board feet, and the price, $130.91 per thousand board feet. It is also true that since the price of western pine in 1972 was 181.946% of the 1967 price, and the production in 1972 was 102.5% of the 1967 production, the value of western pine production in 1972 would be 186.5% of 1967 ($1.81946 \times 1.025 = 1.865$ or 186.5%). After calculating this percentage from the price and the quantity relatives, it can be checked by computing it directly from the value of western pine production in 1967 and 1972. The 1967 value was $732.5 million and the 1972 value of $1,366.2 is 186.5% of the 1967 value. This calculation checks the accuracy of the computation made from the price and the quantity relatives. Such a relationship holds for any commodity used in the construction of an index number.

Since the product of the price relative and the quantity relative for each individual commodity entering into an index number equals the value relative, it follows that the product of a composite price index and a composite quantity index for the same commodities should also equal the value relative computed for the same commodities. If the product of a price index and a quantity index, each computed by a given formula (such as the arithmetic mean of relatives with base-year values as weights), equals the value relative computed directly from the original price and quantity data, the index number formula is said to have met the factors test. If the formula does not meet the factors test, it is subject to criticism on the grounds that if the relatives for individual commodities used in the computation give the correct value relative, the composite price index and the composite quantity index should give the same value relative as computed from the original data.

The factors test can be applied to the price and the quantity indexes for 1972 computed on pages 471 and 472. The price index for 1972 computed by the arithmetic mean of relatives weighted with base-year values is 145.8, and the quantity index computed by the same method is 112.8. The product of the two indexes is 164.5.

$$1.438 \times 1.128 = 1.622 \text{ or } 162.2\%.$$

The value relative computed from the total value of building materials produced in 1972 and 1967 is 162.6,

$$\$3{,}638.9 \div \$2{,}238.5 = 1.626 \text{ or } 162.6\%.$$

In this case the arithmetic mean of relatives weighted with base-year values comes close to meeting the factors test. Indexes computed from this formula usually come close to meeting the test, although it is possible for a rather wide discrepancy to occur.

When the factors test is applied to the geometric mean of relatives weighted by the average values of building materials in 1967 and 1972, computed on page 473, the product of the quantity and the price indexes gives a value relative of 162.9, and comes closer to meeting the factors test.

$$1.453 \times 1.121 = 1.629 \text{ or } 162.9\%.$$

The geometric mean weighted in this manner will come closer to meeting the factors test exactly than the arithmetic mean with base-year values as weights. However, the fact that the latter formula is simple to compute and easy to understand, as well as usually coming reasonably close to meeting the factors test, makes it a very popular one.

Table 17–15 gives the index of prices and the index of quantities produced for each year from 1967 to 1972. It also shows the product of the price and quantity indexes for each year and the true value relative computed from the price and production data.

TABLE 17–15

FACTORS TEST FOR INDEXES OF PRODUCTION AND PRICES OF BUILDING MATERIALS, 1967 TO 1972

Indexes Computed by Arithmetic Mean Weighted with Base-Year Values

Year	Value Relative 1967 = 100	$Q_i \times P_i$
1967	100.0	100.0
1968	114.5	114.4
1969	124.2	125.0
1970	113.0	112.9
1971	134.2	134.5
1972	162.6	162.2

METHOD OF AGGREGATES

The *aggregative method* of computing index numbers is sometimes considered a different method from the method of averaging relatives, but it is in reality an alternative method of computing some of the indexes described previously. Methods using the arithmetic mean of relatives may be expressed as an aggregative formula, and this form of computation is usually used in preference to the average of relatives because the computations are easier. Methods using the geometric mean cannot be computed by the aggregative method.

On pages 50 to 54, in the discussion of averaging ratios, it was pointed out that whenever possible it is better to compute a weighted average of ratios by going back to the basic data from which the ratios were computed rather than to try to weight the ratios. Since index numbers are averages of ratios, this general principle applies to their computation, and it is better to use an aggregative formula if it can be applied. In cases where relatives and values are available but the original data are not, it is necessary to use an average of relatives.

The relationship between the aggregative method and the average of relatives can be illustrated by the computation of the index of building material production in Table 17–10. The amounts in Column 4 were derived for each commodity by multiplying the 1967 value of building materials produced $(p_o q_o)$ by the quantity relative $\frac{q_i}{q_o}$ 100, on $p_o q_o \frac{q_i}{q_o}$ 100. For western pine the computation for 1972 from original data on page 465 was

$$71.95 \times 10{,}180{,}000 \times \frac{10{,}436{,}000}{10{,}180{,}000} \times 100 = 750{,}870{,}200 \times 100.$$

But instead of dividing 10,436,100 by 10,180,000 and then multiplying, it is easier to cancel the 10,180,000 from the numerator and the denominator.

$$71.95 \times \cancel{10{,}180{,}000} \times \frac{10{,}436{,}000}{\cancel{10{,}180{,}000}} \times 100 = 750{,}870{,}200 \times 100.$$

It may, therefore, be written that

$$p_o q_o \frac{q_i}{q_o} 100 = p_o q_i 100$$

and the formula for the quantity index becomes $\frac{\Sigma(p_o q_i 100)}{\Sigma(p_o q_o)}$. Since it is simpler to multiply the final ratio by 100 than to multiply each of the individual products by 100, the formula is usually written

$$Q_i = \frac{\Sigma\,(p_o q_i)}{\Sigma(p_o q_o)}\,100. \qquad \textbf{(17–9)}$$

The simplification of the numerator of the equation gives the result with one multiplication for each commodity instead of a division to secure the relative $\frac{q_i}{q_o}$ and then the multiplication of the relative by the weight $p_o q_o$.

The work sheet for the computation of the index by the method of aggregates must show both the 1967 and 1972 quantities and the 1967 prices. In Table 17–16, $p_o q_i$ is computed for each commodity from the original prices and production data given in Tables 17–1 and 17–2. The resulting index is the same as computed in Table 17–10.

The formula for the price index by the method of aggregates is derived from Formula 17–6 by the same process of simplifying the numerator used

TABLE 17–16

COMPUTATION OF INDEX OF BUILDING MATERIAL PRODUCTION, 1972

1967 = 100

Method of Aggregates

Building Materials (1)	Units (2)	Price per Unit, 1967 p_o (3)	Quantity Produced 1967 q_o (4)	Quantity Produced 1972 q_i (5)	p_oq_o (6)	p_oq_i (7)
Western pine	Thousand board ft.	71.95	10,180,000	10,436,000	732,451,000	750,870,200
Portland cement	Barrels	3.17	374,000,000	440,100,000	1,185,580,000	1,395,117,000
Brick, unglazed	Thousand	38.30	7,117,000	8,397,000	272,581,100	321,605,100
Linseed oil	Pounds	.129	370,600,000	439,700,000	47,807,400	56,721,300
Total					2,238,419,500	2,524,313,600
Index					100.0	112.8

Source: Tables 17–1 and 17–2.

$$Q_i = \frac{\Sigma(p_oq_i)}{\Sigma(p_oq_o)}100$$

$$Q_{72} = \frac{2{,}524{,}313{,}000}{2{,}238{,}419{,}500}100 = 112.8.$$

in deriving Formula 17–9. The quantities in Column 4 of Table 17–11 can be secured directly from the price and quantity data, since $p_oq_o \times \frac{p_i}{p_o}100 = p_iq_o100$. In Table 17–17 the quantities p_iq_o are computed from the original data, and differ from $p_oq_o\frac{p_i}{p_o}100$ computed in Column 4 of Table 17–11 only because of the variations due to rounding. The formula for a price index by the method of aggregates is

$$P_i = \frac{\Sigma(p_iq_o)}{\Sigma(p_oq_o)}100. \qquad \textbf{(17–10)}$$

WEIGHTS DERIVED FROM PERIODS OTHER THAN BASE-YEAR OR CURRENT-YEAR VALUES

The weight schemes discussed to this point have used the prices and the quantities that came from either the base period or from the period being compared with the base period. The use of base-year values has decided practical

TABLE 17-17

COMPUTATION OF INDEX OF BUILDING MATERIAL PRICES, 1968

1963 = 100

Method of Aggregates

Building Materials (1)	Units (2)	Quantity Produced, 1967 q_o (3)	Price per Unit 1967 p_o (4)	Price per Unit 1972 p_i (5)	p_oq_o (6)	p_iq_o (7)
Western pine	Thousand board ft.	10,180,000	71.95	130.91	732,451,000	1,332,663,800
Portland cement	Barrels	374,000,000	3.17	4.18	1,185,580,000	1,563,320,000
Brick	Thousand	7,117,000	38.30	46.76	272,581,100	332,790,920
Linseed oil	Pounds	370,600,000	.129	.092	47,807,400	34,095,200
Total					2,238,419,500	3,262,869,920
Index					100.0	145.8

Source: Tables 17-1 and 17-2.

$$P_i = \frac{\Sigma(p_iq_o)}{\Sigma(p_oq_o)}100$$

$$P_{72} = \frac{3{,}262{,}869{,}920}{2{,}238{,}419{,}500} = 145.8.$$

advantages over using current-year values, since it is to be expected that more complete data would be available for the earlier period than for the current period. If the choice were limited to base-year or current-year weights, it is almost certain that the former would be used. It is possible, however, to use a compromise that consists of computing values for a price index that are based on base-year prices but on some arbitrarily chosen period for quantities. This period might be a year for which census data were available, which would simplify the problem of securing good quantity data. Quantities for periods other than the base year or the current year may be designated q_a to distinguish them from q_o and q_i. The arithmetic mean of relatives weighted by values computed from q_a is

$$P_i = \frac{\sum\left(p_oq_a\frac{p_i}{p_o}\right)}{\Sigma(p_oq_a)}100. \qquad \textbf{(17-11)}$$

Formula 17-11 cancels to the aggregative equivalent

$$P_i = \frac{\Sigma(p_iq_a)}{\Sigma(p_oq_a)}100. \qquad \textbf{(17-12)}$$

When constructing a quantity index, it is possible to use values based on an arbitrarily chosen period for prices and the base period for quantities. In this situation the price data would be designated p_a and the value weights would be $p_a q_o$. The formulas for the weighted arithmetic mean of relatives using $p_a q_o$ as weights and the aggregative equivalent are

$$Q_i = \frac{\Sigma\left(p_a q_o \frac{q_i}{q_o}\right)}{\Sigma(p_a q_o)} 100 \tag{17-13}$$

$$Q_i = \frac{\Sigma(p_a q_i)}{\Sigma(p_a q_o)} 100. \tag{17-14}$$

The formulas above are more commonly used than those with base-year values as weights.

CHAIN INDEXES

The use of chain relatives was discussed in Chapter 13, where it was shown that a relative on a fixed base can be derived from a series of link relatives. It was pointed out that comparable data often do not exist over long periods, but if the data are comparable for two periods it is possible to compute a series of link relatives that may then be chained into a series of relatives expressed as percentages of a fixed base.

In the construction of index numbers, it rarely happens that comparable data are available for long periods; thus, the chain index is an extremely valuable tool. A very popular method of constructing a price index is to compute a series of link indexes based on the arithmetic mean of relatives, weighted with values computed from quantities from an arbitrarily chosen period. If the aggregative form of the index is more convenient, it may be used for making the computations.

In writing a formula for the link index, it is desirable to modify the symbols previously used. Since the base period will always be the preceding period, the price in the base period may be represented by p_{i-1} and the formula for the link index may be written

$$P_{i\,(link)} = \frac{\Sigma(p_i q_a)}{\Sigma(p_{i-1} q_a)} 100. \tag{17-15}$$

On page 350 it was stated that the chain relative may always be computed by multiplying the link relative by the previous chain relative. Since the same principle applies to an index number, the chain index for any period may be computed by multiplying the chain index for the previous period by the link index computed using this period as the base and the following period as the current period. The symbol for the index for the previous period would be $\boldsymbol{P}_{i-1}$. Thus, the formula for the chain index for any current period is

$$P_i = P_{i-1}\left[\frac{\Sigma\left(p_{i-1}q_a\dfrac{p_i}{p_{i-1}}\right)}{\Sigma(p_{i-1}q_a)}\right]100 = P_{i-1}\left[\frac{\Sigma p_i q_a}{\Sigma p_{i-1}q_a}\right]100. \tag{17-16}$$

The chain index may be started with any period desired as 100%, and from this first value of the chain index, each succeeding value may be computed from the current link index.

GENERAL COMMENTS ON INDEX NUMBER CONSTRUCTION

The discussion in this chapter has been devoted to the problems of combining statistical series that cannot properly be added because they are not in comparable units. While this is the central problem of index number construction, it should not be overlooked that the computation of an index number makes use of a wide variety of statistical techniques that have been discussed previously. Although these techniques are not used solely in the construction of index numbers, they will be mentioned to emphasize the fact that the business uses of statistics employ a wide variety of methods.

Sampling

Most index numbers are based on samples rather than on universe data. Thus, the problems of sampling precision enter into the computation of an index number.

Comparability of Data

In comparing prices or volume of production over a period of time, it is essential that a series measure the same factors throughout the period. This gives rise to many difficult problems in constructing index numbers, since the units in which prices and production are measured change with technological improvements. When the price of this year's automobile is compared with the price of the same make in 1926, it is obvious that the units are not the same. Problems of this type, which are present in all types of statistical analysis involving time series, have been discussed in earlier chapters. Securing comparable data for use in constructing an index number is basically no different from the problem of securing comparable data in any other type of analysis, but it must be given careful attention.

Selection of Commodities and Weights

In the construction of an index number, it is necessary to define the problem that is to be solved by the index before deciding which series to include. If the index is to measure the monthly changes in industrial production, for example, it is necessary to include commodities that are produced in factories and that can be measured in monthly volume of production. If an index is to measure the prices received by farmers for their products, it must be based on series that measure the prices paid to farmers. The farm price of cotton is not the same as the price on the cotton exchanges.

When selecting the weights for an index, the use that will be made of the result is an important consideration. If a price index is to measure the changes in the prices paid by consumers, it is necessary for the weights applied to the various commodities to reflect the importance of the individual items to consumers. Likewise, the weights for an index of farm prices should reflect the importance of the various commodities in the income of farmers. Frequently the commodities included in an index and the weights assigned the items do not give the information wanted for the solution of a particular problem. An index may be computed for the specific problem at hand, or the best available index may be used in spite of the fact that it does not measure exactly what the user wants the index to measure.

Choice of Base Period

If an index number is computed for a specific purpose, the base selected may be the period that is the most useful for comparative purposes. However, most indexes used by business are published by public agencies, and since these indexes are used for many purposes it is desirable that the base be a period with which frequent comparisons may be made. The advantages of having all official indexes on the same base are so great that the Bureau of the Budget now recommends a standard base period for all federal general-purpose indexes. In January, 1971, the base period was shifted to 1967; all federal government indexes in the future will be on this new base rather than the old base of 1957–59, which had been used for the preceding nine years.

Some statistical agencies refer to the period q_a (page 479) as the base, without always clarifying the relationship between this period and the period that is given a value of 100. Therefore, the period that is designated as 100% is called the *reference base* when weights from some period other than the base period are being used. In other words, in this text "base period" is used to refer to the period that is given a value of 100. But whenever some misunderstanding might occur "reference base" is used with the same meaning. In some discussions of index numbers, the term $\Sigma p_o q_a$ may be referred to as the sum of base-year values, which is really not correct, since only the prices are from the base year. If q_a refers to 1958 quantities while the base period is 1957–59, $\Sigma p_o q_a$ is more correctly described as 1958 quantities valued at 1957–59 prices.

Since index numbers are expressed as percentage relatives with the base period equal to 100, it is possible to shift the base to any period covered by the index by merely dividing the index for each period by the value for the new period that is to be used as the base. For example, if 1957 is the base with the index number for 1964 equal to 120 and 1965 equal to 132, the year 1964 may be made the base by dividing the 1965 value of 132 by 120 and multiplying by 100, giving a new percentage relative of 110 for 1965 (with 1964 equal to 100). The index number for 1957 on the new base would be found by dividing 100 (the value of the index for 1957) by the value for the new base year, 120. The index number for 1957 on the 1964 base would be 83.

$$\frac{100}{120}100 = 83\%.$$

TWO IMPORTANT INDEXES

The following discussion describes briefly the methods of construction used in two well-known indexes—one an index of prices and the other, a quantity index.

Wholesale Price Index

The *wholesale price index,* also known as the WPI, is compiled by the U.S. Department of Labor, Bureau of Labor Statistics. This index shows the general rate and direction of price movements in the primary markets for commodities. It is designed to measure the change in prices between two periods and attempts to remove from these changes the effects of changes in quality, quantity, and terms of sale. The term "wholesale" is frequently misunderstood, for the prices used are not prices received by wholesalers, jobbers, or distributors, but are the selling prices of representative manufacturers or other producers, or prices quoted on organized exchanges or markets. The policy of the Bureau of Labor Statistics is to revise the wholesale price index weighting structure periodically when data from industrial censuses become available, generally at five-year intervals.

The weights used in the index represent the net selling values of the volume of commodities produced and processed in, or imported into, this country and flowing into primary markets in 1963. Excluded are interplant transfers (where available data permit), military products, and goods sold to household consumers directly by producing establishments. The data are obtained from the 1963 Census of Manufactures and the 1963 Census of Mineral Industries, and from other sources furnished by the U.S. Department of Agriculture, the U.S. Department of Interior's Bureau of Mines and Bureau of Fisheries, and a few other agencies.

The computations are carried out by use of Formula 17–16, with q_a representing the quantities for 1963, using as a reference base the period 1967.

The Bureau of Labor Statistics also compiles an index of purchasing power based on the wholesale price index. It is obtained by computing the reciprocals of the wholesale price index numbers and expressing the reciprocals as percentages of the average of the 1967 base period. As the price index increases, the reciprocal decreases. Thus, if the price index rose from 100 to

200, the purchasing power index would fall from 100 to 50. That is, the price level for the period was double that of the base period; or the purchasing power of the dollar for the period was one half the purchasing power of the base period.

Index of Industrial Production

The *index of industrial production* (IIP), compiled by the Board of Governors of the Federal Reserve System, measures changes in the physical volume or quantity of output of manufacturing, mining, and utilities. The index does not cover agriculture, the construction industry, transportation, or various trade and service industries.

Since the index of industrial production was first introduced in the 1920's, it has been revised from time to time to take account of the growing complexity of the economy, the availability of more data, improvement in statistical processing techniques, and refinements in methods of analysis.

The index is computed by a modification of Formula 17–13.

$$Q_i = \frac{\sum \frac{q_i}{q_o} p_a q_o}{\Sigma p_a q_o} 100 = \sum \left(\frac{q_i}{q_o} p_a q_o \right) \frac{1}{\Sigma p_a q_o} 100$$

$$Q_i = \sum \left(\frac{q_i}{q_o} \frac{p_a q_o}{\Sigma p_a q_o} \right) 100. \quad (17\text{–}17)$$

Since the reference base is the year 1967, q_o represents average quantities produced in 1967. The term p_a represents the value per unit added by manufacture in 1967, computed by subtracting the cost of materials, supplies, containers, fuels, purchased electric energy, and contract work from each industry's gross value of products. Value-added data are used in preference to gross-value figures because they reflect each industry's unduplicated contribution to total output. Gross value of output, which includes material costs, reflects also the contributions made by producers at all earlier stages of fabrication. Formula 17–17 is a weighted average of the relatives $\frac{q_i}{q_o}$, with the weights represented by $\frac{p_a q_o}{\Sigma p_a q_o}$. The total index number or any subgroup can be computed by multiplying the relative for each industry by its weight, summing these products, and dividing by the sum of the weights.

USES OF INDEX NUMBERS

A great many time series are dollar values, such as sales, value of construction contracts awarded, value of building permits issued, national income, wages, and bank deposits. All statistical series that measure aggregate values are the composite result of changes in physical volume and changes in the level of prices. Series such as sales of a company or of all the business concerns in an industry are affected by two factors: (1) *the prices of the items sold,* and (2) *the number of units sold.* If the dollar sales of a business increased during a period when prices were declining, it means that the volume of goods sold increased at a faster rate than the prices declined. If prices rose and volume remained the same, sales in dollars would increase at the same rate as prices. But if prices and physical volume both rise or fall, the dollar sales will show a greater increase or decrease than either of the components.

The sales of Pearl Brewing Company are shown both in barrels and in dollars in Figure 17–1. Both series represent useful information, since the gross income in dollars and the total volume of the product sold are both important to the management of the company. Whenever it is possible to express

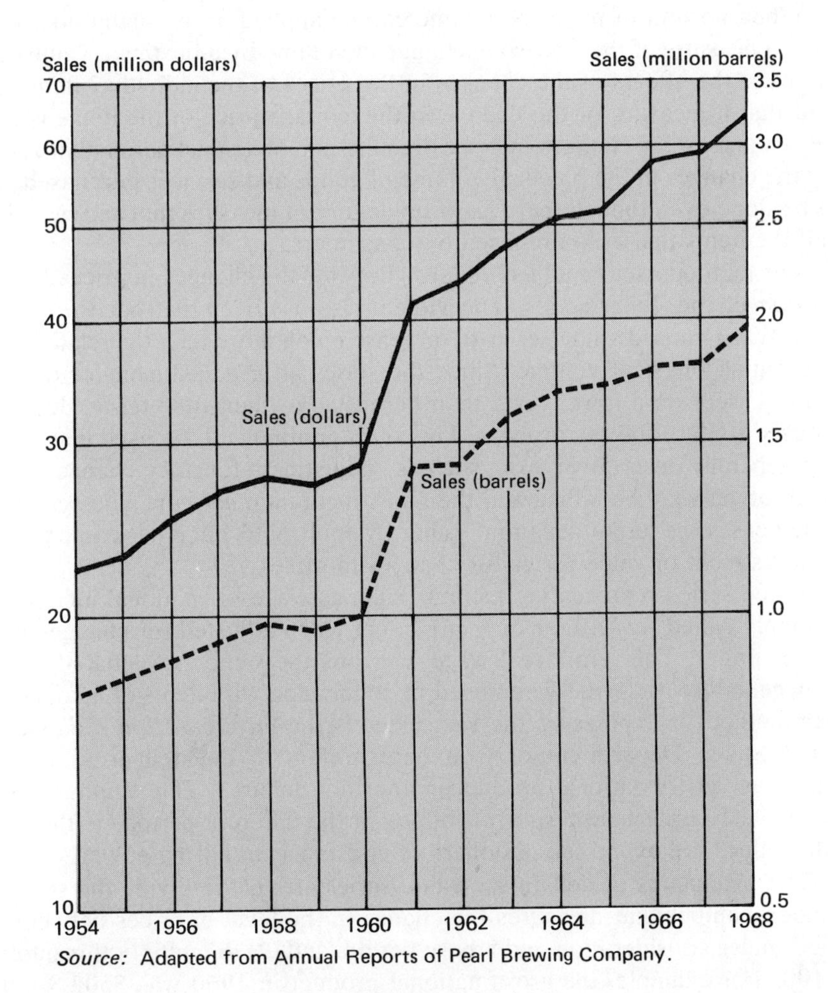

Source: Adapted from Annual Reports of Pearl Brewing Company.

FIGURE 17–1

SALES OF PEARL BREWING COMPANY, 1954 TO 1968

the volume of sales of total production in one physical unit, both of these series may be used; and each one represents a logical, well-defined time series. However, in many situations no adequate physical unit is available. Sales of a department store, for example, are expressed in dollars, but it would be difficult to find a single unit in which to express the physical volume of sales. The total volume of goods and services produced by the economic system of a country is an important item of information, but it is impossible to design a physical unit of measure that can be applied to all kinds of goods and services. Most business concerns sell a variety of products that cannot be reduced to a common physical unit.

Deflation of Time Series

When no unit of physical volume can be applied, it is usually possible to derive a measure of the aggregate change in volume by adjusting a value series to remove the effects of the changes in the prices of the individual items, provided that a measure of the change in the average price of the items is available. Adjusting for changes in prices results in a statistical series that reflects only the changes in the physical volume of goods and services purchased. This can be done even though there is no single unit of measure that can be applied to all the items that are purchased by consumers.

The method used to adjust dollar figures for the changes in prices is called *deflation* of the dollar series. The violent rise in prices that occurred during World War I caused value series to increase much more than the related series representing physical volume. Since the series after adjustment for the level of prices were on a lower level than before adjustment, the term "deflation" was used to describe the process. This term continues to be used instead of a more generally descriptive term, such as "adjustment for price changes," even though prices may be falling and the adjustment *increases* the adjusted series. In such cases the term "deflation" is hardly applicable, but it has come to mean any adjustment of value series for changes in prices.

Value series representing income, such as wages or national income, are commonly called *real wages* or *real income* when adjusted for changes in the level of prices. The term "real wages" means the same as "deflated wages" and is generally preferred when used in connection with items of income. Another method of expressing the same idea is the term *constant dollars*. The United States Department of Commerce refers to deflated gross national product as "gross national product in constant dollars." This emphasizes the fact that the purchasing power of money in the different periods is the same, or the values are expressed in dollars of constant purchasing power.

The mechanics of deflating a value series are: (1) to divide the series by an index number that measures the change in the level of prices between the period under consideration and a base period, and (2) to multiply this quotient by 100. For example, the gross national product in 1960 was $503.8 billion and the prices in 1960 were 103.28% of prices in 1958. The deflated gross national product was $487.8 billion.

$$\text{Deflated gross national product} = \frac{503.8}{103.28} \times 100 = \$487.8.$$

The division by the index number measuring the relation of prices to 1958 expresses gross national product in 1958 dollars, or represents the total value of the gross national product if prices had remained unchanged from the 1958 level. (Multiplying by 100 is necessary since the measure of prices is expressed as a percentage. The price relative is $\frac{103.28}{100}$ and, when dividing by this relative, it is inverted and multiplied by the value being deflated.)

The deflated gross national product might be expressed as "487.8 billions of 1958 dollars" instead of using the term "constant dollars." If another period such as 1939 were used as 100% in expressing the change in the price level, the results would have been in "1939 dollars."

Figure 17–2 and Table 17–18 show the gross national product in the United States for the years 1931 to 1972, expressed both in 1958 dollars and

TABLE 17–18

GROSS NATIONAL PRODUCT, 1931 TO 1972

Current Dollars and 1958 Dollars

Year	Current Dollars (Billions)	1958 Dollars (Billions)	Year	Current Dollars (Billions)	1958 Dollars (Billions)
1931	75.8	169.2	1951	328.4	383.4
1932	58.0	144.1	1952	345.5	395.1
1933	55.6	141.5	1953	364.6	412.8
1934	65.1	154.3	1954	364.8	407.0
1935	72.2	169.6	1955	398.0	438.0
1936	82.5	193.0	1956	419.2	446.1
1937	90.4	203.3	1957	441.1	452.5
1938	84.7	193.0	1958	447.3	447.3
1939	90.5	209.4	1959	483.7	475.9
1940	99.7	227.2	1960	503.8	487.8
1941	124.5	263.7	1961	520.1	497.3
1942	157.9	297.8	1962	560.3	530.0
1943	191.6	337.2	1963	590.5	551.0
1944	210.1	361.3	1964	632.4	581.1
1945	212.0	355.4	1965	684.9	616.7
1946	208.5	312.6	1966	749.9	652.6
1947	231.3	309.9	1967	793.9	669.3
1948	257.6	323.7	1968	864.2	706.6
1949	256.5	324.1	1969	929.1	724.7
1950	284.8	355.3	1970	977.1	722.5
			1971	1,050.5	745.7
			1972	1,155.2	790.7

Source: U.S. Department of Commerce, *Business Statistics* (1973).

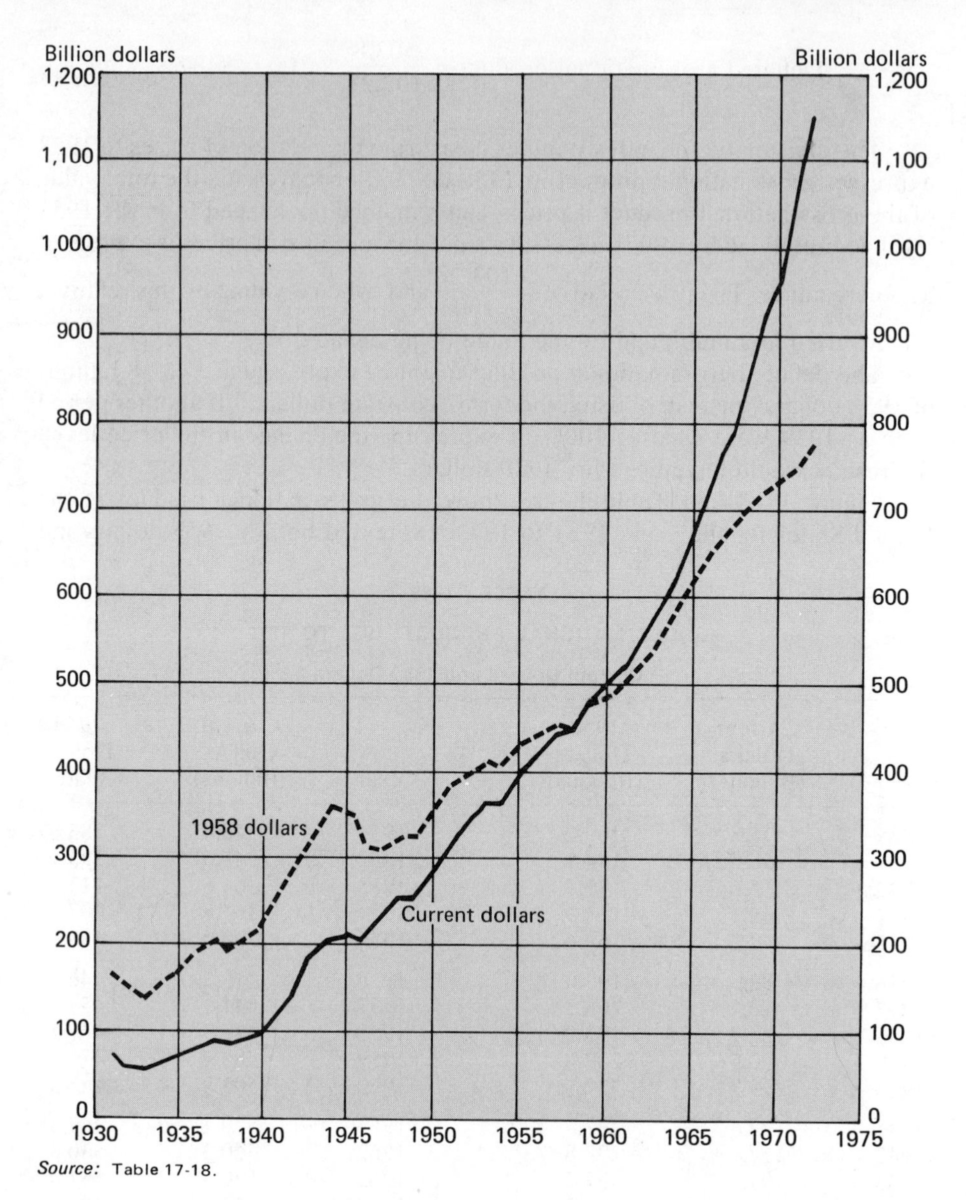

Source: Table 17-18.

FIGURE 17–2

GROSS NATIONAL PRODUCT, 1931 TO 1972

in the current prices prevailing each year. Since 1958 there has been a large rise in the total value of goods and services produced in the United States, but the deflated series has risen much less. This means that a substantial portion of the increase in the value of the total output of goods and services has resulted from rising prices. The series expressed in constant (1958) dollars represents the rise in the physical volume of goods and services produced.

The General Electric Company has constructed an orders price index (1957–59 = 100) to measure the changes in prices of their products. Table 17–19 shows net sales billed in current dollars, the orders price index, and net sales adjusted for changes in the prices of products. This latter series is, in effect, a measure of the physical volume of sales, even though it is impossible to express sales of General Electric Company in comparable physical units as was done for Pearl Brewing Company. Net sales billed increased 26% between 1955 and 1959, but most of this increase resulted from higher prices. After adjustment for price changes, sales rose only 7%. Between 1959 and 1964, net sales billed increased only 14%, but after allowance for declining prices, the adjusted sales rose 32%.

TABLE 17–19

GENERAL ELECTRIC COMPANY

Actual Sales and Sales Adjusted for Changes in Prices
1955 to 1964

Year	Net Sales (Millions of Dollars)	Orders Price Index 1957–59 = 100	Net Sales in 1957–59 Dollars (Millions of Dollars)
1955	3,464	86	4,028
1956	4,090	93	4,398
1957	4,336	99	4,380
1958	4,121	100	4,121
1959	4,350	101	4,307
1960	4,198	98	4,284
1961	4,457	92	4,845
1962	4,793	90	5,326
1963	4,919	87	5,654
1964	4,941	87	5,679

Source: 1964 Annual Report, General Electric Company.

The deflation of a time series can also be used in the computation of real wages. For example, annual hourly earnings in the manufacturing industry in the United States in October, 1973, were $3.95, compared with $.70 in 1939. The increase of 464% did not mean that wage earners had that great an increase in purchasing power. The index of consumer prices was 41.6 in 1939 and 135.5 in October, 1973, an increase of 226%. Each of these indexes was a percentage of the 1967 average index, so the $.70 wage rate in 1939 was $1.68 in 1967 dollars and the October, 1973, wage rate of $3.95 was $2.92 in the same purchasing power.

$$\text{Real wages, 1939} = \frac{.70}{41.6}\,100 = \$1.68.$$

$$\text{Real wages, October, 1973} = \frac{3.89}{132.7}\,100 = \$2.93.$$

The real wages in October, 1973, were 74% more than in 1939. This percentage increase measures the change in the amount of goods and services an hour of work in a factory would buy in 1973 compared with 1939.

Choice of a Price Index. It is essential that the index used to measure the changes in the level of prices accurately reflects the average change in the prices of the commodities or services on which the value series is based. If the measure of the change in prices is not accurate, the resulting deflated series will not be an accurate representation of the changes in physical volume. It is extremely important to insure that the price index used to measure the changes in the prices be comparable with the value series being deflated. For example, if dollar building figures are being deflated, the price index must measure the changes in the cost of building the kind of structures included in the aggregate values. If export figures are being deflated, the price index must measure the prices of export commodities. Frequently the precise index needed is not available, but it should be remembered that when another is substituted, the deflated series will be inaccurate to the degree that the price index used deviated from the correct change in prices.

In spite of the difficulty of securing an accurate measure of the prices included in all aggregate value series, the method of deflating time series is extremely useful. Many series can be compiled only as aggregate values, but when deflated with even an approximate measure of price changes, the series will serve as an approximate measure of physical volume. The concept of real wages and real income is extremely useful, since the fluctuations in total dollar series fail to give a complete picture of the situation in periods of major shifts in the price level.

Contracts in Constant Purchasing Power

One of the serious problems of making long-term contracts to pay specified amounts is the danger that the level of prices will change so much that the amount paid at the time the contract is fulfilled will not buy as many commodities as the contracting parties actually intended. For example, a lease is made to rent property for 50 years. If the general level of prices should double in that time, the result would be that the owner would receive only one half the amount of purchasing power that was intended in the contract. Wage contracts, annuities, bond issues, and agreements for purchases in the future are all subject to serious dislocations when the level of prices changes substantially over a period of time.

One method of eliminating the risk of a shift in the price level is to write contracts for future payments with a clause which provides that the amount paid shall represent a certain amount adjusted for changes in the level of prices. A 50-year lease may provide that the monthly rental be $250 per month, adjusted every five years for any changes in the level of prices. In other words, if the price index specified rose 10% during the five-year period, the rent would be increased by 10%. This would mean that the owner would receive enough rent money to buy the same amount of goods at the end of the lease period as he was able to buy at the beginning, even though prices may have risen by 10%.

An annuity may be made payable as a given number of dollars adjusted to reflect any change in the level of prices. Individuals living on a fixed retirement income find it increasingly difficult to maintain their standard of living as the prices they must pay increase while their monthly income in dollars remains constant. A bond or a long-term note may provide that the amount repaid be adjusted to reflect the amount of change in the level of prices between the date of the loan and the date of repayment. This would insure that the lender would be repaid the same amount of purchasing power that he loaned.

The long-term movement of the price level has been upward during the 20th century, although it does not mean that prices will necessarily always rise. During the last part of the 19th century, the price level declined steadily, with serious effects on long-term contracts written in dollars. Bonds and mortgages were repaid in the amount of dollars borrowed; but since the price level was declining, it meant that the borrower was forced to repay more purchasing power then he had borrowed, resulting in an unexpected gain for the lender. This was a serious burden on the borrower, particularly farmers who had borrowed heavily to buy and improve their land. It is just the reverse of the situation when prices are rising, in which case the borrower can repay in less purchasing power than he received, with a resulting loss in purchasing power to the lender. It seems reasonable to expect that neither party to a long-term contract should gain or lose because of changes in the price level. One simple method of eliminating such losses is to write a contract that provides for repayment in the same amount of purchasing power as borrowed.

Parity Prices

A measure of the purchasing power of farmers' income is furnished by the *parity ratio,* which is computed by the U.S. Department of Agriculture from two index numbers. One index measures the changes in the prices received by farmers for farm products sold, and the other measures the changes in the prices paid by farmers for commodities and services, interest rates, taxes, and wage rates. Since the base of each index is the average of 1910–14, the ratio between the two indexes compares the level of prices received by farmers to prices paid by farmers. For example, in October, 1973, prices paid by farmers were 514% of the level of 1910–14, while prices received by farmers were 468% of the 1910–14 base. The ratio of the index of prices received to the index of prices paid is

$$\text{Parity ratio} = \frac{468}{514} = 91\%.$$

This indicates that the purchasing power of the farmers' production was only 91% of its purchasing power in the five years 1910–14.

Choosing the period 1910 to 1914 as the base for the comparisons assumes that the prices of farm products and the prices of the items farmers buy were in satisfactory balance in that period. In October, 1973, the prices of farm products were 9% below the level that would make it possible for a given amount of farm products to buy the same volume of goods and services as this amount bought in 1910–14. In other words, if the prices of things the farmer buys are three times the level of 1910–14, the prices of farm products must also be three times the level of 1910–14 if the farmers' purchasing power is not to be impaired. When the ratio is above 100, the farmer is relatively better off than he was in the base period; when the ratio is below 100, he is relatively worse off. The Department of Agriculture also computes a parity price for individual commodities, and uses this parity price in its price-support program.

Price-Level Changes and Accounting Statements

Since accounting statements use values based on current prices, assets and profits are expressed as values that are based on different price levels. The level of prices has shown such wide fluctuations during the 20th century that accountants have found it necessary to experiment with methods of measuring the impact of price-level changes on financial reports. There is no general agreement on what adjustments should be made, but all methods make use of an index of prices. If adjustments are to be made with respect to changes in the general level of prices, the problem becomes one of selecting the best index of the general price level. If specific assets are to be adjusted for price changes, special indexes that measure the changes in the prices of specific commodity groups are needed.

STUDY QUESTIONS

17–1. In what important respect are the index numbers in this chapter different from the fixed-base relatives discussed in Chapter 13?

17–2. Explain the logic that underlies the use of the index number technique to combine series that cannot be added.

17–3. Basset Jones in his book *Horses and Apples* argues that it is incorrect to average relatives from series that cannot be averaged. Can you support this argument, or do you consider it proper to average relatives based on data that cannot themselves be properly averaged?

17–4. It was stated in Chapter 3 that the averaging of ratios offers serious problems that must be given careful attention. What is the importance of this principle in the construction of index numbers, since index numbers are averages of relatives?

17–5. Why should the geometric mean not be used with weights computed from base-year values?

17–6. Explain why Formulas 17–5 and 17–9 give the same answer. How would you decide which one to use?

17–7. It was stated in the text that the arithmetic mean is biased when used to average relatives. Does this mean that it is incorrect to use the arithmetic mean to average relatives in constructing index numbers? Give reasons for your answer.

17–8. The index of consumer prices compiled by the Bureau of Labor Statistics includes items that were not for sale in the years before World War II. For example, television sets were not included in the index of 1939. Does this mean that a valid comparison cannot be made between the level of prices in 1973 and 1939, since all the commodities in the two years were not the same?

17–9. Explain the meaning of the "factors test" for an index number. Does it seem reasonable to require that an index number formula give consistent answers when the factors test is applied?

17–10. Formula 17–10 is sometimes considered a significant formula in its own right. The aggregate for the base year represents the value of the goods sold in the base year, computed by multiplying the quantity sold by the price per unit. The aggregate for the current year represents the value of the goods sold in the base year if they were valued at current year prices. The ratio between these two aggregates is a weighted average of the change in prices.

(a) Why would it not be just as logical to construct an index number using quantities in the current year (q_i) as the multiplier in computing the aggregates? Write the formula that would be used in such a calculation.

(b) What is your opinion of using an average of q_o and q_i as the multiplier in constructing an index by the aggregative method? Write the formula that would be used in such a calculation.

17–11. What advantages does a chain index have over a fixed-base index in computing an index of prices? Does the chain index have disadvantages that outweigh its advantages?

17–12. Explain why the wholesale price index uses a chain index with the weights based on quantities in a year that is not the base year (Formula 17–16) instead of using a simpler formula such as Formula 17–10?

17–13. Do you consider the 1910–14 base used in computing the parity ratio a good base period to use? What arguments could you advance for using a more recent base period?

17–14. Why would labor unions be interested in indexes of consumer prices?

17–15. Does the rise in the consumer price index mean that the standard of living in the United States has improved? Explain.

PROBLEMS

17–1. The following data give the price relatives for commodities A and B for the years 1967 and 1973, and the arithmetic mean for the two years.

	1967	1973
Commodity A	100	50
Commodity B	100	200
Arithmetic mean	100	125

If 1973 were chosen as the base instead of 1967, the following relatives would show the relationship of the price of each commodity in 1967 to the price in 1973. The arithmetic mean of the relatives for each year is given in the following data.

	1967	1973
Commodity A	200	100
Commodity B	50	100
Arithmetic mean	125	100

The relationship between the two sets of relatives for commodity A is the same, no matter which year is the base. In each case the 1967 relative is double the 1973 relative. Likewise, the two sets of relatives for commodity B show the same relationship between prices for the two years. In each case the 1967 relative is one half that of 1973.

The averages of the two sets of relatives, however, give two completely different measures of relationship between the two years. When 1967 is taken as the base, the arithmetic mean indicates that the average price of the commodities in 1973 was 25% higher than in 1967. But when 1973 is taken as the base, the average of the relatives indicates that *prices in 1967 were 25% higher than in 1973*. Obviously both of these results cannot be correct. Do you consider one of them to be correct? Explain.

17-2. Compute the geometric mean of the relatives in Problem 17-1 and explain what the results show.

17-3. The following table gives the index of wholesale prices for the years 1950 to 1973, with 1967 = 100.

(a) Shift the base of these index numbers to the year 1950, the year the Korean War started.

(b) Which base shows more clearly the changing prices following the start of the Korean War?

Year	Index	Year	Index	Year	Index
1950	81.8	1958	94.6	1966	99.8
1951	91.1	1959	94.8	1967	100.0
1952	88.6	1960	94.9	1968	102.5
1953	87.4	1961	94.5	1969	106.5
1954	87.6	1962	94.8	1970	110.4
1955	87.8	1963	94.5	1971	113.9
1956	90.7	1964	94.7	1972	119.1
1957	93.3	1965	96.6	1973	135.5

Source: U.S. Council of Joint Economic Advisors for the Joint Economic Committee, *Economic Indicators* (March, 1974).

17-4. A man made a will in 1941 providing that after his death his widow should receive a monthly income of $2,000 from a trust fund. Since he realized that the war was likely to cause a substantial rise in the price level, he provided that the payments should be adjusted each month to insure that the purchasing power of the monthly payment would remain the same as it was in 1941. The will specified that the purchasing power of the dollar should be measured by the cost-of-living index published by the Bureau of Labor Statistics, and if this index were revised or replaced by another index, the purchasing power should then be computed by the use of the revised or new index. The average value of the index for 1941 was 105.2, with 1935–39 = 100.

(a) Compute the amount of income in current dollars to be paid to the widow in January, 1947. The index for January, 1947, was 153.6, with 1935–39 = 100. (By 1947 the name of the index had been changed from cost-of-living index to consumer price index, but the base period was still 1935–39.)

(b) Compute the amount of income in current dollars to be paid to the widow in January, 1960. The index for January, 1960, was 125.4, but the base had been changed to 1947–49 = 100. The average value of the index for 1941 was 62.9, with 1947–49 = 100.

(c) Compute the amount of income in current dollars to be paid to the widow in September, 1973. The index for September, 1973, was 135.5, but in 1971 the base had been changed to 1967 = 100. The average value of the index for 1941 was 48.3, with 1967 = 100.

17–5. The per capita public debt of $2,094 on June 30, 1972, was much larger than in 1940. However, a simple comparison of the size of the per capita debt at different times may not be the most precise method of determining what has happened. The following table gives the consumer price index, and per capita public debt and national income for the years 1940 and 1972.

(a) Compute the percentage increase in the per capita public debt and per capita income between 1940 and 1972.

(b) Compute the ratio of the public debt to per capita national income for each year. What was the percentage change in this ratio? Explain the significance of this change.

(c) Compute the real per capita income in 1940 and 1972, and the percentage increase.

(d) Deflate per capita public debt for 1940 and 1972. What was the percentage increase in deflated per capita public debt? Explain what this percentage means.

(e) Write a brief statement that summarizes the change in the public debt between 1940 and 1972, making use of the above computations.

	1940	1972
Per capita public debt (dollars)[1]	325	2,094
Per capita national income (dollars)[2]	616	4,510
Consumer price index (1967 = 100)[3]	42.0	125.3

[1] As of June 30.
[2] Total for year.
[3] Average for year.

Source: Debt—Bureau of the Budget.
Income—U.S. Department of Commerce.
CIP—U.S. Department of Labor.

17–6. The owner of a lot in the business district of a city is negotiating for its rental on a long-term lease, and he wants to be protected against any substantial decrease in the purchasing power of the dollar. Write a paragraph that he could incorporate in his lease to insure that the purchasing power of his rental income will not diminish over the period of the lease.

17–7. The following table shows quantities sold (in thousands of units) and prices (in dollars and cents) for three commodities over a period of four years.

(a) Set up a table showing quantity relatives and price relatives for each commodity for the four years. Use 1967 = 100 as the base year.

(b) Compute a quantity index for 1970 using an unweighted arithmetic mean of quantity relatives.

Year	Commodity A		Commodity B		Commodity C	
	q_i	p_i	q_i	p_i	q_i	p_i
1967	74	2.20	125	4.80	400	0.75
1968	76	2.35	130	4.85	370	0.72
1969	80	2.41	135	4.75	350	0.70
1970	82	2.50	142	4.60	330	0.65

(c) Compute a price index for 1970 using an unweighted geometric mean of price relatives.

(d) Compute a price index for 1970 using an arithmetic mean of price relatives weighted with base-year values.

(e) Compute a price index for 1970 using a geometric mean of price relatives weighted with base-year values.

(f) Compute a quantity index for 1970 using an arithmetic mean of quantity relatives weighted with 1970 values.

(g) Compute a quantity index for 1970 using a geometric mean of quantity relatives weighted with 1970 values.

(h) Compute a price index for 1970 using the aggregates formula (Formula 17–12) and 1967 base-year values. *Note:* This answer should be same as that in (d).

17–8. If wages for carpenters were \$5.09 per hour when the consumer price index stood at 115.2 and increased to \$6.38 per hour when the consumer price index stood at 135.5, what was the increase, if any, in real wages?

17–9. The data in the following tables serve as the information to be used in Problems 17–9 to 17–15 for the construction of various types of index numbers of prices and volume of production. The first table shows the production of three fuels for the years 1967 to 1971. The second table shows the prices of the same fuels for the same years.

	1967	1968	1969	1970	1971
Bituminous coal (Million tons)	553	545	561	603	552
Crude oil (Million barrels)	3,216	3,316	3,372	3,516	3,454
Natural gas (Billion cubic feet)	18,171	19,322	20,698	21,921	22,493

Source: U.S. Bureau of Mines.

	1967	1968	1969	1970	1971
Bituminous coal (Ton, FOB mine)	4.62	4.67	4.99	6.26	7.07
Crude oil (Barrel)	2.92	2.94	3.09	3.18	3.39
Natural gas (1,000 cubic feet)	.160	.164	.167	.171	.182

Source: U.S. Bureau of Mines.

(a) The first step in the construction of the index numbers is to reduce all the series to percentage relatives, with the year 1967 equal to 100.

(b) Compute the value of production of each fuel in each year. Express the value of production of each fuel as a percentage of total production for each year.

(c) Construct a line chart showing the three price relatives on the same chart for the years 1967 to 1971.

(d) Construct a line chart showing the three production relatives on the same chart for the years 1967 to 1971.

17–10. (a) Construct an index of prices for 1971, using an unweighted arithmetic mean of the relatives.

(b) Construct an index of production for 1971, using an unweighted arithmetic mean of the relatives.

(c) Apply the factors test to the two indexes just computed. What conclusion can you draw from this test?

17–11. (a) Construct an index of prices for 1971, using an unweighted geometric mean of the relatives.

(b) Construct an index of production for 1971, using an unweighted geometric mean of the relatives.

(c) Apply the factors test to the two indexes just computed. How do the results of this test compare with the same test applied to the arithmetic means computed in Problem 17–10?

17–12. (a) Construct an index of prices for 1971, using the arithmetic mean of relatives weighted by the values of the fuels produced in 1967.

(b) Construct an index of production for 1971, using the arithmetic mean of relatives weighted by the values of the fuels produced in 1967.

(c) Apply the factors test to the two indexes just computed.

17–13. (a) Construct an index of prices for 1971, using a weighted geometric mean of the relatives. Use as weights the average value of fuels produced in 1967 and 1971.

(b) Construct an index of production for 1971, using a weighted geometric mean of the relatives. Use as weights the average value of fuels produced in 1967 and 1971.

(c) Apply the factors test to the two indexes just computed. How does the formula used in this problem compare with the use of the factors test in Problem 17–12?

17–14. Construct an index of prices for 1971, using the aggregative Formula 17–10. How does the result compare with the index computed in Problem 17–12(a)?

17–15. Construct an index of production for 1971, using the aggregative Formula 17–9. How does the result compare with the index computed in Problem 17–12(b)?

17–16. The table below shows personal consumption expenditures in the United States for the years 1950 to 1972. During this period prices were rising, with the result that some of the increase in consumer expenditures reflected a price increase rather than an increase in the volume of goods and services bought by consumers. The table also gives the implicit price deflator, which is an index number of the prices of goods and services purchased by consumers.

(a) Compute personal consumption expenditures in 1958 dollars for the years 1950 to 1972.

(b) Plot the two series, consumer expenditures in current dollars and in 1958 dollars, on the same arithmetic chart.

(c) Write a description of what the chart shows concerning the changes in the amount of money consumers spent and the volume of goods and services they secured over the period.

Year	Expenditures (Billions of Dollars)	Deflator (Index Number, 1958 = 100)
1950	191.0	82.9
1951	206.3	88.6
1952	216.7	90.5
1953	230.0	91.7
1954	236.5	92.5
1955	254.4	92.8
1956	266.7	94.8
1957	281.4	97.6
1958	290.1	100.0
1959	311.2	101.3
1960	325.2	102.8
1961	335.2	103.9
1962	355.1	104.9
1963	373.8	106.1
1964	398.9	107.2
1965	431.5	108.9
1966	464.9	111.9
1967	492.2	114.3
1968	533.8	118.4
1969	579.5	123.5
1970	617.6	129.3
1971	667.2	134.4
1972	726.5	137.9

Source: U.S. Department of Commerce, *Survey of Current Business* (July, 1973).

Appendixes

By matching the guides at the edge of this page with the marks opposite them along the edge of the book, you can quickly turn to the section of the appendix you want.

FORMULAS

Part One—Data Analysis

Ch. 3 Averages

Formula Number	Formula		Page
(3–1)	$\mu = \frac{\Sigma X}{N}$	arithmetic mean of the X variable	45
(3–2)	$\mu = \frac{\Sigma fm}{N}$	arithmetic mean of a frequency distribution.	46
(3–3)	$\mu = A + \frac{\Sigma fd'}{N} i$	arithmetic mean of a frequency distribution, using an arbitrary origin	47
(3–4)	$c = \frac{\Sigma fd'}{N} i$	difference between mean (μ) and arbitrary origin (A)	47
(3–5)	$\mu = A + c$	arithmetic mean of a frequency distribution, using an arbitrary origin	47
(3–6)	$\mu = \frac{\Sigma wX}{\Sigma w}$	weighted arithmetic mean, where w = the weight assigned to each item, or X value	48
(3–7)	$H = \frac{N}{\Sigma \frac{1}{X}}$	harmonic mean	54
(3–8)	$G = (X_1 \cdot X_2 \cdot X_3 \cdot \ldots \cdot X_N)^{\frac{1}{N}}$	geometric mean	55
(3–9)	$\log G = \frac{\Sigma \log X}{N}$	geometric mean, in terms of logarithms	55
(3–10)	$Md = L_{Md} + \frac{\frac{N}{2} - F_{L_{Md}}}{f_{Md}} i_{Md}$	median of a frequency distribution	61

Formula Number	Formula	Description	Page
(3–11)	$Mo = L_{Mo} + \frac{d_1}{d_1 + d_2} i_{Mo}$	mode of a frequency distribution	64
(3–12)	$Mo_E = \mu - 3(\mu - Md)$	empirical mode	66

Ch. 4 Dispersion, Skewness, and Kurtosis

Formula Number	Formula	Description	Page
(4–1)	$Q = \frac{Q_3 - Q_1}{2}$	quartile deviation, or semi-interquartile range	79
(4–2)	$Q_1 = L_{Q_1} + \frac{\frac{N}{4} - F_{L_{Q_1}}}{f_{Q_1}} i_{Q_1}$	first quartile of a frequency distribution	80
(4–3)	$Q_3 = L_{Q_3} + \frac{\frac{3N}{4} - F_{L_{Q_3}}}{f_{Q_3}} i_{Q_3}$	third quartile of a frequency distribution	80
(4–4)	$AD = \frac{\Sigma \lvert X - Md \rvert}{N}$	average deviation of ungrouped data	81
(4–5)	$AD = \frac{\Sigma f \lvert m - Md \rvert}{N}$	average deviation of grouped data	83
(4–6)	$\sigma^2 = \frac{\Sigma(X - \mu)^2}{N}$	variance of ungrouped data	84
(4–7)	$\sigma = \sqrt{\frac{\Sigma(X - \mu)^2}{N}}$	standard deviation of ungrouped data	84
(4–8)	$\Sigma(X - \mu)^2 = \Sigma X^2 - \frac{(\Sigma X)^2}{N}$	sum of the squared deviations from the arithmetic mean	86
(4–9)	$\sigma^2 = \frac{\Sigma(X - \mu)^2}{N} = \frac{\Sigma X^2 - \frac{(\Sigma X)^2}{N}}{N}$	variance of ungrouped data	86
(4–10)	$\sigma^2 = \frac{N\Sigma X^2 - (\Sigma X)^2}{N^2}$	variance of ungrouped data	86

Formula Number			Page
(4–11)	$\sigma = \sqrt{\dfrac{N\Sigma X^2 - (\Sigma X)^2}{N^2}}$	standard deviation of ungrouped data	86
(4–12)	$\sigma = \dfrac{\sqrt{N\Sigma X^2 - (\Sigma X)^2}}{N}$	standard deviation of ungrouped data	86
(4–13)	$\sigma^2 = \dfrac{\Sigma f(m - \mu)^2}{N}$	variance of grouped data	87
(4–14)	$\sigma = i\sqrt{\dfrac{\Sigma f(d')^2}{N} - \left(\dfrac{\Sigma fd'}{N}\right)^2}$	standard deviation of grouped data with all classes same size	88
(4–15)	$V = \dfrac{\sigma}{\mu} \cdot 100$	coefficient of variation.	92
(4–16)	$\text{Skewness} = \dfrac{\mu - Mo}{\sigma}$	measure of skewness using the mean, mode, and standard deviation	93
(4–17)	$\text{Skewness} = \dfrac{3(\mu - Md)}{\sigma}$	measure of skewness using the mean, median, and standard deviation	94
(4–18)	$M_3 = \dfrac{\Sigma(X - \mu)^3}{N}$	third moment about the mean of ungrouped data	95
(4–19)	$\beta_1 = \dfrac{M_3^2}{M_2^3}$	measure of relative skewness	95
(4–20)	$M_3 = \dfrac{\Sigma f(d')^3}{N} - 3\dfrac{\Sigma fd'}{N}\dfrac{\Sigma f(d')^2}{N} + 2\left(\dfrac{\Sigma fd'}{N}\right)^3$	third moment about the mean in class interval units	97

Formula Number			Page
(4–21)	$M_2 = \frac{\Sigma f(d')^2}{N} - \left(\frac{\Sigma fd'}{N}\right)^2$	second moment about the mean (also the variance in class-interval units)	97
(4–22)	$M_4 = \frac{\Sigma(X-\mu)^4}{N}$	fourth moment about the mean	98
(4–23)	$\beta_2 = \frac{M_4}{M_2^2}$	measure of relative kurtosis	98
(4–24)	$M_4 = \frac{\Sigma f(d')^4}{N} - 4\frac{\Sigma fd'}{N}\frac{\Sigma f(d')^3}{N} + 6\left(\frac{\Sigma fd'}{N}\right)^2\frac{\Sigma f(d')^2}{N} - 3\left(\frac{\Sigma fd'}{N}\right)^4$	fourth moment about the mean in class interval units	98

Part Two—Statistical Inference

Ch. 5 Probability

(5–1)	$P(A \cup B) = P(A) + P(B) - P(A \cap B)$	general rule of addition for probability of occurrence of two events that may or may not be mutually exclusive	109
(5–2)	$P(A \cup B) = P(A) + P(B)$	special rule of addition given that the two events are mutually exclusive	110
(5–3)	$P(A \cap B) = P(A)P(B \mid A)$	general rule of multiplication for probability of occurrence of *A* and *B*, which may or may not be independent	111

Formula Number			Page
(5–4)	$P(A \cap B) = P(A)P(B)$	special rule of multiplication given that A and B are independent events	113
(5–5)	$P(H_i \mid E) = \dfrac{P(H_i) \cdot P(E \mid H_i)}{\Sigma[P(H_i) \cdot P(E \mid H_i)]}$	Bayes' theorem	114

Ch. 6 Probability Distributions

Formula Number			Page
(6–1)	${}_nC_r = \dfrac{n!}{r!(n-r)!}$	combinations of n items taken r at a time	124
(6–2)	$\mu = n\pi$	arithmetic mean of the hypergeometric	126
(6–3)	$\sigma = \sqrt{n\pi(1-\pi)\dfrac{N-n}{N-1}}$	standard deviation of the hypergeometric	126
(6–4)	$P(r \mid n, \pi) = {}_nC_r\pi^r(1-\pi)^{n-r} = \dfrac{n!}{r!(n-r)!}\pi^r(1-\pi)^{n-r}$	general term of the binomial expansion	127
(6–5)	$\mu = n\pi$	arithmetic mean of the binomial	128
(6–6)	$\sigma = \sqrt{n\pi(1-\pi)}$	standard deviation of the binomial	128
(6–7)	$\mu = \dfrac{n\pi}{n} = \pi$	arithmetic mean of the fraction defective of the binomial	128
(6–8)	$\sigma = \sqrt{\dfrac{\pi(1-\pi)}{n}}$	standard deviation of the fraction defective of the binomial	128
(6–9)	$Y_o = \dfrac{N}{\sigma\sqrt{2\pi}}$ or $Y_o = \dfrac{N}{\sigma}.39894$	maximum ordinate of the normal distribution	132

Formula Number			Page
(6–10)	$Y = Y_0 e^{-\frac{1}{2}\left(\frac{x-\mu}{\sigma}\right)^2}$	................................other ordinates of the normal distribution.....	133
(6–11)	$P(r\|\lambda) = \frac{\lambda^r}{r!} e^{-\lambda}$	................................Poisson probability distribution..............	139

Ch. 7 Survey Sampling

Formula Number			Page
(7–1)	$\sigma_{\bar{x}}^2 = \frac{\sigma^2}{n} \cdot \frac{N-n}{N-1}$	................................variance of the mean...	159
(7–2)	$\sigma_{\bar{x}} = \frac{\sigma}{\sqrt{n}} \sqrt{\frac{N-n}{N-1}}$	................................standard error of the mean	159
(7–3)	$V_{\bar{x}} = \frac{\sigma_{\bar{x}}}{\mu} 100$	................................coefficient of variation	160
(7–4)	$\hat{\sigma}^2 = s^2 \frac{n}{n-1}$	................................estimate of universe variance from sample variance..................	163
(7–5)	$\hat{\sigma} = \sqrt{s^2 \frac{n}{n-1}}$	................................estimate of universe standard deviation from a sample	163
(7–6)	$\hat{\sigma} = \sqrt{\frac{\Sigma(x-\bar{x})^2}{n-1}}$	................................estimate of universe standard deviation from a sample	164
(7–7)	$\hat{\sigma}_{\bar{x}}^2 = \frac{\sigma^2}{n}(1-f)$	................................estimated variance of the mean	165
(7–8)	$\hat{\sigma}_{\bar{x}} = \frac{\hat{\sigma}}{\sqrt{n}} \sqrt{1-f}$	................................estimated standard error of the mean	165
(7–9)	$v_{\bar{x}} = \frac{\hat{\sigma}_{\bar{x}}}{\bar{x}} 100$	................................coefficient of variation.	168
(7–10)	$\hat{\sigma}_p = \sqrt{\frac{\pi(1-\pi)}{n} \frac{N-n}{N-1}}$	................................standard error of a percentage...............	171

Formula Number	Formula	Description	Page
(7–11)	$\hat{\sigma}_p = \sqrt{\frac{pq}{n-1}(1-f)}$	estimate of standard error of a percentage	172
(7–12)	$v_p = \frac{\sigma_p}{p} 100$	relative precision of a percentage	173
(7–13)	$\hat{\sigma}_{Md} = \frac{1.2533\hat{\sigma}}{\sqrt{n}} \sqrt{1-f}$	standard error of the median	173
(7–14)	$\hat{\sigma}_{Q_1} = \frac{1.3626\hat{\sigma}}{\sqrt{n}} \sqrt{1-f}$	standard error of the first quartile	173
(7–15)	$\hat{\sigma}_{Q_3} = \frac{1.3626\hat{\sigma}}{\sqrt{n}} \sqrt{1-f}$	standard error of the third quartile	173
(7–16)	$\hat{\sigma}_s = \sqrt{\frac{m_4 - m_2^2}{4m_2 n}}$	standard error of the standard deviation	173
(7–17)	$\hat{\sigma}_s = \frac{\hat{\sigma}}{\sqrt{2n}}$	standard error of the standard deviation	174
(7–18)	$\Sigma\hat{X} = \frac{\Sigma x}{f}$	unbiased estimate of aggregate value in a universe	174
(7–19)	$\sigma_p = \sqrt{\frac{\pi(1-\pi)}{n}}$	standard error of a percentage, ignoring correction for size of universe	176
(7–20)	$n = \frac{\pi(1-\pi)}{\sigma_p^2}$	estimating the size of sample needed	176
(7–21)	$n = \frac{\sigma^2}{\sigma_{\bar{x}}^2}$	estimating the size of sample needed	177
(7–22)	$n = \left(\frac{3V}{E}\right)^2$	estimating the size of sample needed	179

Formula Number			Page

Ch. 8 Tests of Significance: Parametric Methods

(8–1)	$z = \dfrac{\bar{x} - \mu}{\sigma_{\bar{x}}}$	deviation of sample mean from universe mean in units of the standard error of the mean	191
(8–2)	$z = \dfrac{p - \pi}{\sigma_p}$	deviation of sample percentage from universe percentage in units of the standard error of the percentage	195
(8–3)	$t = \dfrac{\bar{x} - \mu}{\dfrac{s}{\sqrt{n-1}}}$	t distribution	198
(8–4)	$\hat{\sigma}_{\bar{x}_1-\bar{x}_2} = \hat{\sigma}\sqrt{\dfrac{1}{n_1} + \dfrac{1}{n_2}}$	estimated standard deviation of the distribution of differences between two means	199
(8–5)	$t = \dfrac{\bar{x}_1 - \bar{x}_2}{\hat{\sigma}_{\bar{x}_1-\bar{x}_2}}$	t distribution for two sample means	200
(8–6)	$\hat{\sigma} = \sqrt{\dfrac{\Sigma(x_1 - \bar{x}_1)^2 + \Sigma(x_2 - \bar{x}_2)^2}{n_1 + n_2 - 2}}$	estimate of universe standard deviation from two samples	200
(8–7)	$s_{p_1-p_2} = \sqrt{\dfrac{\bar{p}\bar{q}}{n_1} + \dfrac{\bar{p}\bar{q}}{n_2}}$	standard error of the difference between two proportions	202
(8–8)	$z = \dfrac{p_1 - p_2}{s_{p_1-p_2}}$	deviation of differences in two sample percents from an assumed difference of zero in the two universe percents in units of the standard error of the difference.	202

Formula Number			Page
(8–9)	$\Sigma(x-\bar{x})^2 = \Sigma x^2 - \frac{(\Sigma x)^2}{n}$	sum of the squared deviations of the sample x values from the mean of the sample x values	206

Ch. 9 Tests of Significance: Nonparametric Methods

Formula Number			Page
(9–1)	$\chi^2 = \Sigma\left[\frac{(f_o - f_e)^2}{f_e}\right]$	chi-square	223
(9–2)	$\chi^2 = \frac{(ad-bc)^2 n}{(a+b)(c+d)(a+c)(b+d)}$	chi-square	224
(9–3)	$\chi^2 = \Sigma\frac{f_o^2}{f_e} - n$	chi-square	225
(9–4)	$\mu_R = \frac{2n_1n_2}{n_1+n_2} + 1$	average (expected) number of runs	235
(9–5)	$\sigma_R = \sqrt{\frac{2n_1n_2(2n_1n_2 - n_1 - n_2)}{(n_1+n_2)^2\,(n_1+n_2-1)}}$	standard deviation of the number of runs ...	235
(9–6)	$z = \frac{R-\mu_R}{\sigma_R}$	sampling distribution of R	235
(9–7)	$K = \frac{n-1}{2} - (.98)\sqrt{n+1}$	rejection value in sign test when sample is small and tables of binomial distribution are unavailable	238
(9–8)	$z = \frac{s + .5 - n\pi}{\sqrt{n\pi(1-\pi)}}$	normal approximation of the binomial	238

Formula Number			Page
(9–9)	$H = \frac{12}{n(n+1)} \sum_{j=1}^{c} \frac{R_j^2}{n_j} - 3(n+1)$	Kruskal-Wallace H test for analysis of variance	239

Ch. 10 Statistical Quality Control

(10–1)	$c_2 = \sqrt{\frac{2}{n}} \frac{\left(\frac{n-2}{2}\right)!}{\left(\frac{n-3}{2}\right)!}$	computing σ from $\bar{s}$	255
(10–2)	$\hat{\sigma} = \frac{\bar{s}}{c_2}$	estimate of universe standard deviation from the average of several sample standard deviations	255
(10–3)	$A_1 = \frac{3}{c_2\sqrt{n}}$	computing control limits from $\bar{s}$	256
(10–4)	$A_2 = \frac{3}{d_2\sqrt{n}}$	computing control limits from $\bar{R}$	258

Part Three—Analysis of Relationship

Ch. 11 Simple Linear Regression and Correlation

(11–1)	$y_c = a + bx$	straight-line trend	280
(11–2)	$a = \frac{\Sigma x^2\, \Sigma y - \Sigma x \cdot \Sigma xy}{n \cdot \Sigma x^2 - (\Sigma x)^2}$	constants of linear regression equation	281
(11–3)	$b = \frac{n \cdot \Sigma xy - \Sigma x \cdot \Sigma y}{n \cdot \Sigma x^2 - (\Sigma x)^2}$		
(11–4)	$\hat{\sigma}_{yx} = \sqrt{\frac{\Sigma(y - y_c)^2}{n-2}}$	standard error of estimate	287

A

Formula Number	Formula	Description	Page
(11–5)	$\hat{\sigma}_{yx} = \sqrt{\dfrac{\Sigma y^2 - a\Sigma y - b\Sigma xy}{n-2}}$	standard error of estimate	289
(11–6)	$\hat{\kappa}^2 = \dfrac{\hat{\sigma}^2_{yx}}{\hat{\sigma}^2_y}$	coefficient of nondetermination, estimated from a sample	293
(11–7)	$\hat{\rho}^2 = 1 - \hat{\kappa}^2$	coefficient of determination, estimated from a sample	293
(11–8)	$\hat{\rho}^2 = 1 - \dfrac{\hat{\sigma}^2_{yx}}{\hat{\sigma}^2_y}$	coefficient of determination, estimated from a sample	293
(11–9)	$r = \dfrac{n \cdot \Sigma xy - \Sigma x \cdot \Sigma y}{\sqrt{[n \cdot \Sigma x^2 - (\Sigma x)^2][n \cdot \Sigma y^2 - (\Sigma y)^2]}}$	coefficient of correlation, computed from sample data	295
(11–10)	$\hat{\rho}^2 = 1 - (1 - r^2)\left(\dfrac{n-1}{n-2}\right)$	coefficient of correlation, computing $\hat{\rho}^2$ from r	296
(11–11)	$a = \bar{y} - b\bar{x}$	constant of the regression equation	296
(11–12)	$s_{yx} = s_y \sqrt{1 - r^2}$	standard error of estimate for a sample	297
(11–13)	$r_r = 1 - \dfrac{6\Sigma d^2}{n^3 - n}$	coefficient of rank correlation	298
(11–14)	$\sigma_r = \dfrac{1 - \rho^2}{\sqrt{n-1}}$	standard deviation of the distribution of the coefficient of correlation	300
(11–15)	$\hat{\sigma}_r = \dfrac{1 - r^2}{\sqrt{n-1}}$	standard error of the coefficient of correlation	300
(11–16)	$t = r\sqrt{\dfrac{n-2}{1-r^2}}$	t distribution for the coefficient of correlation	301

Formula Number			Page

Ch. 12 Nonlinear, Multiple, and Partial Correlation

(12–1)	$y_c = a + bx + cx^2$	second-degree parabola	313
(12–2)	$\left.\begin{array}{l} \text{I. } \Sigma y = na + b\Sigma x + c\Sigma x^2 \\ \text{II. } \Sigma xy = a\Sigma x + b\Sigma x^2 + c\Sigma x^3 \\ \text{III. } \Sigma x^2 y = a\Sigma x^2 + b\Sigma x^3 + c\Sigma x^4 \end{array}\right\}$	normal equation for computing constants of regression of second-degree parabola equation	313
(12–3)	$\hat{\sigma}_{yx} = \sqrt{\frac{\Sigma(y - y_c)^2}{n - 3}}$	standard error of estimate for nonlinear regression analysis	315
(12–4)	$I^2_{yx} = 1 - \frac{\hat{\sigma}^2_{yx}}{\hat{\sigma}^2_y}$	coefficient of determination	317
(12–5)	$x_{1_c} = a + b_{12.3}x_2 + b_{13.2}x_3$	regression equation with two indcpendent variables	322
(12–6)	$\left.\begin{array}{l} \text{I. } \Sigma x_1 = na + b_{12.3}\Sigma x_2 + b_{13.2}\Sigma x_3 \\ \text{II. } \Sigma x_1 x_2 = a\Sigma x_2 + b_{12.3}\Sigma x_2^2 + b_{13.2}\Sigma x_2 x_3 \\ \text{III. } \Sigma x_1 x_3 = a\Sigma x_3 + b_{12.3}\Sigma x_2 x_3 + b_{13.2}\Sigma x_3^2 \end{array}\right\}$	normal equations for computing constants of regression equation with two independent variables	322
(12–7)	$\hat{\sigma}_{1.23} = \sqrt{\frac{\Sigma(x_1 - x_{1_c})^2}{n - m}}$	standard error of estimate for a multiple regression analysis	325
(12–8)	$\hat{\sigma}_{1.23} = \sqrt{\frac{\Sigma x_1^2 - a\Sigma x_1 - b_{12.3}\Sigma x_1 x_2 - b_{13.2}\Sigma x_1 x_3}{n - m}}$	standard error of estimate for a multiple correlation	326

Formula Number			Page
(12–9)	$\hat{\rho}^2_{1.23} = 1 - \dfrac{\hat{\sigma}^2_{1.23}}{\hat{\sigma}^2_1}$	coefficient of determination for multiple correlation	327
(12–10)	$F = \dfrac{\dfrac{\hat{\rho}^2_A - \hat{\rho}^2_B}{k_A - k_B}}{\dfrac{1 - \hat{\rho}^2_A}{n - k_A}}$	F test for significance of an independent variable in multiple regression	328
(12–11)	$\hat{\rho}^2_{13.2} = 1 - \dfrac{\hat{\sigma}^2_{1.23}}{\hat{\sigma}^2_{12}}$	coefficient of partial correlation	330

Part Four—Analysis of Business Change

Ch. 14 Secular Trend

Formula Number			Page
(14–1)	$y = a + bx$	straight-line equation	373
(14–2)	$b = \dfrac{4(S_2 - S_1)}{N^2}$	slope of a straight line	377
(14–3)	$a = \dfrac{\Sigma y}{N}$	constants of a straight-line equation	380
(14–4)	$b = \dfrac{\Sigma xy}{\Sigma x^2}$		
(14–5)	$\Sigma x^2 = \dfrac{N(N^2 - 1)}{12}$	computing the value of Σx^2 for any odd-numbered value of N when a table is not available.	380
(14–6)	$\Sigma x^2 = \dfrac{N(N^2 - 1)}{3}$	computing the value of Σx^2 for any even-numbered value of N when a table is not available	381

Formula Number			Page
(14–7)	$a = \frac{\Sigma(\log y)}{N}$	y intercept of logarithmic straight line (origin at midrange)	386
(14–8)	$b = \frac{\Sigma(x \cdot \log y)}{\Sigma x^2}$	slope of logarithmic straight line	386
(14–9)	$y_c = a + bx + cx^2$	second-degree parabola	389
(14–10)	$c = \frac{N \cdot \Sigma x^2 y - \Sigma x^2 \Sigma y}{N \cdot \Sigma x^4 - (\Sigma x^2)^2}$	constants of second-degree parabola (origin at midrange)	389
(14–11)	$a = \frac{\Sigma y - c\Sigma x^2}{N}$		
(14–12)	$\Sigma x^4 = \frac{3N^5 - 10N^3 + 7N}{240}$	computing Σx^4 for least-squares trend-fitting	389

Ch. 17 Index Numbers

Formula Number			Page
(17–1)	$Q_i = \frac{\sum \left(\frac{q_i}{q_o} 100\right)}{n}$	quantity index, an unweighted arithmetic mean	466
(17–2)	$P_i = \frac{\sum \left(\frac{p_i}{p_o} 100\right)}{n}$	price index, an unweighted arithmetic mean	467
(17–3)	$\text{Log } Q_i = \frac{\sum \log \left(\frac{q_i}{q_o} 100\right)}{n}$	log of quantity index, an unweighted geometric mean	467

Formula Number			Page
(17–4)	$\text{Log } P_i = \dfrac{\sum \log\left(\dfrac{p_i}{p_o}100\right)}{n}$	log of price index, an unweighted geometric mean	468
(17–5)	$Q_i = \dfrac{\sum\left(p_o q_o \dfrac{q_i}{q_o}100\right)}{\Sigma(p_o q_o)}$	quantity index, arithmetic mean weighted with base-year values.	471
(17–6)	$P_i = \dfrac{\sum\left(p_o q_o \dfrac{p_i}{p_o}100\right)}{\Sigma(p_o q_o)}$	price index, arithmetic mean weighted with base-year values	472
(17–7)	$\text{Log } Q_i = \dfrac{\sum w\left(\log \dfrac{q_i}{q_o}100\right)}{\Sigma w}$	log of quantity index, weighted geometric mean	473
(17–8)	$\text{Log } P_i = \dfrac{\sum w\left(\log \dfrac{p_i}{p_o}100\right)}{\Sigma w}$	log of price index, weighted geometric mean	473
(17–9)	$Q_i = \dfrac{\Sigma(p_o q_i)}{\Sigma(p_o q_o)}100$	quantity index, method of aggregates	477
(17–10)	$P_i = \dfrac{\Sigma(p_i q_o)}{\Sigma(p_o q_o)}100$	price index, method of aggregates	478
(17–11)	$P_i = \dfrac{\sum\left(p_o q_a \dfrac{p_i}{p_o}\right)}{\Sigma(p_o q_a)}100$	price index, arithmetic mean of relatives weighted by values computed from q_a	479
(17–12)	$P_i = \dfrac{\Sigma(p_i q_a)}{\Sigma(p_o q_a)}100$	price index, aggregative equivalent of Formula 17–11	479
(17–13)	$Q_i = \dfrac{\sum\left(p_a q_o \dfrac{q_i}{q_o}\right)}{\Sigma(p_a q_o)}100$	quantity index, arithmetic mean of relatives weighted by values computed from p_a	480

Formula Number			Page
(17–14)	$Q_i = \frac{\Sigma(p_a q_i)}{\Sigma(p_a q_o)} 100$	quantity index, aggregative equivalent of Formula 17–13	480
(17–15)	$P_{i\ (link)} = \frac{\Sigma(p_i q_a)}{\Sigma(p_{i-1} q_a)} 100$	link price index	480
(17–16)	$P_i = P_{i-1}\left[\frac{\sum\left(p_{i-1}q_a \frac{p_i}{p_{i-1}}\right)}{\Sigma(p_{i-1}q_a)}\right] 100$		
	$= P_{i-1}\left[\frac{\Sigma p_i q_a}{\Sigma p_{i-1} q_a}\right] 100$	chain price index	481
(17–17)	$Q_i = \sum\left(\frac{q_i}{q_o}\frac{p_a q_o}{\Sigma p_a q_o}\right) 100$	index of industrial production	484

SQUARES, SQUARE ROOTS, AND RECIPROCALS

The following table gives the values of the squares, square roots, and reciprocals for all sequences of three digits. The values of the squares of the numbers are exact, but the square roots and reciprocals are rounded to six digits.

Finding the square of a number consists merely of finding the value of n in the first column and reading the value of n^2 in the second column. The square of 1.30 is 1.6900, and the square of 13.0 is 169.00. The reciprocal of the number is given in the fifth column. The reciprocal of 1.30 is .769231, and the reciprocal of 13.0 is .0769231. The position of the decimal in the number determines the location of the decimal in the square or reciprocal, but the digits in the square or reciprocal are the same for a given sequence of numbers in n, regardless of the location of the decimal.

The square root of a given sequence, however, is affected by the position of the decimal point. The square root of 1.44 is 1.2, but the square root of 14.4 is 3.79473. Thus, two columns are required for square roots: one for the value of the numbers from 1 to 10, and another for the numbers from 10 to 100, even though the sequence of digits is the same. The square roots of the numbers from 1 to 10 are given in the third column (headed $\sqrt{n}$) and the square roots of the numbers from 10 to 100 are given in the fourth column (headed $\sqrt{10n}$). The column headed $\sqrt{n}$ is used for numbers with an odd number of digits to the left of the decimal, and the column headed $\sqrt{10n}$ is used for numbers with an even number of digits to the left of the decimal.

B

SQUARES — SQUARE ROOTS — RECIPROCALS

n	n^2	$\sqrt{n}$	$\sqrt{10n}$	$1/n$
1.00	1.0000	1.00000	3.16228	1.000000
1.01	1.0201	1.00499	3.17805	.990099
1.02	1.0404	1.00995	3.19374	.980392
1.03	1.0609	1.01489	3.20936	.970874
1.04	1.0816	1.01980	3.22490	.961538
1.05	1.1025	1.02470	3.24037	.952381
1.06	1.1236	1.02956	3.25576	.943396
1.07	1.1449	1.03441	3.27109	.934579
1.08	1.1664	1.03923	3.28634	.925926
1.09	1.1881	1.04403	3.30151	.917431
1.10	1.2100	1.04881	3.31662	.909091
1.11	1.2321	1.05357	3.33167	.900901
1.12	1.2544	1.05830	3.34664	.892857
1.13	1.2769	1.06301	3.36155	.884956
1.14	1.2996	1.06771	3.37639	.877193
1.15	1.3225	1.07238	3.39116	.869565
1.16	1.3456	1.07703	3.40588	.862069
1.17	1.3689	1.08167	3.42053	.854701
1.18	1.3924	1.08628	3.43511	.847458
1.19	1.4161	1.09087	3.44964	.840336
1.20	1.4400	1.09545	3.46410	.833333
1.21	1.4641	1.10000	3.47851	.826446
1.22	1.4884	1.10454	3.49285	.819672
1.23	1.5129	1.10905	3.50714	.813008
1.24	1.5376	1.11355	3.52136	.806452
1.25	1.5625	1.11803	3.53553	.800000
1.26	1.5876	1.12250	3.54965	.793651
1.27	1.6129	1.12694	3.56371	.787402
1.28	1.6384	1.13137	3.57771	.781250
1.29	1.6641	1.13578	3.59166	.775194
1.30	1.6900	1.14018	3.60555	.769231
1.31	1.7161	1.14455	3.61939	.763359
1.32	1.7424	1.14891	3.63318	.757576
1.33	1.7689	1.15326	3.64692	.751880
1.34	1.7956	1.15758	3.66060	.746269
1.35	1.8225	1.16190	3.67423	.740741
1.36	1.8496	1.16619	3.68782	.735294
1.37	1.8769	1.17047	3.70135	.729927
1.38	1.9044	1.17473	3.71484	.724638
1.39	1.9321	1.17898	3.72827	.719424
1.40	1.9600	1.18322	3.74166	.714286
1.41	1.9881	1.18743	3.75500	.709220
1.42	2.0164	1.19164	3.76829	.704225
1.43	2.0449	1.19583	3.78153	.699301
1.44	2.0736	1.20000	3.79473	.694444
1.45	2.1025	1.20416	3.80789	.689655
1.46	2.1316	1.20830	3.82099	.684932
1.47	2.1609	1.21244	3.83406	.680272
1.48	2.1904	1.21655	3.84708	.675676
1.49	2.2201	1.22066	3.86005	.671141
1.50	2.2500	1.22474	3.87298	.666667
n	n^2	$\sqrt{n}$	$\sqrt{10n}$	$1/n$

n	n^2	$\sqrt{n}$	$\sqrt{10n}$	$1/n$
1.50	2.2500	1.22474	3.87298	.666667
1.51	2.2801	1.22882	3.88587	.662252
1.52	2.3104	1.23288	3.89872	.657895
1.53	2.3409	1.23693	3.91152	.653595
1.54	2.3716	1.24097	3.92428	.649351
1.55	2.4025	1.24499	3.93700	.645161
1.56	2.4336	1.24900	3.94968	.641026
1.57	2.4649	1.25300	3.96232	.636943
1.58	2.4964	1.25698	3.97492	.632911
1.59	2.5281	1.26095	3.98748	.628931
1.60	2.5600	1.26491	4.00000	.625000
1.61	2.5921	1.26886	4.01248	.621118
1.62	2.6244	1.27279	4.02492	.617284
1.63	2.6569	1.27671	4.03733	.613497
1.64	2.6896	1.28062	4.04969	.609756
1.65	2.7225	1.28452	4.06202	.606061
1.66	2.7556	1.28841	4.07431	.602410
1.67	2.7889	1.29228	4.08656	.598802
1.68	2.8224	1.29615	4.09878	.595238
1.69	2.8561	1.30000	4.11096	.591716
1.70	2.8900	1.30384	4.12311	.588235
1.71	2.9241	1.30767	4.13521	.584795
1.72	2.9584	1.31149	4.14729	.581395
1.73	2.9929	1.31529	4.15933	.578035
1.74	3.0276	1.31909	4.17133	.574713
1.75	3.0625	1.32288	4.18330	.571429
1.76	3.0976	1.32665	4.19524	.568182
1.77	3.1329	1.33041	4.20714	.564972
1.78	3.1684	1.33417	4.21900	.561798
1.79	3.2041	1.33791	4.23084	.558659
1.80	3.2400	1.34164	4.24264	.555556
1.81	3.2761	1.34536	4.25441	.552486
1.82	3.3124	1.34907	4.26615	.549451
1.83	3.3489	1.35277	4.27785	.546448
1.84	3.3856	1.35647	4.28952	.543478
1.85	3.4225	1.36015	4.30116	.540541
1.86	3.4596	1.36382	4.31277	.537634
1.87	3.4969	1.36748	4.32435	.534759
1.88	3.5344	1.37113	4.33590	.531915
1.89	3.5721	1.37477	4.34741	.529101
1.90	3.6100	1.37840	4.35890	.526316
1.91	3.6481	1.38203	4.37035	.523560
1.92	3.6864	1.38564	4.38178	.520833
1.93	3.7249	1.38924	4.39318	.518135
1.94	3.7636	1.39284	4.40454	.515464
1.95	3.8025	1.39642	4.41588	.512821
1.96	3.8416	1.40000	4.42719	.510204
1.97	3.8809	1.40357	4.43847	.507614
1.98	3.9204	1.40712	4.44972	.505051
1.99	3.9601	1.41067	4.46094	.502513
2.00	4.0000	1.41421	4.47214	.500000
n	n^2	$\sqrt{n}$	$\sqrt{10n}$	$1/n$

SQUARES — SQUARE ROOTS — RECIPROCALS

n	n^2	$\sqrt{n}$	$\sqrt{10n}$	$1/n$
2.00	4.0000	1.41421	4.47214	.500000
2.01	4.0401	1.41774	4.48330	.497512
2.02	4.0804	1.42127	4.49444	.495050
2.03	4.1209	1.42478	4.50555	.492611
2.04	4.1616	1.42829	4.51664	.490196
2.05	4.2025	1.43178	4.52769	.487805
2.06	4.2436	1.43527	4.53872	.485437
2.07	4.2849	1.43875	4.54973	.483092
2.08	4.3264	1.44222	4.56070	.480769
2.09	4.3681	1.44568	4.57165	.478469
2.10	4.4100	1.44914	4.58258	.476190
2.11	4.4521	1.45258	4.59347	.473934
2.12	4.4944	1.45602	4.60435	.471698
2.13	4.5369	1.45945	4.61519	.469484
2.14	4.5796	1.46287	4.62601	.467290
2.15	4.6225	1.46629	4.63681	.465116
2.16	4.6656	1.46969	4.64758	.462963
2.17	4.7089	1.47309	4.65833	.460829
2.18	4.7524	1.47648	4.66905	.458716
2.19	4.7961	1,47986	4.67974	.456621
2.20	4.8400	1.48324	4.69042	.454545
2.21	4.8841	1.48661	4.70106	.452489
2.22	4.9284	1.48997	4.71169	.450450
2.23	4.9729	1.49332	4.72229	.448430
2.24	5.0176	1.49666	4.73286	.446429
2.25	5.0625	1.50000	4.74342	.444444
2.26	5.1076	1.50333	4.75395	.442478
2.27	5.1529	1.50665	4.76445	.440529
2.28	5.1984	1.50997	4.77493	.438596
2.29	5.2441	1.51327	4.78539	.436681
2.30	5.2900	1.51658	4.79583	.434783
2.31	5.3361	1.51987	4.80625	.432900
2.32	5.3824	1.52315	4.81664	.431034
2.33	5.4289	1.52643	4.82701	.429185
2.34	5.4756	1.52971	4.83735	.427350
2.35	5.5225	1.53297	4.84768	.425532
2.36	5.5696	1.53623	4.85798	.423729
2.37	5.6169	1.53948	4.86826	.421941
2.38	5.6644	1.54272	4.87852	.420168
2.39	5.7121	1.54596	4.88876	.418410
2.40	5.7600	1.54919	4.89898	.416667
2.41	5.8081	1.55242	4.90918	.414938
2.42	5.8564	1.55563	4.91935	.413223
2.43	5.9049	1.55885	4.92950	.411523
2.44	5.9536	1.56205	4.93964	.409836
2.45	6.0025	1.56525	4.94975	.408163
2.46	6.0516	1.56844	4.95984	.406504
2.47	6.1009	1.57162	4.96991	.404858
2.48	6.1504	1.57480	4.97996	.403226
2.49	6.2001	1.57797	4.98999	.401606
2.50	6.2500	1.58114	5.00000	.400000
n	n^2	$\sqrt{n}$	$\sqrt{10n}$	$1/n$

n	n^2	$\sqrt{n}$	$\sqrt{10n}$	$1/n$
2.50	6.2500	1.58114	5.00000	.400000
2.51	6.3001	1.58430	5.00999	.398406
2.52	6.3504	1.58745	5.01996	.396825
2.53	6.4009	1.59060	5.02991	.395257
2.54	6.4516	1.59374	5.03984	.393701
2.55	6.5025	1.59687	5.04975	.392157
2.56	6.5536	1.60000	5.05964	.390625
2.57	6.6049	1.60312	5.06952	.389105
2.58	6.6564	1.60624	5.07937	.387597
2.59	6.7081	1.60935	5.08920	.386100
2.60	6.7600	1.61245	5.09902	.384615
2.61	6.8121	1.61555	5.10882	.383142
2.62	6.8644	1.61864	5.11859	.381679
2.63	6.9169	1.62173	5.12835	.380228
2.64	6.9696	1.62481	5.13809	.378788
2.65	7.0225	1.62788	5.14782	.377358
2.66	7.0756	1.63095	5.15752	.375940
2.67	7.1289	1.63401	5.16720	.374532
2.68	7.1824	1.63707	5.17687	.373134
2.69	7.2361	1.64012	5.18652	.371747
2.70	7.2900	1.64317	5.19615	.370370
2.71	7.3441	1.64621	5.20577	.369004
2.72	7.3984	1.64924	5.21536	.367647
2.73	7.4529	1.65227	5.22494	.366300
2.74	7.5076	1.65529	5.23450	.364964
2.75	7.5625	1.65831	5.24404	.363636
2.76	7.6176	1.66132	5.25357	.362319
2.77	7.6729	1.66433	5.26308	.361011
2.78	7.7284	1.66733	5.27257	.359712
2.79	7.7841	1.67033	5.28205	.358423
2.80	7.8400	1.67332	5.29150	.357143
2.81	7.8961	1.67631	5.30094	.355872
2.82	7.9524	1.67929	5.31037	.354610
2.83	8.0089	1.68226	5.31977	.353357
2.84	8.0656	1.68523	5.32917	.352113
2.85	8.1225	1.68819	5.33854	.350877
2.86	8.1796	1.69115	5.34790	.349650
2.87	8.2369	1.69411	5.35724	.348432
2.88	8.2944	1.69706	5.36656	.347222
2.89	8.3521	1.70000	5.37587	.346021
2.90	8.4100	1.70294	5.38516	.344828
2.91	8.4681	1.70587	5.39444	.343643
2.92	8.5264	1.70880	5.40370	.342466
2.93	8.5849	1.71172	5.41295	.341297
2.94	8.6436	1.71464	5.42218	.340136
2.95	8.7025	1.71756	5.43139	.338983
2.96	8.7616	1.72047	5.44059	.337838
2.97	8.8209	1.72337	5.44977	.336700
2.98	8.8804	1.72627	5.45894	.335570
2.99	8.9401	1.72916	5.46809	.334448
3.00	9.0000	1.73205	5.47723	.333333
n	n^2	$\sqrt{n}$	$\sqrt{10n}$	$1/n$

SQUARES — SQUARE ROOTS — RECIPROCALS

n	n^2	$\sqrt{n}$	$\sqrt{10n}$	$1/n$
3.00	9.0000	1.73205	5.47723	.333333
3.01	9.0601	1.73494	5.48635	.332226
3.02	9.1204	1.73781	5.49545	.331126
3.03	9.1809	1.74069	5.50454	.330033
3.04	9.2416	1.74356	5.51362	.328947
3.05	9.3025	1.74642	5.52268	.327869
3.06	9.3636	1.74929	5.53173	.326797
3.07	9.4249	1.75214	5.54076	.325733
3.08	9.4864	1.75499	5.54977	.324675
3.09	9.5481	1.75784	5.55878	.323625
3.10	9.6100	1.76068	5.56776	.322581
3.11	9.6721	1.76352	5.57674	.321543
3.12	9.7344	1.76635	5.58570	.320513
3.13	9.7969	1.76918	5.59464	.319489
3.14	9.8596	1.77200	5.60357	.318471
3.15	9.9225	1.77482	5.61249	.317460
3.16	9.9856	1.77764	5.62139	.316456
3.17	10.0489	1.78045	5.63028	.315457
3.18	10.1124	1.78326	5.63915	.314465
3.19	10.1761	1.78606	5.64801	.313480
3.20	10.2400	1.78885	5.65685	.312500
3.21	10.3041	1.79165	5.66569	.311526
3.22	10.3684	1.79444	5.67450	.310559
3.23	10.4329	1.79722	5.68331	.309598
3.24	10.4976	1.80000	5.69210	.308642
3.25	10.5625	1.80278	5.70088	.307692
3.26	10.6276	1.80555	5.70964	.306748
3.27	10.6929	1.80831	5.71839	.305810
3.28	10.7584	1.81108	5.72713	.304878
3.29	10.8241	1.81384	5.73585	.303951
3.30	10.8900	1.81659	5.74456	.303030
3.31	10.9561	1.81934	5.75326	.302115
3.32	11.0224	1.82209	5.76194	.301205
3.33	11.0889	1.82483	5.77062	.300300
3.34	11.1556	1.82757	5.77927	.299401
3.35	11.2225	1.83030	5.78792	.298507
3.36	11.2896	1.83303	5.79655	.297619
3.37	11.3569	1.83576	5.80517	.296736
3.38	11.4244	1.83848	5.81378	.295858
3.39	11.4921	1.84120	5.82237	.294985
3.40	11.5600	1.84391	5.83095	.294118
3.41	11.6281	1.84662	5.83952	.293255
3.42	11.6964	1.84932	5.84808	.292398
3.43	11.7649	1.85203	5.85662	.291545
3.44	11.8336	1.85472	5.86515	.290698
3.45	11.9025	1.85742	5.87367	.289855
3.46	11.9716	1.86011	5.88218	.289017
3.47	12.0409	1.86279	5.89067	.288184
3.48	12.1104	1.86548	5.89915	.287356
3.49	12.1801	1.86815	5.90762	.286533
3.50	12.2500	1.87083	5.91608	.285714
n	n^2	$\sqrt{n}$	$\sqrt{10n}$	$1/n$

n	n^2	$\sqrt{n}$	$\sqrt{10n}$	$1/n$
3.50	12.2500	1.87083	5.91608	.285714
3.51	12.3201	1.87350	5.92453	.284900
3.52	12.3904	1.87617	5.93296	.284091
3.53	12.4609	1.87883	5.94138	.283286
3.54	12.5316	1.88149	5.94979	.282486
3.55	12.6025	1.88414	5.95819	.281690
3.56	12.6736	1.88680	5.96657	.280899
3.57	12.7449	1.88944	5.97495	.280112
3.58	12.8164	1.89209	5.98331	.279330
3.59	12.8881	1.89473	5.99166	.278552
3.60	12.9600	1.89737	6.00000	.277778
3.61	13.0321	1.90000	6.00833	.277008
3.62	13.1044	1.90263	6.01664	.276243
3.63	13.1769	1.90526	6.02495	.275482
3.64	13.2496	1.90788	6.03324	.274725
3.65	13.3225	1.91050	6.04152	.273973
3.66	13.3956	1.91311	6.04979	.273224
3.67	13.4689	1.91572	6.05805	.272480
3.68	13.5424	1.91833	6.06630	.271739
3.69	13.6161	1.92094	6.07454	.271003
3.70	13.6900	1.92354	6.08276	.270270
3.71	13.7641	1.92614	6.09098	.269542
3.72	13.8384	1.92873	6.09918	.268817
3.73	13.9129	1.93132	6.10737	.268097
3.74	13.9876	1.93391	6.11555	.267380
3.75	14.0625	1.93649	6.12372	.266667
3.76	14.1376	1.93907	6.13188	.265957
3.77	14.2129	1.94165	6.14003	.265252
3.78	14.2884	1.94422	6.14817	.264550
3.79	14.3641	1.94679	6.15630	.263852
3.80	14.4400	1.94936	6.16441	.263158
3.81	14.5161	1.95192	6.17252	.262467
3.82	14.5924	1.95448	6.18061	.261780
3.83	14.6689	1.95704	6.18870	.261097
3.84	14.7456	1.95959	6.19677	.260417
3.85	14.8225	1.96214	6.20484	.259740
3.86	14.8996	1.96469	6.21289	.259067
3.87	14.9769	1.96723	6.22093	.258398
3.88	15.0544	1.96977	6.22896	.257732
3.89	15.1321	1.97231	6.23699	.257069
3.90	15.2100	1.97484	6.24500	.256410
3.91	15.2881	1.97737	6.25300	.255754
3.92	15.3664	1.97990	6.26099	.255102
3.93	15.4449	1.98242	6.26897	.254453
3.94	15.5236	1.98494	6.27694	.253807
3.95	15.6025	1.98746	6.28490	.253165
3.96	15.6816	1.98997	6.29285	.252525
3.97	15.7609	1.99249	6.30079	.251889
3.98	15.8408	1.99499	6.30872	.251256
3.99	15.9201	1.99750	6.31664	.250627
4.00	16.0000	2.00000	6.32456	.250000
n	n^2	$\sqrt{n}$	$\sqrt{10n}$	$1/n$

SQUARES — SQUARE ROOTS — RECIPROCALS

n	n^2	$\sqrt{n}$	$\sqrt{10n}$	$1/n$
4.00	16.0000	2.00000	6.32456	.250000
4.01	16.0801	2.00250	6.33246	.249377
4.02	16.1604	2.00499	6.34035	.248756
4.03	16.2409	2.00749	6.34823	.248139
4.04	16.3216	2.00998	6.35610	.247525
4.05	16.4025	2.01246	6.36396	.246914
4.06	16.4836	2.01494	6.37181	.246305
4.07	16.5649	2.01742	6.37966	.245700
4.08	16.6464	2.01990	6.38749	.245098
4.09	16.7281	2.02237	6.39531	.244499
4.10	16.8100	2.02485	6.40312	.243902
4.11	16.8921	2.02731	6.41093	.243309
4.12	16.9744	2.02978	6.41872	.242718
4.13	17.0569	2.03224	6.42651	.242131
4.14	17.1396	2.03470	6.43428	.241546
4.15	17.2225	2.03715	6.44205	.240964
4.16	17.3056	2.03961	6.44981	.240385
4.17	17.3889	2.04206	6.45755	.239808
4.18	17.4724	2.04450	6.46529	.239234
4.19	17.5561	2.04695	6.47302	.238663
4.20	17.6400	2.04939	6.48074	.238095
4.21	17.7241	2.05183	6.48845	.237530
4.22	17.8084	2.05426	6.49615	.236967
4.23	17.8929	2.05670	6.50384	.236407
4.24	17.9776	2.05913	6.51153	.235849
4.25	18.0625	2.06155	6.51920	.235294
4.26	18.1476	2.06398	6.52687	.234742
4.27	18.2329	2.06640	6.53452	.234192
4.28	18.3184	2.06882	6.54217	.233645
4.29	18.4041	2.07123	6.54981	.233100
4.30	18.4900	2.07364	6.55744	.232558
4.31	18.5761	2.07605	6.56506	.232019
4.32	18.6624	2.07846	6.57267	.231481
4.33	18.7489	2.08087	6.58027	.230947
4.34	18.8356	2.08327	6.58787	.230415
4.35	18.9225	2.08567	6.59545	.229885
4.36	19.0096	2.08806	6.60303	.229358
4.37	19.0969	2.09045	6.61060	.228833
4.38	19.1844	2.09284	6.61816	.228311
4.39	19.2721	2.09523	6.62571	.227790
4.40	19.3600	2.09762	6.63325	.227273
4.41	19.4481	2.10000	6.64078	.226757
4.42	19.5364	2.10238	6.64831	.226244
4.43	19.6249	2.10476	6.65582	.225734
4.44	19.7136	2.10713	6.66333	.225225
4.45	19.8025	2.10950	6.67083	.224719
4.46	19.8916	2.11187	6.67832	.224215
4.47	19.9809	2.11424	6.68581	.223714
4.48	20.0704	2.11660	6.69328	.223214
4.49	20.1601	2.11896	6.70075	.222717
4.50	20.2500	2.12132	6.70820	.222222
n	n^2	$\sqrt{n}$	$\sqrt{10n}$	$1/n$

n	n^2	$\sqrt{n}$	$\sqrt{10n}$	$1/n$
4.50	20.2500	2.12132	6.70820	.222222
4.51	20.3401	2.12368	6.71565	.221729
4.52	20.4304	2.12603	6.72309	.221239
4.53	20.5209	2.12838	6.73053	.220751
4.54	20.6116	2.13073	6.73795	.220264
4.55	20.7025	2.13307	6.74537	.219780
4.56	20.7936	2.13542	6.75278	.219298
4.57	20.8849	2.13776	6.76018	.218818
4.58	20.9764	2.14009	6.76757	.218341
4.59	21.0681	2.14243	6.77495	.217865
4.60	21.1600	2.14476	6.78233	.217391
4.61	21.2521	2.14709	6.78970	.216920
4.62	21.3444	2.14942	6.79706	.216450
4.63	21.4369	2.15174	6.80441	.215983
4.64	21.5296	2.15407	6.81175	.215517
4.65	21.6225	2.15639	6.81909	.215054
4.66	21.7156	2.15870	6.82642	.214592
4.67	21.8089	2.16102	6.83374	.214133
4.68	21.9024	2.16333	6.84105	.213675
4.69	21.9961	2.16564	6.84836	.213220
4.70	22.0900	2.16795	6.85565	.212766
4.71	22.1841	2.17025	6.86294	.212314
4.72	22.2784	2.17256	6.87023	.211864
4.73	22.3729	2.17486	6.87750	.211416
4.74	22.4676	2.17715	6.88477	.210970
4.75	22.5625	2.17945	6.89202	.210526
4.76	22.6576	2.18174	6.89928	.210084
4.77	22.7529	2.18403	6.90652	.209644
4.78	22.8484	2.18632	6.91375	.209205
4.79	22.9441	2.18861	6.92098	.208768
4.80	23.0400	2.19089	6.92820	.208333
4.81	23.1361	2.19317	6.93542	.207900
4.82	23.2324	2.19545	6.94262	.207469
4.83	23.3289	2.19773	6.94982	.207039
4.84	23.4256	2.20000	6.95701	.206612
4.85	23.5225	2.20227	6.96419	.206186
4.86	23.6196	2.20454	6.97137	.205761
4.87	23.7169	2.20681	6.97854	.205339
4.88	23.8144	2.20907	6.98570	.204918
4.89	23.9121	2.21133	6.99285	.204499
4.90	24.0100	2.21359	7.00000	.204082
4.91	24.1081	2.21585	7.00714	.203666
4.92	24.2064	2.21811	7.01427	.203252
4.93	24.3049	2.22036	7.02140	.202840
4.94	24.4036	2.22261	7.02851	.202429
4.95	24.5025	2.22486	7.03562	.202020
4.96	24.6016	2.22711	7.04273	.201613
4.97	24.7009	2.22935	7.04982	.201207
4.98	24.8004	2.23159	7.05691	.200803
4.99	24.9001	2.23383	7.06399	.200401
5.00	25.0000	2.23607	7.07107	.200000
n	n^2	$\sqrt{n}$	$\sqrt{10n}$	$1/n$

SQUARES — SQUARE ROOTS — RECIPROCALS

B

n	n^2	$\sqrt{n}$	$\sqrt{10n}$	$1/n$
5.00	25.0000	2.23607	7.07107	.200000
5.01	25.1001	2.23830	7.07814	.199601
5.02	25.2004	2.24054	7.08520	.199203
5.03	25.3009	2.24277	7.09225	.198807
5.04	25.4016	2.24499	7.09930	.198413
5.05	25.5025	2.24722	7.10634	.198020
5.06	25.6036	2.24944	7.11337	.197628
5.07	25.7049	2.25167	7.12039	.197239
5.08	25.8064	2.25389	7.12741	.196850
5.09	25.9081	2.25610	7.13442	.196464
5.10	26.0100	2.25832	7.14143	.196078
5.11	26.1121	2.26053	7.14843	.195695
5.12	26.2144	2.26274	7.15542	.195312
5.13	26.3169	2.26495	7.16240	.194932
5.14	26.4196	2.26716	7.16938	.194553
5.15	26.5225	2.26936	7.17635	.194175
5.16	26.6256	2.27156	7.18331	.193798
5.17	26.7289	2.27376	7.19027	.193424
5.18	26.8324	2.27596	7.19722	.193050
5.19	26.9361	2.27816	7.20417	.192678
5.20	27.0400	2.28035	7.21110	.192308
5.21	27.1441	2.28254	7.21803	.191939
5.22	27.2484	2.28473	7.22496	.191571
5.23	27.3529	2.28692	7.23187	.191205
5.24	27.4576	2.28910	7.23878	.190840
5.25	27.5625	2.29129	7.24569	.190476
5.26	27.6676	2.29347	7.25259	.190114
5.27	27.7729	2.29565	7.25948	.189753
5.28	27.8784	2.29783	7.26636	.189394
5.29	27.9841	2.30000	7.27324	.189036
5.30	28.0900	2.30217	7.28011	.188679
5.31	28.1961	2.30434	7.28697	.188324
5.32	28.3024	2.30651	7.29383	.187970
5.33	28.4089	2.30868	7.30068	.187617
5.34	28.5156	2.31084	7.30753	.187266
5.35	28.6225	2.31301	7.31437	.186916
5.36	28.7296	2.31517	7.32120	.186567
5.37	28.8369	2.31733	7.32803	.186220
5.38	28.9444	2.31948	7.33485	.185874
5.39	29.0521	2.32164	7.34166	.185529
5.40	29.1600	2.32379	7.34847	.185185
5.41	29.2681	2.32594	7.35527	.184843
5.42	29.3764	2.32809	7.36206	.184502
5.43	29.4849	2.33024	7.36885	.184162
5.44	29.5936	2.33238	7.37564	.183824
5.45	29.7025	2.33452	7.38241	.183486
5.46	29.8116	2.33666	7.38918	.183150
5.47	29.9209	2.33880	7.39594	.182815
5.48	30.0304	2.34094	7.40270	.182482
5.49	30.1401	2.34307	7.40945	.182149
5.50	30.2500	2.34521	7.41620	.181818
n	n^2	$\sqrt{n}$	$\sqrt{10n}$	$1/n$

n	n^2	$\sqrt{n}$	$\sqrt{10n}$	$1/n$
5.50	30.2500	2.34521	7.41620	.181818
5.51	30.3601	2.34734	7.42294	.181488
5.52	30.4704	2.34947	7.42967	.181159
5.53	30.5809	2.35160	7.43640	.180832
5.54	30.6916	2.35372	7.44312	.180505
5.55	30.8025	2.35584	7.44983	.180180
5.56	30.9136	2.35797	7.45654	.179856
5.57	31.0249	2.36008	7.46324	.179533
5.58	31.1364	2.36220	7.46994	.179211
5.59	31.2481	2.36432	7.47663	.178891
5.60	31.3600	2.36643	7.48331	.178571
5.61	31.4721	2.36854	7.48999	.178253
5.62	31.5844	2.37065	7.49667	.177936
5.63	31.6969	2.37276	7.50333	.177620
5.64	31.8096	2.37487	7.50999	.177305
5.65	31.9225	2.37697	7.51665	.176991
5.66	32.0356	2.37908	7.52330	.176678
5.67	32.1489	2.38118	7.52994	.176367
5.68	32.2624	2.38328	7.53658	.176056
5.69	32.3761	2.38537	7.54321	.175747
5.70	32.4900	2.38747	7.54983	.175439
5.71	32.6041	2.38956	7.55645	.175131
5.72	32.7184	2.39165	7.56307	.174825
5.73	32.8329	2.39374	7.56968	.174520
5.74	32.9476	2.39583	7.57628	.174216
5.75	33.0625	2.39792	7.58288	.173913
5.76	33.1776	2.40000	7.58947	.173611
5.77	33.2929	2.40208	7.59605	.173310
5.78	33.4084	2.40416	7.60263	.173010
5.79	33.5241	2.40624	7.60920	.172712
5.80	33.6400	2.40832	7.61577	.172414
5.81	33.7561	2.41039	7.62234	.172117
5.82	33.8724	2.41247	7.62889	.171821
5.83	33.9889	2.41454	7.63544	.171527
5.84	34.1056	2.41661	7.64199	.171233
5.85	34.2225	2.41868	7.64853	.170940
5.86	34.3396	2.42074	7.65506	.170649
5.87	34.4569	2.42281	7.66159	.170358
5.88	34.5744	2.42487	7.66812	.170068
5.89	34.6921	2.42693	7.67463	.169779
5.90	34.8100	2.42899	7.68115	.169492
5.91	34.9281	2.43105	7.68765	.169205
5.92	35.0464	2.43311	7.69415	.168919
5.93	35.1649	2.43516	7.70065	.168634
5.94	35.2836	2.43721	7.70714	.168350
5.95	35.4025	2.43926	7.71362	.168067
5.96	35.5216	2.44131	7.72010	.167785
5.97	35.6409	2.44336	7.72658	.167504
5.98	35.7604	2.44540	7.73305	.167224
5.99	35.8801	2.44745	7.73951	.166945
6.00	36.0000	2.44949	7.74597	.166667
n	n^2	$\sqrt{n}$	$\sqrt{10n}$	$1/n$

SQUARES — SQUARE ROOTS — RECIPROCALS

n	n^2	$\sqrt{n}$	$\sqrt{10n}$	$1/n$
6.00	36.0000	2.44949	7.74597	.166667
6.01	36.1201	2.45153	7.75242	.166389
6.02	36.2404	2.45357	7.75887	.166113
6.03	36.3609	2.45561	7.76531	.165837
6.04	36.4816	2.45764	7.77174	.165563
6.05	36.6025	2.45967	7.77817	.165289
6.06	36.7236	2.46171	7.78460	.165017
6.07	36.8449	2.46374	7.79102	.164745
6.08	36.9664	2.46577	7.79744	.164474
6.09	37.0881	2.46779	7.80385	.164204
6.10	37.2100	2.46982	7.81025	.163934
6.11	37.3321	2.47184	7.81665	.163666
6.12	37.4544	2.47386	7.82304	.163399
6.13	37.5769	2.47588	7.82943	.163132
6.14	37.6996	2.47790	7.83582	.162866
6.15	37.8225	2.47992	7.84219	.162602
6.16	37.9456	2.48193	7.84857	.162338
6.17	38.0689	2.48395	7.85493	.162075
6.18	38.1924	2.48596	7.86130	.161812
6.19	38.3161	2.48797	7.86766	.161551
6.20	38.4400	2.48998	7.87401	.161290
6.21	38.5641	2.49199	7.88036	.161031
6.22	38.6884	2.49399	7.88670	.160772
6.23	38.8129	2.49600	7.89303	.160514
6.24	38.9376	2.49800	7.89937	.160256
6.25	39.0625	2.50000	7.90569	.160000
6.26	39.1876	2.50200	7.91202	.159744
6.27	39.3129	2.50400	7.91833	.159490
6.28	39.4384	2.50599	7.92465	.159236
6.29	39.5641	2.50799	7.93095	.158983
6.30	39.6900	2.50998	7.93725	.158730
6.31	39.8161	2.51197	7.94355	.158479
6.32	39.9424	2.51396	7.94984	.158228
6.33	40.0689	2.51595	7.95613	.157978
6.34	40.1956	2.51794	7.96241	.157729
6.35	40.3225	2.51992	7.96869	.157480
6.36	40.4496	2.52190	7.97496	.157233
6.37	40.5769	2.52389	7.98123	.156986
6.38	40.7044	2.52587	7.98749	.156740
6.39	40.8321	2.52784	7.99375	.156495
6.40	40.9600	2.52982	8.00000	.156250
6.41	41.0881	2.53180	8.00625	.156006
6.42	41.2164	2.53377	8.01249	.155763
6.43	41.3449	2.53574	8.01873	.155521
6.44	41.4736	2.53772	8.02496	.155280
6.45	41.6025	2.53969	8.03119	.155039
6.46	41.7316	2.54165	8.03741	.154799
6.47	41.8609	2.54362	8.04363	.154560
6.48	41.9904	2.54558	8.04984	.154321
6.49	42.1201	2.54755	8.05605	.154083
6.50	42.2500	2.54951	8.06226	.153846
n	n^2	$\sqrt{n}$	$\sqrt{10n}$	$1/n$

n	n^2	$\sqrt{n}$	$\sqrt{10n}$	$1/n$
6.50	42.2500	2.54951	8.06226	.153846
6.51	42.3801	2.55147	8.06846	.153610
6.52	42.5104	2.55343	8.07465	.153374
6.53	42.6409	2.55539	8.08084	.153139
6.54	42.7716	2.55734	8.08703	.152905
6.55	42.9025	2.55930	8.09321	.152672
6.56	43.0336	2.56125	8.09938	.152439
6.57	43.1649	2.56320	8.10555	.152207
6.58	43.2964	2.56515	8.11172	.151976
6.59	43.4281	2.56710	8.11788	.151745
6.60	43.5600	2.56905	8.12404	.151515
6.61	43.6921	2.57099	8.13019	.151286
6.62	43.8244	2.57294	8.13634	.151057
6.63	43.9569	2.57488	8.14248	.150830
6.64	44.0896	2.57682	8.14862	.150602
6.65	44.2225	2.57876	8.15475	.150376
6.66	44.3556	2.58070	8.16088	.150150
6.67	44.4889	2.58263	8.16701	.149925
6.68	44.6224	2.58457	8.17313	.149701
6.69	44.7561	2.58650	8.17924	.149477
6.70	44.8900	2.58844	8.18535	.149254
6.71	45.0241	2.59037	8.19146	.149031
6.72	45.1584	2.59230	8.19756	.148810
6.73	45.2929	2.59422	8.20366	.148588
6.74	45.4276	2.59615	8.20975	.148368
6.75	45.5625	2.59808	8.21584	.148148
6.76	45.6976	2.60000	8.22192	.147929
6.77	45.8329	2.60192	8.22800	.147710
6.78	45.9684	2.60384	8.23408	.147493
6.79	46.1041	2.60576	8.24015	.147275
6.80	46.2400	2.60768	8.24621	.147059
6.81	46.3761	2.60960	8.25227	.146843
6.82	46.5124	2.61151	8.25833	.146628
6.83	46.6489	2.61343	8.26438	.146413
6.84	46.7856	2.61534	8.27043	.146199
6.85	46.9225	2.61725	8.27647	.145985
6.86	47.0596	2.61916	8.28251	.145773
6.87	47.1969	2.62107	8.28855	.145560
6.88	47.3344	2.62298	8.29458	.145349
6.89	47.4721	2.62488	8.30060	.145138
6.90	47.6100	2.62679	8.30662	.144928
6.91	47.7481	2.62869	8.31264	.144718
6.92	47.8864	2.63059	8.31865	.144509
6.93	48.0249	2.63249	8.32466	.144300
6.94	48.1636	2.63439	8.33067	.144092
6.95	48.3025	2.63629	8.33667	.143885
6.96	48.4416	2.63818	8.34266	.143678
6.97	48.5809	2.64008	8.34865	.143472
6.98	48.7204	2.64197	8.35464	.143266
6.99	48.8601	2.64386	8.36062	.143062
7.00	49.0000	2.64575	8.36660	.142857
n	n^2	$\sqrt{n}$	$\sqrt{10n}$	$1/n$

SQUARES — SQUARE ROOTS — RECIPROCALS

B

n	n^2	$\sqrt{n}$	$\sqrt{10n}$	$1/n$
7.00	49.0000	2.64575	8.36660	.142857
7.01	49.1401	2.64764	8.37257	.142653
7.02	49.2804	2.64953	8.37854	.142450
7.03	49.4209	2.65141	8.38451	.142248
7.04	49.5616	2.65330	8.39047	.142045
7.05	49.7025	2.65518	8.39643	.141844
7.06	49.8436	2.65707	8.40238	.141643
7.07	49.9849	2.65895	8.40833	.141443
7.08	50.1264	2.66083	8.41427	.141243
7.09	50.2681	2.66271	8.42021	.141044
7.10	50.4100	2.66458	8.42615	.140845
7.11	50.5521	2.66646	8.43208	.140647
7.12	50.6944	2.66833	8.43801	.140449
7.13	50.8369	2.67021	8.44393	.140252
7.14	50.9796	2.67208	8.44985	.140056
7.15	51.1225	2.67395	8.45577	.139860
7.16	51.2656	2.67582	8.46168	.139665
7.17	51.4089	2.67769	8.46759	.139470
7.18	51.5524	2.67955	8.47349	.139276
7.19	51.6961	2.68142	8.47939	.139082
7.20	51.8400	2.68328	8.48528	.138889
7.21	51.9841	2.68514	8.49117	.138696
7.22	52.1284	2.68701	8.49706	.138504
7.23	52.2729	2.68887	8.50294	.138313
7.24	52.4176	2.69072	8.50882	.138122
7.25	52.5625	2.69258	8.51469	.137931
7.26	52.7076	2.69444	8.52056	.137741
7.27	52.8529	2.69629	8.52643	.137552
7.28	52.9984	2.69815	8.53229	.137363
7.29	53.1441	2.70000	8.53815	.137174
7.30	53.2900	2.70185	8.54400	.136986
7.31	53.4361	2.70370	8.54985	.136799
7.32	53.5824	2.70555	8.55570	.136612
7.33	53.7289	2.70740	8.56154	.136426
7.34	53.8756	2.70924	8.56738	.136240
7.35	54.0225	2.71109	8.57321	.136054
7.36	54.1696	2.71293	8.57904	.135870
7.37	54.3169	2.71477	8.58487	.135685
7.38	54.4644	2.71662	8.59069	.135501
7.39	54.6121	2.71846	8.59651	.135318
7.40	54.7600	2.72029	8.60233	.135135
7.41	54.9081	2.72213	8.60814	.134953
7.42	55.0564	2.72397	8.61394	.134771
7.43	55.2049	2.72580	8.61974	.134590
7.44	55.3536	2.72764	8.62554	.134409
7.45	55.5025	2.72947	8.63134	.134228
7.46	55.6516	2.73130	8.63713	.134048
7.47	55.8009	2.73313	8.64292	.133869
7.48	55.9504	2.73496	8.64870	.133690
7.49	56.1001	2.73679	8.65448	.133511
7.50	56.2500	2.73861	8.66025	.133333
n	n^2	$\sqrt{n}$	$\sqrt{10n}$	$1/n$

n	n^2	$\sqrt{n}$	$\sqrt{10n}$	$1/n$
7.50	56.2500	2.73861	8.66025	.133333
7.51	56.4001	2.74044	8.66603	.133156
7.52	56.5504	2.74226	8.67179	.132979
7.53	56.7009	2.74408	8.67756	.132802
7.54	56.8516	2.74591	8.68332	.132626
7.55	57.0025	2.74773	8.68907	.132450
7.56	57.1536	2.74955	8.69483	.132275
7.57	57.3049	2.75136	8.70057	.132100
7.58	57.4564	2.75318	8.70632	.131926
7.59	57.6081	2.75500	8.71206	.131752
7.60	57.7600	2.75681	8.71780	.131579
7.61	57.9121	2.75862	8.72353	.131406
7.62	58.0644	2.76043	8.72926	.131234
7.63	58.2169	2.76225	8.73499	.131062
7.64	58.3696	2.76405	8.74071	.130890
7.65	58.5225	2.76586	8.74643	.130719
7.66	58.6756	2.76767	8.75214	.130548
7.67	58.8289	2.76948	8.75785	.130378
7.68	58.9824	2.77128	8.76356	.130208
7.69	59.1361	2.77308	8.76926	.130039
7.70	59.2900	2.77489	8.77496	.129870
7.71	59.4441	2.77669	8.78066	.129702
7.72	59.5984	2.77849	8.78635	.129534
7.73	59.7529	2.78029	8.79204	.129366
7.74	59.9076	2.78209	8.79773	.129199
7.75	60.0625	2.78388	8.80341	.129032
7.76	60.2176	2.78568	8.80909	.128866
7.77	60.3729	2.78747	8.81476	.128700
7.78	60.5284	2.78927	8.82043	.128535
7.79	60.6841	2.79106	8.82610	.128370
7.80	60.8400	2.79285	8.83176	.128205
7.81	60.9961	2.79464	8.83742	.128041
7.82	61.1524	2.79643	8.84308	.127877
7.83	61.3089	2.79821	8.84873	.127714
7.84	61.4656	2.80000	8.85438	.127551
7.85	61.6225	2.80179	8.86002	.127389
7.86	61.7796	2.80357	8.86566	.127226
7.87	61.9369	2.80535	8.87130	.127065
7.88	62.0944	2.80713	8.87694	.126904
7.89	62.2521	2.80891	8.88257	.126743
7.90	62.4100	2.81069	8.88819	.126582
7.91	62.5681	2.81247	8.89382	.126422
7.92	62.7264	2.81425	8.89944	.126263
7.93	62.8849	2.81603	8.90505	.126103
7.94	63.0436	2.81780	8.91067	.125945
7.95	63.2025	2.81957	8.91628	.125786
7.96	63.3616	2.82135	8.92188	.125628
7.97	63.5209	2.82312	8.92749	.125471
7.98	63.6804	2.82489	8.93308	.125313
7.99	63.8401	2.82666	8.93868	.125156
8.00	64.0000	2.82843	8.94427	.125000
n	n^2	$\sqrt{n}$	$\sqrt{10n}$	$1/n$

B

SQUARES — SQUARE ROOTS — RECIPROCALS

n	n^2	$\sqrt{n}$	$\sqrt{10n}$	$1/n$
8.00	64.0000	2.82843	8.94427	.125000
8.01	64.1601	2.83019	8.94986	.124844
8.02	64.3204	2.83196	8.95545	.124688
8.03	64.4809	2.83373	8.96103	.124533
8.04	64.6416	2.83549	8.96660	.124378
8.05	64.8025	2.83725	8.97218	.124224
8.06	64.9636	2.83901	8.97775	.124069
8.07	65.1249	2.84077	8.98332	.123916
8.08	65.2864	2.84253	8.98888	.123762
8.09	65.4481	2.84429	8.99444	.123609
8.10	65.6100	2.84605	9.00000	.123457
8.11	65.7721	2.84781	9.00555	.123305
8.12	65.9344	2.84956	9.01110	.123153
8.13	66.0969	2.85132	9.01665	.123001
8.14	66.2596	2.85307	9.02219	.122850
8.15	66.4225	2.85482	9.02774	.122699
8.16	66.5856	2.85657	9.03327	.122549
8.17	66.7489	2.85832	9.03881	.122399
8.18	66.9124	2.86007	9.04434	.122249
8.19	67.0761	2.86182	9.04986	.122100
8.20	67.2400	2.86356	9.05539	.121951
8.21	67.4041	2.86531	9.06091	.121803
8.22	67.5684	2.86705	9.06642	.121655
8.23	67.7329	2.86880	9.07193	.121507
8.24	67.8976	2.87054	9.07744	.121359
8.25	68.0625	2.87228	9.08295	.121212
8.26	68.2276	2.87402	9.08845	.121065
8.27	68.3929	2.87576	9.09395	.120919
8.28	68.5584	2.87750	9.09945	.120773
8.29	68.7241	2.87924	9.10494	.120627
8.30	68.8900	2.88097	9.11043	.120482
8.31	69.0561	2.88271	9.11592	.120337
8.32	69.2224	2.88444	9.12140	.120192
8.33	69.3889	2.88617	9.12688	.120048
8.34	69.5556	2.88791	9.13236	.119904
8.35	69.7225	2.88964	9.13783	.119760
8.36	69.8896	2.89137	9.14330	.119617
8.37	70.0569	2.89310	9.14877	.119474
8.38	70.2244	2.89482	9.15423	.119332
8.39	70.3921	2.89655	9.15969	.119190
8.40	70.5600	2.89828	9.16515	.119048
8.41	70.7281	2.90000	9.17061	.118906
8.42	70.8964	2.90172	9.17606	.118765
8.43	71.0649	2.90345	9.18150	.118624
8.44	71.2336	2.90517	9.18695	.118483
8.45	71.4025	2.90689	9.19239	.118343
8.46	71.5716	2.90861	9.19783	.118203
8.47	71.7409	2.91033	9.20326	.118064
8.48	71.9104	2.91204	9.20869	.117925
8.49	72.0801	2.91376	9.21412	.117786
8.50	72.2500	2.91548	9.21954	.117647
n	n^2	$\sqrt{n}$	$\sqrt{10n}$	$1/n$

n	n^2	$\sqrt{n}$	$\sqrt{10n}$	$1/n$
8.50	72.2500	2.91548	9.21954	.117647
8.51	72.4201	2.91719	9.22497	.117509
8.52	72.5904	2.91890	9.23038	.117371
8.53	72.7609	2.92062	9.23580	.117233
8.54	72.9316	2.92233	9.24121	.117096
8.55	73.1025	2.92404	9.24662	.116959
8.56	73.2736	2.92575	9.25203	.116822
8.57	73.4449	2.92746	9.25743	.116686
8.58	73.6164	2.92916	9.26283	.116550
8.59	73.7881	2.93087	9.26823	.116414
8.60	73.9600	2.93258	9.27362	.116279
8.61	74.1321	2.93428	9.27901	.116144
8.62	74.3044	2.93598	9.28440	.116009
8.63	74.4769	2.93769	9.28978	.115875
8.64	74.6496	2.93939	9.29516	.115741
8.65	74.8225	2.94109	9.30054	.115607
8.66	74.9956	2.94279	9.30591	.115473
8.67	75.1689	2.94449	9.31128	.115340
8.68	75.3424	2.94618	9.31665	.115207
8.69	75.5161	2.94788	9.32202	.115075
8.70	75.6900	2.94958	9.32738	.114943
8.71	75.8641	2.95127	9.33274	.114811
8.72	76.0384	2.95296	9.33809	.114679
8.73	76.2129	2.95466	9.34345	.114548
8.74	76.3876	2.95635	9.34880	.114416
8.75	76.5625	2.95804	9.35414	.114286
8.76	76.7376	2.95973	9.35949	.114155
8.77	76.9129	2.96142	9.36483	.114025
8.78	77.0884	2.96311	9.37017	.113895
8.79	77.2641	2.96479	9.37550	.113766
8.80	77.4400	2.96648	9.38083	.113636
8.81	77.6161	2.96816	9.38616	.113507
8.82	77.7924	2.96985	9.39149	.113379
8.83	77.9689	2.97153	9.39681	.113250
8.84	78.1456	2.97321	9.40213	.113122
8.85	78.3225	2.97489	9.40744	.112994
8.86	78.4996	2.97658	9.41276	.112867
8.87	78.6769	2.97825	9.41807	.112740
8.88	78.8544	2.97993	9.42338	.112613
8.89	79.0321	2.98161	9.42868	.112486
8.90	79.2100	2.98329	9.43398	.112360
8.91	79.3881	2.98496	9.43928	.112233
8.92	79.5664	2.98664	9.44458	.112108
8.93	79.7449	2.98831	9.44987	.111982
8.94	79.9236	2.98998	9.45516	.111857
8.95	80.1025	2.99166	9.46044	.111732
8.96	80.2816	2.99333	9.46573	.111607
8.97	80.4609	2.99500	9.47101	.111483
8.98	80.6404	2.99666	9.47629	.111359
8.99	80.8201	2.99833	9.48156	.111235
9.00	81.0000	3.00000	9.48683	.111111
n	n^2	$\sqrt{n}$	$\sqrt{10n}$	$1/n$

SQUARES — SQUARE ROOTS — RECIPROCALS

n	n^2	$\sqrt{n}$	$\sqrt{10n}$	$1/n$
9.00	81.0000	3.00000	9.48683	.111111
9.01	81.1801	3.00167	9.49210	.110988
9.02	81.3604	3.00333	9.49737	.110865
9.03	81.5409	3.00500	9.50263	.110742
9.04	81.7216	3.00666	9.50789	.110619
9.05	81.9025	3.00832	9.51315	.110497
9.06	82.0836	3.00998	9.51840	.110375
9.07	82.2649	3.01164	9.52365	.110254
9.08	82.4464	3.01330	9.52890	.110132
9.09	82.6281	3.01496	9.53415	.110011
9.10	82.8100	3.01662	9.53939	.109890
9.11	82.9921	3.01828	9.54463	.109769
9.12	83.1744	3.01993	9.54987	.109649
9.13	83.3569	3.02159	9.55510	.109529
9.14	83.5396	3.02324	9.56033	.109409
9.15	83.7225	3.02490	9.56556	.109290
9.16	83.9056	3.02655	9.57079	.109170
9.17	84.0889	3.02820	9.57601	.109051
9.18	84.2724	3.02985	9.58123	.108932
9.19	84.4561	3.03150	9.58645	.108814
9.20	84.6400	3.03315	9.59166	.108696
9.21	84.8241	3.03480	9.59687	.108578
9.22	85.0084	3.03645	9.60208	.108460
9.23	85.1929	3.03809	9.60729	.108342
9.24	85.3776	3.03974	9.61249	.108225
9.25	85.5625	3.04138	9.61769	.108108
9.26	85.7476	3.04302	9.62289	.107991
9.27	85.9329	3.04467	9.62808	.107875
9.28	86.1184	3.04631	9.63328	.107759
9.29	86.3041	3.04795	9.63846	.107643
9.30	86.4900	3.04959	9.64365	.107527
9.31	86.6761	3.05123	9.64883	.107411
9.32	86.8624	3.05287	9.65401	.107296
9.33	87.0489	3.05450	9.65919	.107181
9.34	87.2356	3.05614	9.66437	.107066
9.35	87.4225	3.05778	9.66954	.106952
9.36	87.6096	3.05941	9.67471	.106838
9.37	87.7969	3.06105	9.67988	.106724
9.38	87.9844	3.06268	9.68504	.106610
9.39	88.1721	3.06431	9.69020	.106496
9.40	88.3600	3.06594	9.69536	.106383
9.41	88.5481	3.06757	9.70052	.106270
9.42	88.7364	3.06920	9.70567	.106157
9.43	88.9249	3.07083	9.71082	.106045
9.44	89.1136	3.07246	9.71597	.105932
9.45	89.3025	3.07409	9.72111	.105820
9.46	89.4916	3.07571	9.72625	.105708
9.47	89.6809	3.07734	9.73139	.105597
9.48	89.8704	3.07896	9.73653	.105485
9.49	90.0601	3.08058	9.74166	.105374
9.50	90.2500	3.08221	9.74679	.105263
n	n^2	$\sqrt{n}$	$\sqrt{10n}$	$1/n$

n	n^2	$\sqrt{n}$	$\sqrt{10n}$	$1/n$
9.50	90.2500	3.08221	9.74679	.105263
9.51	90.4401	3.08383	9.75192	.105152
9.52	90.6304	3.08545	9.75705	.105042
9.53	90.8209	3.08707	9.76217	.104932
9.54	91.0116	3.08869	9.76729	.104822
9.55	91.2025	3.09031	9.77241	.104712
9.56	91.3936	3.09192	9.77753	.104603
9.57	91.5849	3.09354	9.78264	.104493
9.58	91.7764	3.09516	9.78775	.104384
9.59	91.9681	3.09677	9.79285	.104275
9.60	92.1600	3.09839	9.79796	.104167
9.61	92.3521	3.10000	9.80306	.104058
9.62	92.5444	3.10161	9.80816	.103950
9.63	92.7369	3.10322	9.81326	.103842
9.64	92.9296	3.10483	9.81835	.103734
9.65	93.1225	3.10644	9.82344	.103627
9.66	93.3156	3.10805	9.82853	.103520
9.67	93.5089	3.10966	9.83362	.103413
9.68	93.7024	3.11127	9.83870	.103306
9.69	93.8961	3.11288	9.84378	.103199
9.70	94.0900	3.11448	9.84886	.103093
9.71	94.2841	3.11609	9.85393	.102987
9.72	94.4784	3.11769	9.85901	.102881
9.73	94.6729	3.11929	9.86408	.102775
9.74	94.8676	3.12090	9.86914	.102669
9.75	95.0625	3.12250	9.87421	.102564
9.76	95.2576	3.12410	9.87927	.102459
9.77	95.4529	3.12570	9.88433	.102354
9.78	95.6484	3.12730	9.88939	.102249
9.79	95.8441	3.12890	9.89444	.102145
9.80	96.0400	3.13050	9.89949	.102041
9.81	96.2361	3.13209	9.90454	.101937
9.82	96.4324	3.13369	9.90959	.101833
9.83	96.6289	3.13528	9.91464	.101729
9.84	96.8256	3.13688	9.91968	.101626
9.85	97.0225	3.13847	9.92472	.101523
9.86	97.2196	3.14006	9.92975	.101420
9.87	97.4169	3.14166	9.93479	.101317
9.88	97.6144	3.14325	9.93982	.101215
9.89	97.8121	3.14484	9.94485	.101112
9.90	98.0100	3.14643	9.94987	.101010
9.91	98.2081	3.14802	9.95490	.100908
9.92	98.4064	3.14960	9.95992	.100806
9.93	98.6049	3.15119	9.96494	.100705
9.94	98.8036	3.15278	9.96995	.100604
9.95	99.0025	3.15436	9.97497	.100503
9.96	99.2016	3.15595	9.97998	.100402
9.97	99.4009	3.15753	9.98499	.100301
9.98	99.6004	3.15911	9.98999	.100200
9.99	99.8001	3.16070	9.99500	.100100
10.00	100.000	3.16228	10.0000	.100000
n	n^2	$\sqrt{n}$	$\sqrt{10n}$	$1/n$

VALUES OF Σx^2 AND Σx^4 FOR COMPUTING LEAST SQUARES TREND LINES

The values of Σx^2 and Σx^4 may be read from the table opposite the value of n (representing the number of years to which the trend is fitted). For values of n greater than 60, use the formulas below. Origin is at mid-range so that $\Sigma x = 0$. For an odd number of years, the x scale is represented by . . . −2, −1, 0, +1, +2 . . . For an even number of years, the x scale is represented by . . . −3, −1, +1, +3 . . .

C

Odd Number of Years

$$\Sigma x^2 = \frac{n(n^2 - 1)}{12}$$

$$\Sigma x^4 = \frac{3n^2 - 7}{20} \Sigma x^2$$

$$\text{or } \Sigma x^4 = \frac{3n^5 - 10n^3 + 7n}{240}$$

Even Number of Years

$$\Sigma x^2 = \frac{n(n^2 - 1)}{3}$$

$$\Sigma x^4 = \frac{3n^2 - 7}{5} \Sigma x^2$$

$$\text{or } \Sigma x^4 = \frac{3n^5 - 10n^3 + 7n}{15}$$

VALUES OF Σx^2 AND Σx^4
FOR COMPUTING LEAST SQUARES TREND LINES

Odd Number of Years x Unit: 1 year			Even Number of Years x Unit: ½ year		
n	Σx^2	Σx^4	n	Σx^2	Σx^4
3	2	2	4	20	164
5	10	34	6	70	1,414
7	28	196	8	168	6,216
9	60	708	10	330	19,338
11	110	1,958	12	572	48,620
13	182	4,550	14	910	105,742
15	280	9,352	16	1,360	206,992
17	408	17,544	18	1,938	374,034
19	570	30,666	20	2,660	634,676
21	770	50,666	22	3,542	1,023,638
23	1,012	79,948	24	4,600	1,583,320
25	1,300	121,420	26	5,850	2,364,570
27	1,638	178,542	28	7,308	3,427,452
29	2,030	255,374	30	8,990	4,842,014
31	2,480	356,624	32	10,912	6,689,056
33	2,992	469,696	34	13,090	9,060,898
35	3,570	654,738	36	15,540	12,062,148
37	4,218	864,690	38	18,278	15,810,470
39	4,940	1,125,332	40	21,320	20,437,352
41	5,740	1,445,332	42	24,682	26,088,874
43	6,622	1,834,294	44	28,380	32,926,476
45	7,590	2,302,806	46	32,430	41,127,726
47	8,628	2,862,488	48	36,848	50,887,088
49	9,800	3,526,040	50	41,650	62,416,690
51	11,050	4,307,290	52	46,852	75,947,092
53	12,402	5,221,242	54	52,470	91,728,054
55	13,860	6,284,124	56	58,520	110,029,304
57	15,428	7,513,436	58	65,018	131,141,306
59	17,110	8,927,998	60	71,980	155,376,028

USE OF LOGARITHMS

Logarithms have been used for more than 300 years to reduce the labor of arithmetic computations. Calculating machines have rather generally replaced logarithms in statistical computations for performing multiplication and division, but logarithms are still used to extract roots and raise to powers. They are practically a necessity in computing the geometric mean and fitting certain types of curves.

A *logarithm* of any positive number to a given base is the power to which the base must be raised to give the number. The number corresponding to the logarithm is the *antilogarithm.* Because of its many advantages, the system of logarithms whose base is 10, known as "common logarithms," is generally used in numerical computation. $\text{Log}_{10} N$ is the common logarithm of N and is written Log N with the base 10 understood.

The common logarithm of 100 is 2, since $10^2 = 100$. Likewise, the common logarithm of 1,000 is 3, since $10^3 = 1{,}000$. Only the exponent "2" is written, since the base is always 10 for common logarithms. For the integral powers of 10, the logarithm is an integer. For all numbers between the integral powers of 10, the logarithm consists of an integer (positive, negative, or zero) plus a positive number less than one, always written as a decimal fraction. The integral portion of the logarithm is called the *characteristic*, and the decimal portion, the *mantissa.* The characteristic may be positive or negative, depending upon the location of the decimal point, but the mantissa is always positive.

Finding the Mantissa

The mantissas of the common logarithms of all numbers having the same sequence of figures are identical regardless of the position of the decimal point. The values of the mantissas for sequences of four digits are given to six decimals in Appendix F. The first three digits of the number are at the left of the page in the column headed N, and the fourth digit is at the top of the table. The mantissa of 1351 is found to be .130655 by reading down the first column to "135" and across the table to the column headed "1."

Determining the Characteristic

The characteristic of the common logarithm of any number may be determined by the following procedure. Locate an imaginary decimal so placed as to make a number between 1 and 10. For all numbers of 10 or more, the imaginary decimal is placed to the left of the real decimal and the characteristic is positive.

For example, in the number 13.51 the imaginary decimal (marked $_\wedge$) is located between 1 and 3 ($1_\wedge3.51$). The number of integers between the decimal and the imaginary decimal is the characteristic of the logarithm.

In this example, the characteristic is 1, the mantissa is .130655, and the logarithm of 13.51 is 1.130655. Other examples of determining the characteristic are given below:

Number	*Characteristic*	*Mantissa*	*Logarithm*
1‸351.	3	.130655	3.130655
1‸35.1	2	.130655	2.130655
1‸3.51	1	.130655	1.130655

For numbers between 1 and 10, the imaginary decimal falls at the same place as the decimal point, making the number of integers between them zero. The logarithm of 1.351 is .130655. For numbers less than 1, the imaginary decimal is to the right of the decimal point and the sign of the characteristic is minus. For example, the characteristic of .001‸351 is − 3 and the logarithm of .001351 is .130655 − 3, or − 2.869345, or 7.130655 − 10.

Interpolation

The table in Appendix F gives the mantissas for numbers of four digits. For a sequence of more than four numbers, the mantissa may be approximated from this table by interpolating between the logarithms of the numbers immediately preceding and following the number for which the logarithm is sought. The logarithm of 13.514 is between 1.130655 and 1.130977, the logarithms of 13.51 and 13.52. The difference between the two mantissas is .000322. Thus, it is assumed that the logarithm of 13.514 will be larger than 1.130655 by 4/10ths of the difference between 1.130655 and 1.130977. This linear interpolation gives the following value for the logarithms, with the last digit approximate:

$$\text{Log } 13.514 = 1.130655 + (.4 \times .000322) = 1.130655 + .000129 = 1.130784$$

The value in Column D, the last column to the right, is the greatest difference between the adjacent mantissas on a given line, and can be used to eliminate the necessity for subtracting to get the difference. Greater accuracy can be obtained by using tables that give the mantissa for a sequence of more than four digits and that carry the values to a greater number of decimal places.

Finding the Antilogarithm

After a calculation has been made with logarithms, it is necessary to find the number represented by the logarithm. This number, called the *antilogarithm*, is determined by a process that is the reverse of finding the logarithm of a number. This may be illustrated by an example in which the problem is to find the value of N when Log N = 2.788546. The digits of the number for which 2.788546 is the logarithm are determined from the mantissa (.788546), and the location of the decimal point is determined by the characteristic (2).

The mantissa is located in the table of logarithms, or if the exact value does not appear in the table, the values just larger and smaller can be found. In this example, 788546 falls between 788522 and 788593 (omitting decimals as in the table). The antilogarithms of these two mantissas are 6145 and 6146. The antilogarithm of 788546 may be approximated by linear interpolation between these two numbers. Since the difference between the two logarithms, 788522 and 788593, is 71, the logarithm of 788546 is located $\frac{24}{71}$ of the distance from 788522 in the direction of 788593. This means that the antilogarithm is approximately 6145 $\frac{24}{71}$, or 61453.

Locating the imaginary decimal point to make a number between 1 and 10 gives 6‸1453. Since the characteristic of the logarithm is +2, the real decimal point is two digits to the right of the imaginary decimal, using the rule given above for finding the characteristic. Therefore, the antilogarithm of 2.788546 is 6‸14.53.

Multiplication

Multiplication is performed by adding the logarithms of the numbers being multiplied and then finding the antilogarithm, as illustrated by the following example.

Find the value of 13.54×94.7

$$\begin{aligned} \log 13.54 &= 1.131619 \\ \text{plus } \log 94.7 &= 1.976350 \\ \hline \log N &= 3.107969 \\ N &= 1_{\wedge}282.2 \end{aligned}$$

The imaginary decimal is located between the first two digits and since the characteristic is +3, the real decimal point is placed three digits to the right of the imaginary decimal.

In using the logarithms of numbers less than 1, it is important to remember that while the characteristic is minus, the mantissa is always positive. The logarithm of .001351 was given before as .130655 − 3, which is, of course, −2.869345. There are advantages, however, in writing this logarithm as 7.130655 − 10. Since 7 was both added and subtracted, the value of the logarithm remains unchanged. (7.130655 − 10 = −2.869345) This device is used in the following example.

Find the value of $64.7 \times .001351$

$$\begin{aligned} \log 64.7 &= 1.810904 \\ \text{plus } \log .001351 &= 7.130655 - 10 \\ \hline \log N &= 8.941559 - 10 \\ N &= .08_{\wedge}741 \end{aligned}$$

Division

Division is performed by subtracting the logarithm of the divisor from the logarithm of the dividend and then finding the antilogarithm, as illustrated in the following examples.

Find the value of $947 \div 13.54$

$$\begin{aligned} \log 947 &= 2.976350 \\ \text{minus } \log 13.54 &= 1.131619 \\ \hline \log N &= 1.844731 \\ N &= 6_{\wedge}9.94 \end{aligned}$$

Find the value of $.00437 \div 64.7$

$$\begin{aligned} \log .00437 &= 7.640481 - 10 \\ \text{minus } \log 64.7 &= 1.810904 \\ \hline \log N &= 5.829577 - 10 \\ N &= .00006_{\wedge}754 \end{aligned}$$

Raising to Powers

To raise a number to a given power, multiply the logarithm of the number by the exponent of the power and find the antilogarithm.

Find the value of N if $N = 1.06^{35}$

$$\begin{aligned} \log N &= 35 \times \log 1.06 \\ \log 1.06 &= .025306 \\ \log N &= 35 \times .025306 = .885710 \\ N &= 7.686 \end{aligned}$$

Extracting Roots

To extract any root of a number, divide the logarithm of the number by the index of the root and find the antilogarithm.

Find the value of N if $N = \sqrt[9]{3.86}$

$$\begin{aligned} \log N &= \frac{\log 3.86}{9} \\ \log 3.86 &= .586587 \\ \log N &= \frac{.586587}{9} = .065176 \\ N &= 1.162 \end{aligned}$$

LOGARITHMS OF NUMBERS 1,000–1,499
Six-Place Mantissas

N	0	1	2	3	4	5	6	7	8	9	D#
100	00 0000	0434	0868	1301	1734	2166	2598	3029	3461	3891	434
01	4321	4751	5181	5609	6038	6466	6894	7321	7748	8174	430
02	00 8600	9026	9451	9876	*0300	*0724	*1147	*1570	*1993	*2415	426
03	01 2837	3259	3680	4100	4521	4940	5360	5779	6197	6616	422
04	01 7033	7451	7868	8284	8700	9116	9532	9947	*0361	*0775	418
05	02 1189	1603	2016	2428	2841	3252	3664	4075	4486	4896	414
06	5306	5715	6125	6533	6942	7350	7757	8164	8571	8978	410
07	02 9384	9789	*0195	*0600	*1004	*1408	*1812	*2216	*2619	*3021	406
08	03 3424	3826	4227	4628	5029	5430	5830	6230	6629	7028	402
09	03 7426	7825	8223	8620	9017	9414	9811	*0207	*0602	*0998	399
110	04 1393	1787	2182	2576	2969	3362	3755	4148	4540	4932	395
11	5323	5714	6105	6495	6885	7275	7664	8053	8442	8830	391
12	04 9218	9606	9993	*0380	*0766	*1153	*1538	*1924	*2309	*2694	388
13	05 3078	3463	3846	4230	4613	4996	5378	5760	6142	6524	385
14	05 6905	7286	7666	8046	8426	8805	9185	9563	9942	*0320	381
15	06 0698	1075	1452	1829	2206	2582	2958	3333	3709	4083	377
16	4458	4832	5206	5580	5953	6326	6699	7071	7443	7815	374
17	06 8186	8557	8928	9298	9668	*0038	*0407	*0776	*1145	*1514	371
18	07 1882	2250	2617	2985	3352	3718	4085	4451	4816	5182	368
19	5547	5912	6276	6640	7004	7368	7731	8094	8457	8819	365
120	07 9181	9543	9904	*0266	*0626	*0987	*1347	*1707	*2067	*2426	362
21	08 2785	3144	3503	3861	4219	4576	4934	5291	5647	6004	359
22	6360	6716	7071	7426	7781	8136	8490	8845	9198	9552	356
23	08 9905	*0258	*0611	*0963	*1315	*1667	*2018	*2370	*2721	*3071	353
24	09 3422	3772	4122	4471	4820	5169	5518	5866	6215	6562	350
25	09 6910	7257	7604	7951	8298	8644	8990	9335	9681	*0026	347
26	10 0371	0715	1059	1403	1747	2091	2434	2777	3119	3462	344
27	3804	4146	4487	4828	5169	5510	5851	6191	6531	6871	342
28	10 7210	7549	7888	8227	8565	8903	9241	9579	9916	*0253	339
29	11 0590	0926	1263	1599	1934	2270	2605	2940	3275	3609	337
130	3943	4277	4611	4944	5278	5611	5943	6276	6608	6940	334
31	11 7271	7603	7934	8265	8595	8926	9256	9586	9915	*0245	332
32	12 0574	0903	1231	1560	1888	2216	2544	2871	3198	3525	329
33	3852	4178	4504	4830	5156	5481	5806	6131	6456	6781	326
34	12 7105	7429	7753	8076	8399	8722	9045	9368	9690	*0012	324
35	13 0334	0655	0977	1298	1619	1939	2260	2580	2900	3219	322
36	3539	3858	4177	4496	4814	5133	5451	5769	6086	6403	319
37	6721	7037	7354	7671	7987	8303	8618	8934	9249	9564	317
38	13 9879	*0194	*0508	*0822	*1136	*1450	*1763	*2076	*2389	*2702	315
39	14 3015	3327	3639	3951	4263	4574	4885	5196	5507	5818	312
140	6128	6438	6748	7058	7367	7676	7985	8294	8603	8911	310
41	14 9219	9527	9835	*0142	*0449	*0756	*1063	*1370	*1676	*1982	308
42	15 2288	2594	2900	3205	3510	3815	4120	4424	4728	5032	306
43	5336	5640	5943	6246	6549	6852	7154	7457	7759	8061	304
44	15 8362	8664	8965	9266	9567	9868	*0168	*0469	*0769	*1068	302
45	16 1368	1667	1967	2266	2564	2863	3161	3460	3758	4055	300
46	4353	4650	4947	5244	5541	5838	6134	6430	6726	7022	297
47	16 7317	7613	7908	8203	8497	8792	9086	9380	9674	9968	296
48	17 0262	0555	0848	1141	1434	1726	2019	2311	2603	2895	293
49	3186	3478	3769	4060	4351	4641	4932	5222	5512	5802	292
N	0	1	2	3	4	5	6	7	8	9	D

*Prefix first two places on next line.
Example: The mantissa for number (N) 1072 is *03* 0195.

#The *highest difference* between adjacent mantissas on the *individual line*. It is also the *lowest difference* between adjacent mantissas on the *preceding line* in many cases.

LOGARITHMS OF NUMBERS 1,500–1,999
Six-Place Mantissas

N	0	1	2	3	4	5	6	7	8	9	D
150	17 6091	6381	6670	6959	7248	7536	7825	8113	8401	8689	290
51	17 8977	9264	9552	9839	*0126	*0413	*0699	*0986	*1272	*1558	288
52	18 1844	2129	2415	2700	2985	3270	3555	3839	4123	4407	286
53	4691	4975	5259	5542	5825	6108	6391	6674	6956	7239	284
54	18 7521	7803	8084	8366	8647	8928	9209	9490	9771	*0051	282
55	19 0332	0612	0892	1171	1451	1730	2010	2289	2567	2846	280
56	3125	3403	3681	3959	4237	4514	4792	5069	5346	5623	278
57	5900	6176	6453	6729	7005	7281	7556	7832	8107	8382	277
58	19 8657	8932	9206	9481	9755	*0029	*0303	*0577	*0850	*1124	275
59	20 1397	1670	1943	2216	2488	2761	3033	3305	3577	3848	273
160	4120	4391	4663	4934	5204	5475	5746	6016	6286	6556	272
61	6826	7096	7365	7634	7904	8173	8441	8710	8979	9247	270
62	20 9515	9783	*0051	*0319	*0586	*0853	*1121	*1388	*1654	*1921	268
63	21 2188	2454	2720	2986	3252	3518	3783	4049	4314	4579	266
64	4844	5109	5373	5638	5902	6166	6430	6694	6957	7221	265
65	21 7484	7747	8010	8273	8536	8798	9060	9323	9585	9846	263
66	22 0108	0370	0631	0892	1153	1414	1675	1936	2196	2456	262
67	2716	2976	3236	3496	3755	4015	4274	4533	4792	5051	260
68	5309	5568	5826	6084	6342	6600	6858	7115	7372	7630	259
69	22 7887	8144	8400	8657	8913	9170	9426	9682	9938	*0193	257
170	23 0449	0704	0960	1215	1470	1724	1979	2234	2488	2742	256
71	2996	3250	3504	3757	4011	4264	4517	4770	5023	5276	254
72	5528	5781	6033	6285	6537	6789	7041	7292	7544	7795	253
73	23 8046	8297	8548	8799	9049	9299	9550	9800	*0050	*0300	251
74	24 0549	0799	1048	1297	1546	1795	2044	2293	2541	2790	250
75	3038	3286	3534	3782	4030	4277	4525	4772	5019	5266	248
76	5513	5759	6006	6252	6499	6745	6991	7237	7482	7728	247
77	24 7973	8219	8464	8709	8954	9198	9443	9687	9932	*0176	246
78	25 0420	0664	0908	1151	1395	1638	1881	2125	2368	2610	244
79	2853	3096	3338	3580	3822	4064	4306	4548	4790	5031	243
180	5273	5514	5755	5996	6237	6477	6718	6958	7198	7439	241
81	25 7679	7918	8158	8398	8637	8877	9116	9355	9594	9833	240
82	26 0071	0310	0548	0787	1025	1263	1501	1739	1976	2214	239
83	2451	2688	2925	3162	3399	3636	3873	4109	4346	4582	237
84	4818	5054	5290	5525	5761	5996	6232	6467	6702	6937	236
85	7172	7406	7641	7875	8110	8344	8578	8812	9046	9279	235
86	26 9513	9746	9980	*0213	*0446	*0679	*0912	*1144	*1377	*1609	234
87	27 1842	2074	2306	2538	2770	3001	3233	3464	3696	3927	232
88	4158	4389	4620	4850	5081	5311	5542	5772	6002	6232	231
89	6462	6692	6921	7151	7380	7609	7838	8067	8296	8525	230
190	27 8754	8982	9211	9439	9667	9895	*0123	*0351	*0578	*0806	229
91	28 1033	1261	1488	1715	1942	2169	2396	2622	2849	3075	228
92	3301	3527	3753	3979	4205	4431	4656	4882	5107	5332	226
93	5557	5782	6007	6232	6456	6681	6905	7130	7354	7578	225
94	28 7802	8026	8249	8473	8696	8920	9143	9366	9589	9812	224
95	29 0035	0257	0480	0702	0925	1147	1369	1591	1813	2034	223
96	2256	2478	2699	2920	3141	3363	3584	3804	4025	4246	222
97	4466	4687	4907	5127	5347	5567	5787	6007	6226	6446	221
98	6665	6884	7104	7323	7542	7761	7979	8198	8416	8635	220
99	29 8853	9071	9289	9507	9725	9943	*0161	*0378	*0595	*0813	218
N	0	1	2	3	4	5	6	7	8	9	D

LOGARITHMS OF NUMBERS 2,000–2,499
Six-Place Mantissas

N	0	1	2	3	4	5	6	7	8	9	D
200	30 1030	1247	1464	1681	1898	2114	2331	2547	2764	2980	217
01	3196	3412	3628	3844	4059	4275	4491	4706	4921	5136	216
02	5351	5566	5781	5996	6211	6425	6639	6854	7068	7282	215
03	7496	7710	7924	8137	8351	8564	8778	8991	9204	9417	214
04	30 9630	9843	*0056	*0268	*0481	*0693	*0906	*1118	*1330	*1542	213
05	31 1754	1966	2177	2389	2600	2812	3023	3234	3445	3656	212
06	3867	4078	4289	4499	4710	4920	5130	5340	5551	5760	211
07	5970	6180	6390	6599	6809	7018	7227	7436	7646	7854	210
08	31 8063	8272	8481	8689	8898	9106	9314	9522	9730	9938	209
09	32 0146	0354	0562	0769	0977	1184	1391	1598	1805	2012	208
210	2219	2426	2633	2839	3046	3252	3458	3665	3871	4077	207
11	4282	4488	4694	4899	5105	5310	5516	5721	5926	6131	206
12	6336	6541	6745	6950	7155	7359	7563	7767	7972	8176	205
13	32 8380	8583	8787	8991	9194	9398	9601	9805	*0008	*0211	204
14	33 0414	0617	0819	1022	1225	1427	1630	1832	2034	2236	203
15	2438	2640	2842	3044	3246	3447	3649	3850	4051	4253	202
16	4454	4655	4856	5057	5257	5458	5658	5859	6059	6260	201
17	6460	6660	6860	7060	7260	7459	7659	7858	8058	8257	200
18	33 8456	8656	8855	9054	9253	9451	9650	9849	*0047	*0246	200
19	34 0444	0642	0841	1039	1237	1435	1632	1830	2028	2225	199
220	2423	2620	2817	3014	3212	3409	3606	3802	3999	4196	198
21	4392	4589	4785	4981	5178	5374	5570	5766	5962	6157	197
22	6353	6549	6744	6939	7135	7330	7525	7720	7915	8110	196
23	34 8305	8500	8694	8889	9083	9278	9472	9666	9860	*0054	195
24	35 0248	0442	0636	0829	1023	1216	1410	1603	1796	1989	194
25	2183	2375	2568	2761	2954	3147	3339	3532	3724	3916	193
26	4108	4301	4493	4685	4876	5068	5260	5452	5643	5834	192
27	6026	6217	6408	6599	6790	6981	7172	7363	7554	7744	191
28	7935	8125	8316	8506	8696	8886	9076	9266	9456	9646	191
29	35 9835	*0025	*0215	*0404	*0593	*0783	*0972	*1161	*1350	*1539	190
230	36 1728	1917	2105	2294	2482	2671	2859	3048	3236	3424	189
31	3612	3800	3988	4176	4363	4551	4739	4926	5113	5301	188
32	5488	5675	5862	6049	6236	6423	6610	6796	6983	7169	187
33	7356	7542	7729	7915	8101	8287	8473	8659	8845	9030	187
34	36 9216	9401	9587	9772	9958	*0143	*0328	*0513	*0698	*0883	186
35	37 1068	1253	1437	1622	1806	1991	2175	2360	2544	2728	185
36	2912	3096	3280	3464	3647	3831	4015	4198	4382	4565	184
37	4748	4932	5115	5298	5481	5664	5846	6029	6212	6394	184
38	6577	6759	6942	7124	7306	7488	7670	7852	8034	8216	183
39	37 8398	8580	8761	8943	9124	9306	9487	9668	9849	*0030	182
240	38 0211	0392	0573	0754	0934	1115	1296	1476	1656	1837	181
41	2017	2197	2377	2557	2737	2917	3097	3277	3456	3636	180
42	3815	3995	4174	4353	4533	4712	4891	5070	5249	5428	180
43	5606	5785	5964	6142	6321	6499	6677	6856	7034	7212	179
44	7390	7568	7746	7923	8101	8279	8456	8634	8811	8989	178
45	38 9166	9343	9520	9698	9875	*0051	*0228	*0405	*0582	*0759	178
46	39 0935	1112	1288	1464	1641	1817	1993	2169	2345	2521	177
47	2697	2873	3048	3224	3400	3575	3751	3926	4101	4277	176
48	4452	4627	4802	4977	5152	5326	5501	5676	5850	6025	175
49	6199	6374	6548	6722	6896	7071	7245	7419	7592	7766	175
N	0	1	2	3	4	5	6	7	8	9	D

LOGARITHMS OF NUMBERS 2,500–2,999
Six-Place Mantissas

N	0	1	2	3	4	5	6	7	8	9	D
250	39 7940	8114	8287	8461	8634	8808	8981	9154	9328	9501	174
51	39 9674	9847	*0020	*0192	*0365	*0538	*0711	*0883	*1056	*1228	173
52	40 1401	1573	1745	1917	2089	2261	2433	2605	2777	2949	172
53	3121	3292	3464	3635	3807	3978	4149	4320	4492	4663	172
54	4834	5005	5176	5346	5517	5688	5858	6029	6199	6370	171
55	6540	6710	6881	7051	7221	7391	7561	7731	7901	8070	171
56	8240	8410	8579	8749	8918	9087	9257	9426	9595	9764	170
57	40 9933	*0102	*0271	*0440	*0609	*0777	*0946	*1114	*1283	*1451	169
58	41 1620	1788	1956	2124	2293	2461	2629	2796	2964	3132	169
59	3300	3467	3635	3803	3970	4137	4305	4472	4639	4806	168
260	4973	5140	5307	5474	5641	5808	5974	6141	6308	6474	167
61	6641	6807	6973	7139	7306	7472	7638	7804	7970	8135	167
62	8301	8467	8633	8798	8964	9129	9295	9460	9625	9791	166
63	41 9956	*0121	*0286	*0451	*0616	*0781	*0945	*1110	*1275	*1439	165
64	42 1604	1768	1933	2097	2261	2426	2590	2754	2918	3082	165
65	3246	3410	3574	3737	3901	4065	4228	4392	4555	4718	164
66	4882	5045	5208	5371	5534	5697	5860	6023	6186	6349	163
67	6511	6674	6836	6999	7161	7324	7486	7648	7811	7973	163
68	8135	8297	8459	8621	8783	8944	9106	9268	9429	9591	162
69	42 9752	9914	*0075	*0236	*0398	*0559	*0720	*0881	*1042	*1203	162
270	43 1364	1525	1685	1846	2007	2167	2328	2488	2649	2809	161
71	2969	3130	3290	3450	3610	3770	3930	4090	4249	4409	161
72	4569	4729	4888	5048	5207	5367	5526	5685	5844	6004	160
73	6163	6322	6481	6640	6799	6957	7116	7275	7433	7592	159
74	7751	7909	8067	8226	8384	8542	8701	8859	9017	9175	159
75	43 9333	9491	9648	9806	9964	*0122	*0279	*0437	*0594	*0752	158
76	44 0909	1066	1224	1381	1538	1695	1852	2009	2166	2323	158
77	2480	2637	2793	2950	3106	3263	3419	3576	3732	3889	157
78	4045	4201	4357	4513	4669	4825	4981	5137	5293	5449	156
79	5604	5760	5915	6071	6226	6382	6537	6692	6848	7003	156
280	7158	7313	7468	7623	7778	7933	8088	8242	8397	8552	155
81	44 8706	8861	9015	9170	9324	9478	9633	9787	9941	*0095	155
82	45 0249	0403	0557	0711	0865	1018	1172	1326	1479	1633	154
83	1786	1940	2093	2247	2400	2553	2706	2859	3012	3165	154
84	3318	3471	3624	3777	3930	4082	4235	4387	4540	4692	153
85	4845	4997	5150	5302	5454	5606	5758	5910	6062	6214	153
86	6366	6518	6670	6821	6973	7125	7276	7428	7579	7731	152
87	7882	8033	8184	8336	8487	8638	8789	8940	9091	9242	152
88	45 9392	9543	9694	9845	9995	*0146	*0296	*0447	*0597	*0748	151
89	46 0898	1048	1198	1348	1499	1649	1799	1948	2098	2248	151
290	2398	2548	2697	2847	2997	3146	3296	3445	3594	3744	150
91	3893	4042	4191	4340	4490	4639	4788	4936	5085	5234	149
92	5383	5532	5680	5829	5977	6126	6274	6423	6571	6719	149
93	6868	7016	7164	7312	7460	7608	7756	7904	8052	8200	148
94	8347	8495	8643	8790	8938	9085	9233	9380	9527	9675	148
95	46 9822	9969	*0116	*0263	*0410	*0557	*0704	*0851	*0998	*1145	147
96	47 1292	1438	1585	1732	1878	2025	2171	2318	2464	2610	147
97	2756	2903	3049	3195	3341	3487	3633	3779	3925	4071	147
98	4216	4362	4508	4653	4799	4944	5090	5235	5381	5526	146
99	5671	5816	5962	6107	6252	6397	6542	6687	6832	6976	146
N	**0**	**1**	**2**	**3**	**4**	**5**	**6**	**7**	**8**	**9**	**D**

E

LOGARITHMS OF NUMBERS 3,000–3,499
Six-Place Mantissas

N	0	1	2	3	4	5	6	7	8	9	D
300	47 7121	7266	7411	7555	7700	7844	7989	8133	8278	8422	145
01	47 8566	8711	8855	8999	9143	9287	9431	9575	9719	9863	145
02	48 0007	0151	0294	0438	0582	0725	0869	1012	1156	1299	144
03	1443	1586	1729	1872	2016	2159	2302	2445	2588	2731	144
04	2874	3016	3159	3302	3445	3587	3730	3872	4015	4157	143
05	4300	4442	4585	4727	4869	5011	5153	5295	5437	5579	143
06	5721	5863	6005	6147	6289	6430	6572	6714	6855	6997	142
07	7138	7280	7421	7563	7704	7845	7986	8127	8269	8410	142
08	8551	8692	8833	8974	9114	9255	9396	9537	9677	9818	141
09	48 9958	*0099	*0239	*0380	*0520	*0661	*0801	*0941	*1081	*1222	141
310	49 1362	1502	1642	1782	1922	2062	2201	2341	2481	2621	140
11	2760	2900	3040	3179	3319	3458	3597	3737	3876	4015	140
12	4155	4294	4433	4572	4711	4850	4989	5128	5267	5406	139
13	5544	5683	5822	5960	6099	6238	6376	6515	6653	6791	139
14	6930	7068	7206	7344	7483	7621	7759	7897	8035	8173	139
15	8311	8448	8586	8724	8862	8999	9137	9275	9412	9550	138
16	49 9687	9824	9962	*0099	*0236	*0374	*0511	*0648	*0785	*0922	138
17	50 1059	1196	1333	1470	1607	1744	1880	2017	2154	2291	137
18	2427	2564	2700	2837	2973	3109	3246	3382	3518	3655	137
19	3791	3927	4063	4199	4335	4471	4607	4743	4878	5014	136
320	5150	5286	5421	5557	5693	5828	5964	6099	6234	6370	136
21	6505	6640	6776	6911	7046	7181	7316	7451	7586	7721	136
22	7856	7991	8126	8260	8395	8530	8664	8799	8934	9068	135
23	50 9203	9337	9471	9606	9740	9874	*0009	*0143	*0277	*0411	135
24	51 0545	0679	0813	0947	1081	1215	1349	1482	1616	1750	134
25	1883	2017	2151	2284	2418	2551	2684	2818	2951	3084	134
26	3218	3351	3484	3617	3750	3883	4016	4149	4282	4415	133
27	4548	4681	4813	4946	5079	5211	5344	5476	5609	5741	133
28	5874	6006	6139	6271	6403	6535	6668	6800	6932	7064	133
29	7196	7328	7460	7592	7724	7855	7987	8119	8251	8382	132
330	8514	8646	8777	8909	9040	9171	9303	9434	9566	9697	132
31	51 9828	9959	*0090	*0221	*0353	*0484	*0615	*0745	*0876	*1007	132
32	52 1138	1269	1400	1530	1661	1792	1922	2053	2183	2314	131
33	2444	2575	2705	2835	2966	3096	3226	3356	3486	3616	131
34	3746	3876	4006	4136	4266	4396	4526	4656	4785	4915	130
35	5045	5174	5304	5434	5563	5693	5822	5951	6081	6210	130
36	6339	6469	6598	6727	6856	6985	7114	7243	7372	7501	130
37	7630	7759	7888	8016	8145	8274	8402	8531	8660	8788	129
38	52 8917	9045	9174	9302	9430	9559	9687	9815	9943	*0072	129
39	53 0200	0328	0456	0584	0712	0840	0968	1096	1223	1351	128
340	1479	1607	1734	1862	1990	2117	2245	2372	2500	2627	128
41	2754	2882	3009	3136	3264	3391	3518	3645	3772	3899	128
42	4026	4153	4280	4407	4534	4661	4787	4914	5041	5167	127
43	5294	5421	5547	5674	5800	5927	6053	6180	6306	6432	127
44	6558	6685	6811	6937	7063	7189	7315	7441	7567	7693	127
45	7819	7945	8071	8197	8322	8448	8574	8699	8825	8951	126
46	53 9076	9202	9327	9452	9578	9703	9829	9954	*0079	*0204	126
47	54 0329	0455	0580	0705	0830	0955	1080	1205	1330	1454	126
48	1579	1704	1829	1953	2078	2203	2327	2452	2576	2701	125
49	2825	2950	3074	3199	3323	3447	3571	3696	3820	3944	125
N	**0**	**1**	**2**	**3**	**4**	**5**	**6**	**7**	**8**	**9**	**D**

LOGARITHMS OF NUMBERS 3,500–3,999
Six-Place Mantissas

N	0	1	2	3	4	5	6	7	8	9	D
350	54 4068	4192	4316	4440	4564	4688	4812	4936	5060	5183	124
51	5307	5431	5555	5678	5802	5925	6049	6172	6296	6419	124
52	6543	6666	6789	6913	7036	7159	7282	7405	7529	7652	124
53	7775	7898	8021	8144	8267	8389	8512	8635	8758	8881	123
54	54 9003	9126	9249	9371	9494	9616	9739	9861	9984	*0106	123
55	55 0228	0351	0473	0595	0717	0840	0962	1084	1206	1328	123
56	1450	1572	1694	1816	1938	2060	2181	2303	2425	2547	122
57	2668	2790	2911	3033	3155	3276	3398	3519	3640	3762	122
58	3883	4004	4126	4247	4368	4489	4610	4731	4852	4973	122
59	5094	5215	5336	5457	5578	5699	5820	5940	6061	6182	121
360	6303	6423	6544	6664	6785	6905	7026	7146	7267	7387	121
61	7507	7627	7748	7868	7988	8108	8228	8349	8469	8589	121
62	8709	8829	8948	9068	9188	9308	9428	9548	9667	9787	120
63	55 9907	*0026	*0146	*0265	*0385	*0504	*0624	*0743	*0863	*0982	120
64	56 1101	1221	1340	1459	1578	1698	1817	1936	2055	2174	120
65	2293	2412	2531	2650	2769	2887	3006	3125	3244	3362	119
66	3481	3600	3718	3837	3955	4074	4192	4311	4429	4548	119
67	4666	4784	4903	5021	5139	5257	5376	5494	5612	5730	119
68	5848	5966	6084	6202	6320	6437	6555	6673	6791	6909	118
69	7026	7144	7262	7379	7497	7614	7732	7849	7967	8084	118
370	8202	8319	8436	8554	8671	8788	8905	9023	9140	9257	118
71	56 9374	9491	9608	9725	9842	9959	*0076	*0193	*0309	*0426	117
72	57 0543	0660	0776	0893	1010	1126	1243	1359	1476	1592	117
73	1709	1825	1942	2058	2174	2291	2407	2523	2639	2755	117
74	2872	2988	3104	3220	3336	3452	3568	3684	3800	3915	116
75	4031	4147	4263	4379	4494	4610	4726	4841	4957	5072	116
76	5188	5303	5419	5534	5650	5765	5880	5996	6111	6226	116
77	6341	6457	6572	6687	6802	6917	7032	7147	7262	7377	116
78	7492	7607	7722	7836	7951	8066	8181	8295	8410	8525	115
79	8639	8754	8868	8983	9097	9212	9326	9441	9555	9669	115
380	57 9784	9898	*0012	*0126	*0241	*0355	*0469	*0583	*0697	*0811	115
81	58 0925	1039	1153	1267	1381	1495	1608	1722	1836	1950	114
82	2063	2177	2291	2404	2518	2631	2745	2858	2972	3085	114
83	3199	3312	3426	3539	3652	3765	3879	3992	4105	4218	114
84	4331	4444	4557	4670	4783	4896	5009	5122	5235	5348	113
85	5461	5574	5686	5799	5912	6024	6137	6250	6362	6475	113
86	6587	6700	6812	6925	7037	7149	7262	7374	7486	7599	113
87	7711	7823	7935	8047	8160	8272	8384	8496	8608	8720	113
88	8832	8944	9056	9167	9279	9391	9503	9615	9726	9838	112
89	58 9950	*0061	*0173	*0284	*0396	*0507	*0619	*0730	*0842	*0953	112
390	59 1065	1176	1287	1399	1510	1621	1732	1843	1955	2066	112
91	2177	2288	2399	2510	2621	2732	2843	2954	3064	3175	111
92	3286	3397	3508	3618	3729	3840	3950	4061	4171	4282	111
93	4393	4503	4614	4724	4834	4945	5055	5165	5276	5386	111
94	5496	5606	5717	5827	5937	6047	6157	6267	6377	6487	111
95	6597	6707	6817	6927	7037	7146	7256	7366	7476	7586	110
96	7695	7805	7914	8024	8134	8243	8353	8462	8572	8681	110
97	8791	8900	9009	9119	9228	9337	9446	9556	9665	9774	110
98	9883	9992	*0101	*0210	*0319	*0428	*0537	*0646	*0755	*0864	109
99	60 0973	1082	1191	1299	1408	1517	1625	1734	1843	1951	109
N	**0**	**1**	**2**	**3**	**4**	**5**	**6**	**7**	**8**	**9**	**D**

LOGARITHMS OF NUMBERS 4,000–4,499

Six-Place Mantissas

N	0	1	2	3	4	5	6	7	8	9	D
400	60 2060	2169	2277	2386	2494	2603	2711	2819	2928	3036	109
01	3144	3253	3361	3469	3577	3686	3794	3902	4010	4118	109
02	4226	4334	4442	4550	4658	4766	4874	4982	5089	5197	108
03	5305	5413	5521	5628	5736	5844	5951	6059	6166	6274	108
04	6381	6489	6596	6704	6811	6919	7026	7133	7241	7348	108
05	7455	7562	7669	7777	7884	7991	8098	8205	8312	8419	108
06	8526	8633	8740	8847	8954	9061	9167	9274	9381	9488	107
07	60 9594	9701	9808	9914	*0021	*0128	*0234	*0341	*0447	*0554	107
08	61 0660	0767	0873	0979	1086	1192	1298	1405	1511	1617	107
09	1723	1829	1936	2042	2148	2254	2360	2466	2572	2678	107
410	2784	2890	2996	3102	3207	3313	3419	3525	3630	3736	106
11	3842	3947	4053	4159	4264	4370	4475	4581	4686	4792	106
12	4897	5003	5108	5213	5319	5424	5529	5634	5740	5845	106
13	5950	6055	6160	6265	6370	6476	6581	6686	6790	6895	106
14	7000	7105	7210	7315	7420	7525	7629	7734	7839	7943	105
15	8048	8153	8257	8362	8466	8571	8676	8780	8884	8989	105
16	61 9093	9198	9302	9406	9511	9615	9719	9824	9928	*0032	105
17	62 0136	0240	0344	0448	0552	0656	0760	0864	0968	1072	104
18	1176	1280	1384	1488	1592	1695	1799	1903	2007	2110	104
19	2214	2318	2421	2525	2628	2732	2835	2939	3042	3146	104
420	3249	3353	3456	3559	3663	3766	3869	3973	4076	4179	104
21	4282	4385	4488	4591	4695	4798	4901	5004	5107	5210	104
22	5312	5415	5518	5621	5724	5827	5929	6032	6135	6238	103
23	6340	6443	6546	6648	6751	6853	6956	7058	7161	7263	103
24	7366	7468	7571	7673	7775	7878	7980	8082	8185	8287	103
25	8389	8491	8593	8695	8797	8900	9002	9104	9206	9308	103
26	62 9410	9512	9613	9715	9817	9919	*0021	*0123	*0224	*0326	102
27	63 0428	0530	0631	0733	0835	0936	1038	1139	1241	1342	102
28	1444	1545	1647	1748	1849	1951	2052	2153	2255	2356	102
29	2457	2559	2660	2761	2862	2963	3064	3165	3266	3367	102
430	3468	3569	3670	3771	3872	3973	4074	4175	4276	4376	101
31	4477	4578	4679	4779	4880	4981	5081	5182	5283	5383	101
32	5484	5584	5685	5785	5886	5986	6087	6187	6287	6388	101
33	6488	6588	6688	6789	6889	6989	7089	7189	7290	7390	101
34	7490	7590	7690	7790	7890	7990	8090	8190	8290	8389	100
35	8489	8589	8689	8789	8888	8988	9088	9188	9287	9387	100
36	63 9486	9586	9686	9785	9885	9984	*0084	*0183	*0283	*0382	100
37	64 0481	0581	0680	0779	0879	0978	1077	1177	1276	1375	100
38	1474	1573	1672	1771	1871	1970	2069	2168	2267	2366	100
39	2465	2563	2662	2761	2860	2959	3058	3156	3255	3354	99
440	3453	3551	3650	3749	3847	3946	4044	4143	4242	4340	99
41	4439	4537	4636	4734	4832	4931	5029	5127	5226	5324	99
42	5422	5521	5619	5717	5815	5913	6011	6110	6208	6306	99
43	6404	6502	6600	6698	6796	6894	6992	7089	7187	7285	98
44	7383	7481	7579	7676	7774	7872	7969	8067	8165	8262	98
45	8360	8458	8555	8653	8750	8848	8945	9043	9140	9237	98
46	64 9335	9432	9530	9627	9724	9821	9919	*0016	*0113	*0210	98
47	65 0308	0405	0502	0599	0696	0793	0890	0987	1084	1181	97
48	1278	1375	1472	1569	1666	1762	1859	1956	2053	2150	97
49	2246	2343	2440	2536	2633	2730	2826	2923	3019	3116	97
N	**0**	**1**	**2**	**3**	**4**	**5**	**6**	**7**	**8**	**9**	**D**

LOGARITHMS OF NUMBERS 4,500–4,999

Six-Place Mantissas

N	0	1	2	3	4	5	6	7	8	9	D
450	65 3213	3309	3405	3502	3598	3695	3791	3888	3984	4080	97
51	4177	4273	4369	4465	4562	4658	4754	4850	4946	5042	97
52	5138	5235	5331	5427	5523	5619	5715	5810	5906	6002	97
53	6098	6194	6290	6386	6482	6577	6673	6769	6864	6960	96
54	7056	7152	7247	7343	7438	7534	7629	7725	7820	7916	96
55	8011	8107	8202	8298	8393	8488	8584	8679	8774	8870	96
56	8965	9060	9155	9250	9346	9441	9536	9631	9726	9821	96
57	65 9916	*0011	*0106	*0201	*0296	*0391	*0486	*0581	*0676	*0771	95
58	66 0865	0960	1055	1150	1245	1339	1434	1529	1623	1718	95
59	1813	1907	2002	2096	2191	2286	2380	2475	2569	2663	95
460	2758	2852	2947	3041	3135	3230	3324	3418	3512	3607	95
61	3701	3795	3889	3983	4078	4172	4266	4360	4454	4548	95
62	4642	4736	4830	4924	5018	5112	5206	5299	5393	5487	94
63	5581	5675	5769	5862	5956	6050	6143	6237	6331	6424	94
64	6518	6612	6705	6799	6892	6986	7079	7173	7266	7360	94
65	7453	7546	7640	7733	7826	7920	8013	8106	8199	8293	94
66	8386	8479	8572	8665	8759	8852	8945	9038	9131	9224	94
67	66 9317	9410	9503	9596	9689	9782	9875	9967	*0060	*0153	93
68	67 0246	0339	0431	0524	0617	0710	0802	0895	0988	1080	93
69	1173	1265	1358	1451	1543	1636	1728	1821	1913	2005	93
470	2098	2190	2283	2375	2467	2560	2652	2744	2836	2929	93
71	3021	3113	3205	3297	3390	3482	3574	3666	3758	3850	93
72	3942	4034	4126	4218	4310	4402	4494	4586	4677	4769	92
73	4861	4953	5045	5137	5228	5320	5412	5503	5595	5687	92
74	5778	5870	5962	6053	6145	6236	6328	6419	6511	6602	92
75	6694	6785	6876	6968	7059	7151	7242	7333	7424	7516	92
76	7607	7698	7789	7881	7972	8063	8154	8245	8336	8427	92
77	8518	8609	8700	8791	8882	8973	9064	9155	9246	9337	91
78	67 9428	9519	9610	9700	9791	9882	9973	*0063	*0154	*0245	91
79	68 0336	0426	0517	0607	0698	0789	0879	0970	1060	1151	91
480	1241	1332	1422	1513	1603	1693	1784	1874	1964	2055	91
81	2145	2235	2326	2416	2506	2596	2686	2777	2867	2957	91
82	3047	3137	3227	3317	3407	3497	3587	3677	3767	3857	90
83	3947	4037	4127	4217	4307	4396	4486	4576	4666	4756	90
84	4845	4935	5025	5114	5204	5294	5383	5473	5563	5652	90
85	5742	5831	5921	6010	6100	6189	6279	6368	6458	6547	90
86	6636	6726	6815	6904	6994	7083	7172	7261	7351	7440	90
87	7529	7618	7707	7796	7886	7975	8064	8153	8242	8331	90
88	8420	8509	8598	8687	8776	8865	8953	9042	9131	9220	89
89	68 9309	9398	9486	9575	9664	9753	9841	9930	*0019	*0107	89
490	69 0196	0285	0373	0462	0550	0639	0728	0816	0905	0993	89
91	1081	1170	1258	1347	1435	1524	1612	1700	1789	1877	89
92	1965	2053	2142	2230	2318	2406	2494	2583	2671	2759	89
93	2847	2935	3023	3111	3199	3287	3375	3463	3551	3639	88
94	3727	3815	3903	3991	4078	4166	4254	4342	4430	4517	88
95	4605	4693	4781	4868	4956	5044	5131	5219	5307	5394	88
96	5482	5569	5657	5744	5832	5919	6007	6094	6182	6269	88
97	6356	6444	6531	6618	6706	6793	6880	6968	7055	7142	88
98	7229	7317	7404	7491	7578	7665	7752	7839	7926	8014	88
99	8101	8188	8275	8362	8449	8535	8622	8709	8796	8883	87
N	0	1	2	3	4	5	6	7	8	9	D

LOGARITHMS OF NUMBERS 5,000–5,499

Six-Place Mantissas

N	0	1	2	3	4	5	6	7	8	9	D
500	69 8970	9057	9144	9231	9317	9404	9491	9578	9664	9751	87
01	69 9838	9924	*0011	*0098	*0184	*0271	*0358	*0444	*0531	*0617	87
02	70 0704	0790	0877	0963	1050	1136	1222	1309	1395	1482	87
03	1568	1654	1741	1827	1913	1999	2086	2172	2258	2344	87
04	2431	2517	2603	2689	2775	2861	2947	3033	3119	3205	86
05	3291	3377	3463	3549	3635	3721	3807	3893	3979	4065	86
06	4151	4236	4322	4408	4494	4579	4665	4751	4837	4922	86
07	5008	5094	5179	5265	5350	5436	5522	5607	5693	5778	86
08	5864	5949	6035	6120	6206	6291	6376	6462	6547	6632	86
09	6718	6803	6888	6974	7059	7144	7229	7315	7400	7485	86
510	7570	7655	7740	7826	7911	7996	8081	8166	8251	8336	86
11	8421	8506	8591	8676	8761	8846	8931	9015	9100	9185	85
12	70 9270	9355	9440	9524	9609	9694	9779	9863	9948	*0033	85
13	71 0117	0202	0287	0371	0456	0540	0625	0710	0794	0879	85
14	0963	1048	1132	1217	1301	1385	1470	1554	1639	1723	85
15	1807	1892	1976	2060	2144	2229	2313	2397	2481	2566	85
16	2650	2734	2818	2902	2986	3070	3154	3238	3323	3407	85
17	3491	3575	3659	3742	3826	3910	3994	4078	4162	4246	84
18	4330	4414	4497	4581	4665	4749	4833	4916	5000	5084	84
19	5167	5251	5335	5418	5502	5586	5669	5753	5836	5920	84
520	6003	6087	6170	6254	6337	6421	6504	6588	6671	6754	84
21	6838	6921	7004	7088	7171	7254	7338	7421	7504	7587	84
22	7671	7754	7837	7920	8003	8086	8169	8253	8336	8419	84
23	8502	8585	8668	8751	8834	8917	9000	9083	9165	9248	83
24	71 9331	9414	9497	9580	9663	9745	9828	9911	9994	*0077	83
25	72 0159	0242	0325	0407	0490	0573	0655	0738	0821	0903	83
26	0986	1068	1151	1233	1316	1398	1481	1563	1646	1728	83
27	1811	1893	1975	2058	2140	2222	2305	2387	2469	2552	83
28	2634	2716	2798	2881	2963	3045	3127	3209	3291	3374	83
29	3456	3538	3620	3702	3784	3866	3948	4030	4112	4194	82
530	4276	4358	4440	4522	4604	4685	4767	4849	4931	5013	82
31	5095	5176	5258	5340	5422	5503	5585	5667	5748	5830	82
32	5912	5993	6075	6156	6238	6320	6401	6483	6564	6646	82
33	6727	6809	6890	6972	7053	7134	7216	7297	7379	7460	82
34	7541	7623	7704	7785	7866	7948	8029	8110	8191	8273	82
35	8354	8435	8516	8597	8678	8759	8841	8922	9003	9084	82
36	9165	9246	9327	9408	9489	9570	9651	9732	9813	9893	81
37	72 9974	*0055	*0136	*0217	*0298	*0378	*0459	*0540	*0621	*0702	81
38	73 0782	0863	0944	1024	1105	1186	1266	1347	1428	1508	81
39	1589	1669	1750	1830	1911	1991	2072	2152	2233	2313	81
540	2394	2474	2555	2635	2715	2796	2876	2956	3037	3117	81
41	3197	3278	3358	3438	3518	3598	3679	3759	3839	3919	81
42	3999	4079	4160	4240	4320	4400	4480	4560	4640	4720	81
43	4800	4880	4960	5040	5120	5200	5279	5359	5439	5519	80
44	5599	5679	5759	5838	5918	5998	6078	6157	6237	6317	80
45	6397	6476	6556	6635	6715	6795	6874	6954	7034	7113	80
46	7193	7272	7352	7431	7511	7590	7670	7749	7829	7908	80
47	7987	8067	8146	8225	8305	8384	8463	8543	8622	8701	80
48	8781	8860	8939	9018	9097	9177	9256	9335	9414	9493	80
49	73 9572	9651	9731	9810	9889	9968	*0047	*0126	*0205	*0284	80
N	**0**	**1**	**2**	**3**	**4**	**5**	**6**	**7**	**8**	**9**	**D**

E

LOGARITHMS OF NUMBERS 5,500–5,999
Six-Place Mantissas

N	0	1	2	3	4	5	6	7	8	9	D
550	74 0363	0442	0521	0600	0678	0757	0836	0915	0994	1073	79
51	1152	1230	1309	1388	1467	1546	1624	1703	1782	1860	79
52	1939	2018	2096	2175	2254	2332	2411	2489	2568	2647	79
53	2725	2804	2882	2961	3039	3118	3196	3275	3353	3431	79
54	3510	3588	3667	3745	3823	3902	3980	4058	4136	4215	79
55	4293	4371	4449	4528	4606	4684	4762	4840	4919	4997	79
56	5075	5153	5231	5309	5387	5465	5543	5621	5699	5777	78
57	5855	5933	6011	6089	6167	6245	6323	6401	6479	6556	78
58	6634	6712	6790	6868	6945	7023	7101	7179	7256	7334	78
59	7412	7489	7567	7645	7722	7800	7878	7955	8033	8110	78
560	8188	8266	8343	8421	8498	8576	8653	8731	8808	8885	78
61	8963	9040	9118	9195	9272	9350	9427	9504	9582	9659	78
62	74 9736	9814	9891	9968	*0045	*0123	*0200	*0277	*0354	*0431	78
63	75 0508	0586	0663	0740	0817	0894	0971	1048	1125	1202	78
64	1279	1356	1433	1510	1587	1664	1741	1818	1895	1972	77
65	2048	2125	2202	2279	2356	2433	2509	2586	2663	2740	77
66	2816	2893	2970	3047	3123	3200	3277	3353	3430	3506	77
67	3583	3660	3736	3813	3889	3966	4042	4119	4195	4272	77
68	4348	4425	4501	4578	4654	4730	4807	4883	4960	5036	77
69	5112	5189	5265	5341	5417	5494	5570	5646	5722	5799	77
570	5875	5951	6027	6103	6180	6256	6332	6408	6484	6560	77
71	6636	6712	6788	6864	6940	7016	7092	7168	7244	7320	76
72	7396	7472	7548	7624	7700	7775	7851	7927	8003	8079	76
73	8155	8230	8306	8382	8458	8533	8609	8685	8761	8836	76
74	8912	8988	9063	9139	9214	9290	9366	9441	9517	9592	76
75	75 9668	9743	9819	9894	9970	*0045	*0121	*0196	*0272	*0347	76
76	76 0422	0498	0573	0649	0724	0799	0875	0950	1025	1101	76
77	1176	1251	1326	1402	1477	1552	1627	1702	1778	1853	76
78	1928	2003	2078	2153	2228	2303	2378	2453	2529	2604	76
79	2679	2754	2829	2904	2978	3053	3128	3203	3278	3353	75
580	3428	3503	3578	3653	3727	3802	3877	3952	4027	4101	75
81	4176	4251	4326	4400	4475	4550	4624	4699	4774	4848	75
82	4923	4998	5072	5147	5221	5296	5370	5445	5520	5594	75
83	5669	5743	5818	5892	5966	6041	6115	6190	6264	6338	75
84	6413	6487	6562	6636	6710	6785	6859	6933	7007	7082	75
85	7156	7230	7304	7379	7453	7527	7601	7675	7749	7823	75
86	7898	7972	8046	8120	8194	8268	8342	8416	8490	8564	74
87	8638	8712	8786	8860	8934	9008	9082	9156	9230	9303	74
88	76 9377	9451	9525	9599	9673	9746	9820	9894	9968	*0042	74
89	77 0115	0189	0263	0336	0410	0484	0557	0631	0705	0778	74
590	0852	0926	0999	1073	1146	1220	1293	1367	1440	1514	74
91	1587	1661	1734	1808	1881	1955	2028	2102	2175	2248	74
92	2322	2395	2468	2542	2615	2688	2762	2835	2908	2981	74
93	3055	3128	3201	3274	3348	3421	3494	3567	3640	3713	74
94	3786	3860	3933	4006	4079	4152	4225	4298	4371	4444	74
95	4517	4590	4663	4736	4809	4882	4955	5028	5100	5173	73
96	5246	5319	5392	5465	5538	5610	5683	5756	5829	5902	73
97	5974	6047	6120	6193	6265	6338	6411	6483	6556	6629	73
98	6701	6774	6846	6919	6992	7064	7137	7209	7282	7354	73
99	7427	7499	7572	7644	7717	7789	7862	7934	8006	8079	73
N	**0**	**1**	**2**	**3**	**4**	**5**	**6**	**7**	**8**	**9**	**D**

LOGARITHMS OF NUMBERS 6,000–6,499
Six-Place Mantissas

N	0	1	2	3	4	5	6	7	8	9	D
600	77 8151	8224	8296	8368	8441	8513	8585	8658	8730	8802	73
01	8874	8947	9019	9091	9163	9236	9308	9380	9452	9524	73
02	77 9596	9669	9741	9813	9885	9957	*0029	*0101	*0173	*0245	73
03	78 0317	0389	0461	0533	0605	0677	0749	0821	0893	0965	72
04	1037	1109	1181	1253	1324	1396	1468	1540	1612	1684	72
05	1755	1827	1899	1971	2042	2114	2186	2258	2329	2401	72
06	2473	2544	2616	2688	2759	2831	2902	2974	3046	3117	72
07	3189	3260	3332	3403	3475	3546	3618	3689	3761	3832	72
08	3904	3975	4046	4118	4189	4261	4332	4403	4475	4546	72
09	4617	4689	4760	4831	4902	4974	5045	5116	5187	5259	72
610	5330	5401	5472	5543	5615	5686	5757	5828	5899	5970	72
11	6041	6112	6183	6254	6325	6396	6467	6538	6609	6680	71
12	6751	6822	6893	6964	7035	7106	7177	7248	7319	7390	71
13	7460	7531	7602	7673	7744	7815	7885	7956	8027	8098	71
14	8168	8239	8310	8381	8451	8522	8593	8663	8734	8804	71
15	8875	8946	9016	9087	9157	9228	9299	9369	9440	9510	71
16	78 9581	9651	9722	9792	9863	9933	*0004	*0074	*0144	*0215	71
17	79 0285	0356	0426	0496	0567	0637	0707	0778	0848	0918	71
18	0988	1059	1129	1199	1269	1340	1410	1480	1550	1620	71
19	1691	1761	1831	1901	1971	2041	2111	2181	2252	2322	71
620	2392	2462	2532	2602	2672	2742	2812	2882	2952	3022	70
21	3092	3162	3231	3301	3371	3441	3511	3581	3651	3721	70
22	3790	3860	3930	4000	4070	4139	4209	4279	4349	4418	70
23	4488	4558	4627	4697	4767	4836	4906	4976	5045	5115	70
24	5185	5254	5324	5393	5463	5532	5602	5672	5741	5811	70
25	5880	5949	6019	6088	6158	6227	6297	6366	6436	6505	70
26	6574	6644	6713	6782	6852	6921	6990	7060	7129	7198	70
27	7268	7337	7406	7475	7545	7614	7683	7752	7821	7890	70
28	7960	8029	8098	8167	8236	8305	8374	8443	8513	8582	70
29	8651	8720	8789	8858	8927	8996	9065	9134	9203	9272	69
630	79 9341	9409	9478	9547	9616	9685	9754	9823	9892	9961	69
31	80 0029	0098	0167	0236	0305	0373	0442	0511	0580	0648	69
32	0717	0786	0854	0923	0992	1061	1129	1198	1266	1335	69
33	1404	1472	1541	1609	1678	1747	1815	1884	1952	2021	69
34	2089	2158	2226	2295	2363	2432	2500	2568	2637	2705	69
35	2774	2842	2910	2979	3047	3116	3184	3252	3321	3389	69
36	3457	3525	3594	3662	3730	3798	3867	3935	4003	4071	69
37	4139	4208	4276	4344	4412	4480	4548	4616	4685	4753	69
38	4821	4889	4957	5025	5093	5161	5229	5297	5365	5433	68
39	5501	5569	5637	5705	5773	5841	5908	5976	6044	6112	68
640	6180	6248	6316	6384	6451	6519	6587	6655	6723	6790	68
41	6858	6926	6994	7061	7129	7197	7264	7332	7400	7467	68
42	7535	7603	7670	7738	7806	7873	7941	8008	8076	8143	68
43	8211	8279	8346	8414	8481	8549	8616	8684	8751	8818	68
44	8886	8953	9021	9088	9156	9223	9290	9358	9425	9492	68
45	80 9560	9627	9694	9762	9829	9896	9964	*0031	*0098	*0165	68
46	81 0233	0300	0367	0434	0501	0569	0636	0703	0770	0837	68
47	0904	0971	1039	1106	1173	1240	1307	1374	1441	1508	68
48	1575	1642	1709	1776	1843	1910	1977	2044	2111	2178	67
49	2245	2312	2379	2445	2512	2579	2646	2713	2780	2847	67
N	**0**	**1**	**2**	**3**	**4**	**5**	**6**	**7**	**8**	**9**	**D**

LOGARITHMS OF NUMBERS 6,500–6,999
Six-Place Mantissas

N	0	1	2	3	4	5	6	7	8	9	D
650	81 2913	2980	3047	3114	3181	3247	3314	3381	3448	3514	67
51	3581	3648	3714	3781	3848	3914	3981	4048	4114	4181	67
52	4248	4314	4381	4447	4514	4581	4647	4714	4780	4847	67
53	4913	4980	5046	5113	5179	5246	5312	5378	5445	5511	67
54	5578	5644	5711	5777	5843	5910	5976	6042	6109	6175	67
55	6241	6308	6374	6440	6506	6573	6639	6705	6771	6838	67
56	6904	6970	7036	7102	7169	7235	7301	7367	7433	7499	67
57	7565	7631	7698	7764	7830	7896	7962	8028	8094	8160	67
58	8226	8292	8358	8424	8490	8556	8622	8688	8754	8820	66
59	8885	8951	9017	9083	9149	9215	9281	9346	9412	9478	66
660	81 9544	9610	9676	9741	9807	9873	9939	*0004	*0070	*0136	66
61	82 0201	0267	0333	0399	0464	0530	0595	0661	0727	0792	66
62	0858	0924	0989	1055	1120	1186	1251	1317	1382	1448	66
63	1514	1579	1645	1710	1775	1841	1906	1972	2037	2103	66
64	2168	2233	2299	2364	2430	2495	2560	2626	2691	2756	66
65	2822	2887	2952	3018	3083	3148	3213	3279	3344	3409	66
66	3474	3539	3605	3670	3735	3800	3865	3930	3996	4061	66
67	4126	4191	4256	4321	4386	4451	4516	4581	4646	4711	65
68	4776	4841	4906	4971	5036	5101	5166	5231	5296	5361	65
69	5426	5491	5556	5621	5686	5751	5815	5880	5945	6010	65
670	6075	6140	6204	6269	6334	6399	6464	6528	6593	6658	65
71	6723	6787	6852	6917	6981	7046	7111	7175	7240	7305	65
72	7369	7434	7499	7563	7628	7692	7757	7821	7886	7951	65
73	8015	8080	8144	8209	8273	8338	8402	8467	8531	8595	65
74	8660	8724	8789	8853	8918	8982	9046	9111	9175	9239	65
75	9304	9368	9432	9497	9561	9625	9690	9754	9818	9882	65
76	82 9947	*0011	*0075	*0139	*0204	*0268	*0332	*0396	*0460	*0525	65
77	83 0589	0653	0717	0781	0845	0909	0973	1037	1102	1166	65
78	1230	1294	1358	1422	1486	1550	1614	1678	1742	1806	64
79	1870	1934	1998	2062	2126	2189	2253	2317	2381	2445	64
680	2509	2573	2637	2700	2764	2828	2892	2956	3020	3083	64
81	3147	3211	3275	3338	3402	3466	3530	3593	3657	3721	64
82	3784	3848	3912	3975	4039	4103	4166	4230	4294	4357	64
83	4421	4484	4548	4611	4675	4739	4802	4866	4929	4993	64
84	5056	5120	5183	5247	5310	5373	5437	5500	5564	5627	64
85	5691	5754	5817	5881	5944	6007	6071	6134	6197	6261	64
86	6324	6387	6451	6514	6577	6641	6704	6767	6830	6894	64
87	6957	7020	7083	7146	7210	7273	7336	7399	7462	7525	64
88	7588	7652	7715	7778	7841	7904	7967	8030	8093	8156	64
89	8219	8282	8345	8408	8471	8534	8597	8660	8723	8786	63
690	8849	8912	8975	9038	9101	9164	9227	9289	9352	9415	63
91	83 9478	9541	9604	9667	9729	9792	9855	9918	9981	*0043	63
92	84 0106	0169	0232	0294	0357	0420	0482	0545	0608	0671	63
93	0733	0796	0859	0921	0984	1046	1109	1172	1234	1297	63
94	1359	1422	1485	1547	1610	1672	1735	1797	1860	1922	63
95	1985	2047	2110	2172	2235	2297	2360	2422	2484	2547	63
96	2609	2672	2734	2796	2859	2921	2983	3046	3108	3170	63
97	3233	3295	3357	3420	3482	3544	3606	3669	3731	3793	63
98	3855	3918	3980	4042	4104	4166	4229	4291	4353	4415	63
99	4477	4539	4601	4664	4726	4788	4850	4912	4974	5036	63
N	0	1	2	3	4	5	6	7	8	9	D

E

LOGARITHMS OF NUMBERS 7,000–7,499
Six-Place Mantissas

N	0	1	2	3	4	5	6	7	8	9	D
700	84 5098	5160	5222	5284	5346	5408	5470	5532	5594	5656	62
01	5718	5780	5842	5904	5966	6028	6090	6151	6213	6275	62
02	6337	6399	6461	6523	6585	6646	6708	6770	6832	6894	62
03	6955	7017	7079	7141	7202	7264	7326	7388	7449	7511	62
04	7573	7634	7696	7758	7819	7881	7943	8004	8066	8128	62
05	8189	8251	8312	8374	8435	8497	8559	8620	8682	8743	62
06	8805	8866	8928	8989	9051	9112	9174	9235	9297	9358	62
07	84 9419	9481	9542	9604	9665	9726	9788	9849	9911	9972	62
08	85 0033	0095	0156	0217	0279	0340	0401	0462	0524	0585	62
09	0646	0707	0769	0830	0891	0952	1014	1075	1136	1197	62
710	1258	1320	1381	1442	1503	1564	1625	1686	1747	1809	62
11	1870	1931	1992	2053	2114	2175	2236	2297	2358	2419	61
12	2480	2541	2602	2663	2724	2785	2846	2907	2968	3029	61
13	3090	3150	3211	3272	3333	3394	3455	3516	3577	3637	61
14	3698	3759	3820	3881	3941	4002	4063	4124	4185	4245	61
15	4306	4367	4428	4488	4549	4610	4670	4731	4792	4852	61
16	4913	4974	5034	5095	5156	5216	5277	5337	5398	5459	61
17	5519	5580	5640	5701	5761	5822	5882	5943	6003	6064	61
18	6124	6185	6245	6306	6366	6427	6487	6548	6608	6668	61
19	6729	6789	6850	6910	6970	7031	7091	7152	7212	7272	61
720	7332	7393	7453	7513	7574	7634	7694	7755	7815	7875	61
21	7935	7995	8056	8116	8176	8236	8297	8357	8417	8477	61
22	8537	8597	8657	8718	8778	8838	8898	8958	9018	9078	61
23	9138	9198	9258	9318	9379	9439	9499	9559	9619	9679	61
24	85 9739	9799	9859	9918	9978	*0038	*0098	*0158	*0218	*0278	60
25	86 0338	0398	0458	0518	0578	0637	0697	0757	0817	0877	60
26	0937	0996	1056	1116	1176	1236	1295	1355	1415	1475	60
27	1534	1594	1654	1714	1773	1833	1893	1952	2012	2072	60
28	2131	2191	2251	2310	2370	2430	2489	2549	2608	2668	60
29	2728	2787	2847	2906	2966	3025	3085	3144	3204	3263	60
730	3323	3382	3442	3501	3561	3620	3680	3739	3799	3858	60
31	3917	3977	4036	4096	4155	4214	4274	4333	4392	4452	60
32	4511	4570	4630	4689	4748	4808	4867	4926	4985	5045	60
33	5104	5163	5222	5282	5341	5400	5459	5519	5578	5637	60
34	5696	5755	5814	5874	5933	5992	6051	6110	6169	6228	60
35	6287	6346	6405	6465	6524	6583	6642	6701	6760	6819	60
36	6878	6937	6996	7055	7114	7173	7232	7291	7350	7409	59
37	7467	7526	7585	7644	7703	7762	7821	7880	7939	7998	59
38	8056	8115	8174	8233	8292	8350	8409	8468	8527	8586	59
39	8644	8703	8762	8821	8879	8938	8997	9056	9114	9173	59
740	9232	9290	9349	9408	9466	9525	9584	9642	9701	9760	59
41	86 9818	9877	9935	9994	*0053	*0111	*0170	*0228	*0287	*0345	59
42	87 0404	0462	0521	0579	0638	0696	0755	0813	0872	0930	59
43	0989	1047	1106	1164	1223	1281	1339	1398	1456	1515	59
44	1573	1631	1690	1748	1806	1865	1923	1981	2040	2098	59
45	2156	2215	2273	2331	2389	2448	2506	2564	2622	2681	59
46	2739	2797	2855	2913	2972	3030	3088	3146	3204	3262	59
47	3321	3379	3437	3495	3553	3611	3669	3727	3785	3844	59
48	3902	3960	4018	4076	4134	4192	4250	4308	4366	4424	58
49	4482	4540	4598	4656	4714	4772	4830	4888	4945	5003	58
N	0	1	2	3	4	5	6	7	8	9	D

LOGARITHMS OF NUMBERS 7,500–7,999

Six-Place Mantissas

N	0	1	2	3	4	5	6	7	8	9	D
750	87 5061	5119	5177	5235	5293	5351	5409	5466	5524	5582	58
51	5640	5698	5756	5813	5871	5929	5987	6045	6102	6160	58
52	6218	6276	6333	6391	6449	6507	6564	6622	6680	6737	58
53	6795	6853	6910	6968	7026	7083	7141	7199	7256	7314	58
54	7371	7429	7487	7544	7602	7659	7717	7774	7832	7889	58
55	7947	8004	8062	8119	8177	8234	8292	8349	8407	8464	58
56	8522	8579	8637	8694	8752	8809	8866	8924	8981	9039	58
57	9096	9153	9211	9268	9325	9383	9440	9497	9555	9612	58
58	87 9669	9726	9784	9841	9898	9956	*0013	*0070	*0127	*0185	58
59	88 0242	0299	0356	0413	0471	0528	0585	0642	0699	0756	58
760	0814	0871	0928	0985	1042	1099	1156	1213	1271	1328	58
61	1385	1442	1499	1556	1613	1670	1727	1784	1841	1898	57
62	1955	2012	2069	2126	2183	2240	2297	2354	2411	2468	57
63	2525	2581	2638	2695	2752	2809	2866	2923	2980	3037	57
64	3093	3150	3207	3264	3321	3377	3434	3491	3548	3605	57
65	3661	3718	3775	3832	3888	3945	4002	4059	4115	4172	57
66	4229	4285	4342	4399	4455	4512	4569	4625	4682	4739	57
67	4795	4852	4909	4965	5022	5078	5135	5192	5248	5305	57
68	5361	5418	5474	5531	5587	5644	5700	5757	5813	5870	57
69	5926	5983	6039	6096	6152	6209	6265	6321	6378	6434	57
770	6491	6547	6604	6660	6716	6773	6829	6885	6942	6998	57
71	7054	7111	7167	7223	7280	7336	7392	7449	7505	7561	57
72	7617	7674	7730	7786	7842	7898	7955	8011	8067	8123	57
73	8179	8236	8292	8348	8404	8460	8516	8573	8629	8685	57
74	8741	8797	8853	8909	8965	9021	9077	9134	9190	9246	57
75	9302	9358	9414	9470	9526	9582	9638	9694	9750	9806	56
76	88 9862	9918	9974	*0030	*0086	*0141	*0197	*0253	*0309	*0365	56
77	89 0421	0477	0533	0589	0645	0700	0756	0812	0868	0924	56
78	0980	1035	1091	1147	1203	1259	1314	1370	1426	1482	56
79	1537	1593	1649	1705	1760	1816	1872	1928	1983	2039	56
780	2095	2150	2206	2262	2317	2373	2429	2484	2540	2595	56
81	2651	2707	2762	2818	2873	2929	2985	3040	3096	3151	56
82	3207	3262	3318	3373	3429	3484	3540	3595	3651	3706	56
83	3762	3817	3873	3928	3984	4039	4094	4150	4205	4261	56
84	4316	4371	4427	4482	4538	4593	4648	4704	4759	4814	56
85	4870	4925	4980	5036	5091	5146	5201	5257	5312	5367	56
86	5423	5478	5533	5588	5644	5699	5754	5809	5864	5920	56
87	5975	6030	6085	6140	6195	6251	6306	6361	6416	6471	56
88	6526	6581	6636	6692	6747	6802	6857	6912	6967	7022	56
89	7077	7132	7187	7242	7297	7352	7407	7462	7517	7572	55
790	7627	7682	7737	7792	7847	7902	7957	8012	8067	8122	55
91	8176	8231	8286	8341	8396	8451	8506	8561	8615	8670	55
92	8725	8780	8835	8890	8944	8999	9054	9109	9164	9218	55
93	9273	9328	9383	9437	9492	9547	9602	9656	9711	9766	55
94	89 9821	9875	9930	9985	*0039	*0094	*0149	*0203	*0258	*0312	55
95	90 0367	0422	0476	0531	0586	0640	0695	0749	0804	0859	55
96	0913	0968	1022	1077	1131	1186	1240	1295	1349	1404	55
97	1458	1513	1567	1622	1676	1731	1785	1840	1894	1948	55
98	2003	2057	2112	2166	2221	2275	2329	2384	2438	2492	55
99	2547	2601	2655	2710	2764	2818	2873	2927	2981	3036	55
N	**0**	**1**	**2**	**3**	**4**	**5**	**6**	**7**	**8**	**9**	**D**

LOGARITHMS OF NUMBERS 8,000–8,499
Six-Place Mantissas

N	0	1	2	3	4	5	6	7	8	9	D
800	90 3090	3144	3199	3253	3307	3361	3416	3470	3524	3578	55
01	3633	3687	3741	3795	3849	3904	3958	4012	4066	4120	55
02	4174	4229	4283	4337	4391	4445	4499	4553	4607	4661	55
03	4716	4770	4824	4878	4932	4986	5040	5094	5148	5202	54
04	5256	5310	5364	5418	5472	5526	5580	5634	5688	5742	54
05	5796	5850	5904	5958	6012	6066	6119	6173	6227	6281	54
06	6335	6389	6443	6497	6551	6604	6658	6712	6766	6820	54
07	6874	6927	6981	7035	7089	7143	7196	7250	7304	7358	54
08	7411	7465	7519	7573	7626	7680	7734	7787	7841	7895	54
09	7949	8002	8056	8110	8163	8217	8270	8324	8378	8431	54
810	8485	8539	8592	8646	8699	8753	8807	8860	8914	8967	54
11	9021	9074	9128	9181	9235	9289	9342	9396	9449	9503	54
12	90 9556	9610	9663	9716	9770	9823	9877	9930	9984	*0037	54
13	91 0091	0144	0197	0251	0304	0358	0411	0464	0518	0571	54
14	0624	0678	0731	0784	0838	0891	0944	0998	1051	1104	54
15	1158	1211	1264	1317	1371	1424	1477	1530	1584	1637	54
16	1690	1743	1797	1850	1903	1956	2009	2063	2116	2169	54
17	2222	2275	2328	2381	2435	2488	2541	2594	2647	2700	54
18	2753	2806	2859	2913	2966	3019	3072	3125	3178	3231	54
19	3284	3337	3390	3443	3496	3549	3602	3655	3708	3761	53
820	3814	3867	3920	3973	4026	4079	4132	4184	4237	4290	53
21	4343	4396	4449	4502	4555	4608	4660	4713	4766	4819	53
22	4872	4925	4977	5030	5083	5136	5189	5241	5294	5347	53
23	5400	5453	5505	5558	5611	5664	5716	5769	5822	5875	53
24	5927	5980	6033	6085	6138	6191	6243	6296	6349	6401	53
25	6454	6507	6559	6612	6664	6717	6770	6822	6875	6927	53
26	6980	7033	7085	7138	7190	7243	7295	7348	7400	7453	53
27	7506	7558	7611	7663	7716	7768	7820	7873	7925	7978	53
28	8030	8083	8135	8188	8240	8293	8345	8397	8450	8502	53
29	8555	8607	8659	8712	8764	8816	8869	8921	8973	9026	53
830	9078	9130	9183	9235	9287	9340	9392	9444	9496	9549	53
31	91 9601	9653	9706	9758	9810	9862	9914	9967	*0019	*0071	53
32	92 0123	0176	0228	0280	0332	0384	0436	0489	0541	0593	53
33	0645	0697	0749	0801	0853	0906	0958	1010	1062	1114	53
34	1166	1218	1270	1322	1374	1426	1478	1530	1582	1634	52
35	1686	1738	1790	1842	1894	1946	1998	2050	2102	2154	52
36	2206	2258	2310	2362	2414	2466	2518	2570	2622	2674	52
37	2725	2777	2829	2881	2933	2985	3037	3089	3140	3192	52
38	3244	3296	3348	3399	3451	3503	3555	3607	3658	3710	52
39	3762	3814	3865	3917	3969	4021	4072	4124	4176	4228	52
840	4279	4331	4383	4434	4486	4538	4589	4641	4693	4744	52
41	4796	4848	4899	4951	5003	5054	5106	5157	5209	5261	52
42	5312	5364	5415	5467	5518	5570	5621	5673	5725	5776	52
43	5828	5879	5931	5982	6034	6085	6137	6188	6240	6291	52
44	6342	6394	6445	6497	6548	6600	6651	6702	6754	6805	52
45	6857	6908	6959	7011	7062	7114	7165	7216	7268	7319	52
46	7370	7422	7473	7524	7576	7627	7678	7730	7781	7832	52
47	7883	7935	7986	8037	8088	8140	8191	8242	8293	8345	52
48	8396	8447	8498	8549	8601	8652	8703	8754	8805	8857	52
49	8908	8959	9010	9061	9112	9163	9215	9266	9317	9368	52
N	0	1	2	3	4	5	6	7	8	9	D

E

LOGARITHMS OF NUMBERS 8,500–8,999

Six-Place Mantissas

N	0	1	2	3	4	5	6	7	8	9	D
850	92 9419	9470	9521	9572	9623	9674	9725	9776	9827	9879	52
51	92 9930	9981	*0032	*0083	*0134	*0185	*0236	*0287	*0338	*0389	51
52	93 0440	0491	0542	0592	0643	0694	0745	0796	0847	0898	51
53	0949	1000	1051	1102	1153	1204	1254	1305	1356	1407	51
54	1458	1509	1560	1610	1661	1712	1763	1814	1865	1915	51
55	1966	2017	2068	2118	2169	2220	2271	2322	2372	2423	51
56	2474	2524	2575	2626	2677	2727	2778	2829	2879	2930	51
57	2981	3031	3082	3133	3183	3234	3285	3335	3386	3437	51
58	3487	3538	3589	3639	3690	3740	3791	3841	3892	3943	51
59	3993	4044	4094	4145	4195	4246	4296	4347	4397	4448	51
860	4498	4549	4599	4650	4700	4751	4801	4852	4902	4953	51
61	5003	5054	5104	5154	5205	5255	5306	5356	5406	5457	51
62	5507	5558	5608	5658	5709	5759	5809	5860	5910	5960	51
63	6011	6061	6111	6162	6212	6262	6313	6363	6413	6463	51
64	6514	6564	6614	6665	6715	6765	6815	6865	6916	6966	51
65	7016	7066	7117	7167	7217	7267	7317	7367	7418	7468	51
66	7518	7568	7618	7668	7718	7769	7819	7869	7919	7969	51
67	8019	8069	8119	8169	8219	8269	8320	8370	8420	8470	51
68	8520	8570	8620	8670	8720	8770	8820	8870	8920	8970	50
69	9020	9070	9120	9170	9220	9270	9320	9369	9419	9469	50
870	93 9519	9569	9619	9669	9719	9769	9819	9869	9918	9968	50
71	94 0018	0068	0118	0168	0218	0267	0317	0367	0417	0467	50
72	0516	0566	0616	0666	0716	0765	0815	0865	0915	0964	50
73	1014	1064	1114	1163	1213	1263	1313	1362	1412	1462	50
74	1511	1561	1611	1660	1710	1760	1809	1859	1909	1958	50
75	2008	2058	2107	2157	2207	2256	2306	2355	2405	2455	50
76	2504	2554	2603	2653	2702	2752	2801	2851	2901	2950	50
77	3000	3049	3099	3148	3198	3247	3297	3346	3396	3445	50
78	3495	3544	3593	3643	3692	3742	3791	3841	3890	3939	50
79	3989	4038	4088	4137	4186	4236	4285	4335	4384	4433	50
880	4483	4532	4581	4631	4680	4729	4779	4828	4877	4927	50
81	4976	5025	5074	5124	5173	5222	5272	5321	5370	5419	50
82	5469	5518	5567	5616	5665	5715	5764	5813	5862	5912	50
83	5961	6010	6059	6108	6157	6207	6256	6305	6354	6403	50
84	6452	6501	6551	6600	6649	6698	6747	6796	6845	6894	50
85	6943	6992	7041	7090	7140	7189	7238	7287	7336	7385	50
86	7434	7483	7532	7581	7630	7679	7728	7777	7826	7875	49
87	7924	7973	8022	8070	8119	8168	8217	8266	8315	8364	49
88	8413	8462	8511	8560	8609	8657	8706	8755	8804	8853	49
89	8902	8951	8999	9048	9097	9146	9195	9244	9292	9341	49
890	9390	9439	9488	9536	9585	9634	9683	9731	9780	9829	49
91	94 9878	9926	9975	*0024	*0073	*0121	*0170	*0219	*0267	*0316	49
92	95 0365	0414	0462	0511	0560	0608	0657	0706	0754	0803	49
93	0851	0900	0949	0997	1046	1095	1143	1192	1240	1289	49
94	1338	1386	1435	1483	1532	1580	1629	1677	1726	1775	49
95	1823	1872	1920	1969	2017	2066	2114	2163	2211	2260	49
96	2308	2356	2405	2453	2502	2550	2599	2647	2696	2744	49
97	2792	2841	2889	2938	2986	3034	3083	3131	3180	3228	49
98	3276	3325	3373	3421	3470	3518	3566	3615	3663	3711	49
99	3760	3808	3856	3905	3953	4001	4049	4098	4146	4194	49
N	0	1	2	3	4	5	6	7	8	9	D

LOGARITHMS OF NUMBERS 9,000–9,499
Six-Place Mantissas

N	0	1	2	3	4	5	6	7	8	9	D
900	95 4243	4291	4339	4387	4435	4484	4532	4580	4628	4677	49
01	4725	4773	4821	4869	4918	4966	5014	5062	5110	5158	49
02	5207	5255	5303	5351	5399	5447	5495	5543	5592	5640	49
03	5688	5736	5784	5832	5880	5928	5976	6024	6072	6120	48
04	6168	6216	6265	6313	6361	6409	6457	6505	6553	6601	48
05	6649	6697	6745	6793	6840	6888	6936	6984	7032	7080	48
06	7128	7176	7224	7272	7320	7368	7416	7464	7512	7559	48
07	7607	7655	7703	7751	7799	7847	7894	7942	7990	8038	48
08	8086	8134	8181	8229	8277	8325	8373	8421	8468	8516	48
09	8564	8612	8659	8707	8755	8803	8850	8898	8946	8994	48
910	9041	9089	9137	9185	9232	9280	9328	9375	9423	9471	48
11	9518	9566	9614	9661	9709	9757	9804	9852	9900	9947	48
12	95 9995	*0042	*0090	*0138	*0185	*0233	*0280	*0328	*0376	*0423	48
13	96 0471	0518	0566	0613	0661	0709	0756	0804	0851	0899	48
14	0946	0994	1041	1089	1136	1184	1231	1279	1326	1374	48
15	1421	1469	1516	1563	1611	1658	1706	1753	1801	1848	48
16	1895	1943	1990	2038	2085	2132	2180	2227	2275	2322	48
17	2369	2417	2464	2511	2559	2606	2653	2701	2748	2795	48
18	2843	2890	2937	2985	3032	3079	3126	3174	3221	3268	48
19	3316	3363	3410	3457	3504	3552	3599	3646	3693	3741	48
920	3788	3835	3882	3929	3977	4024	4071	4118	4165	4212	48
21	4260	4307	4354	4401	4448	4495	4542	4590	4637	4684	48
22	4731	4778	4825	4872	4919	4966	5013	5061	5108	5155	48
23	5202	5249	5296	5343	5390	5437	5484	5531	5578	5625	47
24	5672	5719	5766	5813	5860	5907	5954	6001	6048	6095	47
25	6142	6189	6236	6283	6329	6376	6423	6470	6517	6564	47
26	6611	6658	6705	6752	6799	6845	6892	6939	6986	7033	47
27	7080	7127	7173	7220	7267	7314	7361	7408	7454	7501	47
28	7548	7595	7642	7688	7735	7782	7829	7875	7922	7969	47
29	8016	8062	8109	8156	8203	8249	8296	8343	8390	8436	47
930	8483	8530	8576	8623	8670	8716	8763	8810	8856	8903	47
31	8950	8996	9043	9090	9136	9183	9229	9276	9323	9369	47
32	9416	9463	9509	9556	9602	9649	9695	9742	9789	9835	47
33	96 9882	9928	9975	*0021	*0068	*0114	*0161	*0207	*0254	*0300	47
34	97 0347	0393	0440	0486	0533	0579	0626	0672	0719	0765	47
35	0812	0858	0904	0951	0997	1044	1090	1137	1183	1229	47
36	1276	1322	1369	1415	1461	1508	1554	1601	1647	1693	47
37	1740	1786	1832	1879	1925	1971	2018	2064	2110	2157	47
38	2203	2249	2295	2342	2388	2434	2481	2527	2573	2619	47
39	2666	2712	2758	2804	2851	2897	2943	2989	3035	3082	47
940	3128	3174	3220	3266	3313	3359	3405	3451	3497	3543	47
41	3590	3636	3682	3728	3774	3820	3866	3913	3959	4005	47
42	4051	4097	4143	4189	4235	4281	4327	4374	4420	4466	47
43	4512	4558	4604	4650	4696	4742	4788	4834	4880	4926	46
44	4972	5018	5064	5110	5156	5202	5248	5294	5340	5386	46
45	5432	5478	5524	5570	5616	5662	5707	5753	5799	5845	46
46	5891	5937	5983	6029	6075	6121	6167	6212	6258	6304	46
47	6350	6396	6442	6488	6533	6579	6625	6671	6717	6763	46
48	6808	6854	6900	6946	6992	7037	7083	7129	7175	7220	46
49	7266	7312	7358	7403	7449	7495	7541	7586	7632	7678	46
N	0	1	2	3	4	5	6	7	8	9	D

LOGARITHMS OF NUMBERS 9,500–9,999
Six-Place Mantissas

N	0	1	2	3	4	5	6	7	8	9	D
950	97 7724	7769	7815	7861	7906	7952	7998	8043	8089	8135	46
51	8181	8226	8272	8317	8363	8409	8454	8500	8546	8591	46
52	8637	8683	8728	8774	8819	8865	8911	8956	9002	9047	46
53	9093	9138	9184	9230	9275	9321	9366	9412	9457	9503	46
54	97 9548	9594	9639	9685	9730	9776	9821	9867	9912	9958	46
55	98 0003	0049	0094	0140	0185	0231	0276	0322	0367	0412	46
56	0458	0503	0549	0594	0640	0685	0730	0776	0821	0867	46
57	0912	0957	1003	1048	1093	1139	1184	1229	1275	1320	46
58	1366	1411	1456	1501	1547	1592	1637	1683	1728	1773	46
59	1819	1864	1909	1954	2000	2045	2090	2135	2181	2226	46
960	2271	2316	2362	2407	2452	2497	2543	2588	2633	2678	46
61	2723	2769	2814	2859	2904	2949	2994	3040	3085	3130	46
62	3175	3220	3265	3310	3356	3401	3446	3491	3536	3581	46
63	3626	3671	3716	3762	3807	3852	3897	3942	3987	4032	46
64	4077	4122	4167	4212	4257	4302	4347	4392	4437	4482	45
65	4527	4572	4617	4662	4707	4752	4797	4842	4887	4932	45
66	4977	5022	5067	5112	5157	5202	5247	5292	5337	5382	45
67	5426	5471	5516	5561	5606	5651	5696	5741	5786	5830	45
68	5875	5920	5965	6010	6055	6100	6144	6189	6234	6279	45
69	6324	6369	6413	6458	6503	6548	6593	6637	6682	6727	45
970	6772	6817	6861	6906	6951	6996	7040	7085	7130	7175	45
71	7219	7264	7309	7353	7398	7443	7488	7532	7577	7622	45
72	7666	7711	7756	7800	7845	7890	7934	7979	8024	8068	45
73	8113	8157	8202	8247	8291	8336	8381	8425	8470	8514	45
74	8559	8604	8648	8693	8737	8782	8826	8871	8916	8960	45
75	9005	9049	9094	9138	9183	9227	9272	9316	9361	9405	45
76	9450	9494	9539	9583	9628	9672	9717	9761	9806	9850	45
77	98 9895	9939	9983	*0028	*0072	*0117	*0161	*0206	*0250	*0294	45
78	99 0339	0383	0428	0472	0516	0561	0605	0650	0694	0738	45
79	0783	0827	0871	0916	0960	1004	1049	1093	1137	1182	45
980	1226	1270	1315	1359	1403	1448	1492	1536	1580	1625	45
81	1669	1713	1758	1802	1846	1890	1935	1979	2023	2067	45
82	2111	2156	2200	2244	2288	2333	2377	2421	2465	2509	45
83	2554	2598	2642	2686	2730	2774	2819	2863	2907	2951	45
84	2995	3039	3083	3127	3172	3216	3260	3304	3348	3392	45
85	3436	3480	3524	3568	3613	3657	3701	3745	3789	3833	45
86	3877	3921	3965	4009	4053	4097	4141	4185	4229	4273	44
87	4317	4361	4405	4449	4493	4537	4581	4625	4669	4713	44
88	4757	4801	4845	4889	4933	4977	5021	5065	5108	5152	44
89	5196	5240	5284	5328	5372	5416	5460	5504	5547	5591	44
990	5635	5679	5723	5767	5811	5854	5898	5942	5986	6030	44
91	6074	6117	6161	6205	6249	6293	6337	6380	6424	6468	44
92	6512	6555	6599	6643	6687	6731	6774	6818	6862	6906	44
93	6949	6993	7037	7080	7124	7168	7212	7255	7299	7343	44
94	7386	7430	7474	7517	7561	7605	7648	7692	7736	7779	44
95	7823	7867	7910	7954	7998	8041	8085	8129	8172	8216	44
96	8259	8303	8347	8390	8434	8477	8521	8564	8608	8652	44
97	8695	8739	8782	8826	8869	8913	8956	9000	9043	9087	44
98	9131	9174	9218	9261	9305	9348	9392	9435	9479	9522	44
99	99 9565	9609	9652	9696	9739	9783	9826	9870	9913	9957	44
N	0	1	2	3	4	5	6	7	8	9	D

BINOMIAL PROBABILITY DISTRIBUTION

$$P(r \mid n, p) = \binom{n}{r} p^{r} q^{n-r}$$

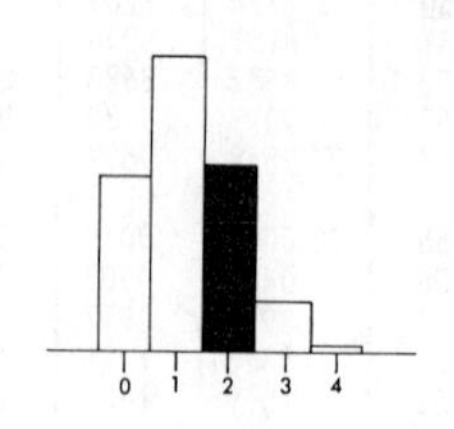

$P(r = 2 \mid n = 4, p = 0.3) = 0.2646$

n = 1

r \ p	.01	.02	.03	.04	.05	.06	.07	.08	.09	.10
0	.9900	.9800	.9700	.9600	.9500	.9400	.9300	.9200	.9100	.9000
1	.0100	.0200	.0300	.0400	.0500	.0600	.0700	.0800	.0900	.1000
	.11	.12	.13	.14	.15	.16	.17	.18	.19	.20
0	.8900	.8800	.8700	.8600	.8500	.8400	.8300	.8200	.8100	.8000
1	.1100	.1200	.1300	.1400	.1500	.1600	.1700	.1800	.1900	.2000
	.21	.22	.23	.24	.25	.26	.27	.28	.29	.30
0	.7900	.7800	.7700	.7600	.7500	.7400	.7300	.7200	.7100	.7000
1	.2100	.2200	.2300	.2400	.2500	.2600	.2700	.2800	.2900	.3000
	.31	.32	.33	.34	.35	.36	.37	.38	.39	.40
0	.6900	.6800	.6700	.6600	.6500	.6400	.6300	.6200	.6100	.6000
1	.3100	.3200	.3300	.3400	.3500	.3600	.3700	.3800	.3900	.4000
	.41	.42	.43	.44	.45	.46	.47	.48	.49	.50
0	.5900	.5800	.5700	.5600	.5500	.5400	.5300	.5200	.5100	.5000
1	.4100	.4200	.4300	.4400	.4500	.4600	.4700	.4800	.4900	.5000

n = 2

r \ p	.01	.02	.03	.04	.05	.06	.07	.08	.09	.10
0	.9801	.9604	.9409	.9216	.9025	.8836	.8649	.8464	.8281	.8100
1	.0198	.0392	.0582	.0768	.0950	.1128	.1302	.1472	.1638	.1800
2	.0001	.0004	.0009	.0016	.0025	.0036	.0049	.0064	.0081	.0100
	.11	.12	.13	.14	.15	.16	.17	.18	.19	.20
0	.7921	.7744	.7569	.7396	.7225	.7056	.6889	.6724	.6561	.6400
1	.1958	.2112	.2262	.2408	.2550	.2688	.2822	.2952	.3078	.3200
2	.0121	.0144	.0169	.0196	.0225	.0256	.0289	.0324	.0361	.0400
	.21	.22	.23	.24	.25	.26	.27	.28	.29	.30
0	.6241	.6084	.5929	.5776	.5625	.5476	.5329	.5184	.5041	.4900
1	.3318	.3432	.3542	.3648	.3750	.3848	.3942	.4032	.4118	.4200
2	.0441	.0484	.0529	.0576	.0625	.0676	.0729	.0784	.0841	.0900
	.31	.32	.33	.34	.35	.36	.37	.38	.39	.40
0	.4761	.4624	.4489	.4356	.4225	.4096	.3969	.3844	.3721	.3600
1	.4278	.4352	.4422	.4488	.4550	.4608	.4662	.4712	.4758	.4800
2	.0961	.1024	.1089	.1156	.1225	.1296	.1369	.1444	.1521	.1600
	.41	.42	.43	.44	.45	.46	.47	.48	.49	.50
0	.3481	.3364	.3249	.3136	.3025	.2916	.2809	.2704	.2601	.2500
1	.4838	.4872	.4902	.4928	.4950	.4968	.4982	.4992	.4998	.5000
2	.1681	.1764	.1849	.1936	.2025	.2116	.2209	.2304	.2401	.2500

n = 3

r \ p	.01	.02	.03	.04	.05	.06	.07	.08	.09	.10
0	.9704	.9412	.9127	.8847	.8574	.8306	.8044	.7787	.7536	.7290
1	.0294	.0576	.0847	.1106	.1354	.1590	.1816	.2031	.2236	.2430
2	.0003	.0012	.0026	.0046	.0071	.0102	.0137	.0177	.0221	.0270
3	.0000	.0000	.0000	.0001	.0001	.0002	.0003	.0005	.0007	.0010
	.11	.12	.13	.14	.15	.16	.17	.18	.19	.20
0	.7050	.6815	.6585	.6361	.6141	.5927	.5718	.5514	.5314	.5120
1	.2614	.2788	.2952	.3106	.3251	.3387	.3513	.3631	.3740	.3840
2	.0323	.0380	.0441	.0506	.0574	.0645	.0720	.0797	.0877	.0960
3	.0013	.0017	.0022	.0027	.0034	.0041	.0049	.0058	.0069	.0080
	.21	.22	.23	.24	.25	.26	.27	.28	.29	.30
0	.4930	.4746	.4565	.4390	.4219	.4052	.3890	.3732	.3579	.3430
1	.3932	.4015	.4091	.4159	.4219	.4271	.4316	.4355	.4386	.4410
2	.1045	.1133	.1222	.1313	.1406	.1501	.1597	.1693	.1791	.1890
3	.0093	.0106	.0122	.0138	.0156	.0176	.0197	.0220	.0244	.0270
	.31	.32	.33	.34	.35	.36	.37	.38	.39	.40
0	.3285	.3144	.3008	.2875	.2746	.2621	.2500	.2383	.2270	.2160
1	.4428	.4439	.4444	.4443	.4436	.4424	.4406	.4382	.4354	.4320
2	.1989	.2089	.2189	.2289	.2389	.2488	.2587	.2686	.2783	.2880
3	.0298	.0328	.0359	.0393	.0429	.0467	.0507	.0549	.0593	.0640
	.41	.42	.43	.44	.45	.46	.47	.48	.49	.50
0	.2054	.1951	.1852	.1756	.1664	.1575	.1489	.1406	.1327	.1250
1	.4282	.4239	.4191	.4140	.4084	.4024	.3961	.3894	.3823	.3750
2	.2975	.3069	.3162	.3252	.3341	.3428	.3512	.3594	.3674	.3750
3	.0689	.0741	.0795	.0852	.0911	.0973	.1038	.1106	.1176	.1250

n = 4

r \ p	.01	.02	.03	.04	.05	.06	.07	.08	.09	.10
0	.9606	.9224	.8853	.8493	.8145	.7807	.7481	.7164	.6857	.6561
1	.0388	.0753	.1095	.1416	.1715	.1993	.2252	.2492	.2713	.2916
2	.0006	.0023	.0051	.0088	.0135	.0191	.0254	.0325	.0402	.0486
3	.0000	.0000	.0001	.0002	.0005	.0008	.0013	.0019	.0027	.0036
4	.0000	.0000	.0000	.0000	.0000	.0000	.0000	.0000	.0001	.0001
	.11	.12	.13	.14	.15	.16	.17	.18	.19	.20
0	.6274	.5997	.5729	.5470	.5220	.4979	.4746	.4521	.4305	.4096
1	.3102	.3271	.3424	.3562	.3685	.3793	.3888	.3970	.4039	.4096
2	.0575	.0669	.0767	.0870	.0975	.1084	.1195	.1307	.1421	.1536
3	.0047	.0061	.0076	.0094	.0115	.0138	.0163	.0191	.0222	.0256
4	.0001	.0002	.0003	.0004	.0005	.0007	.0008	.0010	.0013	.0016
	.21	.22	.23	.24	.25	.26	.27	.28	.29	.30
0	.3895	.3702	.3515	.3336	.3164	.2999	.2840	.2687	.2541	.2401
1	.4142	.4176	.4200	.4214	.4219	.4214	.4201	.4180	.4152	.4116
2	.1651	.1767	.1882	.1996	.2109	.2221	.2331	.2439	.2544	.2646
3	.0293	.0332	.0375	.0420	.0469	.0520	.0575	.0632	.0693	.0756
4	.0019	.0023	.0028	.0033	.0039	.0046	.0053	.0061	.0071	.0081
	.31	.32	.33	.34	.35	.36	.37	.38	.39	.40
0	.2267	.2138	.2015	.1897	.1785	.1678	.1575	.1478	.1385	.1296
1	.4074	.4025	.3970	.3910	.3845	.3775	.3701	.3623	.3541	.3456
2	.2745	.2841	.2933	.3021	.3105	.3185	.3260	.3330	.3396	.3456
3	.0822	.0891	.0963	.1038	.1115	.1194	.1276	.1361	.1447	.1536
4	.0092	.0105	.0119	.0134	.0150	.0168	.0187	.0209	.0231	.0256
	.41	.42	.43	.44	.45	.46	.47	.48	.49	.50
0	.1212	.1132	.1056	.0983	.0915	.0850	.0789	.0731	.0677	.0625
1	.3368	.3278	.3185	.3091	.2995	.2897	.2799	.2700	.2600	.2500
2	.3511	.3560	.3604	.3643	.3675	.3702	.3723	.3738	.3747	.3750
3	.1627	.1719	.1813	.1908	.2005	.2102	.2201	.2300	.2400	.2500
4	.0283	.0311	.0342	.0375	.0410	.0448	.0488	.0531	.0576	.0625

n = 5

r \ p	.01	.02	.03	.04	.05	.06	.07	.08	.09	.10
0	.9510	.9039	.8587	.8154	.7738	.7339	.6957	.6591	.6240	.5905
1	.0480	.0922	.1328	.1699	.2036	.2342	.2618	.2866	.3086	.3280
2	.0010	.0038	.0082	.0142	.0214	.0299	.0394	.0498	.0610	.0729
3	.0000	.0001	.0003	.0006	.0011	.0019	.0030	.0043	.0060	.0081
4	.0000	.0000	.0000	.0000	.0000	.0001	.0001	.0002	.0003	.0004

r \ p	.11	.12	.13	.14	.15	.16	.17	.18	.19	.20
0	.5584	.5277	.4984	.4704	.4437	.4182	.3939	.3707	.3487	.3277
1	.3451	.3598	.3724	.3829	.3915	.3983	.4034	.4069	.4089	.4096
2	.0853	.0981	.1113	.1247	.1382	.1517	.1652	.1786	.1919	.2048
3	.0105	.0134	.0166	.0203	.0244	.0289	.0338	.0392	.0450	.0512
4	.0007	.0009	.0012	.0017	.0022	.0028	.0035	.0043	.0053	.0064
5	.0000	.0000	.0000	.0001	.0001	.0001	.0001	.0002	.0002	.0003

r \ p	.21	.22	.23	.24	.25	.26	.27	.28	.29	.30
0	.3077	.2887	.2707	.2536	.2373	.2219	.2073	.1935	.1804	.1681
1	.4090	.4072	.4043	.4003	.3955	.3898	.3834	.3762	.3685	.3602
2	.2174	.2297	.2415	.2529	.2637	.2739	.2836	.2926	.3010	.3087
3	.0578	.0648	.0721	.0798	.0879	.0962	.1049	.1138	.1229	.1323
4	.0077	.0091	.0108	.0126	.0146	.0169	.0194	.0221	.0251	.0284
5	.0004	.0005	.0006	.0008	.0010	.0012	.0014	.0017	.0021	.0024

r \ p	.31	.32	.33	.34	.35	.36	.37	.38	.39	.40
0	.1564	.1454	.1350	.1252	.1160	.1074	.0992	.0916	.0845	.0778
1	.3513	.3421	.3325	.3226	.3124	.3020	.2914	.2808	.2700	.2592
2	.3157	.3220	.3275	.3323	.3364	.3397	.3423	.3441	.3452	.3456
3	.1418	.1515	.1613	.1712	.1811	.1911	.2010	.2109	.2207	.2304
4	.0319	.0357	.0397	.0441	.0488	.0537	.0590	.0646	.0706	.0768
5	.0029	.0034	.0039	.0045	.0053	.0060	.0069	.0079	.0090	.0102

r \ p	.41	.42	.43	.44	.45	.46	.47	.48	.49	.50
0	.0715	.0656	.0602	.0551	.0503	.0459	.0418	.0380	.0345	.0312
1	.2484	.2376	.2270	.2164	.2059	.1956	.1854	.1755	.1657	.1562
2	.3452	.3442	.3424	.3400	.3369	.3332	.3289	.3240	.3185	.3125
3	.2399	.2492	.2583	.2671	.2757	.2838	.2916	.2990	.3060	.3125
4	.0834	.0902	.0974	.1049	.1128	.1209	.1293	.1380	.1470	.1562
5	.0116	.0131	.0147	.0165	.0185	.0206	.0229	.0255	.0282	.0312

n = 6

r \ p	.01	.02	.03	.04	.05	.06	.07	.08	.09	.10
0	.9415	.8858	.8330	.7828	.7351	.6899	.6470	.6064	.5679	.5314
1	.0571	.1085	.1546	.1957	.2321	.2642	.2922	.3164	.3370	.3543
2	.0014	.0055	.0120	.0204	.0305	.0422	.0550	.0688	.0833	.0984
3	.0000	.0002	.0005	.0011	.0021	.0036	.0055	.0080	.0110	.0146
4	.0000	.0000	.0000	.0000	.0001	.0002	.0003	.0005	.0008	.0012
5	.0000	.0000	.0000	.0000	.0000	.0000	.0000	.0000	.0000	.0001

r \ p	.11	.12	.13	.14	.15	.16	.17	.18	.19	.20
0	.4970	.4644	.4336	.4046	.3771	.3513	.3269	.3040	.2824	.2621
1	.3685	.3800	.3888	.3952	.3993	.4015	.4018	.4004	.3975	.3932
2	.1139	.1295	.1452	.1608	.1762	.1912	.2057	.2197	.2331	.2458
3	.0188	.0236	.0289	.0349	.0415	.0486	.0562	.0643	.0729	.0819
4	.0017	.0024	.0032	.0043	.0055	.0069	.0086	.0106	.0128	.0154
5	.0001	.0001	.0002	.0003	.0004	.0005	.0007	.0009	.0012	.0015
6	.0000	.0000	.0000	.0000	.0000	.0000	.0000	.0000	.0000	.0001

r \ p	.21	.22	.23	.24	.25	.26	.27	.28	.29	.30
0	.2431	.2252	.2084	.1927	.1780	.1642	.1513	.1393	.1281	.1176
1	.3877	.3811	.3735	.3651	.3560	.3462	.3358	.3251	.3139	.3025
2	.2577	.2687	.2789	.2882	.2966	.3041	.3105	.3160	.3206	.3241
3	.0913	.1011	.1111	.1214	.1318	.1424	.1531	.1639	.1746	.1852
4	.0182	.0214	.0249	.0287	.0330	.0375	.0425	.0478	.0535	.0595
5	.0019	.0024	.0030	.0036	.0044	.0053	.0063	.0074	.0087	.0102
6	.0001	.0001	.0001	.0002	.0002	.0003	.0004	.0005	.0006	.0007

n = 6 (Continued)

r \ p	.31	.32	.33	.34	.35	.36	.37	.38	.39	.40
0	.1079	.0989	.0905	.0827	.0754	.0687	.0625	.0568	.0515	.0467
1	.2909	.2792	.2673	.2555	.2437	.2319	.2203	.2089	.1976	.1866
2	.3267	.3284	.3292	.3290	.3280	.3261	.3235	.3201	.3159	.3110
3	.1957	.2061	.2162	.2260	.2355	.2446	.2533	.2616	.2693	.2765
4	.0660	.0727	.0799	.0873	.0951	.1032	.1116	.1202	.1291	.1382
5	.0119	.0137	.0157	.0180	.0205	.0232	.0262	.0295	.0330	.0369
6	.0009	.0011	.0013	.0015	.0018	.0022	.0026	.0030	.0035	.0041

r \ p	.41	.42	.43	.44	.45	.46	.47	.48	.49	.50
0	.0422	.0381	.0343	.0308	.0277	.0248	.0222	.0198	.0176	.0156
1	.1759	.1654	.1552	.1454	.1359	.1267	.1179	.1095	.1014	.0938
2	.3055	.2994	.2928	.2856	.2780	.2699	.2615	.2527	.2436	.2344
3	.2831	.2891	.2945	.2992	.3032	.3065	.3091	.3110	.3121	.3125
4	.1475	.1570	.1666	.1763	.1861	.1958	.2056	.2153	.2249	.2344
5	.0410	.0455	.0503	.0554	.0609	.0667	.0729	.0795	.0864	.0938
6	.0048	.0055	.0063	.0073	.0083	.0095	.0108	.0122	.0138	.0156

n = 7

r \ p	.01	.02	.03	.04	.05	.06	.07	.08	.09	.10
0	.9321	.8681	.8080	.7514	.6983	.6485	.6017	.5578	.5168	.4783
1	.0659	.1240	.1749	.2192	.2573	.2897	.3170	.3396	.3578	.3720
2	.0020	.0076	.0162	.0274	.0406	.0555	.0716	.0886	.1061	.1240
3	.0000	.0003	.0008	.0019	.0036	.0059	.0090	.0128	.0175	.0230
4	.0000	.0000	.0000	.0001	.0002	.0004	.0007	.0011	.0017	.0026
5	.0000	.0000	.0000	.0000	.0000	.0000	.0000	.0001	.0001	.0002

r \ p	.11	.12	.13	.14	.15	.16	.17	.18	.19	.20
0	.4423	.4087	.3773	.3479	.3206	.2951	.2714	.2493	.2288	.2097
1	.3827	.3901	.3946	.3965	.3960	.3935	.3891	.3830	.3756	.3670
2	.1419	.1596	.1769	.1936	.2097	.2248	.2391	.2523	.2643	.2753
3	.0292	.0363	.0441	.0525	.0617	.0714	.0816	.0923	.1033	.1147
4	.0036	.0049	.0066	.0086	.0109	.0136	.0167	.0203	.0242	.0287
5	.0003	.0004	.0006	.0008	.0012	.0016	.0021	.0027	.0034	.0043
6	.0000	.0000	.0000	.0000	.0001	.0001	.0001	.0002	.0003	.0004

r \ p	.21	.22	.23	.24	.25	.26	.27	.28	.29	.30
0	.1920	.1757	.1605	.1465	.1335	.1215	.1105	.1003	.0910	.0824
1	.3573	.3468	.3356	.3237	.3115	.2989	.2860	.2731	.2600	.2471
2	.2850	.2935	.3007	.3067	.3115	.3150	.3174	.3186	.3186	.3177
3	.1263	.1379	.1497	.1614	.1730	.1845	.1956	.2065	.2169	.2269
4	.0336	.0389	.0447	.0510	.0577	.0648	.0724	.0803	.0886	.0972
5	.0054	.0066	.0080	.0097	.0115	.0137	.0161	.0187	.0217	.0250
6	.0005	.0006	.0008	.0010	.0013	.0016	.0020	.0024	.0030	.0036
7	.0000	.0000	.0000	.0000	.0001	.0001	.0001	.0001	.0002	.0002

r \ p	.31	.32	.33	.34	.35	.36	.37	.38	.39	.40
0	.0745	.0672	.0606	.0546	.0490	.0440	.0394	.0352	.0314	.0280
1	.2342	.2215	.2090	.1967	.1848	.1732	.1619	.1511	.1407	.1306
2	.3156	.3127	.3088	.3040	.2985	.2922	.2853	.2778	.2698	.2613
3	.2363	.2452	.2535	.2610	.2679	.2740	.2793	.2838	.2875	.2903
4	.1062	.1154	.1248	.1345	.1442	.1541	.1640	.1739	.1838	.1935
5	.0286	.0326	.0369	.0416	.0466	.0520	.0578	.0640	.0705	.0774
6	.0043	.0051	.0061	.0071	.0084	.0098	.0113	.0131	.0150	.0172
7	.0003	.0003	.0004	.0005	.0006	.0008	.0009	.0011	.0014	.0016

r \ p	.41	.42	.43	.44	.45	.46	.47	.48	.49	.50
0	.0249	.0221	.0195	.0173	.0152	.0134	.0117	.0103	.0090	.0078
1	.1211	.1119	.1032	.0950	.0872	.0798	.0729	.0664	.0604	.0547
2	.2524	.2431	.2336	.2239	.2140	.2040	.1940	.1840	.1740	.1641
3	.2923	.2934	.2937	.2932	.2918	.2897	.2867	.2830	.2786	.2734
4	.2031	.2125	.2216	.2304	.2388	.2468	.2543	.2612	.2676	.2734
5	.0847	.0923	.1003	.1086	.1172	.1261	.1353	.1447	.1543	.1641
6	.0196	.0223	.0252	.0284	.0320	.0358	.0400	.0445	.0494	.0547
7	.0019	.0023	.0027	.0032	.0037	.0044	.0051	.0059	.0068	.0078

n = 8

r \ p	.01	.02	.03	.04	.05	.06	.07	.08	.09	.10
0	.9227	.8508	.7837	.7214	.6634	.6096	.5596	.5132	.4703	.4305
1	.0746	.1389	.1939	.2405	.2793	.3113	.3370	.3570	.3721	.3826
2	.0026	.0099	.0210	.0351	.0515	.0695	.0888	.1087	.1288	.1488
3	.0001	.0004	.0013	.0029	.0054	.0089	.0134	.0189	.0255	.0331
4	.0000	.0000	.0001	.0002	.0004	.0007	.0013	.0021	.0031	.0046
5	.0000	.0000	.0000	.0000	.0000	.0000	.0001	.0001	.0002	.0004
	.11	.12	.13	.14	.15	.16	.17	.18	.19	.20
0	.3937	.3596	.3282	.2992	.2725	.2479	.2252	.2044	.1853	.1678
1	.3892	.3923	.3923	.3897	.3847	.3777	.3691	.3590	.3477	.3355
2	.1684	.1872	.2052	.2220	.2376	.2518	.2646	.2758	.2855	.2936
3	.0416	.0511	.0613	.0723	.0839	.0959	.1084	.1211	.1339	.1468
4	.0064	.0087	.0115	.0147	.0185	.0228	.0277	.0332	.0393	.0459
5	.0006	.0009	.0014	.0019	.0026	.0035	.0045	.0058	.0074	.0092
6	.0000	.0001	.0001	.0002	.0002	.0003	.0005	.0006	.0009	.0011
7	.0000	.0000	.0000	.0000	.0000	.0000	.0000	.0000	.0001	.0001
	.21	.22	.23	.24	.25	.26	.27	.28	.29	.30
0	.1517	.1370	.1236	.1113	.1001	.0899	.0806	.0722	.0646	.0576
1	.3226	.3092	.2953	.2812	.2670	.2527	.2386	.2247	.2110	.1977
2	.3002	.3052	.3087	.3108	.3115	.3108	.3089	.3058	.3017	.2965
3	.1596	.1722	.1844	.1963	.2076	.2184	.2285	.2379	.2464	.2541
4	.0530	.0607	.0689	.0775	.0865	.0959	.1056	.1156	.1258	.1361
5	.0113	.0137	.0165	.0196	.0231	.0270	.0313	.0360	.0411	.0467
6	.0015	.0019	.0025	.0031	.0038	.0047	.0058	.0070	.0084	.0100
7	.0001	.0002	.0002	.0003	.0004	.0005	.0006	.0008	.0010	.0012
8	.0000	.0000	.0000	.0000	.0000	.0000	.0000	.0000	.0001	.0001
	.31	.32	.33	.34	.35	.36	.37	.38	.39	.40
0	.0514	.0457	.0406	.0360	.0319	.0281	.0248	.0218	.0192	.0168
1	.1847	.1721	.1600	.1484	.1373	.1267	.1166	.1071	.0981	.0896
2	.2904	.2835	.2758	.2675	.2587	.2494	.2397	.2297	.2194	.2090
3	.2609	.2668	.2717	.2756	.2786	.2805	.2815	.2815	.2806	.2787
4	.1465	.1569	.1673	.1775	.1875	.1973	.2067	.2157	.2242	.2322
5	.0527	.0591	.0659	.0732	.0808	.0888	.0971	.1058	.1147	.1239
6	.0118	.0139	.0162	.0188	.0217	.0250	.0285	.0324	.0367	.0413
7	.0015	.0019	.0023	.0028	.0033	.0040	.0048	.0057	.0067	.0079
8	.0001	.0001	.0001	.0002	.0002	.0003	.0004	.0004	.0005	.0007
	.41	.42	.43	.44	.45	.46	.47	.48	.49	.50
0	.0147	.0128	.0111	.0097	.0084	.0072	.0062	.0053	.0046	.0039
1	.0816	.0742	.0672	.0608	.0548	.0493	.0442	.0395	.0352	.0312
2	.1985	.1880	.1776	.1672	.1569	.1469	.1371	.1275	.1183	.1094
3	.2759	.2723	.2679	.2627	.2568	.2503	.2431	.2355	.2273	.2188
4	.2397	.2465	.2526	.2580	.2627	.2665	.2695	.2717	.2730	.2734
5	.1332	.1428	.1525	.1622	.1719	.1816	.1912	.2006	.2098	.2188
6	.0463	.0517	.0575	.0637	.0703	.0774	.0848	.0926	.1008	.1094
7	.0092	.0107	.0124	.0143	.0164	.0188	.0215	.0244	.0277	.0312
8	.0008	.0010	.0012	.0014	.0017	.0020	.0024	.0028	.0033	.0039

n = 9

r \ p	.01	.02	.03	.04	.05	.06	.07	.08	.09	.10
0	.9135	.8337	.7602	.6925	.6302	.5730	.5204	.4722	.4279	.3874
1	.0830	.1531	.2116	.2597	.2985	.3292	.3525	.3695	.3809	.3874
2	.0034	.0125	.0262	.0433	.0629	.0840	.1061	.1285	.1507	.1722
3	.0001	.0006	.0019	.0042	.0077	.0125	.0186	.0261	.0348	.0446
4	.0000	.0000	.0001	.0003	.0006	.0012	.0021	.0034	.0052	.0074
5	.0000	.0000	.0000	.0000	.0000	.0001	.0002	.0003	.0005	.0008
6	.0000	.0000	.0000	.0000	.0000	.0000	.0000	.0000	.0000	.0001

F

n = 9 (Continued)

r \ p	.11	.12	.13	.14	.15	.16	.17	.18	.19	.20
0	.3504	.3165	.2855	.2573	.2316	.2082	.1869	.1676	.1501	.1342
1	.3897	.3884	.3840	.3770	.3679	.3569	.3446	.3312	.3169	.3020
2	.1927	.2119	.2295	.2455	.2597	.2720	.2823	.2908	.2973	.3020
3	.0556	.0674	.0800	.0933	.1069	.1209	.1349	.1489	.1627	.1762
4	.0103	.0138	.0179	.0228	.0283	.0345	.0415	.0490	.0573	.0661
5	.0013	.0019	.0027	.0037	.0050	.0066	.0085	.0108	.0134	.0165
6	.0001	.0002	.0003	.0004	.0006	.0008	.0012	.0016	.0021	.0028
7	.0000	.0000	.0000	.0000	.0000	.0001	.0001	.0001	.0002	.0003
	.21	.22	.23	.24	.25	.26	.27	.28	.29	.30
0	.1199	.1069	.0952	.0846	.0751	.0665	.0589	.0520	.0458	.0404
1	.2867	.2713	.2558	.2404	.2253	.2104	.1960	.1820	.1685	.1556
2	.3049	.3061	.3056	.3037	.3003	.2957	.2899	.2831	.2754	.2668
3	.1891	.2014	.2130	.2238	.2336	.2424	.2502	.2569	.2624	.2668
4	.0754	.0852	.0954	.1060	.1168	.1278	.1388	.1499	.1608	.1715
5	.0200	.0240	.0285	.0335	.0389	.0449	.0513	.0583	.0657	.0735
6	.0036	.0045	.0057	.0070	.0087	.0105	.0127	.0151	.0179	.0210
7	.0004	.0005	.0007	.0010	.0012	.0016	.0020	.0025	.0031	.0039
8	.0000	.0000	.0001	.0001	.0001	.0001	.0002	.0002	.0003	.0004
	.31	.32	.33	.34	.35	.36	.37	.38	.39	.40
0	.0355	.0311	.0272	.0238	.0207	.0180	.0156	.0135	.0117	.0101
1	.1433	.1317	.1206	.1102	.1004	.0912	.0826	.0747	.0673	.0605
2	.2576	.2478	.2376	.2270	.2162	.2052	.1941	.1831	.1721	.1612
3	.2701	.2721	.2731	.2729	.2716	.2693	.2660	.2618	.2567	.2508
4	.1820	.1921	.2017	.2109	.2194	.2272	.2344	.2407	.2462	.2508
5	.0818	.0904	.0994	.1086	.1181	.1278	.1376	.1475	.1574	.1672
6	.0245	.0284	.0326	.0373	.0424	.0479	.0539	.0603	.0671	.0743
7	.0047	.0057	.0069	.0082	.0098	.0116	.0136	.0158	.0184	.0212
8	.0005	.0007	.0008	.0011	.0013	.0016	.0020	.0024	.0029	.0035
9	.0000	.0000	.0000	.0001	.0001	.0001	.0001	.0002	.0002	.0003
	.41	.42	.43	.44	.45	.46	.47	.48	.49	.50
0	.0087	.0074	.0064	.0054	.0046	.0039	.0033	.0028	.0023	.0020
1	.0542	.0484	.0431	.0383	.0339	.0299	.0263	.0231	.0202	.0176
2	.1506	.1402	.1301	.1204	.1110	.1020	.0934	.0853	.0776	.0703
3	.2442	.2369	.2291	.2207	.2119	.2027	.1933	.1837	.1739	.1641
4	.2545	.2573	.2592	.2601	.2600	.2590	.2571	.2543	.2506	.2461
5	.1769	.1863	.1955	.2044	.2128	.2207	.2280	.2347	.2408	.2461
6	.0819	.0900	.0983	.1070	.1160	.1253	.1348	.1445	.1542	.1641
7	.0244	.0279	.0318	.0360	.0407	.0458	.0512	.0571	.0635	.0703
8	.0042	.0051	.0060	.0071	.0083	.0097	.0114	.0132	.0153	.0176
9	.0003	.0004	.0005	.0006	.0008	.0009	.0011	.0014	.0016	.0020

n = 10

r \ p	.01	.02	.03	.04	.05	.06	.07	.08	.09	.10
0	.9044	.8171	.7374	.6648	.5987	.5386	.4840	.4344	.3894	.3487
1	.0914	.1667	.2281	.2770	.3151	.3438	.3643	.3777	.3851	.3874
2	.0042	.0153	.0317	.0519	.0746	.0988	.1234	.1478	.1714	.1937
3	.0001	.0008	.0026	.0058	.0105	.0168	.0248	.0343	.0452	.0574
4	.0000	.0000	.0001	.0004	.0010	.0019	.0033	.0052	.0078	.0112
5	.0000	.0000	.0000	.0000	.0001	.0001	.0003	.0005	.0009	.0015
6	.0000	.0000	.0000	.0000	.0000	.0000	.0000	.0000	.0001	.0001
	.11	.12	.13	.14	.15	.16	.17	.18	.19	.20
0	.3118	.2785	.2484	.2213	.1969	.1749	.1552	.1374	.1216	.1074
1	.3854	.3798	.3712	.3603	.3474	.3331	.3178	.3017	.2852	.2684
2	.2143	.2330	.2496	.2639	.2759	.2856	.2929	.2980	.3010	.3020
3	.0706	.0847	.0995	.1146	.1298	.1450	.1600	.1745	.1883	.2013
4	.0153	.0202	.0260	.0326	.0401	.0483	.0573	.0670	.0773	.0881
5	.0023	.0033	.0047	.0064	.0085	.0111	.0141	.0177	.0218	.0264
6	.0002	.0004	.0006	.0009	.0012	.0018	.0024	.0032	.0043	.0055
7	.0000	.0000	.0000	.0001	.0001	.0002	.0003	.0004	.0006	.0008
8	.0000	.0000	.0000	.0000	.0000	.0000	.0000	.0000	.0001	.0001

n = 10 (Continued)

r \ p	.21	.22	.23	.24	.25	.26	.27	.28	.29	.30
0	.0947	.0834	.0733	.0643	.0563	.0492	.0430	.0374	.0326	.0282
1	.2517	.2351	.2188	.2030	.1877	.1730	.1590	.1456	.1330	.1211
2	.3011	.2984	.2942	.2885	.2816	.2735	.2646	.2548	.2444	.2335
3	.2134	.2244	.2343	.2429	.2503	.2563	.2609	.2642	.2662	.2668
4	.0993	.1108	.1225	.1343	.1460	.1576	.1689	.1798	.1903	.2001
5	.0317	.0375	.0439	.0509	.0584	.0664	.0750	.0839	.0933	.1029
6	.0070	.0088	.0109	.0134	.0162	.0195	.0231	.0272	.0317	.0368
7	.0011	.0014	.0019	.0024	.0031	.0039	.0049	.0060	.0074	.0090
8	.0001	.0002	.0002	.0003	.0004	.0005	.0007	.0009	.0011	.0014
9	.0000	.0000	.0000	.0000	.0000	.0000	.0001	.0001	.0001	.0001

r \ p	.31	.32	.33	.34	.35	.36	.37	.38	.39	.40
0	.0245	.0211	.0182	.0157	.0135	.0115	.0098	.0084	.0071	.0060
1	.1099	.0995	.0898	.0808	.0725	.0649	.0578	.0514	.0456	.0403
2	.2222	.2107	.1990	.0873	.1757	.1642	.1529	.1419	.1312	.1209
3	.2662	.2644	.2614	.2573	.2522	.2462	.2394	.2319	.2237	.2150
4	.2093	.2177	.2253	.2320	.2377	.2424	.2461	.2487	.2503	.2508
5	.1128	.1229	.1332	.1434	.1536	.1636	.1734	.1829	.1920	.2007
6	.0422	.0482	.0547	.0616	.0689	.0767	.0849	.0934	.1023	.1115
7	.0108	.0130	.0154	.0181	.0212	.0247	.0285	.0327	.0374	.0425
8	.0018	.0023	.0028	.0035	.0043	.0052	.0063	.0075	.0090	.0106
9	.0002	.0002	.0003	.0004	.0005	.0006	.0008	.0010	.0013	.0016
10	.0000	.0000	.0000	.0000	.0000	.0000	.0000	.0001	.0001	.0001

r \ p	.41	.42	.43	.44	.45	.46	.47	.48	.49	.50
0	.0051	.0043	.0036	.0030	.0025	.0021	.0017	.0014	.0012	.0010
1	.0355	.0312	.0273	.0238	.0207	.0180	.0155	.0133	.0114	.0098
2	.1111	.1017	.0927	.0843	.0763	.0688	.0619	.0554	.0494	.0439
3	.2058	.1963	.1865	.1765	.1665	.1564	.1464	.1364	.1267	.1172
4	.2503	.2488	.2462	.2427	.2384	.2331	.2271	.2204	.2130	.2051
5	.2087	.2162	.2229	.2289	.2340	.2383	.2417	.2441	.2456	.2461
6	.1209	.1304	.1401	.1499	.1596	.1692	.1786	.1878	.1966	.2051
7	.0480	.0540	.0604	.0673	.0746	.0824	.0905	.0991	.1080	.1172
8	.0125	.0147	.0171	.0198	.0229	.0263	.0301	.0343	.0389	.0439
9	.0019	.0024	.0029	.0035	.0042	.0050	.0059	.0070	.0083	.0098
10	.0001	.0002	.0002	.0003	.0003	.0004	.0005	.0006	.0008	.0010

n = 11

r \ p	.01	.02	.03	.04	.05	.06	.07	.08	.09	.10
0	.8953	.8007	.7153	.6382	.5688	.5063	.4501	.3996	.3544	.3138
1	.0995	.1798	.2433	.2925	.3293	.3555	.3727	.3823	.3855	.3835
2	.0050	.0183	.0376	.0609	.0867	.1135	.1403	.1662	.1906	.2131
3	.0002	.0011	.0035	.0076	.0137	.0217	.0317	.0434	.0566	.0710
4	.0000	.0000	.0002	.0006	.0014	.0028	.0048	.0075	.0112	.0158
5	.0000	.0000	.0000	.0000	.0001	.0002	.0005	.0009	.0015	.0025
6	.0000	.0000	.0000	.0000	.0000	.0000	.0000	.0001	.0002	.0003

r \ p	.11	.12	.13	.14	.15	.16	.17	.18	.19	.20
0	.2775	.2451	.2161	.1903	.1673	.1469	.1288	.1127	.0985	.0859
1	.3773	.3676	.3552	.3408	.3248	.3078	.2901	.2721	.2541	.2362
2	.2332	.2507	.2654	.2774	.2866	.2932	.2971	.2987	.2980	.2953
3	.0865	.1025	.1190	.1355	.1517	.1675	.1826	.1967	.2097	.2215
4	.0214	.0280	.0356	.0441	.0536	.0638	.0748	.0864	.0984	.1107
5	.0037	.0053	.0074	.0101	.0132	.0170	.0214	.0265	.0323	.0388
6	.0005	.0007	.0011	.0016	.0023	.0032	.0044	.0058	.0076	.0097
7	.0000	.0001	.0001	.0002	.0003	.0004	.0006	.0009	.0013	.0017
8	.0000	.0000	.0000	.0000	.0000	.0000	.0001	.0001	.0001	.0002

n = 11 (Continued)

r \ p	.21	.22	.23	.24	.25	.26	.27	.28	.29	.30
0	.0748	.0650	.0564	.0489	.0422	.0364	.0314	.0270	.0231	.0198
1	.2187	.2017	.1854	.1697	.1549	.1408	.1276	.1153	.1038	.0932
2	.2907	.2845	.2768	.2680	.2581	.2474	.2360	.2242	.2121	.1998
3	.2318	.2407	.2481	.2539	.2581	.2608	.2619	.2616	.2599	.2568
4	.1232	.1358	.1482	.1603	.1721	.1832	.1937	.2035	.2123	.2201
5	.0459	.0536	.0620	.0709	.0803	.0901	.1003	.1108	.1214	.1321
6	.0122	.0151	.0185	.0224	.0268	.0317	.0371	.0431	.0496	.0566
7	.0023	.0030	.0039	.0050	.0064	.0079	.0098	.0120	.0145	.0173
8	.0003	.0004	.0006	.0008	.0011	.0014	.0018	.0023	.0030	.0037
9	.0000	.0000	.0001	.0001	.0001	.0002	.0002	.0003	.0004	.0005

r \ p	.31	.32	.33	.34	.35	.36	.37	.38	.39	.40
0	.0169	.0144	.0122	.0104	.0088	.0074	.0062	.0052	.0044	.0036
1	.0834	.0744	.0662	.0587	.0518	.0457	.0401	.0351	.0306	.0266
2	.1874	.1751	.1630	.1511	.1395	.1284	.1177	.1075	.0978	.0887
3	.2526	.2472	.2408	.2335	.2254	.2167	.2074	.1977	.1876	.1774
4	.2269	.2326	.2372	.2406	.2428	.2438	.2436	.2423	.2399	.2365
5	.1427	.1533	.1636	.1735	.1830	.1920	.2003	.2079	.2148	.2207
6	.0641	.0721	.0806	.0894	.0985	.1080	.1176	.1274	.1373	.1471
7	.0206	.0242	.0283	.0329	.0379	.0434	.0494	.0558	.0627	.0701
8	.0046	.0057	.0070	.0085	.0102	.0122	.0145	.0171	.0200	.0234
9	.0007	.0009	.0011	.0015	.0018	.0023	.0028	.0035	.0043	.0052
10	.0001	.0001	.0001	.0001	.0002	.0003	.0003	.0004	.0005	.0007

r \ p	.41	.42	.43	.44	.45	.46	.47	.48	.49	.50
0	.0030	.0025	.0021	.0017	.0014	.0011	.0009	.0008	.0006	.0005
1	.0231	.0199	.0171	.0147	.0125	.0107	.0090	.0076	.0064	.0054
2	.0801	.0721	.0646	.0577	.0513	.0454	.0401	.0352	.0308	.0269
3	.1670	.1566	.1462	.1359	.1259	.1161	.1067	.0976	.0888	.0806
4	.2321	.2267	.2206	.2136	.2060	.1978	.1892	.1801	.1707	.1611
5	.2258	.2299	.2329	.2350	.2360	.2359	.2348	.2327	.2296	.2256
6	.1569	.1664	.1757	.1846	.1931	.2010	.2083	.2148	.2206	.2256
7	.0779	.0861	.0947	.1036	.1128	.1223	.1319	.1416	.1514	.1611
8	.0271	.0312	.0357	.0407	.0462	.0521	.0585	.0654	.0727	.0806
9	.0063	.0075	.0090	.0107	.0126	.0148	.0173	.0201	.0233	.0269
10	.0009	.0011	.0014	.0017	.0021	.0025	.0031	.0037	.0045	.0054
11	.0001	.0001	.0001	.0001	.0002	.0002	.0002	.0003	.0004	.0005

n = 12

r \ p	.01	.02	.03	.04	.05	.06	.07	.08	.09	.10
0	.8864	.7847	.6938	.6127	.5404	.4759	.4186	.3677	.3225	.2824
1	.1074	.1922	.2575	.3064	.3413	.3645	.3781	.3837	.3827	.3766
2	.0060	.0216	.0438	.0702	.0988	.1280	.1565	.1835	.2082	.2301
3	.0002	.0015	.0045	.0098	.0173	.0272	.0393	.0532	.0686	.0852
4	.0000	.0001	.0003	.0009	.0021	.0039	.0067	.0104	.0153	.0213
5	.0000	.0000	.0000	.0001	.0002	.0004	.0008	.0014	.0024	.0038
6	.0000	.0000	.0000	.0000	.0000	.0000	.0001	.0001	.0003	.0005

r \ p	.11	.12	.13	.14	.15	.16	.17	.18	.19	.20
0	.2470	.2157	.1880	.1637	.1422	.1234	.1069	.0924	.0798	.0687
1	.3663	.3529	.3372	.3197	.3012	.2821	.2627	.2434	.2245	.2062
2	.2490	.2647	.2771	.2863	.2924	.2955	.2960	.2939	.2897	.2835
3	.1026	.1203	.1380	.1553	.1720	.1876	.2021	.2151	.2265	.2362
4	.0285	.0369	.0464	.0569	.0683	.0804	.0931	.1062	.1195	.1329
5	.0056	.0081	.0111	.0148	.0193	.0245	.0305	.0373	.0449	.0532
6	.0008	.0013	.0019	.0028	.0040	.0054	.0073	.0096	.0123	.0155
7	.0001	.0001	.0002	.0004	.0006	.0009	.0013	.0018	.0025	.0033
8	.0000	.0000	.0000	.0000	.0001	.0001	.0002	.0002	.0004	.0005
9	.0000	.0000	.0000	.0000	.0000	.00000	.0000	.0000	.0000	.0001

n = 12 (Continued)

r \ p	.21	.22	.23	.24	.25	.26	.27	.28	.29	.30
0	.0591	.0507	.0434	.0371	.0317	.0270	.0229	.0194	.0164	.0138
1	.1885	.1717	.1557	.1407	.1267	.1137	.1016	.0906	.0804	.0712
2	.2756	.2663	.2558	.2444	.2323	.2197	.2068	.1937	.1807	.1678
3	.2442	.2503	.2547	.2573	.2581	.2573	.2549	.2511	.2460	.2397
4	.1460	.1589	.1712	.1828	.1936	.2034	.2122	.2197	.2261	.2311
5	.0621	.0717	.0818	.0924	.1032	.1143	.1255	.1367	.1477	.1585
6	.0193	.0236	.0285	.0340	.0401	.0469	.0542	.0620	.0704	.0792
7	.0044	.0057	.0073	.0092	.0115	.0141	.0172	.0207	.0246	.0291
8	.0007	.0010	.0014	.0018	.0024	.0031	.0040	.0050	.0063	.0078
9	.0001	.0001	.0002	.0003	.0004	.0005	.0007	.0009	.0011	.0015
10	.0000	.0000	.0000	.0000	.0000	.0001	.0001	.0001	.0001	.0002

r \ p	.31	.32	.33	.34	.35	.36	.37	.38	.39	.40
0	.0116	.0098	.0082	.0068	.0057	.0047	.0039	.0032	.0027	.0022
1	.0628	.0552	.0484	.0422	.0368	.0319	.0276	.0237	.0204	.0174
2	.1552	.1429	.1310	.1197	.1088	.0986	.0890	.0800	.0716	.0639
3	.2324	.2241	.2151	.2055	.1954	.1849	.1742	.1634	.1526	.1419
4	.2349	.2373	.2384	.2382	.2367	.2340	.2302	.2254	.2195	.2128
5	.1688	.1787	.1879	.1963	.2039	.2106	.2163	.2210	.2246	.2270
6	.0885	.0981	.1079	.1180	.1281	.1382	.1482	.1580	.1675	.1766
7	.0341	.0396	.0456	.0521	.0591	.0666	.0746	.0830	.0918	.1009
8	.0096	.0116	.0140	.0168	.0199	.0234	.0274	.0318	.0367	.0420
9	.0019	.0024	.0031	.0038	.0048	.0059	.0071	.0087	.0104	.0125
10	.0003	.0003	.0005	.0006	.0008	.0010	.0013	.0016	.0020	.0025
11	.0000	.0000	.0000	.0001	.0001	.0001	.0001	.0002	.0002	.0003

r \ p	.41	.42	.43	.44	.45	.46	.47	.48	.49	.50
0	.0018	.0014	.0012	.0010	.0008	.0006	.0005	.0004	.0003	.0002
1	.0148	.0126	.0106	.0090	.0075	.0063	.0052	.0043	.0036	.0029
2	.0567	.0502	.0442	.0388	.0339	.0294	.0255	.0220	.0189	.0161
3	.1314	.1211	.1111	.1015	.0923	.0836	.0754	.0676	.0604	.0537
4	.2054	.1973	.1886	.1794	.1700	.1602	.1504	.1405	.1306	.1208
5	.2284	.2285	.2276	.2256	.2225	.2184	.2134	.2075	.2008	.1934
6	.1851	.1931	.2003	.2068	.2124	.2171	.2208	.2234	.2250	.2256
7	.1103	.1198	.1295	.1393	.1489	.1585	.1678	.1768	.1853	.1934
8	.0479	.0542	.0611	.0684	.0762	.0844	.0930	.1020	.1113	.1208
9	.0148	.0175	.0205	.0239	.0277	.0319	.0367	.0418	.0475	.0537
10	.0031	.0038	.0046	.0056	.0068	.0082	.0098	.0116	.0137	.0161
11	.0004	.0005	.0006	.0008	.0010	.0013	.0016	.0019	.0024	.0029
12	.0000	.0000	.0000	.0001	.0001	.0001	.0001	.0001	.0002	.0002

n = 13

r \ p	.01	.02	.03	.04	.05	.06	.07	.08	.09	.10
0	.8775	.7690	.6730	.5882	.5133	.4474	.3893	.3383	.2935	.2542
1	.1152	.2040	.2706	.3186	.3512	.3712	.3809	.3824	.3773	.3672
2	.0070	.0250	.0502	.0797	.1109	.1422	.1720	.1995	.2239	.2448
3	.0003	.0019	.0057	.0122	.0214	.0333	.0475	.0636	.0812	.0997
4	.0000	.0001	.0004	.0013	.0028	.0053	.0089	.0138	.0201	.0277
5	.0000	.0000	.0000	.0001	.0003	.0006	.0012	.0022	.0036	.0055
6	.0000	.0000	.0000	.0000	.0000	.0001	.0001	.0003	.0005	.0008
7	.0000	.0000	.0000	.0000	.0000	.0000	.0000	.0000	.0000	.0001

r \ p	.11	.12	.13	.14	.15	.16	.17	.18	.19	.20
0	.2198	.1898	.1636	.1408	.1209	.1037	.0887	.0758	.0646	.0550
1	.3532	.3364	.3178	.2979	.2774	.2567	.2362	.2163	.1970	.1787
2	.2619	.2753	.2849	.2910	.2937	.2934	.2903	.2848	.2773	.2680
3	.1187	.1376	.1561	.1737	.1900	.2049	.2180	.2293	.2385	.2457
4	.0367	.0469	.0583	.0707	.0838	.0976	.1116	.1258	.1399	.1535
5	.0082	.0115	.0157	.0207	.0266	.0335	.0412	.0497	.0591	.0691
6	.0013	.0021	.0031	.0045	.0063	.0085	.0112	.0145	.0185	.0230
7	.0002	.0003	.0005	.0007	.0011	.0016	.0023	.0032	.0043	.0058
8	.0000	.0000	.0001	.0001	.0001	.0002	.0004	.0005	.0008	.0011
9	.0000	.0000	.0000	.0000	.0000	.0000	.0000	.0001	.0001	.0001

n = 13 (Continued)

r \ p	.21	.22	.23	.24	.25	.26	.27	.28	.29	.30
0	.0467	.0396	.0334	.0282	.0238	.0200	.0167	.0140	.0117	.0097
1	.1613	.1450	.1299	.1159	.1029	.0911	.0804	.0706	.0619	.0540
2	.2573	.2455	.2328	.2195	.2059	.1921	.1784	.1648	.1516	.1388
3	.2508	.2539	.2550	.2542	.2517	.2475	.2419	.2351	.2271	.2181
4	.1667	.1790	.1904	.2007	.2097	.2174	.2237	.2285	.2319	.2337
5	.0797	.0909	.1024	.1141	.1258	.1375	.1489	.1600	.1705	.1803
6	.0283	.0342	.0408	.0480	.0559	.0644	.0734	.0829	.0928	.1030
7	.0075	.0096	.0122	.0152	.0186	.0226	.0272	.0323	.0379	.0442
8	.0015	.0020	.0027	.0036	.0047	.0060	.0075	.0094	.0116	.0142
9	.0002	.0003	.0005	.0006	.0009	.0012	.0015	.0020	.0026	.0034
10	.0000	.0000	.0001	.0001	.0001	.0002	.0002	.0003	.0004	.0006
11	.0000	.0000	.0000	.0000	.0000	.0000	.0000	.0000	.0000	.0001

r \ p	.31	.32	.33	.34	.35	.36	.37	.38	.39	.40
0	.0080	.0066	.0055	.0045	.0037	.0030	.0025	.0020	.0016	.0013
1	.0469	.0407	.0351	.0302	.0259	.0221	.0188	.0159	.0135	.0113
2	.1265	.1148	.1037	.0933	.0836	.0746	.0663	.0586	.0516	.0453
3	.2084	.1981	.1874	.1763	.1651	.1538	.1427	.1317	.1210	.1107
4	.2341	.2331	.2307	.2270	.2222	.2163	.2095	.2018	.1934	.1845
5	.1893	.1974	.2045	.2105	.2154	.2190	.2215	.2227	.2226	.2214
6	.1134	.1239	.1343	.1446	.1546	.1643	.1734	.1820	.1898	.1968
7	.0509	.0583	.0662	.0745	.0833	.0924	.1019	.1115	.1213	.1312
8	.0172	.0206	.0244	.0288	.0336	.0390	.0449	.0513	.0582	.0656
9	.0043	.0054	.0067	.0082	.0101	.0122	.0146	.0175	.0207	.0243
10	.0008	.0010	.0013	.0017	.0022	.0027	.0034	.0043	.0053	.0065
11	.0001	.0001	.0002	.0002	.0003	.0004	.0006	.0007	.0009	.0012
12	.0000	.0000	.0000	.0000	.0000	.0000	.0001	.0001	.0001	.0001

r \ p	.41	.42	.43	.44	.45	.46	.47	.48	.49	.50
0	.0010	.0008	.0007	.0005	.0004	.0003	.0003	.0002	.0002	.0001
1	.0095	.0079	.0066	.0054	.0045	.0037	.0030	.0024	.0020	.0016
2	.0395	.0344	.0298	.0256	.0220	.0188	.0160	.0135	.0114	.0095
3	.1007	.0913	.0823	.0739	.0660	.0587	.0519	.0457	.0401	.0349
4	.1750	.1653	.1553	.1451	.1350	.1250	.1151	.1055	.0962	.0873
5	.2189	.2154	.2108	.2053	.1989	.1917	.1838	.1753	.1664	.1571
6	.2029	.2080	.2121	.2151	.2169	.2177	.2173	.2158	.2131	.2095
7	.1410	.1506	.1600	.1690	.1775	.1854	.1927	.1992	.2048	.2095
8	.0735	.0818	.0905	.0996	.1089	.1185	.1282	.1379	.1476	.1571
9	.0284	.0329	.0379	.0435	.0495	.0561	.0631	.0707	.0788	.0873
10	.0079	.0095	.0114	.0137	.0162	.0191	.0224	.0261	.0303	.0349
11	.0015	.0019	.0024	.0029	.0036	.0044	.0054	.0066	.0079	.0095
12	.0002	.0002	.0003	.0004	.0005	.0006	.0008	.0010	.0013	.0016
13	.0000	.0000	.0000	.0000	.0000	.0000	.0001	.0001	.0001	.0001

n = 14

r \ p	.01	.02	.03	.04	.05	.06	.07	.08	.09	.10
0	.8687	.7536	.6528	.5647	.4877	.4205	.3620	.3112	.2670	.2288
1	.1229	.2153	.2827	.3294	.3593	.3758	.3815	.3788	.3698	.3559
2	.0081	.0286	.0568	.0892	.1229	.1559	.1867	.2141	.2377	.2570
3	.0003	.0023	.0070	.0149	.0259	.0398	.0562	.0745	.0940	.1142
4	.0000	.0001	.0006	.0017	.0037	.0070	.0116	.0178	.0256	.0349
5	.0000	.0000	.0000	.0001	.0004	.0009	.0018	.0031	.0051	.0078
6	.0000	.0000	.0000	.0000	.0000	.0001	.0002	.0004	.0008	.0013
7	.0000	.0000	.0000	.0000	.0000	.0000	.0000	.0000	.0001	.0002

r \ p	.11	.12	.13	.14	.15	.16	.17	.18	.19	.20
0	.1956	.1670	.1423	.1211	.1028	.0871	.0736	.0621	.0523	.0440
1	.3385	.3188	.2977	.2759	.2539	.2322	.2112	.1910	.1719	.1539
2	.2720	.2826	.2892	.2919	.2912	.2875	.2811	.2725	.2620	.2501
3	.1345	.1542	.1728	.1901	.2056	.2190	.2303	.2393	.2459	.2501
4	.0457	.0578	.0710	.0851	.0998	.1147	.1297	.1444	.1586	.1720
5	.0113	.0158	.0212	.0277	.0352	.0437	.0531	.0634	.0744	.0860
6	.0021	.0032	.0048	.0068	.0093	.0125	.0163	.0209	.0262	.0322
7	.0003	.0005	.0008	.0013	.0019	.0027	.0038	.0052	.0070	.0092
8	.0000	.0001	.0001	.0002	.0003	.0005	.0007	.0010	.0014	.0020
9	.0000	.0000	.0000	.0000	.0000	.0001	.0001	.0001	.0002	.0003

n = 14 (Continued)

r \ p	.21	.22	.23	.24	.25	.26	.27	.28	.29	.30
0	.0369	.0309	.0258	.0214	.0178	.0148	.0122	.0101	.0083	.0068
1	.1372	.1218	.1077	.0948	.0832	.0726	.0632	.0548	.0473	.0407
2	.2371	.2234	.2091	.1946	.1802	.1659	.1519	.1385	.1256	.1134
3	.2521	.2520	.2499	.2459	.2402	.2331	.2248	.2154	.2052	.1943
4	.1843	.1955	.2052	.2135	.2202	.2252	.2286	.2304	.2305	.2290
5	.0980	.1103	.1226	.1348	.1468	.1583	.1691	.1792	.1883	.1963
6	.0391	.0466	.0549	.0639	.0734	.0834	.0938	.1045	.1153	.1262
7	.0119	.0150	.0188	.0231	.0280	.0335	.0397	.0464	.0538	.0618
8	.0028	.0037	.0049	.0064	.0082	.0103	.0128	.0158	.0192	.0232
9	.0005	.0007	.0010	.0013	.0018	.0024	.0032	.0041	.0052	.0066
10	.0001	.0001	.0001	.0002	.0003	.0004	.0006	.0008	.0011	.0014
11	.0000	.0000	.0000	.0000	.0000	.0001	.0001	.0001	.0002	.0002

r \ p	.31	.32	.33	.34	.35	.36	.37	.38	.39	.40
0	.0055	.0045	.0037	.0030	.0024	.0019	.0016	.0012	.0010	.0008
1	.0349	.0298	.0253	.0215	.0181	.0152	.0128	.0106	.0088	.0073
2	.1018	.0911	.0811	.0719	.0634	.0557	.0487	.0424	.0367	.0317
3	.1830	.1715	.1598	.1481	.1366	.1253	.1144	.1039	.0940	.0845
4	.2261	.2219	.2164	.2098	.2022	.1938	.1848	.1752	.1652	.1549
5	.2032	.2088	.2132	.2161	.2178	.2181	.2170	.2147	.2112	.2066
6	.1369	.1474	.1575	.1670	.1759	.1840	.1912	.1974	.2026	.2066
7	.0703	.0793	.0886	.0983	.1082	.1183	.1283	.1383	.1480	.1574
8	.0276	.0326	.0382	.0443	.0510	.0582	.0659	.0742	.0828	.0918
9	.0083	.0102	.0125	.0152	.0183	.0218	.0258	.0303	.0353	.0408
10	.0019	.0024	.0031	.0039	.0049	.0061	.0076	.0093	.0113	.0136
11	.0003	.0004	.0006	.0007	.0010	.0013	.0016	.0021	.0026	.0033
12	.0000	.0000	.0001	.0001	.0001	.0002	.0002	.0003	.0004	.0005
13	.0000	.0000	.0000	.0000	.0000	.0000	.0000	.0000	.0000	.0001

r \ p	.41	.42	.43	.44	.45	.46	.47	.48	.49	.50
0	.0006	.0005	.0004	.0003	.0002	.0002	.0001	.0001	.0001	.0001
1	.0060	.0049	.0040	.0033	.0027	.0021	.0017	.0014	.0011	.0009
2	.0272	.0233	.0198	.0168	.0141	.0118	.0099	.0082	.0068	.0056
3	.0757	.0674	.0597	.0527	.0462	.0403	.0350	.0303	.0260	.0222
4	.1446	.1342	.1239	.1138	.1040	.0945	.0854	.0768	.0687	.0611
5	.2009	.1943	.1869	.1788	.1701	.1610	.1515	.1418	.1320	.1222
6	.2094	.2111	.2115	.2108	.2088	.2057	.2015	.1963	.1902	.1833
7	.1663	.1747	.1824	.1892	.1952	.2003	.2043	.2071	.2089	.2095
8	.1011	.1107	.1204	.1301	.1398	.1493	.1585	.1673	.1756	.1833
9	.0469	.0534	.0605	.0682	.0762	.0848	.0937	.1030	.1125	.1222
10	.0163	.0193	.0228	.0268	.0312	.0361	.0415	.0475	.0540	.0611
11	.0041	.0051	.0063	.0076	.0093	.0112	.0134	.0160	.0189	.0222
12	.0007	.0009	.0012	.0015	.0019	.0024	.0030	.0037	.0045	.0056
13	.0001	.0001	.0001	.0002	.0002	.0003	.0004	.0005	.0007	.0009
14	.0000	.0000	.0000	.0000	.0000	.0000	.0000	.0000	.0000	.0001

n = 15

r \ p	.01	.02	.03	.04	.05	.06	.07	.08	.09	.10
0	.8601	.7386	.6333	.5421	.4633	.3953	.3367	.2863	.2430	.2059
1	.1303	.2261	.2938	.3388	.3658	.3785	.3801	.3734	.3605	.3432
2	.0092	.0323	.0636	.0988	.1348	.1691	.2003	.2273	.2496	.2669
3	.0004	.0029	.0085	.0178	.0307	.0468	.0653	.0857	.1070	.1285
4	.0000	.0002	.0008	.0022	.0049	.0090	.0148	.0223	.0317	.0428
5	.0000	.0000	.0001	.0002	.0006	.0013	.0024	.0043	.0069	.0105
6	.0000	.0000	.0000	.0000	.0000	.0001	.0003	.0006	.0011	.0019
7	.0000	.0000	.0000	.0000	.0000	.0000	.0000	.0001	.0001	.0003

F

n = 15 (Continued)

r \ p	.11	.12	.13	.14	.15	.16	.17	.18	.19	.20
0	.1741	.1470	.1238	.1041	.0874	.0731	.0611	.0510	.0424	.0352
1	.3228	.3006	.2775	.2542	.2312	.2090	.1878	.1678	.1492	.1319
2	.2793	.2870	.2903	.2897	.2856	.2787	.2692	.2578	.2449	.2309
3	.1496	.1696	.1880	.2044	.2184	.2300	.2389	.2452	.2489	.2501
4	.0555	.0694	.0843	.0998	.1156	.1314	.1468	.1615	.1752	.1876
5	.0151	.0208	.0277	.0357	.0449	.0551	.0662	.0780	.0904	.1032
6	.0031	.0047	.0069	.0097	.0132	.0175	.0226	.0285	.0353	.0430
7	.0005	.0008	.0013	.0020	.0030	.0043	.0059	.0081	.0107	.0138
8	.0001	.0001	.0002	.0003	.0005	.0008	.0012	.0018	.0025	.0035
9	.0000	.0000	.0000	.0000	.0001	.0001	.0002	.0003	.0005	.0007
10	.0000	.0000	.0000	.0000	.0000	.0000	.0000	.0000	.0001	.0001

r \ p	.21	.22	.23	.24	.25	.26	.27	.28	.29	.30
0	.0291	.0241	.0198	.0163	.0134	.0109	.0089	.0072	.0059	.0047
1	.1162	.1018	.0889	.0772	.0668	.0576	.0494	.0423	.0360	.0305
2	.2162	.2010	.1858	.1707	.1559	.1416	.1280	.1150	.1029	.0916
3	.2490	.2457	.2405	.2336	.2252	.2156	.2051	.1939	.1821	.1700
4	.1986	.2079	.2155	.2213	.2252	.2273	.2276	.2262	.2231	.2186
5	.1161	.1290	.1416	.1537	.1651	.1757	.1852	.1935	.2005	.2061
6	.0514	.0606	.0705	.0809	.0917	.1029	.1142	.1254	.1365	.1472
7	.0176	.0220	.0271	.0329	.0393	.0465	.0543	.0627	.0717	.0811
8	.0047	.0062	.0081	.0104	.0131	.0163	.0201	.0244	.0293	.0348
9	.0010	.0014	.0019	.0025	.0034	.0045	.0058	.0074	.0093	.0116
10	.0002	.0002	.0003	.0005	.0007	.0009	.0013	.0017	.0023	.0030
11	.0000	.0000	.0000	.0001	.0001	.0002	.0002	.0003	.0004	.0006
12	.0000	.0000	.0000	.0000	.0000	.0000	.0000	.0000	.0001	.0001

r \ p	.31	.32	.33	.34	.35	.36	.37	.38	.39	.40
0	.0038	.0031	.0025	.0020	.0016	.0012	.0010	.0008	.0006	.0005
1	.0258	.0217	.0182	.0152	.0126	.0104	.0086	.0071	.0058	.0047
2	.0811	.0715	.0627	.0547	.0476	.0411	.0354	.0303	.0259	.0219
3	.1579	.1457	.1338	.1222	.1110	.1002	.0901	.0805	.0716	.0634
4	.2128	.2057	.1977	.1888	.1792	.1692	.1587	.1481	.1374	.1268
5	.210	.2130	.2142	.2140	.2123	.2093	.2051	.1997	.1933	.1859
6	.1575	.1671	.1759	.1837	.1906	.1963	.2008	.2040	.2059	.2066
7	.0910	.1011	.1114	.1217	.1319	.1419	.1516	.1608	.1693	.1771
8	.0409	.0476	.0549	.0627	.0710	.0798	.0890	.0985	.1082	.1181
9	.0143	.0174	.0210	.0251	.0298	.0349	.0407	.0470	.0538	.0612
10	.0038	.0049	.0062	.0078	.0096	.0118	.0143	.0173	.0206	.0245
11	.0008	.0011	.0014	.0018	.0024	.0030	.0038	.0048	.0060	.0074
12	.0001	.0002	.0002	.0003	.0004	.0006	.0007	.0010	.0013	.0016
13	.0000	.0000	.0000	.0000	.0001	.0001	.0001	.0001	.0002	.0003

r \ p	.41	.42	.43	.44	.45	.46	.47	.48	.49	.50
0	.0004	.0003	.0002	.0002	.0001	.0001	.0001	.0001	.0000	.0000
1	.0038	.0031	.0025	.0020	.0016	.0012	.0010	.0008	.0006	.0005
2	.0185	.0156	.0130	.0108	.0090	.0074	.0060	.0049	.0040	.0032
3	.0558	.0489	.0426	.0369	.0318	.0272	.0232	.0197	.0166	.0139
4	.1163	.1061	.0963	.0869	.0780	.0696	.0617	.0545	.0478	.0417
5	.1778	.1691	.1598	.1502	.1404	.1304	.1204	.1106	.1010	.0916
6	.2060	.2041	.2010	.1967	.1914	.1851	.1780	.1702	.1617	.1527
7	.1840	.1900	.1949	.1987	.2013	.2028	.2030	.2020	.1997	.1964
8	.1279	.1376	.1470	.1561	.1647	.1727	.1800	.1864	.1919	.1964
9	.0691	.0775	.0863	.0954	.1048	.1144	.1241	.1338	.1434	.1527
10	.0288	.0337	.0390	.0450	.0515	.0585	.0661	.0741	.0827	.0916
11	.0091	.0111	.0134	.0161	.0191	.0226	.0266	.0311	.0361	.0417
12	.0021	.0027	.0034	.0042	.0052	.0064	.0079	.0096	.0116	.0139
13	.0003	.0004	.0006	.0008	.0010	.0013	.0016	.0020	.0026	.0032
14	.0000	.0000	.0001	.0001	.0001	.0002	.0002	.0003	.0004	.0005

F

n = 16

r \ p	.01	.02	.03	.04	.05	.06	.07	.08	.09	.10
0	.8515	.7238	.6143	.5204	.4401	.3716	.3131	.2634	.2211	.1853
1	.1376	.2363	.3040	.3469	.3706	.3795	.3771	.3665	.3499	.3294
2	.0104	.0362	.0705	.1084	.1463	.1817	.2129	.2390	.2596	.2745
3	.0005	.0034	.0102	.0211	.0359	.0541	.0748	.0970	.1198	.1423
4	.0000	.0002	.0010	.0029	.0061	.0112	.0183	.0274	.0385	.0514
5	.0000	.0000	.0001	.0003	.0008	.0017	.0033	.0057	.0091	.0137
6	.0000	.0000	.0000	.0000	.0001	.0002	.0005	.0009	.0017	.0028
7	.0000	.0000	.0000	.0000	.0000	.0000	.0000	.0001	.0002	.0004
8	.0000	.0000	.0000	.0000	.0000	.0000	.0000	.0000	.0000	.0001
	.11	.12	.13	.14	.15	.16	.17	.18	.19	.20
0	.1550	.1293	.1077	.0895	.0743	.0614	.0507	.0418	.0343	.0281
1	.3065	.2822	.2575	.2332	.2097	.1873	.1662	.1468	.1289	.1126
2	.2841	.2886	.2886	.2847	.2775	.2675	.2554	.2416	.2267	.2111
3	.1638	.1837	.2013	.2163	.2285	.2378	.2441	.2475	.2482	.2463
4	.0658	.0814	.0977	.1144	.1311	.1472	.1625	.1766	.1892	.2001
5	.0195	.0266	.0351	.0447	.0555	.0673	.0799	.0930	.1065	.1201
6	.0044	.0067	.0096	.0133	.0180	.0235	.0300	.0374	.0458	.0550
7	.0008	.0013	.0020	.0031	.0045	.0064	.0088	.0117	.0153	.0197
8	.0001	.0002	.0003	.0006	.0009	.0014	.0020	.0029	.0041	.0055
9	.0000	.0000	.0000	.0001	.0001	.0002	.0004	.0006	.0008	.0012
10	.0000	.0000	.0000	.0000	.0000	.0000	.0001	.0001	.0001	.0002
	.21	.22	.23	.24	.25	.26	.27	.28	.29	.30
0	.0230	.0188	.0153	.0124	.0100	.0081	.0065	.0052	.0042	.0033
1	.0979	.0847	.0730	.0626	.0535	.0455	.0385	.0325	.0273	.0228
2	.1952	.1792	.1635	.1482	.1336	.1198	.1068	.0947	.0835	.0732
3	.2421	.2359	.2279	.2185	.2079	.1964	.1843	.1718	.1591	.1465
4	.2092	.2162	.2212	.2242	.2252	.2243	.2215	.2171	.2112	.2040
5	.1334	.1464	.1586	.1699	.1802	.1891	.1966	.2026	.2071	.2099
6	.0650	.0757	.0869	.0984	.1101	.1218	.1333	.1445	.1551	.1649
7	.0247	.0305	.0371	.0444	.0524	.0611	.0704	.0803	.0905	.1010
8	.0074	.0097	.0125	.0158	.0197	.0242	.0293	.0351	.0416	.0487
9	.0017	.0024	.0033	.0044	.0058	.0075	.0096	.0121	.0151	.0185
10	.0003	.0005	.0007	.0010	.0014	.0019	.0025	.0033	.0043	.0056
11	.0000	.0001	.0001	.0002	.0002	.0004	.0005	.0007	.0010	.0013
12	.0000	.0000	.0000	.0000	.0000	.0001	.0001	.0001	.0002	.0002
	.31	.32	.33	.34	.35	.36	.37	.38	.39	.40
0	.0026	.0021	.0016	.0013	.0010	.0008	.0006	.0005	.0004	.0003
1	.0190	.0157	.0130	.0107	.0087	.0071	.0058	.0047	.0038	.0030
2	.0639	.0555	.0480	.0413	.0353	.0301	.0255	.0215	.0180	.0150
3	.1341	.1220	.1103	.0992	.0888	.0790	.0699	.0615	.0538	.0468
4	.1958	.1865	.1766	.1662	.1553	.1444	.1333	.1224	.1118	.1014
5	.2111	.2107	.2088	.2054	.2008	.1949	.1879	.1801	.1715	.1623
6	.1739	.1818	.1885	.1940	.1982	.2010	.2024	.2024	.2010	.1983
7	.1116	.1222	.1326	.1428	.1524	.1615	.1698	.1772	.1836	.1889
8	.0564	.0647	.0735	.0827	.0923	.1022	.1122	.1222	.1320	.1417
9	.0225	.0271	.0322	.0379	.0442	.1511	.0586	.0666	.0750	.0840
10	.0071	.0089	.0111	.0137	.0167	.0201	.0241	.0286	.0336	.0392
11	.0017	.0023	.0030	.0038	.0049	.0062	.0077	.0095	.0117	.0142
12	.0003	.0004	.0006	.0008	.0011	.0014	.0019	.0024	.0031	.0040
13	.0000	.0001	.0001	.0001	.0002	.0003	.0003	.0005	.0006	.0008
14	.0000	.0000	.0000	.0000	.0000	.0000	.0000	.0001	.0001	.0001
	.41	.42	.43	.44	.45	.46	.47	.48	.49	.50
0	.0002	.0002	.0001	.0001	.0001	.0001	.0000	.0000	.0000	.0000
1	.0024	.0019	.0015	.0012	.0009	.0007	.0005	.0004	.0003	.0002
2	.0125	.0103	.0085	.0069	.0056	.0046	.0037	.0029	.0023	.0018
3	.0405	.0349	.0299	.0254	.0215	.0181	.0151	.0126	.0104	.0085
4	.0915	.0821	.0732	.0649	.0572	.0501	.0436	.0378	.0325	.0278
5	.1526	.1426	.1325	.1224	.1123	.1024	.0929	.0837	.0749	.0667
6	.1944	.1894	.1833	.1762	.1684	.1600	.1510	.1416	.1319	.1222
7	.1930	.1959	.1975	.1978	.1969	.1947	.1912	.1867	.1811	.1746
8	.1509	.1596	.1676	.1749	.1812	.1865	.1908	.1939	.1958	.1964
9	.0932	.1027	.1124	.1221	.1318	.1413	.1504	.1591	.1672	.1746
10	.0453	.0521	.0594	.0672	.0755	.0842	.0934	.1028	.1124	.1222
11	.0172	.0206	.0244	.0288	.0337	.0391	.0452	.0518	.0589	.0667
12	.0050	.0062	.0077	.0094	.0115	.0139	.0167	.0199	.0236	.0278
13	.0011	.0014	.0018	.0023	.0029	.0036	.0046	.0057	.0070	.0085
14	.0002	.0002	.0003	.0004	.0005	.0007	.0009	.0011	.0014	.0018
15	.0000	.0000	.0000	.0000	.0001	.0001	.0001	.0001	.0002	.0002

n = 17

r \ p	.01	.02	.03	.04	.05	.06	.07	.08	.09	.10
0	.8429	.7093	.5958	.4996	.4181	.3493	.2912	.2423	.2012	.1668
1	.1447	.2461	.3133	.3539	.3741	.3790	.3726	.3582	.3383	.3150
2	.0117	.0402	.0775	.1180	.1575	.1935	.2244	.2492	.2677	.2800
3	.0006	.0041	.0120	.0246	.0415	.0618	.0844	.1083	.1324	.1556
4	.0000	.0003	.0013	.0036	.0076	.0138	.0222	.0330	.0458	.0605
5	.0000	.0000	.0001	.0004	.0010	.0023	.0044	.0075	.0118	.0175
6	.0000	.0000	.0000	.0000	.0001	.0003	.0007	.0013	.0023	.0039
7	.0000	.0000	.0000	.0000	.0000	.0000	.0001	.0002	.0004	.0007
8	.0000	.0000	.0000	.0000	.0000	.0000	.0000	.0000	.0000	.0001

r \ p	.11	.12	.13	.14	.15	.16	.17	.18	.19	.20
0	.1379	.1138	.0937	.0770	.0631	.0516	.0421	.0343	.0278	.0225
1	.2898	.2638	.2381	.2131	.1893	.1671	.1466	.1279	.1109	.0957
2	.2865	.2878	.2846	.2775	.2673	.2547	.2402	.2245	.2081	.1914
3	.1771	.1963	.2126	.2259	.2359	.2425	.2460	.2464	.2441	.2393
4	.0766	.0937	.1112	.1287	.1457	.1617	.1764	.1893	.2004	.2093
5	.0246	.0332	.0432	.0545	.0668	.0801	.0939	.1081	.1222	.1361
6	.0061	.0091	.0129	.0177	.0236	.0305	.0385	.0474	.0573	.0680
7	.0012	.0019	.0030	.0045	.0065	.0091	.0124	.0164	.0211	.0267
8	.0002	.0003	.0006	.0009	.0014	.0022	.0032	.0045	.0062	.0084
9	.0000	.0000	.0001	.0002	.0003	.0004	.0006	.0010	.0015	.0021
10	.0000	.0000	.0000	.0000	.0000	.0001	.0001	.0002	.0003	.0004
11	.0000	.0000	.0000	.0000	.0000	.0000	.0000	.0000	.0000	.0001

r \ p	.21	.22	.23	.24	.25	.26	.27	.28	.29	.30
0	.0182	.0146	.0118	.0094	.0075	.0060	.0047	.0038	.0030	.0023
1	.0822	.0702	.0597	.0505	.0426	.0357	.0299	.0248	.0206	.0169
2	.1747	.1584	.1427	.1277	.1136	.1005	.0883	.0772	.0672	.0581
3	.2322	.2234	.2131	.2016	.1893	.1765	.1634	.1502	.1372	.1245
4	.2161	.2205	.2228	.2228	.2209	.2170	.2115	.2044	.1961	.1868
5	.1493	.1617	.1730	.1830	.1914	.1982	.2033	.2067	.2083	.2081
6	.0794	.0912	.1034	.1156	.1276	.1393	.1504	.1608	.1701	.1784
7	.0332	.0404	.0485	.0573	.0668	.0769	.0874	.0982	.1092	.1201
8	.0110	.0143	.0181	.0226	.0279	.0338	.0404	.0478	.0558	.0644
9	.0029	.0040	.0054	.0071	.0093	.0119	.0150	.0186	.0228	.0276
10	.0006	.0009	.0013	.0018	.0025	.0033	.0044	.0058	.0074	.0095
11	.0001	.0002	.0002	.0004	.0005	.0007	.0010	.0014	.0019	.0026
12	.0000	.0000	.0000	.0001	.0001	.0001	.0002	.0003	.0004	.0006
13	.0000	.0000	.0000	.0000	.0000	.0000	.0000	.0000	.0001	.0001

r \ p	.31	.32	.33	.34	.35	.36	.37	.38	.39	.40
0	.0018	.0014	.0011	.0009	.0007	.0005	.0004	.0003	.0002	.0002
1	.0139	.0114	.0093	.0075	.0060	.0048	.0039	.0031	.0024	.0019
2	.0500	.0428	.0364	.0309	.0260	.0218	.0182	.0151	.0125	.0102
3	.1123	.1007	.0898	.0795	.0701	.0614	.0534	.0463	.0398	.0341
4	.1766	.1659	.1547	.1434	.1320	.1208	.1099	.0993	.0892	.0796
5	.2063	.2030	.1982	.1921	.1849	.1767	.1677	.1582	.1482	.1379
6	.1854	.1910	.1952	.1979	.1991	.1988	.1970	.1939	.1895	.1839
7	.1309	.1413	.1511	.1602	.1685	.1757	.1818	.1868	.1904	.1927
8	.0735	.0831	.0930	.1032	.1134	.1235	.1335	.1431	.1521	.1606
9	.0330	.0391	.0458	.0531	.0611	.0695	.0784	.0877	.0973	.1070
10	.0119	.0147	.0181	.0219	.0263	.0313	.0368	.0430	.0498	.0571
11	.0034	.0044	.0057	.0072	.0090	.0112	.0138	.0168	.0202	.0242
12	.0008	.0010	.0014	.0018	.0024	.0031	.0040	.0051	.0065	.0081
13	.0001	.0002	.0003	.0004	.0005	.0007	.0009	.0012	.0016	.0021
14	.0000	.0000	.0000	.0001	.0001	.0001	.0002	.0002	.0003	.0004
15	.0000	.0000	.0000	.0000	.0000	.0000	.0000	.0000	.0000	.0001

n = 17 (Continued)

r \ p	.41	.42	.43	.44	.45	.46	.47	.48	.49	.50
0	.0001	.0001	.0001	.0001	.0000	.0000	.0000	.0000	.0000	.0000
1	.0015	.0012	.0009	.0007	.0005	.0004	.0003	.0002	.0002	.0001
2	.0084	.0068	.0055	.0044	.0035	.0028	.0022	.0017	.0013	.0010
3	.0290	.0246	.0207	.0173	.0144	.0119	.0097	.0079	.0064	.0052
4	.0706	.0622	.0546	.0475	.0411	.0354	.0302	.0257	.0217	.0182
5	.1276	.1172	.1070	.0971	.0875	.0784	.0697	.0616	.0541	.0472
6	.1773	.1697	.1614	.1525	.1432	.1335	.1237	.1138	.1040	.0944
7	.1936	.1932	.1914	.1883	.1841	.1787	.1723	.1650	.1570	.1484
8	.1682	.1748	.1805	.1850	.1883	.1903	.1910	.1904	.1886	.1855
9	.1169	.1266	.1361	.1453	.1540	.1621	.1694	.1758	.1812	.1855
10	.0650	.0733	.0822	.0914	.1008	.1105	.1202	.1298	.1393	.1484
11	.0287	.0338	.0394	.0457	.0525	.0599	.0678	.0763	.0851	.0944
12	.0100	.0122	.0149	.0179	.0215	.0255	.0301	.0352	.0409	.0472
13	.0027	.0034	.0043	.0054	.0068	.0084	.0103	.0125	.0151	.0182
14	.0005	.0007	.0009	.0012	.0016	.0020	.0026	.0033	.0041	.0052
15	.0001	.0001	.0001	.0002	.0003	.0003	.0005	.0006	.0008	.0010
16	.0000	.0000	.0000	.0000	.0000	.0000	.0001	.0001	.0001	.0001

n = 18

r \ p	.01	.02	.03	.04	.05	.06	.07	.08	.09	.10
0	.8345	.6951	.5780	.4796	.3972	.3283	.2708	.2229	.1831	.1501
1	.1517	.2554	.3217	.3597	.3763	.3772	.3669	.3489	.3260	.3002
2	.0130	.0443	.0846	.1274	.1683	.2047	.2348	.2579	.2741	.2835
3	.0007	.0048	.0140	.0283	.0473	.0697	.0942	.1196	.1446	.1680
4	.0000	.0004	.0016	.0044	.0093	.0167	.0266	.0390	.0536	.0700
5	.0000	.0000	.0001	.0005	.0014	.0030	.0056	.0095	.0148	.0218
6	.0000	.0000	.0000	.0000	.0002	.0004	.0009	.0018	.0032	.0052
7	.0000	.0000	.0000	.0000	.0000	.0000	.0001	.0003	.0005	.0010
8	.0000	.0000	.0000	.0000	.0000	.0000	.0000	.0000	.0001	.0002

r \ p	.11	.12	.13	.14	.15	.16	.17	.18	.19	.20
0	.1227	.1002	.0815	.0662	.0536	.0434	.0349	.0281	.0225	.0180
1	.2731	.2458	.2193	.1940	.1704	.1486	.1288	.1110	.0951	.0811
2	.2869	.2850	.2785	.2685	.2556	.2407	.2243	.2071	.1897	.1723
3	.1891	.2072	.2220	.2331	.2406	.2445	.2450	.2425	.2373	.2297
4	.0877	.1060	.1244	.1423	.1592	.1746	.1882	.1996	.2087	.2153
5	.0303	.0405	.0520	.0649	.0787	.0931	.1079	.1227	.1371	.1507
6	.0081	.0120	.0168	.0229	.0301	.0384	.0479	.0584	.0697	.0816
7	.0017	.0028	.0043	.0064	.0091	.0126	.0168	.0220	.0280	.0350
8	.0003	.0005	.0009	.0014	.0022	.0033	.0047	.0066	.0090	.0120
9	.0000	.0001	.0001	.0003	.0004	.0007	.0011	.0016	.0024	.0033
10	.0000	.0000	.0000	.0000	.0001	.0001	.0002	.0003	.0005	.0008
11	.0000	.0000	.0000	.0000	.0000	.0000	.0000	.0001	.0001	.0001

r \ p	.21	.22	.23	.24	.25	.26	.27	.28	.29	.30
0	.0144	.0114	.0091	.0072	.0056	.0044	.0035	.0027	.0021	.0016
1	.0687	.0580	.0487	.0407	.0338	.0280	.0231	.0189	.0155	.0126
2	.1553	.1390	.1236	.1092	.0958	.0836	.0725	.0626	.0537	.0458
3	.2202	.2091	.1969	.1839	.1704	.1567	.1431	.1298	.1169	.1046
4	.2195	.2212	.2205	.2177	.2130	.2065	.1985	.1892	.1790	.1681
5	.1634	.1747	.1845	.1925	.1988	.2031	.2055	.2061	.2048	.2017
6	.0941	.1067	.1194	.1317	.1436	.1546	.1647	.1736	.1812	.1873
7	.0429	.0516	.0611	.0713	.0820	.0931	.1044	.1157	.1269	.1376
8	.0157	.0200	.0251	.0310	.0376	.0450	.0531	.0619	.0713	.0811
9	.0046	.0063	.0083	.0109	.0139	.0176	.0218	.0267	.0323	.0386
10	.0011	.0016	.0022	.0031	.0042	.0056	.0073	.0094	.0119	.0149
11	.0002	.0003	.0005	.0007	.0010	.0014	.0020	.0026	.0035	.0046
12	.0000	.0001	.0001	.0001	.0002	.0003	.0004	.0006	.0008	.0012
13	.0000	.0000	.0000	.0000	.0000	.0000	.0001	.0001	.0002	.0002

n = 18 (Continued)

r \ p	.31	.32	.33	.34	.35	.36	.37	.38	.39	.40
0	.0013	.0010	.0007	.0006	.0004	.0003	.0002	.0002	.0001	.0001
1	.0102	.0082	.0066	.0052	.0042	.0033	.0026	.0020	.0016	.0012
2	.0388	.0327	.0275	.0229	.0190	.0157	.0129	.0105	.0086	.0069
3	.0930	.0822	.0722	.0630	.0547	.0471	.0404	.0344	.0292	.0246
4	.1567	.1450	.1333	.1217	.1104	.0994	.0890	.0791	.0699	.0614
5	.1971	.1911	.1838	.1755	.1664	.1566	.1463	.1358	.1252	.1146
6	.1919	.1948	.1962	.1959	.1941	.1908	.1862	.1803	.1734	.1655
7	.1478	.1572	.1656	.1730	.1792	.1840	.1875	.1895	.1900	.1892
8	.0913	.1017	.1122	.1226	.1327	.1423	.1514	.1597	.1671	.1734
9	.0456	.0532	.0614	.0701	.0794	.0890	.0988	.1087	.1187	.1284
10	.0184	.0225	.0272	.0325	.0385	.0450	.0522	.0600	.0683	.0771
11	.0060	.0077	.0097	.0122	.0151	.0184	.0223	.0267	.0318	.0374
12	.0016	.0021	.0028	.0037	.0047	.0060	.0076	.0096	.0118	.0145
13	.0003	.0005	.0006	.0009	.0012	.0016	.0021	.0027	.0035	.0045
14	.0001	.0001	.0001	.0002	.0002	.0003	.0004	.0006	.0008	.0011
15	.0000	.0000	.0000	.0000	.0000	.0000	.0001	.0001	.0001	.0002

r \ p	.41	.42	.43	.44	.45	.46	.47	.48	.49	.50
0	.0001	.0001	.0000	.0000	.0000	.0000	.0000	.0000	.0000	.0000
1	.0009	.0007	.0005	.0004	.0003	.0002	.0002	.0001	.0001	.0001
2	.0055	.0044	.0035	.0028	.0022	.0017	.0013	.0010	.0008	.0006
3	.0206	.0171	.0141	.0116	.0095	.0077	.0062	.0050	.0039	.0031
4	.0536	.0464	.0400	.0342	.0291	.0246	.0206	.0172	.0142	.0117
5	.1042	.0941	.0844	.0753	.0666	.0586	.0512	.0444	.0382	.0327
6	.1569	.1477	.1380	.1281	.1181	.1081	.0983	.0887	.0796	.0708
7	.1869	.1833	.1785	.1726	.1657	.1579	.1494	.1404	.1310	.1214
8	.1786	.1825	.1852	.1864	.1864	.1850	.1822	.1782	.1731	.1669
9	.1379	.1469	.1552	.1628	.1694	.1751	.1795	.1828	.1848	.1855
10	.0862	.0957	.1054	.1151	.1248	.1342	.1433	.1519	.1598	.1669
11	.0436	.0504	.0578	.0658	.0742	.0831	.0924	.1020	.1117	.1214
12	.0177	.0213	.0254	.0301	.0354	.0413	.0478	.1549	.0626	.0708
13	.0057	.0071	.0089	.0109	.0134	.0162	.0196	.0234	.0278	.0327
14	.0014	.0018	.0024	.0031	.0039	.0049	.0062	.0077	.0095	.0117
15	.0003	.0004	.0005	.0006	.0009	.0011	.0015	.0019	.0024	.0031
16	.0000	.0000	.0001	.0001	.0001	.0002	.0002	.0003	.0004	.0006
17	.0000	.0000	.0000	.0000	.0000	.0000	.0000	.0000	.0000	.0001

n = 19

r \ p	.01	.02	.03	.04	.05	.06	.07	.08	.09	.10
0	.8262	.6812	.5606	.4604	.3774	.3086	.2519	.2051	.1666	.1351
1	.1586	.2642	.3294	.3645	.3774	.3743	.3602	.3389	.3131	.2852
2	.0144	.0485	.0917	.1367	.1787	.2150	.2440	.2652	.2787	.2852
3	.0008	.0056	.0161	.0323	.0533	.0778	.1041	.1307	.1562	.1796
4	.0000	.0005	.0020	.0054	.0112	.0199	.0313	.0455	.0618	.0798
5	.0000	.0000	.0002	.0007	.0018	.0038	.0071	.0119	.0183	.0266
6	.0000	.0000	.0000	.0001	.0002	.0006	.0012	.0024	.0042	.0069
7	.0000	.0000	.0000	.0000	.0000	.0001	.0002	.0004	.0008	.0014
8	.0000	.0000	.0000	.0000	.0000	.0000	.0000	.0001	.0001	.0002

r \ p	.11	.12	.13	.14	.15	.16	.17	.18	.19	.20
0	.1092	.0881	.0709	.0569	.0456	.0364	.0290	.0230	.0182	.0144
1	.2565	.2284	.2014	.1761	.1529	.1318	.1129	.0961	.0813	.0685
2	.2854	.2803	.2708	.2581	.2428	.2259	.2081	.1898	.1717	.1540
3	.1999	.2166	.2293	.2381	.2428	.2439	.2415	.2361	.2282	.2182
4	.0988	.1181	.1371	.1550	.1714	.1858	.1979	.2073	.2141	.2182
5	.0366	.0483	.0614	.0757	.0907	.1062	.1216	.1365	.1507	.1636
6	.0106	.0154	.0214	.0288	.0374	.0472	.0581	.0699	.0825	.0955
7	.0024	.0039	.0059	.0087	.0122	.0167	.0221	.0285	.0359	.0443
8	.0004	.0008	.0013	.0021	.0032	.0048	.0068	.0094	.0126	.0166
9	.0001	.0001	.0002	.0004	.0007	.0011	.0017	.0025	.0036	.0051
10	.0000	.0000	.0000	.0001	.0001	.0002	.0003	.0006	.0009	.0013
11	.0000	.0000	.0000	.0000	.0000	.0000	.0001	.0001	.0002	.0003

F

n = 19 (Continued)

r \ p	.21	.22	.23	.24	.25	.26	.27	.28	.29	.30
0	.0113	.0089	.0070	.0054	.0042	.0033	.0025	.0019	.0015	.0011
1	.0573	.0477	.0396	.0326	.0268	.0219	.0178	.0144	.0116	.0093
2	.1371	.1212	.1064	.0927	.0803	.0692	.0592	.0503	.0426	.0358
3	.2065	.1937	.1800	.1659	.1517	.1377	.1240	.1109	.0985	.0869
4	.2196	.2185	.2151	.2096	.2023	.1935	.1835	.1726	.1610	.1491
5	.1751	.1849	.1928	.1986	.2023	.2040	.2036	.2013	.1973	.1916
6	.1086	.1217	.1343	.1463	.1574	.1672	.1757	.1827	.1880	.1916
7	.0536	.0637	.0745	.0858	.0974	.1091	.1207	.1320	.1426	.1525
8	.0214	.0270	.0334	.0406	.0487	.0575	.0670	.0770	.0874	.0981
9	.0069	.0093	.0122	.0157	.0198	.0247	.0303	.0366	.0436	.0514
10	.0018	.0026	.0036	.0050	.0066	.0087	.0112	.0142	.0178	.0220
11	.0004	.0006	.0009	.0013	.0018	.0025	.0034	.0045	.0060	.0077
12	.0001	.0001	.0002	.0003	.0004	.0006	.0008	.0012	.0016	.0022
13	.0000	.0000	.0000	.0000	.0001	.0001	.0002	.0002	.0004	.0005
14	.0000	.0000	.0000	.0000	.0000	.0000	.0000	.0000	.0001	.0001

r \ p	.31	.32	.33	.34	.35	.36	.37	.38	.39	.40
0	.0009	.0007	.0005	.0004	.0003	.0002	.0002	.0001	.0001	.0001
1	.0074	.0059	.0046	.0036	.0029	.0022	.0017	.0013	.0010	.0008
2	.0299	.0249	.0206	.0169	.0138	.0112	.0091	.0073	.0058	.0046
3	.0762	.0664	.0574	.0494	.0422	.0358	.0302	.0253	.0211	.0175
4	.1370	.1249	.1131	.1017	.0909	.0806	.0710	.0621	.0540	.0467
5	.1846	.1764	.1672	.1572	.1468	.1360	.1251	.1143	.1036	.0933
6	.1935	.1936	.1921	.1890	.1844	.1785	.1714	.1634	.1546	.1451
7	.1615	.1692	.1757	.1808	.1844	.1865	.1870	.1860	.1835	.1797
8	.1088	.1195	.1298	.1397	.1489	.1573	.1647	.1710	.1760	.1797
9	.0597	.0687	.0782	.0880	.0980	.1082	.1182	.1281	.1375	.1464
10	.0268	.0323	.0385	.0453	.0528	.0608	.0694	.0785	.0879	.0976
11	.0099	.0124	.0155	.0191	.0233	.0280	.0334	.0394	.0460	.0532
12	.0030	.0039	.0051	.0066	.0083	.0105	.0131	.0161	.0196	.0237
13	.0007	.0010	.0014	.0018	.0024	.0032	.0041	.0053	.0067	.0085
14	.0001	.0002	.0003	.0004	.0006	.0008	.0010	.0014	.0018	.0024
15	.0000	.0000	.0000	.0001	.0001	.0001	.0002	.0003	.0004	.0005
16	.0000	.0000	.0000	.0000	.0000	.0000	.0000	.0000	.0001	.0001

r \ p	.41	.42	.43	.44	.45	.46	.47	.48	.49	.50
0	.0000	.0000	.0000	.0000	.0000	.0000	.0000	.0000	.0000	.0000
1	.0006	.0004	.0003	.0002	.0002	.0001	.0001	.0001	.0001	.0000
2	.0037	.0029	.0022	.0017	.0013	.0010	.0008	.0006	.0004	.0003
3	.0144	.0118	.0096	.0077	.0062	.0049	.0039	.0031	.0024	.0018
4	.0400	.0341	.0289	.0243	.0203	.0168	.0138	.0113	.0092	.0074
5	.0834	.0741	.0653	.0572	.0497	.0429	.0368	.0313	.0265	.0222
6	.1353	.1252	.1150	.1049	.0949	.0853	.0751	.0674	.0593	.0518
7	.1746	.1683	.1611	.1530	.1443	.1350	.1254	.1156	.1058	.0961
8	.1820	.1829	.1823	.1803	.1771	.1725	.1668	.1601	.1525	.1442
9	.1546	.1618	.1681	.1732	.1771	.1796	.1808	.1806	.1791	.1762
10	.1074	.1172	.1268	.1361	.1449	.1530	.1603	.1667	.1721	.1762
11	.0611	.0694	.0783	.0875	.0970	.1066	.1163	.1259	.1352	.1442
12	.0283	.0335	.0394	.0458	.0529	.0606	.0688	.0775	.0866	.0961
13	.0106	.0131	.0160	.0194	.0233	.0278	.0328	.0385	.0448	.0518
14	.0032	.0041	.0052	.0065	.0082	.0101	.0125	.0152	.0185	.0222
15	.0007	.0010	.0013	.0017	.0022	.0029	.0037	.0047	.0059	.0074
16	.0001	.0002	.0002	.0003	.0005	.0006	.0008	.0011	.0014	.0018
17	.0000	.0000	.0000	.0000	.0001	.0001	.0001	.0002	.0002	.0003

n = 20

r \ p	.01	.02	.03	.04	.05	.06	.07	.08	.09	.10
0	.8179	.6676	.5438	.4420	.3585	.2901	.2342	.1887	.1516	.1216
1	.1652	.2725	.3364	.3683	.3774	.3703	.3526	.3282	.3000	.2702
2	.0159	.0528	.0988	.1458	.1887	.2246	.2521	.2711	.2818	.2852
3	.0010	.0065	.0183	.0364	.0596	.0860	.1139	.1414	.1672	.1901
4	.0000	.0006	.0024	.0065	.0133	.0233	.0364	.0523	.0703	.0898
5	.0000	.0000	.0002	.0009	.0022	.0048	.0088	.0145	.0222	.0319
6	.0000	.0000	.0000	.0001	.0003	.0008	.0017	.0032	.0055	.0089
7	.0000	.0000	.0000	.0000	.0000	.0001	.0002	.0005	.0011	.0020
8	.0000	.0000	.0000	.0000	.0000	.0000	.0000	.0001	.0002	.0004
9	.0000	.0000	.0000	.0000	.0000	.0000	.0000	.0000	.0000	.0001

n = 20 (Continued)

r \ p	.11	.12	.13	.14	.15	.16	.17	.18	.19	.20
0	.0972	.0776	.0617	.0490	.0388	.0306	.0241	.0189	.0148	.0115
1	.2403	.2115	.1844	.1595	.1368	.1165	.0986	.0829	.0693	.0576
2	.2822	.2740	.2618	.2466	.2293	.2109	.1919	.1730	.1545	.1369
3	.2093	.2242	.2347	.2409	.2428	.2410	.2358	.2278	.2175	.2054
4	.1099	.1299	.1491	.1666	.1821	.1951	.2053	.2125	.2168	.2182
5	.0435	.0567	.0713	.1868	.1028	.1189	.1345	.1493	.1627	.1746
6	.0134	.0193	.0266	.0353	.0454	.0566	.0689	.0819	.0954	.1091
7	.0033	.0053	.0080	.0115	.0160	.0216	.0282	.0360	.0448	.0545
8	.0007	.0012	.0019	.0030	.0046	.0067	.0094	.0128	.0171	.0222
9	.0001	.0002	.0004	.0007	.0011	.0017	.0026	.0038	.0053	.0074
10	.0000	.0000	.0001	.0001	.0002	.0004	.0006	.0009	.0014	.0020
11	.0000	.0000	.0000	.0000	.0000	.0001	.0001	.0002	.0003	.0005
12	.0000	.0000	.0000	.0000	.0000	.0000	.0000	.0000	.0001	.0001

r \ p	.21	.22	.23	.24	.25	.26	.27	.28	.29	.30
0	.0090	.0069	.0054	.0041	.0032	.0024	.0018	.0014	.0011	.0008
1	.0477	.0392	.0321	.0261	.0211	.0170	.0137	.0109	.0087	.0068
2	.1204	.1050	.0910	.0783	.0669	.0569	.0480	.0403	.0336	.0278
3	.1920	.1777	.1631	.1484	.1339	.1199	.1065	.0940	.0823	.0716
4	.2169	.2131	.2070	.1991	.1897	.1790	.1675	.1553	.1429	.1304
5	.1845	.1923	.1979	.2012	.2023	.2013	.1982	.1933	.1868	.1789
6	.1226	.1356	.1478	.1589	.1686	.1768	.1833	.1879	.1907	.1916
7	.0652	.0765	.0883	.1003	.1124	.1242	.1356	.1462	.1558	.1643
8	.0282	.0351	.0429	.0515	.0609	.0709	.0815	.0924	.1034	.1144
9	.0100	.0132	.0171	.0217	.0271	.0332	.0402	.0479	.0563	.0654
10	.0029	.0041	.0056	.0075	.0099	.0128	.0163	.0205	.0253	.0308
11	.0007	.0010	.0015	.0022	.0030	.0041	.0055	.0072	.0094	.0120
12	.0001	.0002	.0003	.0005	.0008	.0011	.0015	.0021	.0029	.0039
13	.0000	.0000	.0001	.0001	.0002	.0002	.0003	.0005	.0007	.0010
14	.0000	.0000	.0000	.0000	.0000	.0000	.0001	.0001	.0001	.0002

r \ p	.31	.32	.33	.34	.35	.36	.37	.38	.39	.40
0	.0006	.0004	.0003	.0002	.0002	.0001	.0001	.0001	.0001	.0000
1	.0054	.0042	.0033	.0025	.0020	.0015	.0011	.0009	.0007	.0005
2	.0229	.0188	.0153	.0124	.0100	.0080	.0064	.0050	.0040	.0031
3	.0619	.0531	.0453	.0383	.0323	.0270	.0224	.0185	.0152	.0123
4	.1181	.1062	.0947	.0839	.0738	.0645	.0559	.0482	.0412	.0350
5	.1698	.1599	.1493	.1384	.1272	.1161	.1051	.0945	.0843	.0746
6	.1907	.1881	.1839	.1782	.1712	.1632	.1543	.1447	.1347	.1244
7	.1714	.1770	.1811	.1836	.1844	.1836	.1812	.1774	.1722	.1659
8	.1251	.1354	.1450	.1537	.1614	.1678	.1730	.1767	.1790	.1797
9	.0750	.0849	.0952	.1056	.1158	.1259	.1354	.1444	.1526	.1597
10	.0370	.0440	.0516	.0598	.0686	.0779	.0875	.0974	.1073	.1171
11	.0151	.0188	.0231	.0280	.0336	.0398	.0467	.0542	.0624	.0710
12	.0051	.0066	.0085	.0108	.0136	.0168	.0206	.0249	.0299	.0355
13	.0014	.0019	.0026	.0034	.0045	.0058	.0074	.0094	.0118	.0146
14	.0003	.0005	.0006	.0009	.0012	.0016	.0022	.0029	.0038	.0049
15	.0001	.0001	.0001	.0002	.0003	.0004	.0005	.0007	.0010	.0013
16	.0000	.0000	.0000	.0000	.0000	.0001	.0001	.0001	.0002	.0003

r \ p	.41	.42	.43	.44	.45	.46	.47	.48	.49	.50
0	.0000	.0000	.0000	.0000	.0000	.0000	.0000	.0000	.0000	.0000
1	.0004	.0003	.0002	.0001	.0001	.0001	.0001	.0000	.0000	.0000
2	.0024	.0018	.0014	.0011	.0008	.0006	.0005	.0003	.0002	.0002
3	.0100	.0080	.0064	.0051	.0040	.0031	.0024	.0019	.0014	.0011
4	.0295	.0247	.0206	.0170	.0139	.0113	.0092	.0074	.0059	.0046
5	.0656	.0573	.0496	.0427	.0365	.0309	.0260	.0217	.0180	.0148
6	.1140	.1037	.0936	.0839	.0746	.0658	.0577	.0501	.0432	.0370
7	.1585	.1502	.1413	.1318	.1221	.1122	.1023	.0925	.0830	.0739
8	.1790	.1768	.1732	.1683	.1623	.1553	.1474	.1388	.1296	.1201
9	.1658	.1707	.1742	.1763	.1771	.1763	.1742	.1708	.1661	.1602
10	.1268	.1359	.1446	.1524	.1593	.1652	.1700	.1734	.1755	.1762
11	.0801	.0895	.0991	.1089	.1185	.1280	.1370	.1455	.1533	.1602
12	.0417	.0486	.0561	.0642	.0727	.0818	.0911	.1007	.1105	.1201
13	.0178	.0217	.0260	.0310	.0366	.0429	.0497	.0572	.0653	.0739
14	.0062	.0078	.0098	.0122	.0150	.0183	.0221	.0264	.0314	.0370
15	.0017	.0023	.0030	.0038	.0049	.0062	.0078	.0098	.0121	.0148
16	.0004	.0005	.0007	.0009	.0013	.0017	.0022	.0028	.0036	.0046
17	.0001	.0001	.0001	.0002	.0002	.0003	.0005	.0006	.0008	.0011
18	.0000	.0000	.0000	.0000	.0000	.0000	.0001	.0001	.0001	.0002

F

ORDINATES OF THE NORMAL DISTRIBUTION

Values of the ordinates of the normal distribution may be computed by multiplying the term $\frac{N}{\sigma}$ for the distribution by the values in the table. (The values in this table represent the ordinates of a normal distribution for which $\frac{N}{\sigma} = 1$.) In a distribution for which $N = 1{,}000$, $\mu = \$400$, and $\sigma = \$20$, the maximum ordinate ($Y_o$) is computed as follows, since Y_o is the ordinate when $\frac{X - \mu}{\sigma} = 0$:

$$Y_o = \frac{N}{\sigma}.39894 = \frac{1{,}000}{20}.39894 = 19.947$$

The ordinate for any value of $X - \mu$ may be computed in a similar manner. When $X = 440$, $z = \frac{X - \mu}{\sigma} = \frac{440 - 400}{20} = 2$. The table gives the value of the ordinate when $\frac{X - \mu}{\sigma} = 2$ as .05399, which is the ordinate of a distribution for which $\frac{N}{\sigma} = 1$. When $\frac{N}{\sigma} = \frac{1{,}000}{20}$, $Y = \frac{1{,}000}{20}.05399 = 2.70$. If the normal distribution is to be plotted on the grid with a histogram or frequency polygon, the standard deviation is expressed in class-interval units $\left(\frac{\sigma}{i}\right)$.

G

ORDINATES OF THE NORMAL DISTRIBUTION

z or $\frac{X - \mu}{\sigma}$	Ordinate at z or $\frac{X - \mu}{\sigma}$	z or $\frac{X - \mu}{\sigma}$	Ordinate at z or $\frac{X - \mu}{\sigma}$
.0	.39894	2.0	.05399
.1	.39695	2.1	.04398
.2	.39104	2.2	.03547
.3	.38139	2.3	.02833
.4	.36827	2.4	.02239
.5	.35207	2.5	.01753
.6	.33322	2.6	.01358
.7	.31225	2.7	.01042
.8	.28969	2.8	.00792
.9	.26609	2.9	.00595
1.0	.24197	3.0	.00443
1.1	.21785	3.1	.00327
1.2	.19419	3.2	.00238
1.3	.17137	3.3	.00172
1.4	.14973	3.4	.00123
1.5	.12952	3.5	.00087
1.6	.11092	3.6	.00061
1.7	.09405	3.7	.00042
1.8	.07895	3.8	.00029
1.9	.06562	3.9	.00020
		4.0	.00014

AREAS OF THE NORMAL CURVE BETWEEN MAXIMUM ORDINATE AND ORDINATE AT z

The values in the table show the fraction of the area of the normal curve that lies between the maximum ordinate (Y_o) and the ordinate at various distances from the maximum ordinate, measured by $\frac{X-\mu}{\sigma}$, or z. Reading down the table to $z = 1.00$, the fraction of the curve is .34134. Since the normal curve is symmetrical, slightly more than 68% of the area of the normal curve lies within the range of $+1\,\sigma$ and $-1\,\sigma$. This means that 68% of the individual values of a normal distribution fall within this range.

The percentage of items falling within any range expressed in standard deviation units can be computed in the same manner by doubling the fraction in the table. For example, 95% of the items in a normal distribution fall between $\pm 1.96\,\sigma$, and 99% fall within the range $\pm 2.576\,\sigma$.

AREAS OF THE NORMAL CURVE BETWEEN MAXIMUM ORDINATE AND ORDINATE AT z

z or $\frac{X-\mu}{\sigma}$	.00	.01	.02	.03	.04	.05	.06	.07	.08	.09
0.0	.00000	.00399	.00798	.01197	.01595	.01994	.02392	.02790	.03188	.03586
0.1	.03983	.04380	.04776	.05172	.05567	.05962	.06356	.06749	.07142	.07535
0.2	.07926	.08317	.08706	.09095	.09483	.09871	.10257	.10642	.11026	.11409
0.3	.11791	.12172	.12552	.12930	.13307	.13683	.14058	.14431	.14803	.15173
0.4	.15542	.15910	.16276	.16640	.17003	.17364	.17724	.18082	.18439	.18793
0.5	.19146	.19497	.19847	.20194	.20540	.20884	.21226	.21566	.21904	.22240
0.6	.22575	.22907	.23237	.23565	.23891	.24215	.24537	.24857	.25175	.25490
0.7	.25804	.26115	.26424	.26730	.27035	.27337	.27637	.27935	.28230	.28524
0.8	.28814	.29103	.29389	.29673	.29955	.30234	.30511	.30785	.31057	.31327
0.9	.31594	.31859	.32121	.32381	.32639	.32894	.33147	.33398	.33646	.33891
1.0	.34134	.34375	.34614	.34850	.35083	.35314	.35543	.35769	.35993	.36214
1.1	.36433	.36650	.36864	.37076	.37286	.37493	.37698	.37900	.38100	.38298
1.2	.38493	.38686	.38877	.39065	.39251	.39435	.39617	.39796	.39973	.40147
1.3	.40320	.40490	.40658	.40824	.40988	.41149	.41309	.41466	.41621	.41774
1.4	.41924	.42073	.42220	.42364	.42507	.42647	.42786	.42922	.43056	.43189
1.5	.43319	.43448	.43574	.43699	.43822	.43943	.44062	.44179	.44295	.44408
1.6	.44520	.44630	.44738	.44845	.44950	.45053	.45154	.45254	.45352	.45449
1.7	.45543	.45637	.45728	.45818	.45907	.45994	.46080	.46164	.46246	.46327
1.8	.46407	.46485	.46562	.46638	.46712	.46784	.46856	.46926	.46995	.47062
1.9	.47128	.47193	.47257	.47320	.47381	.47441	.47500	.47558	.47615	.47670
2.0	.47725	.47778	.47831	.47882	.47932	.47982	.48030	48077	.48124	.48169
2.1	.48214	.48257	.48300	.48341	.48382	.48422	.48461	.48500	.48537	.48574
2.2	.48610	.48645	.48679	.48713	.48745	.48778	.48809	.48840	.48870	.48899
2.3	.48928	.48956	.48983	.49010	.49036	.49061	.49086	.49111	.49134	.49158
2.4	.49180	.49202	.49224	.49245	.49266	.49286	.49305	.49324	.49343	.49361
2.5	.49379	.49396	.49413	.49430	.49446	.49461	.49477	.49492	.49506	.49520
2.6	.49534	.49547	.49560	.49573	.49585	.49598	.49609	.49621	.49632	.49643
2.7	.49653	.49664	.49674	.49683	.49693	.49702	.49711	.49720	.49728	.49736
2.8	.49744	.49752	.49760	.49767	.49774	.49781	.49788	.49795	.49801	.49807
2.9	.49813	.49819	.49825	.49831	.49386	.49841	.49846	.49851	.49856	.49861
3.0	.49865	.49869	.49874	.49878	.49882	.49886	.49889	.49893	.49897	.49900
3.1	.49903	.49906	.49910	.49913	.49916	.49918	.49921	.49924	.49926	.49929
3.2	.49931	.49934	.49936	.49938	.49940	.49942	.49944	.49946	.49948	.49950
3.3	.49952	.49953	.49955	.49957	.49958	.49960	.49961	.49962	.49964	.49965
3.4	.49966	.49968	.49969	.49970	.49971	.49972	.49973	.49974	.49975	.49976
3.5	.49977									
3.6	.49984									
3.7	.49989									
3.8	.49993									
3.9	.49995									
4.0	.49997									

VALUES OF $e^{-\lambda}$

λ	$e^{-\lambda}$	λ	$e^{-\lambda}$	λ	$e^{-\lambda}$	λ	$e^{-\lambda}$	λ	$e^{-\lambda}$
.01	.990050	.28	.755784	.75	.472367	3.20	.0407622	5.90	.00273945
.02	.980199	.29	.748264	.80	.449329	3.30	.0368832	6.00	.00247875
.03	.970446	.30	.740818	.85	.427415	3.40	.0333733	6.10	.00224287
.04	.960789	.31	.733467	.90	.406570	3.50	.0301974	6.20	.00202943
.05	.951229	.32	.726149	.95	.386741	3.60	.0273237	6.30	.00183631
.06	.941765	.33	.718924	1.00	.367879	3.70	.0247235	6.40	.00166156
.07	.932394	.34	.711770	1.10	.332871	3.80	.0223708	6.50	.00150344
.08	.923116	.35	.704688	1.20	.301194	3.90	.0202419	6.60	.00136037
.09	.913931	.36	.697676	1.30	.272532	4.00	.0183156	6.70	.00123091
.10	.904837	.37	.690743	1.40	.246597	4.10	.0165727	6.80	.00111378
.11	.895834	.38	.683861	1.50	.223130	4.20	.0149956	6.90	.00100779
.12	.886920	.39	.677057	1.60	.201897	4.30	.0135686	7.00	.00091188
.13	.878095	.40	.670320	1.70	.182684	4.40	.0122773	7.50	.00055308
.14	.869358	.41	.663650	1.80	.165299	4.50	.0111090	8.00	.00033546
.15	.860708	.42	.657047	1.90	.149569	4.60	.0100518	8.50	.00020347
.16	.852144	.43	.650509	2.00	.135335	4.70	.00909528	9.00	.00012341
.17	.843665	.44	.644036	2.10	.122456	4.80	.00822975	9.50	.00007485
.18	.835270	.45	.637628	2.20	.110803	4.90	.00744658	10.00	.00004540
.19	.826959	.46	.631284	2.30	.100259	5.00	.00673795	10.50	.00002754
.20	.818731	.47	.625002	2.40	.0907180	5.10	.00609675	11.00	.00001670
.21	.810584	.48	.618783	2.50	.0820850	5.20	.00551656	11.50	.00001013
.22	.802519	.49	.612626	2.60	.0742736	5.30	.00499159	12.00	.00000614
.23	.794534	.50	.606531	2.70	.0672055	5.40	.00451658	12.50	.00000373
.24	.786628	.55	.576950	2.80	.0608101	5.50	.00408677	13.00	.00000226
.25	.778801	.60	.548812	2.90	.0550232	5.60	.00369786		
.26	.771052	.65	.522046	3.00	.0497871	5.70	.00334597		
.27	.763379	.70	.496585	3.10	.0450492	5.80	.00302756		

I

POISSON PROBABILITY DISTRIBUTION

$$P(r \mid \lambda) = \frac{\lambda^r}{r!} e^{-\lambda}$$

$P(r = 1 \mid \lambda = 0.7) = 0.3476$

r	λ 0.10	0.20	0.30	0.40	0.50	0.60	0.70	0.80	0.90	1.00
0	.9048	.8187	.7408	.6703	.6065	.5488	.4966	.4493	.4066	.3679
1	.0905	.1637	.2222	.2681	.3033	.3293	.3476	.3595	.3659	.3679
2	.0045	.0164	.0333	.0536	.0758	.0988	.1217	.1438	.1647	.1839
3	.0002	.0011	.0033	.0072	.0126	.0198	.0284	.0383	.0494	.0613
4	.0000	.0001	.0003	.0007	.0016	.0030	.0050	.0077	.0111	.0153
5	.0000	.0000	.0000	.0001	.0002	.0004	.0007	.0012	.0020	.0031
6	.0000	.0000	.0000	.0000	.0000	.0000	.0001	.0002	.0003	.0005
7	.0000	.0000	.0000	.0000	.0000	.0000	.0000	.0000	.0000	.0001

r	λ 1.10	1.20	1.30	1.40	1.50	1.60	1.70	1.80	1.90	2.00
0	.3329	.3012	.2725	.2466	.2231	.2019	.1827	.1653	.1496	.1353
1	.3662	.3614	.3543	.3452	.3347	.3230	.3106	.2975	.2842	.2707
2	.2014	.2169	.2303	.2417	.2510	.2584	.2640	.2678	.2700	.2707
3	.0738	.0867	.0998	.1128	.1255	.1378	.1496	.1607	.1710	.1804
4	.0203	.0260	.0324	.0395	.0471	.0551	.0636	.0723	.0812	.0902
5	.0045	.0062	.0084	.0111	.0141	.0176	.0216	.0260	.0309	.0361
6	.0008	.0012	.0018	.0026	.0035	.0047	.0061	.0078	.0098	.0120
7	.0001	.0002	.0003	.0005	.0008	.0011	.0015	.0020	.0027	.0034
8	.0000	.0000	.0001	.0001	.0001	.0002	.0003	.0005	.0006	.0009
9	.0000	.0000	.0000	.0000	.0000	.0000	.0001	.0001	.0001	.0002

r	λ 2.10	2.20	2.30	2.40	2.50	2.60	2.70	2.80	2.90	3.00
0	.1225	.1108	.1003	.0907	.0821	.0743	.0672	.0608	.0550	.0498
1	.2572	.2438	.2306	.2177	.2052	.1931	.1815	.1703	.1596	.1494
2	.2700	.2681	.2652	.2613	.2565	.2510	.2450	.2384	.2314	.2240
3	.1890	.1966	.2033	.2090	.2138	.2176	.2205	.2225	.2237	.2240
4	.0992	.1082	.1169	.1254	.1336	.1414	.1488	.1557	.1622	.1680
5	.0417	.0476	.0538	.0602	.0668	.0735	.0804	.0872	.0940	.1008
6	.0146	.0174	.0206	.0241	.0278	.0319	.0362	.0407	.0455	.0504
7	.0044	.0055	.0068	.0083	.0099	.0118	.0139	.0163	.0188	.0216
8	.0011	.0015	.0019	.0025	.0031	.0038	.0047	.0057	.0068	.0081
9	.0003	.0004	.0005	.0007	.0009	.0011	.0014	.0018	.0022	.0027
10	.0001	.0001	.0001	.0002	.0002	.0003	.0004	.0005	.0006	.0008
11	.0000	.0000	.0000	.0000	.0000	.0001	.0001	.0001	.0002	.0002
12	.0000	.0000	.0000	.0000	.0000	.0000	.0000	.0000	.0000	.0001

r	λ 3.10	3.20	3.30	3.40	3.50	3.60	3.70	3.80	3.90	4.00
0	.0450	.0408	.0369	.0334	.0302	.0273	.0247	.0224	.0202	.0183
1	.1397	.1304	.1217	.1135	.1057	.0984	.0915	.0850	.0789	.0733
2	.2165	.2087	.2008	.1929	.1850	.1771	.1692	.1615	.1539	.1465
3	.2237	.2226	.2209	.2186	.2158	.2125	.2087	.2046	.2001	.1954
4	.1733	.1781	.1823	.1858	.1888	.1912	.1931	.1944	.1951	.1954

POISSON PROBABILITY DISTRIBUTION

r	λ 3.10	3.20	3.30	3.40	3.50	3.60	3.70	3.80	3.90	4.00
5	.1075	.1140	.1203	.1264	.1322	.1377	.1429	.1477	.1522	.1563
6	.0555	.0608	.0662	.0716	.0771	.0826	.0881	.0936	.0989	.1042
7	.0246	.0278	.0312	.0348	.0385	.0425	.0466	.0508	.0551	.0595
8	.0095	.0111	.0129	.0148	.0169	.0191	.0215	.0241	.0269	.0298
9	.0033	.0040	.0047	.0056	.0066	.0076	.0089	.0102	.0116	.0132
10	.0010	.0013	.0016	.0019	.0023	.0028	.0033	.0039	.0045	.0053
11	.0003	.0004	.0005	.0006	.0007	.0009	.0011	.0013	.0016	.0019
12	.0001	.0001	.0001	.0002	.0002	.0003	.0003	.0004	.0005	.0006
13	.0000	.0000	.0000	.0000	.0001	.0001	.0001	.0001	.0002	.0002
14	.0000	.0000	.0000	.0000	.0000	.0000	.0000	.0000	.0000	.0001

r	λ 4.10	4.20	4.30	4.40	4.50	4.60	4.70	4.80	4.90	5.00
0	.0166	.0150	.0136	.0123	.0111	.0101	.0091	.0082	.0074	.0067
1	.0679	.0630	.0583	.0540	.0500	.0462	.0427	.0395	.0365	.0337
2	.1393	.1323	.1254	.1188	.1125	.1063	.1005	.0948	.0894	.0842
3	.1904	.1852	.1798	.1743	.1687	.1631	.1574	.1517	.1460	.1404
4	.1951	.1944	.1933	.1917	.1898	.1875	.1849	.1820	.1789	.1755
5	.1600	.1633	.1662	.1687	.1708	.1725	.1738	.1747	.1753	.1755
6	.1093	.1143	.1191	.1237	.1281	.1323	.1362	.1398	.1432	.1462
7	.0640	.0686	.0732	.0778	.0824	.0869	.0914	.0959	.1002	.1044
8	.0328	.0360	.0393	.0428	.0463	.0500	.0537	.0575	.0614	.0653
9	.0150	.0168	.0188	.0209	.0232	.0255	.0281	.0307	.0334	.0363
10	.0061	.0071	.0081	.0092	.0104	.0118	.0132	.0147	.0164	.0181
11	.0023	.0027	.0032	.0037	.0043	.0049	.0056	.0064	.0073	.0082
12	.0008	.0009	.0011	.0013	.0016	.0019	.0022	.0026	.0030	.0034
13	.0002	.0003	.0004	.0005	.0006	.0007	.0008	.0009	.0011	.0013
14	.0001	.0001	.0001	.0001	.0002	.0002	.0003	.0003	.0004	.0005
15	.0000	.0000	.0000	.0000	.0001	.0001	.0001	.0001	.0001	.0002

r	λ 5.10	5.20	5.30	5.40	5.50	5.60	5.70	5.80	5.90	6.00
0	.0061	.0055	.0050	.0045	.0041	.0037	.0033	.0030	.0027	.0025
1	.0311	.0287	.0265	.0244	.0225	.0207	.0191	.0176	.0162	.0149
2	.0793	.0746	.0701	.0659	.0618	.0580	.0544	.0509	.0477	.0446
3	.1348	.1293	.1239	.1185	.1133	.1082	.1033	.0985	.0938	.0892
4	.1719	.1681	.1641	.1600	.1558	.1515	.1472	.1428	.1383	.1339
5	.1753	.1748	.1740	.1728	.1714	.1697	.1678	.1656	.1632	.1606
6	.1490	.1515	.1537	.1555	.1571	.1584	.1594	.1601	.1605	.1606
7	.1086	.1125	.1163	.1200	.1234	.1267	.1298	.1326	.1353	.1377
8	.0692	.0731	.0771	.0810	.0849	.0887	.0925	.0962	.0998	.1033
9	.0392	.0423	.0454	.0486	.0519	.0552	.0586	.0620	.0654	.0688
10	.0200	.0220	.0241	.0262	.0285	.0309	.0334	.0359	.0386	.0413
11	.0093	.0104	.0116	.0129	.0143	.0157	.0173	.0190	.0207	.0225
12	.0039	.0045	.0051	.0058	.0065	.0073	.0082	.0092	.0102	.0113
13	.0015	.0018	.0021	.0024	.0028	.0032	.0036	.0041	.0046	.0052
14	.0006	.0007	.0008	.0009	.0011	.0013	.0015	.0017	.0019	.0022
15	.0002	.0002	.0003	.0003	.0004	.0005	.0006	.0007	.0008	.0009
16	.0001	.0001	.0001	.0001	.0001	.0002	.0002	.0002	.0003	.0003
17	.0000	.0000	.0000	.0000	.0000	.0001	.0001	.0001	.0001	.0001

r	λ 6.10	6.20	6.30	6.40	6.50	6.60	6.70	6.80	6.90	7.00
0	.0022	.0020	.0018	.0017	.0015	.0014	.0012	.0011	.0010	.0009
1	.0137	.0126	.0116	.0106	.0098	.0090	.0082	.0076	.0070	.0064
2	.0417	.0390	.0364	.0340	.0318	.0296	.0276	.0258	.0240	.0223
3	.0848	.0806	.0765	.0726	.0688	.0652	.0617	.0584	.0552	.0521
4	.1294	.1249	.1205	.1161	.1118	.1076	.1034	.0992	.0952	.0912
5	.1579	.1549	.1519	.1487	.1454	.1420	.1385	.1349	.1314	.1277
6	.1605	.1601	.1595	.1586	.1575	.1562	.1546	.1529	.1511	.1490
7	.1399	.1418	.1435	.1450	.1462	.1472	.1480	.1486	.1489	.1490
8	.1066	.1099	.1130	.1160	.1188	.1215	.1240	.1263	.1284	.1304
9	.0723	.0757	.0791	.0825	.0858	.0891	.0923	.0954	.0985	.1014
10	.0441	.0469	.0498	.0528	.0558	.0588	.0618	.0649	.0679	.0710
11	.0244	.0265	.0285	.0307	.0330	.0353	.0377	.0401	.0426	.0452
12	.0124	.0137	.0150	.0164	.0179	.0194	.0210	.0227	.0245	.0263
13	.0058	.0065	.0073	.0081	.0089	.0099	.0108	.0119	.0130	.0142
14	.0025	.0029	.0033	.0037	.0041	.0046	.0052	.0058	.0064	.0071

POISSON PROBABILITY DISTRIBUTION

r	6.10	6.20	6.30	6.40	λ 6.50	6.60	6.70	6.80	6.90	7.00
15	.0010	.0012	.0014	.0016	.0018	.0020	.0023	.0026	.0029	.0033
16	.0004	.0005	.0005	.0006	.0007	.0008	.0010	.0011	.0013	.0014
17	.0001	.0002	.0002	.0002	.0003	.0003	.0004	.0004	.0005	.0006
18	.0000	.0001	.0001	.0001	.0001	.0001	.0001	.0002	.0002	.0002
19	.0000	.0000	.0000	.0000	.0000	.0000	.0001	.0001	.0001	.0001

r	7.10	7.20	7.30	7.40	λ 7.50	7.60	7.70	7.80	7.90	8.00
0	.0008	.0007	.0007	.0006	.0006	.0005	.0005	.0004	.0004	.0003
1	.0059	.0054	.0049	.0045	.0041	.0038	.0035	.0032	.0029	.0027
2	.0208	.0194	.0180	.0167	.0156	.0145	.0134	.0125	.0116	.0107
3	.0492	.0464	.0438	.0413	.0389	.0366	.0345	.0324	.0305	.0286
4	.0874	.0836	.0799	.0764	.0729	.0696	.0663	.0632	.0602	.0573
5	.1241	.1204	.1167	.1130	.1094	.1057	.1021	.0986	.0951	.0916
6	.1468	.1445	.1420	.1394	.1367	.1339	.1311	.1282	.1252	.1221
7	.1489	.1486	.1481	.1474	.1465	.1454	.1442	.1428	.1413	.1396
8	.1321	.1337	.1351	.1363	.1373	.1381	.1388	.1392	.1395	.1396
9	.1042	.1070	.1096	.1121	.1144	.1167	.1187	.1207	.1224	.1241
10	.0740	.0770	.0800	.0829	.0858	.0887	.0914	.0941	.0967	.0993
11	.0478	.0504	.0531	.0558	.0585	.0613	.0640	.0667	.0695	.0722
12	.0283	.0303	.0323	.0344	.0366	.0388	.0411	.0434	.0457	.0481
13	.0154	.0168	.0181	.0196	.0211	.0227	.0243	.0260	.0278	.0296
14	.0078	.0086	.0095	.0104	.0113	.0123	.0134	.0145	.0157	.0169
15	.0037	.0041	.0046	.0051	.0057	.0062	.0069	.0075	.0083	.0090
16	.0016	.0019	.0021	.0024	.0026	.0030	.0033	.0037	.0041	.0045
17	.0007	.0008	.0009	.0010	.0012	.0013	.0015	.0017	.0019	.0021
18	.0003	.0003	.0004	.0004	.0005	.0006	.0006	.0007	.0008	.0009
19	.0001	.0001	.0001	.0002	.0002	.0002	.0003	.0003	.0003	.0004
20	.0000	.0000	.0001	.0001	.0001	.0001	.0001	.0001	.0001	.0002
21	.0000	.0000	.0000	.0000	.0000	.0000	.0000	.0000	.0001	.0001

r	8.10	8.20	8.30	8.40	λ 8.50	8.60	8.70	8.80	8.90	9.00
0	.0003	.0003	.0002	.0002	.0002	.0002	.0002	.0002	.0001	.0001
1	.0025	.0023	.0021	.0019	.0017	.0016	.0014	.0013	.0012	.0011
2	.0100	.0092	.0086	.0079	.0074	.0068	.0063	.0058	.0054	.0050
3	.0269	.0252	.0237	.0222	.0208	.0195	.0183	.0171	.0160	.0150
4	.0544	.0517	.0491	.0466	.0443	.0420	.0398	.0377	.0357	.0337
5	.0882	.0849	.0816	.0784	.0752	.0722	.0692	.0663	.0635	.0607
6	.1191	.1160	.1128	.1097	.1066	.1034	.1003	.0972	.0941	.0911
7	.1378	.1358	.1338	.1317	.1294	.1271	.1247	.1222	.1197	.1171
8	.1395	.1392	.1388	.1382	.1375	.1366	.1356	.1344	.1332	.1318
9	.1256	.1269	.1280	.1290	.1299	.1306	.1311	.1315	.1317	.1318
10	.1017	.1040	.1063	.1084	.1104	.1123	.1140	.1157	.1172	.1186
11	.0749	.0776	.0802	.0828	.0853	.0878	.0902	.0925	.0948	.0970
12	.0505	.0530	.0555	.0579	.0604	.0629	.0654	.0679	.0703	.0728
13	.0315	.0334	.0354	.0374	.0395	.0416	.0438	.0459	.0481	.0504
14	.0182	.0196	.0210	.0225	.0240	.0256	.0272	.0289	.0306	.0324
15	.0098	.0107	.0116	.0126	.0136	.0147	.0158	.0169	.0182	.0194
16	.0050	.0055	.0060	.0066	.0072	.0079	.0086	.0093	.0101	.0109
17	.0024	.0026	.0029	.0033	.0036	.0040	.0044	.0048	.0053	.0058
18	.0011	.0012	.0014	.0015	.0017	.0019	.0021	.0024	.0026	.0029
19	.0005	.0005	.0006	.0007	.0008	.0009	.0010	.0011	.0012	.0014
20	.0002	.0002	.0002	.0003	.0003	.0004	.0004	.0005	.0005	.0006
21	.0001	.0001	.0001	.0001	.0001	.0002	.0002	.0002	.0002	.0003
22	.0000	.0000	.0000	.0000	.0001	.0001	.0001	.0001	.0001	.0001

r	9.10	9.20	9.30	9.40	λ 9.50	9.60	9.70	9.80	9.90	10.00
0	.0001	.0001	.0001	.0001	.0001	.0001	.0001	.0001	.0001	.0000
1	.0010	.0009	.0009	.0008	.0007	.0007	.0006	.0005	.0005	.0005
2	.0046	.0043	.0040	.0037	.0034	.0031	.0029	.0027	.0025	.0023
3	.0140	.0131	.0123	.0115	.0107	.0100	.0093	.0087	.0081	.0076
4	.0319	.0302	.0285	.0269	.0254	.0240	.0226	.0213	.0201	.0189
5	.0581	.0555	.0530	.0506	.0483	.0460	.0439	.0418	.0398	..0378
6	.0881	.0851	.0822	.0793	.0764	.0736	.0709	.0682	.0656	.0631
7	.1145	.1118	.1091	.1064	.1037	.1010	.0982	.0955	.0928	.0901
8	.1302	.1286	.1269	.1251	.1232	.1212	.1191	.1170	.1148	.1126
9	.1317	.1315	.1311	.1306	.1300	.1293	.1284	.1274	.1263	.1251

POISSON PROBABILITY DISTRIBUTION

r	λ 9.10	9.20	9.30	9.40	9.50	9.60	9.70	9.80	9.90	10.00
10	.1198	.1210	.1219	.1228	.1235	.1241	.1245	.1249	.1250	.1251
11	.0991	.1012	.1031	.1049	.1067	.1083	.1098	.1112	.1125	.1137
12	.0752	.0776	.0799	.0822	.0844	.0866	.0888	.0908	.0928	.0948
13	.0526	.0549	.0572	.0594	.0617	.0640	.0662	.0685	.0707	.0729
14	.0342	.0361	.0380	.0399	.0419	.0439	.0459	.0479	.0500	.0521
15	.0208	.0221	.0235	.0250	.0265	.0281	.0297	.0313	.0330	.0347
16	.0118	.0127	.0137	.0147	.0157	.0168	.0180	.0192	.0204	.0217
17	.0063	.0069	.0075	.0081	.0088	.0095	.0103	.0111	.0119	.0128
18	.0032	.0035	.0039	.0042	.0046	.0051	.0055	.0060	.0065	.0071
19	.0015	.0017	.0019	.0021	.0023	.0026	.0028	.0031	.0034	.0037
20	.0007	.0008	.0009	.0010	.0011	.0012	.0014	.0015	.0017	.0019
21	.0003	.0003	.0004	.0004	.0005	.0006	.0006	.0007	.0008	.0009
22	.0001	.0001	.0002	.0002	.0002	.0002	.0003	.0003	.0004	.0004
23	.0000	.0001	.0001	.0001	.0001	.0001	.0001	.0001	.0002	.0002
24	.0000	.0000	.0000	.0000	.0000	.0000	.0000	.0001	.0001	.0001

r	λ 11.	12.	13.	14.	15.	16.	17.	18.	19.	20.
0	.0000	.0000	.0000	.0000	.0000	.0000	.0000	.0000	.0000	.0000
1	.0002	.0001	.0000	.0000	.0000	.0000	.0000	.0000	.0000	.0000
2	.0010	.0004	.0002	.0001	.0000	.0000	.0000	.0000	.0000	.0000
3	.0037	.0018	.0008	.0004	.0002	.0001	.0000	.0000	.0000	.0000
4	.0102	.0053	.0027	.0013	.0006	.0003	.0001	.0001	.0000	.0000
5	.0224	.0127	.0070	.0037	.0019	.0010	.0005	.0002	.0001	.0001
6	.0411	.0255	.0152	.0087	.0048	.0026	.0014	.0007	.0004	.0002
7	.0646	.0437	.0281	.0174	.0104	.0060	.0034	.0019	.0010	.0005
8	.0888	.0655	.0457	.0304	.0194	.0120	.0072	.0042	.0024	.0013
9	.1085	.0874	.0661	.0473	.0324	.0213	.0135	.0083	.0050	.0029
10	.1194	.1048	.0859	.0663	.0486	.0341	.0230	.0150	.0095	.0058
11	.1194	.1144	.1015	.0844	.0663	.0496	.0355	.0245	.0164	.0106
12	.1094	.1144	.1099	.0984	.0829	.0661	.0504	.0368	.0259	.0176
13	.0926	.1056	.1099	.1060	.0956	.0814	.0658	.0509	.0378	.0271
14	.0728	.0905	.1021	.1060	.1024	.0930	.0800	.0655	.0514	.0387
15	.0534	.0724	.0885	.0989	.1024	.0992	.0906	.0786	.0650	.0516
16	.0367	.0543	.0719	.0866	.0960	.0992	.0963	.0884	.0772	.0646
17	.0237	.0383	.0550	.0713	.0847	.0934	.0963	.0936	.0863	.0760
18	.0145	.0256	.0397	.0554	.0706	.0830	.0909	.0936	.0911	.0844
19	.0084	.0161	.0272	.0409	.0557	.0699	.0814	.0887	.0911	.0888
20	.0046	.0097	.0177	.0286	.0418	.0559	.0692	.0798	.0866	.0888
21	.0024	.0055	.0109	.0191	.0299	.0426	.0560	.0684	.0783	.0846
22	.0012	.0030	.0065	.0121	.0204	.0310	.0433	.0560	.0676	.0769
23	.0006	.0016	.0037	.0074	.0133	.0216	.0320	.0438	.0559	.0669
24	.0003	.0008	.0020	.0043	.0083	.0144	.0226	.0329	.0442	.0557
25	.0001	.0004	.0010	.0024	.0050	.0092	.0154	.0237	.0336	.0446
26	.0000	.0002	.0005	.0013	.0029	.0057	.0101	.0164	.0246	.0343
27	.0000	.0001	.0002	.0007	.0016	.0034	.0063	.0109	.0173	.0254
28	.0000	.0000	.0001	.0003	.0009	.0019	.0038	.0070	.0117	.0181
29	.0000	.0000	.0001	.0002	.0004	.0011	.0023	.0044	.0077	.0125
30	.0000	.0000	.0000	.0001	.0002	.0006	.0013	.0026	.0049	.0083
31	.0000	.0000	.0000	.0000	.0001	.0003	.0007	.0015	.0030	.0054
32	.0000	.0000	.0000	.0000	.0001	.0001	.0004	.0009	.0018	.0034
33	.0000	.0000	.0000	.0000	.0000	.0001	.0002	.0005	.0010	.0020
34	.0000	.0000	.0000	.0000	.0000	.0000	.0001	.0002	.0006	.0012
35	.0000	.0000	.0000	.0000	.0000	.0000	.0000	.0001	.0003	.0007
36	.0000	.0000	.0000	.0000	.0000	.0000	.0000	.0001	.0002	.0004
37	.0000	.0000	.0000	.0000	.0000	.0000	.0000	.0000	.0001	.0002
38	.0000	.0000	.0000	.0000	.0000	.0000	.0000	.0000	.0000	.0001
39	.0000	.0000	.0000	.0000	.0000	.0000	.0000	.0000	.0000	.0001

DISTRIBUTION OF t

Degrees of Freedom	Probability .50	.30	.20	.10	.05	.02	.01
1	1.000	1.963	3.078	6.314	12.706	31.821	63.657
2	.816	1.386	1.886	2.920	4.303	6.965	9.925
3	.765	1.250	1.638	2.353	3.182	4.541	5.841
4	.741	1.190	1.533	2.132	2.776	3.747	4.604
5	.727	1.156	1.476	2.015	2.571	3.365	4.032
6	.718	1.134	1.440	1.943	2.447	3.143	3.707
7	.711	1.119	1.415	1.895	2.365	2.998	3.499
8	.706	1.108	1.397	1.860	2.306	2.896	3.355
9	.703	1.100	1.383	1.833	2.262	2.821	3.250
10	.700	1.093	1.372	1.812	2.228	2.764	3.169
11	.697	1.088	1.363	1.796	2.201	2.718	3.106
12	.695	1.083	1.356	1.782	2.179	2.681	3.055
13	.694	1.079	1.350	1.771	2.160	2.650	3.012
14	.692	1.076	1.345	1.761	2.145	2.624	2.977
15	.691	1.074	1.341	1.753	2.131	2.602	2.947
16	.690	1.071	1.337	1.746	2.120	2.583	2.921
17	.689	1.069	1.333	1.740	2.110	2.567	2.898
18	.688	1.067	1.330	1.734	2.101	2.552	2.878
19	.688	1.066	1.328	1.729	2.093	2.539	2.861
20	.687	1.064	1.325	1.725	2.086	2.528	2.845
21	.686	1.063	1.323	1.721	2.080	2.518	2.831
22	.686	1.061	1.321	1.717	2.074	2.508	2.819
23	.685	1.060	1.319	1.714	2.069	2.500	2.807
24	.685	1.059	1.318	1.711	2.064	2.492	2.797
25	.684	1.058	1.316	1.708	2.060	2.485	2.787
26	.684	1.058	1.315	1.706	2.056	2.479	2.779
27	.684	1.057	1.314	1.703	2.052	2.473	2.771
28	.683	1.056	1.313	1.701	2.048	2.467	2.763
29	.683	1.055	1.311	1.699	2.045	2.462	2.756
30	.683	1.055	1.310	1.697	2.042	2.457	2.750
40	.681	1.050	1.303	1.684	2.021	2.423	2.704
60	.679	1.046	1.296	1.671	2.000	2.390	2.660
120	.677	1.041	1.289	1.658	1.980	2.358	2.617
∞	.674	1.036	1.282	1.645	1.960	2.326	2.576

Appendix K is abridged from Table III of Fisher and Yates: *Statistical Tables for Biological, Agricultural, and Medical Research*, published by Oliver and Boyd Ltd., Edinburgh, and by permission of the authors and publishers.

DISTRIBUTION OF χ^2

Degrees of Freedom	Probability						
	.50.	.30.	.20.	.10.	.05.	.02.	.01.
1	.455	1.074	1.642	2.706	3.841	5.412	6.635
2	1.386	2.408	3.219	4.605	5.991	7.824	9.210
3	2.366	3.665	4.642	6.251	7.815	9.837	11.345
4	3.357	4.878	5.989	7.779	9.488	11.668	13.277
5	4.351	6.064	7.289	9.236	11.070	13.388	15.086
6	5.348	7.231	8.558	10.645	12.592	15.033	16.812
7	6.346	8.383	9.803	12.017	14.067	16.622	18.475
8	7.344	9.524	11.030	13.362	15.507	18.168	20.090
9	8.343	10.656	12.242	14.684	16.919	19.679	21.666
10	9.342	11.781	13.442	15.987	18.307	21.161	23.209
11	10.341	12.899	14.631	17.275	19.675	22.618	24.725
12	11.340	14.011	15.812	18.549	21.026	24.054	26.217
13	12.340	15.119	16.985	19.812	22.362	25.472	27.688
14	13.339	16.222	18.151	21.064	23.685	26.873	29.141
15	14.339	17.322	19.311	22.307	24.996	28.259	30.578
16	15.338	18.418	20.465	23.542	26.296	29.633	32.000
17	16.338	19.511	21.615	24.769	27.587	30.995	33.409
18	17.338	20.601	22.760	25.989	28.869	33.346	34.805
19	18.338	21.689	23.900	27.204	30.144	33.687	36.191
20	19.337	22.775	25.038	28.412	31.410	35.020	37.566
21	20.337	23.858	26.171	29.615	32.671	36.343	38.932
22	21.337	24.939	27.301	30.813	33.924	37.659	40.289
23	22.337	26.018	28.429	32.007	35.172	38.968	41.638
24	23.337	27.096	29.553	33.196	36.415	40.270	42.980
25	24.337	28.172	30.675	34.382	37.652	41.566	44.314
26	25.336	29.246	31.795	35.563	38.885	42.856	45.642
27	26.336	30.319	32.912	36.741	40.113	44.140	46.963
28	27.336	31.391	34.027	37.916	41.337	45.419	48.278
29	28.336	32.461	35.139	39.087	42.557	46.693	49.588
30	29.336	33.530	36.250	40.256	43.773	47.962	50.892

Appendix L is abridged from Table IV of Fisher and Yates: *Statistical Tables for Biological, Agricultural, and Medical Research*, published by Oliver and Boyd Ltd., Edinburgh, and by permission of the authors and publishers.

DISTRIBUTION OF F

5% (Roman Type) and 1% (**Bold Face Type**) Points for the Distribution of F

$d.f._2$	$d.f._1$ Degrees of Freedom (for greater mean square) 1	2	3	4	5	6	7	8	9	10	11	12	14	16	20	24	30	40	50	75	100	200	500	∞	$d.f._2$
1	161	200	216	225	230	234	237	239	241	242	243	244	245	246	248	249	250	251	252	253	253	254	254	254	1
	4,052	**4,999**	**5,403**	**5,625**	**5,764**	**5,859**	**5,928**	**5,981**	**6,022**	**6,056**	**6,082**	**6,106**	**6,142**	**6,169**	**6,208**	**6,234**	**6,261**	**6,286**	**6,302**	**6,323**	**6,334**	**6,352**	**6,361**	**6,366**	
2	18.51	19.00	19.16	19.25	19.30	19.33	19.36	19.37	19.38	19.39	19.40	19.41	19.42	19.43	19.44	19.45	19.46	19.47	19.47	19.48	19.49	19.49	19.50	19.50	2
	98.49	**99.00**	**99.17**	**99.25**	**99.30**	**99.33**	**99.36**	**99.37**	**99.39**	**99.40**	**99.41**	**99.42**	**99.43**	**99.44**	**99.45**	**99.46**	**99.47**	**99.48**	**99.48**	**99.49**	**99.49**	**99.49**	**99.50**	**99.50**	
3	10.13	9.55	9.28	9.12	9.01	8.94	8.88	8.84	8.81	8.78	8.76	8.74	8.71	8.69	8.66	8.64	8.62	8.60	8.58	8.57	8.56	8.54	8.54	8.53	3
	34.12	**30.82**	**29.46**	**28.71**	**28.24**	**27.91**	**27.67**	**27.49**	**27.34**	**27.23**	**27.13**	**27.05**	**26.92**	**26.83**	**26.69**	**26.60**	**26.50**	**26.41**	**26.35**	**26.27**	**26.23**	**26.18**	**26.14**	**26.12**	
4	7.71	6.94	6.59	6.39	6.26	6.16	6.09	6.04	6.00	5.96	5.93	5.91	5.87	5.84	5.80	5.77	5.74	5.71	5.70	5.68	5.66	5.65	5.64	5.63	4
	21.20	**18.00**	**16.69**	**15.98**	**15.52**	**15.21**	**14.98**	**14.80**	**14.66**	**14.54**	**14.45**	**14.37**	**14.24**	**14.15**	**14.02**	**13.93**	**13.83**	**13.74**	**13.69**	**13.61**	**13.57**	**13.52**	**13.48**	**13.46**	
5	6.61	5.79	5.41	5.19	5.05	4.95	4.88	4.82	4.78	4.74	4.70	4.68	4.64	4.60	4.56	4.53	4.50	4.46	4.44	4.42	4.40	4.38	4.37	4.36	5
	16.26	**13.27**	**12.06**	**11.39**	**10.97**	**10.67**	**10.45**	**10.29**	**10.15**	**10.05**	**9.96**	**9.89**	**9.77**	**9.68**	**9.55**	**9.47**	**9.38**	**9.29**	**9.24**	**9.17**	**9.13**	**9.07**	**9.04**	**9.02**	
6	5.99	5.14	4.76	4.53	4.39	4.28	4.21	4.15	4.10	4.06	4.03	4.00	3.96	3.92	3.87	3.84	3.81	3.77	3.75	3.72	3.71	3.69	3.68	3.67	6
	13.74	**10.92**	**9.78**	**9.15**	**8.75**	**8.47**	**8.26**	**8.10**	**7.98**	**7.87**	**7.79**	**7.72**	**7.60**	**7.52**	**7.39**	**7.31**	**7.23**	**7.14**	**7.09**	**7.02**	**6.99**	**6.94**	**6.90**	**6.88**	
7	5.59	4.74	4.35	4.12	3.97	3.87	3.79	3.73	3.68	3.63	3.60	3.57	3.52	3.49	3.44	3.41	3.38	3.34	3.32	3.29	3.28	3.25	3.24	3.23	7
	12.25	**9.55**	**8.45**	**7.85**	**7.46**	**7.19**	**7.00**	**6.84**	**6.71**	**6.62**	**6.54**	**6.47**	**6.35**	**6.27**	**6.15**	**6.07**	**5.98**	**5.90**	**5.85**	**5.78**	**5.75**	**5.70**	**5.67**	**5.65**	
8	5.32	4.46	4.07	3.84	3.69	3.58	3.50	3.44	3.39	3.34	3.31	3.28	3.23	3.20	3.15	3.12	3.08	3.05	3.03	3.00	2.98	2.96	2.94	2.93	8
	11.26	**8.65**	**7.59**	**7.01**	**6.63**	**6.37**	**6.19**	**6.03**	**5.91**	**5.82**	**5.74**	**5.67**	**5.56**	**5.48**	**5.36**	**5.28**	**5.20**	**5.11**	**5.06**	**5.00**	**4.96**	**4.91**	**4.88**	**4.86**	
9	5.12	4.26	3.86	3.63	3.48	3.37	3.29	3.23	3.18	3.13	3.10	3.07	3.02	2.98	2.93	2.90	2.86	2.82	2.80	2.77	2.76	2.73	2.72	2.71	9
	10.56	**8.02**	**6.99**	**6.42**	**6.06**	**5.80**	**5.62**	**5.47**	**5.35**	**5.26**	**5.18**	**5.11**	**5.00**	**4.92**	**4.80**	**4.73**	**4.64**	**4.56**	**4.51**	**4.45**	**4.41**	**4.36**	**4.33**	**4.31**	
10	4.96	4.10	3.71	3.48	3.33	3.22	3.14	3.07	3.02	2.97	2.94	2.91	2.86	2.82	2.77	2.74	2.70	2.67	2.64	2.61	2.59	2.56	2.55	2.54	10
	10.04	**7.56**	**6.55**	**5.99**	**5.64**	**5.39**	**5.21**	**5.06**	**4.95**	**4.85**	**4.78**	**4.71**	**4.60**	**4.52**	**4.41**	**4.33**	**4.25**	**4.17**	**4.12**	**4.05**	**4.01**	**3.96**	**3.93**	**3.91**	
11	4.84	3.98	3.59	3.36	3.20	3.09	3.01	2.95	2.90	2.86	2.82	2.79	2.74	2.70	2.65	2.61	2.57	2.53	2.50	2.47	2.45	2.42	2.41	2.40	11
	9.65	**7.20**	**6.22**	**5.67**	**5.32**	**5.07**	**4.88**	**4.74**	**4.63**	**4.54**	**4.46**	**4.40**	**4.29**	**4.21**	**4.10**	**4.02**	**3.94**	**3.86**	**3.80**	**3.74**	**3.70**	**3.66**	**3.62**	**3.60**	
12	4.75	3.88	3.49	3.26	3.11	3.00	2.92	2.85	2.80	2.76	2.72	2.69	2.64	2.60	2.54	2.50	2.46	2.42	2.40	2.36	2.35	2.32	2.31	2.30	12
	9.33	**6.93**	**5.95**	**5.41**	**5.06**	**4.82**	**4.65**	**4.50**	**4.39**	**4.30**	**4.22**	**4.16**	**4.05**	**3.98**	**3.86**	**3.78**	**3.70**	**3.61**	**3.56**	**3.49**	**3.46**	**3.41**	**3.38**	**3.36**	
13	4.67	3.80	3.41	3.18	3.02	2.92	2.84	2.77	2.72	2.67	2.63	2.60	2.55	2.51	2.46	2.42	2.38	2.34	2.32	2.28	2.26	2.24	2.22	2.21	13
	9.07	**6.70**	**5.74**	**5.20**	**4.86**	**4.62**	**4.44**	**4.30**	**4.19**	**4.10**	**4.02**	**3.96**	**3.85**	**3.78**	**3.67**	**3.59**	**3.51**	**3.42**	**3.37**	**3.30**	**3.27**	**3.21**	**3.18**	**3.16**	

M

DISTRIBUTION OF F

$d.f._2$	$d.f._1$ Degrees of Freedom (for greater mean square)																								$d.f._2$
	1	2	3	4	5	6	7	8	9	10	11	12	14	16	20	24	30	40	50	75	100	200	500	∞	
14	4.60	3.74	3.34	3.11	2.96	2.85	2.77	2.70	2.65	2.60	2.56	2.53	2.48	2.44	2.39	2.35	2.31	2.27	2.24	2.21	2.19	2.16	2.14	2.13	14
	8.86	**6.51**	**5.56**	**5.03**	**4.69**	**4.46**	**4.28**	**4.14**	**4.03**	**3.94**	**3.86**	**3.80**	**3.70**	**3.62**	**3.51**	**3.43**	**3.34**	**3.26**	**3.21**	**3.14**	**3.11**	**3.06**	**3.02**	**3.00**	
15	4.54	3.68	3.29	3.06	2.90	2.79	2.70	2.64	2.59	2.55	2.51	2.48	2.43	2.39	2.33	2.29	2.25	2.21	2.18	2.15	2.12	2.10	2.08	2.07	15
	8.68	**6.36**	**5.42**	**4.89**	**4.56**	**4.32**	**4.14**	**4.00**	**3.89**	**3.80**	**3.73**	**3.67**	**3.56**	**3.48**	**3.36**	**3.29**	**3.20**	**3.12**	**3.07**	**3.00**	**2.97**	**2.92**	**2.89**	**2.87**	
16	4.49	3.63	3.24	3.01	2.85	2.74	2.66	2.59	2.54	2.49	2.45	2.42	2.37	2.33	2.28	2.24	2.20	2.16	2.13	2.09	2.07	2.04	2.02	2.01	16
	8.53	**6.23**	**5.29**	**4.77**	**4.44**	**4.20**	**4.03**	**3.89**	**3.78**	**3.69**	**3.61**	**3.55**	**3.45**	**3.37**	**3.25**	**3.18**	**3.10**	**3.01**	**2.96**	**2.98**	**2.86**	**2.80**	**2.77**	**2.75**	
17	4.45	3.59	3.20	2.96	2.81	2.70	2.62	2.55	2.50	2.45	2.41	2.38	2.33	2.29	2.23	2.19	2.15	2.11	2.08	2.04	2.02	1.99	1.97	1.96	17
	8.40	**6.11**	**5.18**	**4.67**	**4.34**	**4.10**	**3.93**	**3.79**	**3.68**	**3.59**	**3.52**	**3.45**	**3.35**	**3.27**	**3.16**	**3.08**	**3.00**	**2.92**	**2.86**	**2.79**	**2.76**	**2.70**	**2.67**	**2.65**	
18	4.41	3.55	3.16	2.93	2.77	3.66	2.58	2.51	2.46	2.41	2.37	2.34	2.29	2.25	2.19	2.15	2.11	2.07	2.04	2.00	1.98	1.95	1.93	1.92	18
	8.28	**6.01**	**5.09**	**4.58**	**4.25**	**4.01**	**3.85**	**3.71**	**3.60**	**3.51**	**3.44**	**3.37**	**3.27**	**3.19**	**3.07**	**3.00**	**2.91**	**2.83**	**2.78**	**2.71**	**2.68**	**2.62**	**2.59**	**2.57**	
19	4.38	3.52	3.13	2.90	2.74	2.63	2.55	2.48	2.43	2.38	2.34	2.31	2.26	2.21	2.15	2.11	2.07	2.02	2.00	1.96	1.94	1.91	1.90	1.88	19
	8.18	**5.93**	**5.01**	**4.50**	**4.17**	**3.94**	**3.77**	**3.63**	**3.52**	**3.43**	**3.36**	**3.30**	**3.19**	**3.12**	**3.00**	**2.92**	**2.84**	**2.76**	**2.70**	**2.63**	**2.60**	**2.54**	**2.51**	**2.49**	
20	4.35	3.49	3.10	2.87	2.71	2.60	2.52	2.45	2.40	2.35	2.31	2.28	2.23	2.18	2.12	2.08	2.04	1.99	1.96	1.92	1.90	1.87	1.85	1.84	20
	8.10	**5.85**	**4.94**	**4.43**	**4.10**	**3.87**	**3.71**	**3.56**	**3.45**	**3.37**	**3.30**	**3.23**	**3.13**	**3.05**	**2.94**	**2.86**	**2.77**	**2.69**	**2.63**	**2.56**	**2.53**	**2.47**	**2.44**	**2.42**	
21	4.32	3.47	3.07	2.84	2.68	2.57	2.49	2.42	2.37	2.32	2.28	2.25	2.20	2.15	2.09	2.05	2.00	1.96	1.93	1.89	1.87	1.84	1.82	1.81	21
	8.02	**5.78**	**4.87**	**4.37**	**4.04**	**3.81**	**3.65**	**3.51**	**3.40**	**3.31**	**3.24**	**3.17**	**3.07**	**2.99**	**2.88**	**2.80**	**2.72**	**2.63**	**2.58**	**2.51**	**2.47**	**2.42**	**2.38**	**2.36**	
22	4.30	3.44	3.05	2.82	2.66	2.55	2.47	2.40	2.35	2.30	2.26	2.23	2.18	2.13	2.07	2.03	1.98	1.93	1.91	1.87	1.84	1.81	1.80	1.78	22
	7.94	**5.72**	**4.82**	**4.31**	**3.99**	**3.76**	**3.59**	**3.45**	**3.35**	**3.26**	**3.18**	**3.12**	**3.02**	**2.94**	**2.83**	**2.75**	**2.67**	**2.58**	**2.53**	**2.46**	**2.42**	**2.37**	**2.33**	**2.31**	
23	4.28	3.42	3.03	2.80	2.64	2.53	2.45	2.38	2.32	2.28	2.24	2.20	2.14	2.10	2.04	2.00	1.96	1.91	1.88	1.84	1.82	1.79	1.77	1.76	23
	7.88	**5.66**	**4.76**	**4.26**	**3.94**	**3.71**	**3.54**	**3.41**	**3.30**	**3.21**	**3.14**	**3.07**	**2.97**	**2.89**	**2.78**	**2.70**	**2.62**	**2.53**	**2.48**	**2.41**	**2.37**	**2.32**	**2.28**	**2.26**	
24	4.26	3.40	3.01	2.78	2.62	2.51	2.43	2.36	2.30	2.26	2.22	2.18	2.13	2.09	2.02	1.98	1.94	1.89	1.86	1.82	1.80	1.76	1.74	1.73	24
	7.82	**5.61**	**4.72**	**4.22**	**3.90**	**3.67**	**3.50**	**3.36**	**3.25**	**3.17**	**3.09**	**3.03**	**2.93**	**2.85**	**2.74**	**2.66**	**2.58**	**2.49**	**2.44**	**2.36**	**2.33**	**2.27**	**2.23**	**2.21**	
25	4.24	3.38	2.99	2.76	2.60	2.49	2.41	2.34	2.28	2.24	2.20	2.16	2.11	2.06	2.00	1.96	1.92	1.87	1.84	1.80	1.77	1.74	1.72	1.71	25
	7.77	**5.57**	**4.68**	**4.18**	**3.86**	**3.63**	**3.46**	**3.32**	**3.21**	**3.13**	**3.05**	**2.99**	**2.89**	**2.81**	**2.70**	**2.62**	**2.54**	**2.45**	**2.40**	**2.32**	**2.29**	**2.23**	**2.19**	**2.17**	
26	4.22	3.37	2.98	2.74	2.59	2.47	2.39	2.32	2.27	2.22	2.18	2.15	2.10	2.05	1.99	1.95	1.90	1.85	1.82	1.78	1.76	1.72	1.70	1.69	26
	7.72	**5.53**	**4.64**	**4.14**	**3.82**	**3.59**	**3.42**	**3.29**	**3.17**	**3.09**	**3.02**	**2.96**	**2.86**	**2.77**	**2.66**	**2.58**	**2.50**	**2.41**	**2.36**	**2.28**	**2.25**	**2.19**	**2.15**	**2.13**	

The function, $F = e$ with exponent $2z$, is computed in part from Fisher's table VI (7). Additional entries are by interpolation, mostly graphical.

DISTRIBUTION OF *F*

$d.f._2$	$d.f._1$ Degrees of Freedom (for greater mean square)																								$d.f._2$
	1	2	3	4	5	6	7	8	9	10	11	12	14	16	20	24	30	40	50	75	100	200	500	∞	
27	4.21	3.35	2.96	2.73	2.57	2.46	2.37	2.30	2.25	2.20	2.16	2.13	2.08	2.03	1.97	1.93	1.88	1.84	1.80	1.76	1.74	1.71	1.68	1.67	27
	7.68	**5.49**	**4.60**	**4.11**	**3.79**	**3.56**	**3.39**	**3.26**	**3.14**	**3.06**	**2.98**	**2.93**	**2.83**	**2.74**	**2.63**	**2.55**	**2.47**	**2.38**	**2.33**	**2.25**	**2.21**	**2.16**	**2.12**	**2.10**	
28	4.20	3.34	2.95	2.71	2.56	2.44	2.36	2.29	2.24	2.19	2.15	2.12	2.06	2.02	1.96	1.91	1.87	1.81	1.78	1.75	1.72	1.69	1.67	1.65	28
	7.64	**5.45**	**4.57**	**4.07**	**3.76**	**3.53**	**3.36**	**3.23**	**3.11**	**3.03**	**2.95**	**2.90**	**2.80**	**2.71**	**2.60**	**2.52**	**2.44**	**2.35**	**2.30**	**2.22**	**2.18**	**2.13**	**2.09**	**2.06**	
29	4.18	3.33	2.93	2.70	2.54	2.43	2.35	2.28	2.22	2.18	2.14	2.10	2.05	2.00	1.94	1.90	1.85	1.80	1.77	1.73	1.71	1.68	1.65	1.64	29
	7.60	**5.42**	**4.54**	**4.04**	**3.73**	**3.50**	**3.33**	**3.20**	**3.08**	**3.00**	**2.92**	**2.87**	**2.77**	**2.68**	**2.57**	**2.49**	**2.41**	**2.32**	**2.27**	**2.19**	**2.15**	**2.10**	**2.06**	**2.03**	
30	4.17	3.32	2.92	2.69	2.53	2.42	2.34	2.27	2.21	2.16	2.12	2.09	2.04	1.99	1.93	1.89	1.84	1.79	1.76	1.72	1.69	1.66	1.64	1.62	30
	7.56	**5.39**	**4.51**	**4.02**	**3.70**	**3.47**	**3.30**	**3.17**	**3.06**	**2.98**	**2.90**	**2.84**	**2.74**	**2.66**	**2.55**	**2.47**	**2.38**	**2.29**	**2.24**	**2.16**	**2.13**	**2.07**	**2.03**	**2.01**	
32	4.15	3.30	2.90	2.67	2.51	2.40	2.32	2.25	2.19	2.14	2.10	2.07	2.02	1.97	1.91	1.86	1.82	1.76	1.74	1.69	1.67	1.64	1.61	1.59	32
	7.50	**5.34**	**4.46**	**3.97**	**3.66**	**3.42**	**3.25**	**3.12**	**3.01**	**2.94**	**2.86**	**2.80**	**2.70**	**2.62**	**2.51**	**2.42**	**2.34**	**2.25**	**2.20**	**2.12**	**2.08**	**2.02**	**1.98**	**1.96**	
34	4.13	3.28	2.88	2.65	2.49	2.38	2.30	2.23	2.17	2.12	2.08	2.05	2.00	1.95	1.89	1.84	1.80	1.74	1.71	1.67	1.64	1.61	1.59	1.57	34
	7.44	**5.29**	**4.42**	**3.93**	**3.61**	**3.38**	**3.21**	**3.08**	**2.97**	**2.89**	**2.82**	**2.76**	**2.66**	**2.58**	**2.47**	**2.38**	**2.30**	**2.21**	**2.15**	**2.08**	**2.04**	**1.98**	**1.94**	**1.91**	
36	4.11	3.26	2.86	2.63	2.48	2.36	2.28	2.21	2.15	2.10	2.06	2.03	1.98	1.93	1.87	1.82	1.78	1.72	1.69	1.65	1.62	1.59	1.56	1.55	36
	7.39	**5.25**	**4.38**	**3.89**	**3.58**	**3.35**	**3.18**	**3.04**	**2.94**	**2.86**	**2.78**	**2.72**	**2.62**	**2.54**	**2.43**	**2.35**	**2.26**	**2.17**	**2.12**	**2.04**	**2.00**	**1.94**	**1.90**	**1.87**	
38	4.10	3.25	2.85	2.62	2.46	2.35	2.26	2.19	2.14	2.09	2.05	2.02	1.96	1.92	1.85	1.80	1.76	1.71	1.67	1.63	1.60	1.57	1.54	1.53	38
	7.35	**5.21**	**4.34**	**3.86**	**3.54**	**3.32**	**3.15**	**3.02**	**2.91**	**2.82**	**2.75**	**2.69**	**2.59**	**2.51**	**2.40**	**2.32**	**2.22**	**2.14**	**2.08**	**2.00**	**1.97**	**1.90**	**1.86**	**1.84**	
40	4.08	3.23	2.84	2.61	2.45	2.34	2.25	2.18	2.12	2.07	2.04	2.00	1.95	1.90	1.84	1.79	1.74	1.69	1.66	1.61	1.59	1.55	1.53	1.51	40
	7.31	**5.18**	**4.31**	**3.83**	**3.51**	**3.29**	**3.12**	**2.99**	**2.88**	**2.80**	**2.73**	**2.66**	**2.56**	**2.49**	**2.37**	**2.29**	**2.20**	**2.11**	**2.05**	**1.97**	**1.94**	**1.88**	**1.84**	**1.81**	
42	4.07	3.22	2.83	2.59	2.44	2.32	2.24	2.17	2.11	2.06	2.02	1.99	1.94	1.89	1.82	1.78	1.73	1.68	1.64	1.60	1.57	1.54	1.51	1.49	42
	7.27	**5.15**	**4.29**	**3.80**	**3.49**	**3.26**	**3.10**	**2.96**	**2.86**	**2.77**	**2.70**	**2.64**	**2.54**	**2.46**	**2.35**	**2.26**	**2.17**	**2.08**	**2.02**	**1.94**	**1.91**	**1.85**	**1.80**	**1.78**	
44	4.06	3.21	2.82	2.58	2.43	2.31	2.23	2.16	2.10	2.05	2.01	1.98	1.92	1.88	1.81	1.76	1.72	1.66	1.63	1.58	1.56	1.52	1.50	1.48	44
	7.24	**5.12**	**4.26**	**3.78**	**3.46**	**3.24**	**3.07**	**2.94**	**2.84**	**2.75**	**2.68**	**2.62**	**2.52**	**2.44**	**2.32**	**2.24**	**2.15**	**2.06**	**2.00**	**1.92**	**1.88**	**1.82**	**1.78**	**1.75**	
46	4.05	3.20	2.81	2.57	2.42	2.30	2.22	2.14	2.09	2.04	2.00	1.97	1.91	1.87	1.80	1.75	1.71	1.65	1.62	1.57	1.54	1.51	1.48	1.46	46
	7.21	**5.10**	**4.24**	**3.76**	**3.44**	**3.22**	**3.05**	**2.92**	**2.82**	**2.73**	**2.66**	**2.60**	**2.50**	**2.42**	**2.30**	**2.22**	**2.13**	**2.04**	**1.98**	**1.90**	**1.86**	**1.80**	**1.76**	**1.72**	
48	4.04	3.19	2.80	2.56	2.41	2.30	2.21	2.14	2.08	2.03	1.99	1.96	1.90	1.86	1.79	1.74	1.70	1.64	1.61	1.56	1.53	1.50	1.47	1.45	48
	7.19	**5.08**	**4.22**	**3.74**	**3.42**	**3.20**	**3.04**	**2.90**	**2.80**	**2.71**	**2.64**	**2.58**	**2.48**	**2.40**	**2.28**	**2.20**	**2.11**	**2.02**	**1.96**	**1.88**	**1.84**	**1.78**	**1.73**	**1.70**	

M

DISTRIBUTION OF F

$d.f._2$	$d.f._1$ Degrees of Freedom (for greater mean square) 1	2	3	4	5	6	7	8	9	10	11	12	14	16	20	24	30	40	50	75	100	200	500	∞	$d.f._2$
50	4.03 **7.17**	3.18 **5.06**	2.79 **4.20**	2.56 **3.72**	2.40 **3.41**	2.29 **3.18**	2.20 **3.02**	2.13 **2.88**	2.07 **2.78**	2.02 **2.70**	1.98 **2.62**	1.95 **2.56**	1.90 **2.46**	1.85 **2.39**	1.78 **2.26**	1.74 **2.18**	1.69 **2.10**	1.63 **2.00**	1.60 **1.94**	1.55 **1.86**	1.52 **1.82**	1.48 **1.76**	1.46 **1.71**	1.44 **1.68**	50
55	4.02 **7.12**	3.17 **5.01**	2.78 **4.16**	2.54 **3.68**	2.38 **3.37**	2.27 **3.15**	2.18 **2.98**	2.11 **2.85**	2.05 **2.75**	2.00 **2.66**	1.97 **2.59**	1.93 **2.53**	1.88 **2.43**	1.83 **2.35**	1.76 **2.23**	1.72 **2.15**	1.67 **2.06**	1.61 **1.96**	1.58 **1.90**	1.52 **1.82**	1.50 **1.78**	1.46 **1.71**	1.43 **1.66**	1.41 **1.64**	55
60	4.00 **7.08**	3.15 **4.98**	2.76 **4.13**	2.52 **3.65**	2.37 **3.34**	2.25 **3.12**	2.17 **2.95**	2.10 **2.82**	2.04 **2.72**	1.99 **2.63**	1.95 **2.56**	1.92 **2.50**	1.86 **2.40**	1.81 **2.32**	1.75 **2.20**	1.70 **2.12**	1.65 **2.03**	1.59 **1.93**	1.56 **1.87**	1.50 **1.79**	1.48 **1.74**	1.44 **1.68**	1.41 **1.63**	1.39 **1.60**	60
65	3.99 **7.04**	3.14 **4.95**	2.75 **4.10**	2.51 **3.62**	2.36 **3.31**	2.24 **3.09**	2.15 **2.93**	2.08 **2.79**	2.02 **2.70**	1.98 **2.61**	1.94 **2.54**	1.90 **2.47**	1.85 **2.37**	1.80 **2.30**	1.73 **2.18**	1.68 **2.09**	1.63 **2.00**	1.57 **1.90**	1.54 **1.84**	1.49 **1.76**	1.46 **1.71**	1.42 **1.64**	1.39 **1.60**	1.37 **1.56**	65
70	3.98 **7.01**	3.13 **4.92**	2.74 **4.08**	2.50 **3.60**	2.35 **3.29**	2.23 **3.07**	2.14 **2.91**	2.07 **2.77**	2.01 **2.67**	1.97 **2.59**	1.93 **2.51**	1.89 **2.45**	1.84 **2.35**	1.79 **2.28**	1.72 **2.15**	1.67 **2.07**	1.62 **1.98**	1.56 **1.88**	1.53 **1.82**	1.47 **1.74**	1.45 **1.69**	1.40 **1.62**	1.37 **1.56**	1.35 **1.53**	70
80	3.96 **6.96**	3.11 **4.88**	2.72 **4.04**	2.48 **3.56**	2.33 **3.25**	2.21 **3.04**	2.12 **2.87**	2.05 **2.74**	1.99 **2.64**	1.95 **2.55**	1.91 **2.48**	1.88 **2.41**	1.82 **2.32**	1.77 **2.24**	1.70 **2.11**	1.65 **2.03**	1.60 **1.94**	1.54 **1.84**	1.51 **1.78**	1.45 **1.70**	1.42 **1.65**	1.38 **1.57**	1.35 **1.52**	1.32 **1.49**	80
100	3.94 **6.90**	3.09 **4.82**	2.70 **3.98**	2.46 **3.51**	2.30 **3.20**	2.19 **2.99**	2.10 **2.82**	2.03 **2.69**	1.97 **2.59**	1.92 **2.51**	1.88 **2.43**	1.85 **2.36**	1.79 **2.26**	1.75 **2.19**	1.68 **2.06**	1.63 **1.98**	1.57 **1.89**	1.51 **1.79**	1.48 **1.73**	1.42 **1.64**	1.39 **1.59**	1.34 **1.51**	1.30 **1.46**	1.28 **1.43**	100
125	3.92 **6.84**	3.07 **4.78**	2.68 **3.94**	2.44 **3.47**	2.29 **3.17**	2.17 **2.95**	2.08 **2.79**	2.01 **2.65**	1.95 **2.56**	1.90 **2.47**	1.86 **2.40**	1.83 **2.33**	1.77 **2.23**	1.72 **2.15**	1.65 **2.03**	1.60 **1.94**	1.55 **1.85**	1.49 **1.75**	1.45 **1.68**	1.39 **1.59**	1.36 **1.54**	1.31 **1.46**	1.27 **1.40**	1.25 **1.37**	125
150	3.91 **6.81**	3.06 **4.75**	2.67 **3.91**	2.43 **3.44**	2.27 **3.14**	2.16 **2.92**	2.07 **2.76**	2.00 **2.62**	1.94 **2.53**	1.89 **2.44**	1.85 **2.37**	1.82 **2.30**	1.76 **2.20**	1.71 **2.12**	1.64 **2.00**	1.59 **1.91**	1.54 **1.83**	1.47 **1.72**	1.44 **1.66**	1.37 **1.56**	1.34 **1.51**	1.29 **1.43**	1.25 **1.37**	1.22 **1.33**	150
200	3.89 **6.76**	3.04 **4.71**	2.65 **3.88**	2.41 **3.41**	2.26 **3.11**	2.14 **2.90**	2.05 **2.73**	1.98 **2.60**	1.92 **2.50**	1.87 **2.41**	1.83 **2.34**	1.80 **2.28**	1.74 **2.17**	1.69 **2.09**	1.62 **1.97**	1.57 **1.88**	1.52 **1.79**	1.45 **1.69**	1.42 **1.62**	1.35 **1.53**	1.32 **1.48**	1.26 **1.39**	1.22 **1.33**	1.19 **1.28**	200
400	3.86 **6.70**	3.02 **4.66**	2.62 **3.83**	2.39 **3.36**	2.23 **3.06**	2.12 **2.85**	2.03 **2.69**	1.96 **2.55**	1.90 **2.46**	1.85 **2.37**	1.81 **2.29**	1.78 **2.23**	1.72 **2.12**	1.67 **2.04**	1.60 **1.92**	1.54 **1.84**	1.49 **1.74**	1.42 **1.64**	1.38 **1.57**	1.32 **1.47**	1.28 **1.42**	1.22 **1.32**	1.16 **1.24**	1.13 **1.19**	400
1000	3.85 **6.66**	3.00 **4.62**	2.61 **3.80**	2.38 **3.34**	2.22 **3.04**	2.10 **2.82**	2.02 **2.66**	1.95 **2.53**	1.89 **2.43**	1.84 **2.34**	1.80 **2.26**	1.76 **2.20**	1.70 **2.09**	1.65 **2.01**	1.58 **1.89**	1.53 **1.81**	1.47 **1.71**	1.41 **1.61**	1.36 **1.54**	1.30 **1.44**	1.26 **1.38**	1.19 **1.28**	1.13 **1.19**	1.08 **1.11**	1000
∞	3.84 **6.64**	2.99 **4.60**	2.60 **3.78**	2.37 **3.32**	2.21 **3.02**	2.09 **2.80**	2.01 **2.64**	1.94 **2.51**	1.88 **2.41**	1.83 **2.32**	1.79 **2.24**	1.75 **2.18**	1.69 **2.07**	1.64 **1.99**	1.57 **1.87**	1.52 **1.79**	1.46 **1.69**	1.40 **1.59**	1.35 **1.52**	1.28 **1.41**	1.24 **1.36**	1.17 **1.25**	1.11 **1.15**	1.00 **1.00**	∞

RANDOM SAMPLING NUMBERS

FIRST 1,512 RANDOM DIGITS

	A	B	C	D	E	F	G
1	345769	953810	627280	423578	353511	899906	827008
	549075	004410	059309	271243	403382	248735	972383
	423480	950812	197145	556566	655917	046169	363201
	554518	514280	950974	482196	058868	474936	724289
	797165	670995	791954	188521	950156	086813	033365
	062730	163375	602168	908350	360861	152201	966097
2	356756	519371	679389	371912	502903	936741	636775
	700770	781547	916968	136999	801855	605975	295802
	279584	733750	487151	116069	274869	416181	610911
	862434	481154	391464	021094	761599	474456	582253
	199585	167701	170778	934765	761328	275799	323046
	048736	514507	977406	158840	846761	198016	933522
3	815218	609732	629295	517386	824505	676788	304971
	643021	527212	492869	261844	914505	354436	355772
	164332	245407	517804	422658	751712	583087	286872
	174303	085157	308590	535846	503131	266915	465641
	136325	414066	452293	649359	844625	674828	953396
	117780	407444	426115	108970	621527	601599	652376
4	435697	245510	946158	934221	824917	509832	362638
	912252	579474	848845	824321	049853	151126	052643
	754438	658573	717914	040054	630638	264060	594641
	322053	924909	048177	957012	801464	833319	978384
	897199	125506	708669	408374	737887	906201	599469
	046637	642050	435779	502427	027842	515775	811203
5.	721653	260190	842505	797017	157497	179041	979346
	202312	011976	373248	374293	802292	646914	171322
	354014	356787	511271	904434	068589	329862	829316
	682909	809290	793392	098004	120575	469925	112743
	897690	572456	871574	465543	486529	507767	608677
	139029	160636	417690	191242	625269	104858	020808
6	341447	723998	905614	519309	926345	240082	395043
	415603	129727	894956	780924	227496	134056	023014
	014881	496311	750082	707823	738906	157591	072396
	827235	783798	324650	485324	568156	098331	768720
	261607	730824	341940	259028	253973	145183	658110
	527920	834376	972906	627959	654790	342497	593779

Appendix N is abridged with the permission of the author and publisher from "New Random Sampling Numbers," *Baylor Business Studies* No. 1, by H. N. Broom, published by The School of Business, Baylor University, Waco, Texas.

SECOND 1,512 RANDOM DIGITS

	A	B	C	D	E	F	G
1	835431	206253	467521	029822	700399	554652	450184
	512651	743206	118787	587401	921517	015407	206860
	376187	189133	154812	828785	667020	998697	579598
	092530	869028	483691	165063	847894	041617	762973
	238036	016856	290105	538530	079931	412195	838814
	308168	717698	919814	092230	215657	469994	805803
2	773429	915639	900911	276895	149505	540379	224349
	171626	601259	009905	572567	441960	299704	313987
	180570	665625	424048	713009	830314	664642	521021
	558715	965963	494210	875287	488595	898691	713010
	345067	361180	989224	138905	355519	045847	746266
	583819	310956	174728	099164	118461	758000	496302
3	615026	599459	722322	555090	572720	826686	456517
	812358	389535	166779	441968	105639	632418	340890
	784592	003651	279275	055646	341897	510689	026160
	094619	636747	934082	787345	772825	603866	565688
	450908	919891	157771	114333	710179	062848	615156
	593546	728768	984323	290410	970562	906724	315005
4	873778	491131	209695	604075	783895	862911	772026
	965705	317845	169619	921361	315606	990029	745251
	311163	943589	540958	556212	760508	129963	236556
	454554	284761	269019	924179	670780	389869	519229
	124330	819763	596075	064570	495169	030185	866211
	920765	122124	423205	596357	469969	072245	359269
5	183002	540547	312909	389818	464023	768381	377241
	600135	865974	929756	162716	415598	878513	994633
	235787	023117	895285	027055	943962	381112	530492
	953379	655834	283102	836259	437761	391976	940853
	009658	521970	537626	806052	715247	808585	252503
	176570	849057	387097	311529	893745	450267	182626
6	747456	304530	931013	678688	270736	355032	400713
	486876	631985	368395	154273	959983	672523	210456
	987193	268135	867829	025419	301168	409545	131960
	358155	950977	170562	246987	884126	785621	467942
	021394	182615	049084	942153	278313	872709	693590
	735047	428941	630704	893281	716045	267529	427605

THIRD 1,512 RANDOM DIGITS

	A	B	C	D	E	F	G
1	133877	894168	670664	007673	436272	479568	247014
	909935	172305	428979	775425	004071	896108	519806
	204092	380210	589306	421798	273014	842846	750253
	906975	390605	040857	206293	173991	258115	043825
	387430	513087	738318	344565	465609	416995	943451
	045890	563165	460571	633567	481740	951614	668403
2	837159	143979	698357	219259	924875	691935	843585
	796578	982105	540570	724307	369621	562203	757320
	509998	316652	678549	468115	387469	316301	013153
	067045	238296	042458	275413	499300	680274	026351
	207634	540337	350587	013692	412939	274513	984596
	980620	875228	496017	581165	251684	275169	588760
3	347609	157545	919210	690074	532650	922600	693037
	475802	466358	379889	594832	514118	205292	371756
	818821	932102	628457	533138	655279	704197	584316
	362078	838671	765113	410097	138149	701956	928874
	072228	759522	791735	398202	162345	294805	828520
	147935	014193	536872	552021	693458	018447	788748
4	419843	160700	338910	184107	235002	024298	449135
	825546	648481	916364	607857	436970	438087	798960
	082314	418158	781469	991818	721194	358904	450970
	915221	233704	129127	767232	098851	584646	353870
	765613	354681	367568	496453	308935	131432	204643
	036236	196087	690273	453073	595160	410830	466051
5	607104	305543	705229	623194	613727	696054	758402
	308792	376543	027151	165422	560769	814957	589180
	857280	462801	434761	324058	482908	294374	976175
	959721	758687	456782	568719	404563	154205	663418
	207153	231920	518416	804920	932735	082468	322964
	403778	187984	157069	719462	053157	953043	416342
6	286108	108539	428918	149527	723573	636055	737916
	411295	291930	481424	871000	172070	273030	456317
	679313	787369	159935	164716	835268	174221	959886
	405323	376852	057589	437497	357398	838285	098772
	917458	429205	795610	905859	676942	294087	791952
	659514	078457	711589	690730	104700	912369	848269

FOURTH 1,512 RANDOM DIGITS

	A	B	C	D	E	F	G
1	142582	838531	948535	204547	621651	329695	014694
	097086	190024	666521	170674	144070	124008	702818
	358324	034739	403012	692427	208539	381841	432976
	257091	654023	191287	731088	259167	352640	004388
	928731	264667	956546	240744	769932	574832	914694
	729816	278812	119374	895490	818386	267958	560523
2	781062	721128	169905	290611	024176	160727	856247
	093549	401262	079175	117813	842686	246713	649987
	829708	656390	804223	434596	134518	401187	589048
	550416	658096	352864	576572	178144	051421	836509
	072934	572971	564253	950363	656948	923152	790087
	646941	109528	073147	354187	771592	647850	086352
3	905725	867727	033964	579862	045061	896494	589268
	727696	156430	671765	127312	335860	407661	709388
	742859	985436	487786	403118	684839	561387	985352
	095217	375204	659737	001286	046025	616072	224715
	791020	730765	212021	763149	590401	433554	462302
	824304	754426	728896	070857	137631	634735	426189
4	194358	810596	051443	917458	855114	808348	568628
	364029	285129	651482	180425	166024	465370	467021
	675894	027149	802421	058779	786349	597533	917864
	009913	955754	981235	888191	437609	131287	967580
	605600	593586	200254	365462	154578	179723	358203
	792003	304109	298794	661389	720132	741928	088924
5	357790	028381	163072	758986	302348	248362	909435
	482034	980395	236510	516007	654864	890157	740017
	319302	713745	057612	027685	180265	981029	237304
	622343	241778	137067	061429	489784	439401	438854
	870846	008446	490322	136989	703895	591878	506804
	603242	818115	746069	437465	507246	713641	936584
6	143632	031587	688275	170345	823659	049277	970129
	954182	040683	787002	775349	571341	854167	020533
	283056	857426	252542	404561	546734	822595	481604
	891730	027420	126799	821731	195465	709433	240637
	497562	204798	343671	124740	855713	016508	282359
	862867	628369	179980	851292	200332	260919	634484

Index

C

D

E

J

K

L

M

N

O

S

T

U

V

W

X

Y